Waste Water Technology
Origin, Collection, Treatment
and Analysis of Waste Water

Edited by

Institut Fresenius GmbH
Taunusstein-Neuhof
W. Fresenius and **W. Schneider**

and by

Forschungsinstitut für
Wassertechnologie
an der RWTH Aachen (FIW)
B. Böhnke and **K. Pöppinghaus**

Commissioned by the
Deutsche Gesellschaft für Technische Zusammenarbeit (GTZ) GmbH

Springer-Verlag Berlin Heidelberg New York
London Paris Tokyo

Deutsche Gesellschaft für Technische Zusammenarbeit (GTZ) GmbH
Dag-Hammarskjöld-Weg 1, 6236 Eschborn 1, near Frankfurt/M.

Institut Fresenius GmbH, 6204 Taunusstein-Neuhof
W. Fresenius and W. Schneider

Authors: W. Czysz   W. Schneider
          A. Denne   E. Staudte
          H. Rump   W. Supperl

Forschungsinstitut für Wassertechnologie an der RWTH Aachen – FiW
Mies-van-der-Rohe-Straße 17, 5100 Aachen
B. Böhnke and K. Pöppinghaus

Authors: E. Blitz †   K. Pöppinghaus
          B. Böhnke   K. Siekmann
          P. Doetsch   S. Thomas
          P. Dreschmann

ISBN 3-540-17450-8 Springer-Verlag Berlin Heidelberg NewYork
ISBN 0-387-17450-8 Springer-Verlag NewYork Berlin Heidelberg

Library of Congress Cataloging-in-Publication Data
Waste water technology:
origin, collection, treatment, and analysis of waste water / edited by W. Fresenius ... [et al.];
commissioned by Deutsche Gesellschaft für Technische Zusammenarbeit.
Includes index.
Bibliography: p.
  ISBN 0-387-17450-8 (U.S.)
1. Sewage disposal. 2. Sewage.
I. Fresenius, Wilhelm. II. Deutsche Gesellschaft für Technische Zusammenarbeit.
TD741.W33 1989
628.3'72--dc20   89-6196

This work is subject to copyright. All rights are reserved, whether the whole or part of the material is concerned, specifically the rights of translation, reprinting, re-use of illustrations, recitation, broadcasting, reproduction on microfilms or in other ways, and storage in data banks. Duplication of this publication or parts thereof is only permitted under the provision of the German Copyright Law of September 9, 1965, in its version of June 24, 1985, and a copyright fee must always be paid. Violations fall under the prosecution act of the German Copyright Law.

© Springer-Verlag Berlin Heidelberg 1989
Printed in Germany

The use of registered names, trademarks, etc. in this publication does not imply, even in the absence of a specific statement, that such names are exempt from the relevant protective laws and regulations and therefore free for general use.

Offsetprinting: Mercedes-Druck, Berlin; Bookbinding: Lüderitz & Bauer, Berlin
2161/3020 543210 – Printed on acid-free paper

Preface

Waste water disposal is one of the major environmental problems of today. This is not only true for the industrial countries of Europe and overseas, but also for those countries that are still on the threshold of industrial development. In order to encourage the education and training of experts in this field, the Institut Fresenius, Taunusstein, and the Forschungsinstitut für Wassertechnologie at the RWTH Aachen were commissioned by the Deutsche Gesellschaft für Technische Zusammenarbeit (GTZ) GmbH to make an effort to describe all the problems of waste water disposal that are deemed most urgent and how to solve them.

The book deals with the formation, the collection, the analysis, and the treatment of wastes of various origins. It describes numerous experiences which the authors have had in their practical activities in various countries. So the book does not only impart the basic knowledge of waste water disposal but offers at the same time working examples and suggestions.

However, the authors are aware of the fact that the information given cannot cover all the specific conditions of waste water disposal prevailing in the respective countries.

The information presented here will enable the user of this study to solve specific waste water problems as they occur in municipal and industrial areas and to identify which measures should be taken to ensure orderly disposal.

Despite the clearly stated objective of the Water Decade 1980-1990 to give priority to water supply and disposal in rural areas, it cannot be denied that the main problems arise in urban districts and on the outskirts of major conurbations. Waste water disposal in small townships and rural areas is, therefore, not the subject of this book.

The editors thank the authors and the editorial staff for their careful and engaged work. They feel very much obliged, too, to the colleagues in the respective departments and the GTZ, on whose behalf the work was undertaken, for their many helpful suggestions and their never failing willingness to offer support.

Institut Fresenius GmbH
Taunusstein-Neuhof
W. Fresenius and W. Schneider

Forschungsinstitut für
Wassertechnologie
an der RWTH Aachen (FiW)
B. Böhnke and K. Pöppinghaus

Foreword

One of the chief goals of the United Nations International Drinking Water Supply and Sanitation Decade (1981 - 1990) is to reduce the prevalence of water-induced disease and associated mortality in the Third World. The achievement of this goal demands an integrated concept linking the supply of hygienic potable water to complementary sewage disposal and sanitation measures.

A coordinated and appropriate procedure in these three fields requires access to systematic compilations of existing expertise, data and methods which are as comprehensive as possible. Publications of this nature, like the American "state-of-the-art reports", are intended to serve as a general guiding principle or as a source of current information to assist the engineer, the chemist, the technician and the student.

In the projects which it has supported, the Deutsche Gesellschaft für Technische Zusammenarbeit (GTZ) GmbH has found that professionals working in developing countries who are confronted with a broad spectrum of tasks, and the counterparts being trained or advised there, are the very people who often miss such compilations and reference works.

The GTZ has therefore commissioned a series of publications directly related to specific projects, which are designed to meet this need. The titles published to date are:
- "Technologie des Trinkwassers" (Drinking Water Technology). Eds.: Fresenius and Schneider, 1980, also available in French.
- "Methodensammlung zur Wasseruntersuchung" (A Compilation of Methods for Water Analysis). Eds.: Fresenius and Schneider, 1977, 3 volumes, also available in French.
- "La construction des canalisations - exécution et contrôle" (Water Conduit Construction - Execution and Monitoring). Forschungsinstitut für Wassertechnologie, Aachen and the Deutsche Gesellschaft für Technische Zusammenarbeit (GTZ) GmbH, Eschborn; workshop handbook, 1980.

- "Traitement des eaux résiduaires industrielles" (Treatment of Industrial Waste Water). Forschungsinstitut für Wassertechnologie, Aachen and the Deutsche Gesellschaft für Technische Zusammenarbeit (GTZ) GmbH, Eschborn; workshop handbook, 1981.

This publication is conceived as a further component which, in conjunction with the existing literature, provides a complete overview as to the state of the art in the water supply and sanitation sector. It focuses on problems relating to the diversion, analysis and treatment of industrial and municipal sewage.

The authors have consciously avoided descriptions of procedures and technologies relevant to smaller settlements and rural regions in the Third World. There are already special publications available in this field (including those published by the World Bank and WHO; see bibliography), which explain the special project strategies and appropriate technologies applicable to decentralized water supply and sanitation concepts. Technologies of this nature - such as low-cost sanitation - are currently at the top of the agenda in discussions within projects of Technical Cooperation with developing countries, and rightly so. Nevertheless, the inadequate sanitation and the associated strain on the environment which occur in urban agglomerations in the Third World, as a result of socioeconomic factors (rural exodus and urbanization, industrialization, formation of slums), should not be dismissed as being of lesser importance: here too there is a growing desire amongst local decision-making bodies, training institutions and target groups for effective support. With the right development measures, the quality of life of large sections of the population can be improved.

This work was produced jointly by practicians from the Institut Fresenius, Taunusstein and the Forschungsinstitut für Wassertechnologie at the Rheinisch-Westfälische Technische Hochschule, Aachen.

The GTZ wishes to thank the authors for their dedicated work, and express its hope that the publication will meet with a broad cross-section of readers. It has been designed both as a textbook to familiarize those undergoing training with knowledge of the fundamentals, and for use as an aid by those providing that training, or acting as advisers. Naturally, the work is primarily aimed at the practician who, in the course of project planning and design, construction, operational back-up, running of the laboratory, and maintenance and rehabilitation measures needs a manual to assist him in solving problems.

Eschborn, 1988

Dr. Klaus Erbel
(Head of the GTZ Hydraulic Engineering
and Water Resources Development Division)

Contents

1	INTRODUCTION	1
1.1	CLASSIFICATION OF WASTE WATER	1
1.2	COMPOSITION OF WASTE WATERS	3
1.3	HARMFUL EFFECTS OF WASTE WATER ON NATURAL WATERS	6
2	TYPES AND AMOUNT OF WASTE WATER	12
2.1	DOMESTIC SEWAGE WATER	12
2.1.1	Origin and Types of Domestic Sewage	12
2.1.2	Amount and Composition of Sewage Water in Private Households	13
2.1.2.1	Amount of Domestic Sewage	13
2.1.2.2	Composition of Domestic Sewage	17
2.1.3	Amount and Composition of Sewage Water from Public Buildings	19
2.1.4	Amount and Composition of Sewage Water from Tourist Facilities	21
2.1.4.1	Restaurants	21
2.1.4.2	Hotels	21
2.1.4.3	Camping Facilities	22
2.1.4.4	Organized Beaches	22
2.2	MUNICIPAL SEWAGE WATER	23
2.2.1	Amount of Municipal Sewage Water	23
2.2.2	Fluctuations in the Amount of Sewage Water	29
2.2.3	Sewage Water per Inhabitant	31
2.2.4	Waste Concentrations	40
2.2.5	Fluctuations in the Waste Concentration and Waste Load	41
2.3	FOREIGN WATER	42
2.3.1	Types of Foreign Water	42
2.3.2	Origin and Amount of Foreign Water	43
2.4	COMMERCIAL AND INDUSTRIAL WASTE WATER AMOUNTS	44
2.4.1	Inorganic Industrial Waste Water	60
2.4.1.1	Waste Waters from Nonmetallic Minerals Industries	61

2.4.1.2	Waste Waters from the Metal Processing Industry	62
2.4.1.2.1	Waste Waters from Iron Works	64
2.4.1.2.2	Waste Waters from Steel and Rolling Mills	64
2.4.1.2.3	Waste Waters from Machinery Production and Mechanical Workshops	66
2.4.1.2.4	Waste Waters from Metal Pickling Plants	66
2.4.1.2.5	Waste Waters from Eloxal Works	70
2.4.1.2.6	Waste Waters from Electroplating Plants	71
2.4.1.3	Waste Waters from Mines and Ore Dressing Plants	74
2.4.1.3.1	Pit Waters	74
2.4.1.3.2	Pithead Bath Waste Waters	75
2.4.1.3.3	Waste Waters from Coal Washing and Dressing	76
2.4.1.3.4	Waste Waters from Hard-Coal Coking Plants	77
2.4.1.3.5	Coke Quenching Water	78
2.4.1.3.6	Waste Waters from Charcoal Manufacture	78
2.4.1.4	Waste Waters from the Chemical Industry	79
2.4.1.4.1	Waste Waters from Mineral Acid Production	79
2.4.1.4.2	Waste Waters from the Potash Industry	80
2.4.1.4.3	Waste Waters from Soda Factories	81
2.4.1.4.4	Waste Waters from Fertilizer Production	82
2.4.1.4.5	Waste Waters from Antimony Pentasulfide Production	85
2.4.2	Organic Industrial Waste Water	85
2.4.2.1	Waste Waters from the Pharmaceutical and Cosmetic Industries	86
2.4.2.1.1	Waste Waters from the Pharmaceutical Industry	86
2.4.2.1.2	Waste Waters from the Cosmetic Industry	87
2.4.2.2	Waste Waters from Dyestuff Manufacturing	87
2.4.2.2.1	Waste Waters from Inorganic Dyestuffs	88
2.4.2.2.2	Waste Waters from Organic Dyestuffs	89
2.4.2.3	Waste Waters from Soap and Synthetic Detergents Production	91
2.4.2.3.1	Waste Waters from Soap Production with Fatty Acids	91
2.4.2.3.2	Waste Waters from Synthetic Detergent Production	93
2.4.2.4	Waste Waters from Plastics Industries	98
2.4.2.4.1	Waste Waters from Cellulose-based Plastics Industry	98
2.4.2.4.2	Waste Waters from Plastics Production (Condensation Products)	99
2.4.2.4.3	Waste Waters from Plastics Production (Polymerization Products)	101
2.4.2.5	Waste Waters from Tanneries and Leather Producing Plants	103
2.4.2.5.1	Waste Waters from Tanneries	103
2.4.2.5.2	Waste Waters from Leather Producing Plants	109
2.4.2.6	Waste Waters from the Textile Industry	109
2.4.2.6.1	Waste Waters from Spinning Mills	110
2.4.2.6.1.1	Waste Waters from Flax and Hemp Retteries	110
2.4.2.6.1.2	Waste Waters from Silk Boiling	114
2.4.2.6.1.3	Waste Waters from Artificial Silk and Rayon Staple Production	116
2.4.2.6.2	Waste Waters from Cotton Bleaching	120

2.4.2.6.3	Waste Waters from Wool Washing	121
2.4.2.6.4	Waste Waters from Cloth Production	122
2.4.2.7	Waste Waters from Wood Processing, Pulp, Paper and Cardboard Mills	126
2.4.2.7.1	Waste Waters from Wood Processing and Mechanical Wood Pulp Mills	127
2.4.2.7.2	Waste Waters from Chemical Pulp Production	130
2.4.2.7.2.1	Waste Waters from Sulfite Pulp Production	130
2.4.2.7.2.2	Waste Waters from Sulfate Pulp Production	132
2.4.2.7.3	Waste Waters from Straw Pulp Production	132
2.4.2.7.4	Waste Waters from Paper Mills	132
2.4.2.7.5	Waste Waters from Strawboard Production	135
2.4.2.7.6	Waste Waters from Wood-Fiber Board Production by Wet Processing	136
2.4.2.8	Waste Waters from Oil Refining Industry	138
2.4.2.8.1	Waste Waters from Oilfields	138
2.4.2.8.2	Waste Waters from Oil Refineries	139
2.4.2.8.3	Waste Waters from Petrochemical Plants	142
2.4.2.8.4	Waste Waters Containing Oil from Oil Storage, Gasoline Stations and Workshops	144
2.4.2.9	Waste Waters from the Food Processing Industry	145
2.4.2.9.1	Waste Waters from Sugar Refineries	154
2.4.2.9.1.1	Beet Sugar	154
2.4.2.9.1.2	Sugar from Sugar Cane	156
2.4.2.9.2	Waste Waters from Milk Processing Plants	159
2.4.2.9.3	Waste Waters from Non-alcoholic Beverage Production	164
2.4.2.9.4	Waste Waters from Breweries	166
2.4.2.9.5	Waste Waters from Margarine, Edible Fat, and Edible Oil Production	170
2.4.2.9.6	Waste Waters from Slaughterhouses and Meat Processing Plants	173
2.4.2.9.7	Waste Waters from Fish Canning Factories	176
2.4.2.9.8	Waste Waters from Fruit and Vegetable Canning	179
2.4.2.9.9	Waste Waters from Starch Production and Potato Processing	183
2.4.2.9.9.1	Waste Waters from Potato Starch Production	183
2.4.2.9.9.2	Waste Waters from Wheat Starch Production	186
2.4.2.9.9.3	Waste Waters from Maize Starch Production	188
2.4.2.9.9.4	Waste Waters from Rice Starch Production	189
2.4.2.9.9.5	Waste Waters from Starch Sugar and Syrup Production	189
2.4.2.9.9.6	Waste Waters from Dried Potato Products	189
2.4.2.9.9.7	Waste Waters from Potato Chip Production	190
2.4.2.9.10	Waste Waters from Wine Production	190
2.4.2.10	Waste Waters from Processing Vegetable and Animal Waste Products	193
2.4.2.10.1	Agricultural Waste Waters	193
2.4.2.10.1.1	Waste Waters from Pectin Production	196
2.4.2.10.1.2	Waste Waters from Gut String and Gut Processing Plants	198
2.4.2.10.1.3	Utilization of Animal Excreta	199
2.4.2.10.2	Waste Waters from Carcass Disposal Plants	200
2.4.2.10.3	Waste Waters from Fish Meal Production	202

2.5	STORM WATER	204
2.5.1	Types of Precipitation	204
2.5.2	Contamination of Rainwater	206
2.5.2.1	Atmospheric Impurities	206
2.5.2.2	Ground Surface Impurities	207
2.5.3	Contamination of Wet Weather Flow in the Sewerage System	211
2.5.3.1	Contamination of Effluent in Combined Sewerage Systems	211
2.5.3.2	Contamination of Effluent in Separate Storm Drainage Systems	212
2.6	OTHER INDUSTRIAL WATERS	213
2.6.1	Sea Water	213
2.6.2	Reutilized Water	214
2.6.2.1	Multiple Use and Recirculation of Industrial Waste Waters	217
2.6.2.1.1	Reutilization of Waste Waters within the Plant	217
2.6.2.2	Utilization of Waste Waters for Other Purposes	222
2.6.2.2.1	Reutilization of Waste Waters for Drinking and Bathing Purposes	222
2.6.2.2.2	Indirect Utilization of Waste Water for Obtaining Drinking Water	223
2.6.2.2.3	Utilization of Waste Waters in Dual Purpose Plants	224
2.6.2.2.3.1	Utilization of Waste Waters for Recharging Groundwater	224
2.6.2.2.4	Utilization of Waste Waters in Recreation Areas	225
2.6.2.2.5	Utilization of Waste Waters for Agricultural Purposes	226
2.6.2.2.6	Recovery of Valuable Residues from Industrial Waste Waters	228
3	WASTE WATER DISPOSAL IN RURAL AREAS AND SMALL VILLAGES	233
4	COLLECTION AND DRAINAGE OF WASTE WATER	234
4.1	DEFINITIONS	235
4.1.1	Methods of Waste Water Disposal	235
4.1.2	Parts of the Sewerage System	236
4.1.3	Waste Water Disposal Scheme	238
4.2	TYPES OF DRAINAGE SYSTEMS AND DRAINAGE METHODS	240
4.2.1	Drainage or Waste Water Disposal Systems	240
4.2.2	Drainage Methods	241
4.2.2.1	Combined and Separate Sewage Systems	241
4.2.2.2	Pressure Drainage and Vacuum Drainage	244
4.2.2.2.1	Pressure Drainage Systems	245
4.2.2.2.2	Vacuum Drainage Systems	246
4.2.2.3	Other Methods	248

4.3	PLANNING OF DRAINAGE SYSTEMS	249
4.3.1	Preliminary Design	250
4.3.2	Criteria for the Design of a Waste Water Disposal System	251
4.3.2.1	Choice of Drainage System	251
4.3.2.2	Choice of Drainage Method	254
4.3.2.3	Planning a Municipal Sewerage System	255
4.3.2.4	Removal of Deposits from the Sewers	257
4.3.2.5	Choice of Receiving Water	258
4.3.2.6	Choice of Scheme for the Sewer Network	259
4.3.2.7	Allocation and Delimitation of Drainage Area for Each Sewer	263
4.3.3	Calculation of Sewage Volume	264
4.3.3.1	Calculation of Domestic Sewage Volume	265
4.3.3.2	Calculation of Foreign Water	266
4.3.3.3	Calculation of Commercial and Industrial Waste Waters	267
4.3.3.4	Calculation of Dry Weather Flow	267
4.3.3.5	Calculation of Storm Water	270
4.3.3.6	Calculation of Flow from Surface Waters	275
4.3.3.7	Completing the Calculations	276
4.3.4	Hydraulic Calculation of the Sewer Network	278
4.3.4.1	General Characteristic Parameters	278
4.3.4.2	Theoretical Foundations	281
4.3.4.3	Calculation of Pipelines	284
4.3.4.3.1	Calculation of Pipes and Channels	284
4.3.4.3.1.1	Determining Frictional Losses	284
4.3.4.3.1.2	Determining the Total Discharge	300
4.3.4.3.1.3	Determining the Closed Sewer Cross Sections with Natural Water Head	300
4.3.4.3.1.4	Determining the Open Sewer Cross Sections	306
4.3.4.3.1.5	Determining the Cross Sections for Pressure Pipes	307
4.3.4.3.1.6	Calculation of Overflows	309
4.3.4.3.1.7	Hydraulic Waterhammer	318
4.3.5	Static Calculation of Sewer Lines	320
4.3.6	Layout of Sewers, Gradients, and Depths	322
4.3.6.1	Routing of Sewers	322
4.3.6.2	Gradient	324
4.3.6.3	Hydraulic Conditions	327
4.3.6.4	Graphic Representation	330
4.4	BUILDING COMPONENTS OF DRAINAGE SYSTEMS	335
4.4.1	Conduits	337
4.4.1.1	Open Channels	337
4.4.1.2	Closed Conduits	338
4.4.1.2.1	Concrete and Reinforced Concrete Pipes	338
4.4.1.2.2	Vitrified Clay Pipes	340
4.4.1.2.3	In situ Concrete Sewers and Special Construction	341
4.4.1.2.4	Asbestos Cement Pipes	342
4.4.1.2.5	Plastic Pipes and Plastic Coated Pipes	342

4.4.1.2.6	Cast Iron Pipes	343
4.4.1.2.7	Steel Pipes	344
4.4.2	Structures	344
4.4.2.1	House Connections	344
4.4.2.2	Road Gulleys	345
4.4.2.3	Intake Structures	347
4.4.2.4	Manholes	348
4.4.2.5	Junctions	349
4.4.2.6	Drop Manholes	352
4.4.2.7	Flushing Devices	352
4.4.2.8	Storm Water Surplusing Works	353
4.4.2.8.1	Storm Overflows	353
4.4.2.8.2	Storm-Water Tanks	354
4.4.2.8.3	Regulators	355
4.4.2.9	Pumping Stations	356
4.4.2.10	Sewer Crossings	363
4.5	SEWER CONSTRUCTION	363
4.5.1	First Stage of Construction - Setting Out Sewer Lines	364
4.5.2	Excavation	369
4.5.3	Sheeting and Bracing	371
4.5.3.1	Horizontal Sheeting	371
4.5.3.2	Vertical Sheeting	378
4.5.4	Pipe Bedding	385
4.5.4.1	Prefabricated Pipes	385
4.5.5	Special Structures	397
4.5.5.1	Sewer Outfalls	397
4.5.5.2	Sheeting of Large Trenches	398
4.5.5.3	Weirs	398
4.5.5.4	Special Formwork	400
4.6	OPERATION AND MAINTENANCE OF SEWER NETWORKS	400
4.6.1	Official Acceptance of the Sewer System	403
4.6.2	Operation and Maintenance of Sewers and Auxiliary Facilities	406
4.6.2.1	Operation	406
4.6.2.2	Safety Measures in the Sewer System	408
5	CONDITIONS FOR THE DISPOSAL OF INDUSTRIAL AND MUNICIPAL WASTE WATER AND SLUDGES	410
5.1	GENERAL REQUIREMENTS FOR DRAINAGE OF INDUSTRIAL EFFLUENTS IN PUBLIC SEWERAGE SYSTEMS (Indirect Discharge)	410
5.1.1	General Requirements and Regulations for Sewerage Systems	413
5.1.2	General Requirements and Regulations for Public Sewage Treatment Plants	414
5.2	GENERAL REQUIREMENTS AND REGULATIONS FOR THE DISCHARGE OF INDUSTRIAL WASTES INTO WATER BODIES (Direct Discharge)	419

5.2.1	Classification System for Water Quality	420
5.2.2	Limit Values for Immission	423
5.2.3	Limit Values for Emission	424
5.2.4	National and International Standard Values	425
5.2.4.1	LAWA Emission Values (Recommended)	425
5.2.4.2	Minimum Requirements (Emission Values)	444
5.2.4.3	EC Standards	448
5.2.4.4	French Guidelines	457
5.3	CHECKING OF PRESCRIBED PARAMETERS	460
5.3.1	Checking the Emission Values at the Treatment Plant Inlet	460
5.3.2	Checking the Emission Values at the Treatment Plant Outlet	462
5.3.3	Checking the Immission Values	462
5.4	STATUTORY REGULATIONS FOR SLUDGE USAGE IN AGRICULTURE	463
6	TREATMENT OF WASTE WATER	467
6.1	METHODS OF WASTE WATER TREATMENT	470
6.2	MECHANICAL PURIFICATION	477
6.2.1	Coarse Separation	479
6.2.1.1	Grating	480
6.2.1.2	Screening	484
6.2.1.3	Filters	488
6.2.1.3.1	Filters with Coarse Sand or Fine Gravel Fillings	489
6.2.1.3.2	Drum Filters	492
6.2.1.3.3	Suction Filters	494
6.2.1.3.4	Disk Filters	494
6.2.1.3.5	Pressure Filters	495
6.2.2	Gravity Separation	495
6.2.2.1	Flotation Systems	498
6.2.2.1.1	Fat and Oil Separators	503
6.2.2.1.2	Mineral Oil Separators	506
6.2.2.1.3	Gasoline Separators	510
6.2.2.2	Sedimentation Systems	512
6.2.2.2.1	Sand Traps	515
6.2.2.2.2	Sedimentation Basins	521
6.2.2.2.2.1	Rectangular Basins	525
6.2.2.2.2.2	Round Basins	529
6.2.2.2.2.3	Two-level Settling Systems	532
6.2.2.2.2.4	Lamella Sedimentation Basins	536
6.2.2.2.2.5	Vertical Flow Sedimentation Basins	537
6.2.2.2.2.6	Other Types of Sedimentation Basins	538
6.2.3	Equilibration Basins	539
6.3	CHEMICAL PROCESSES OF WASTE WATER TREATMENT	541
6.3.1	Methods of Treatment	541
6.3.1.1	Neutralization	542

6.3.1.2	Flocculation	544
6.3.1.2.1	The General Flocculation Process	544
6.3.1.2.2	Precipitation by Flocculation as a Single Stage Waste Water Treatment, Preliminary Precipitation	550
6.3.1.2.3	Secondary Precipitation	551
6.3.1.2.4	Simultaneous Precipitation	553
6.3.1.3	Adsorption	555
6.3.1.4	Extraction by Solvents	558
6.3.1.5	Membrane Separation (Ultrafiltration, Reverse Osmosis)	559
6.3.1.6	Distillation	562
6.3.1.7	Oxidation	562
6.3.1.8	Reduction	563
6.3.1.9	Ion Exchange Process	564
6.3.1.10	Other Methods of Physico-Chemical Treatment	566
6.3.1.10.1	Stripping	566
6.3.1.10.2	Aeration	566
6.3.1.10.3	Gassing with Carbonic Acid or Flue Gases	567
6.3.1.10.4	Evaporation	567
6.3.1.10.5	The Freezing Method	567
6.3.1.10.6	Crystallization	567
6.3.1.10.7	Electrolysis	567
6.3.1.10.8	Dialysis	568
6.3.2	Utilization of Chemicals in Waste Water Treatment	568
6.3.2.1	Commercial Aluminum and Iron Salts	569
6.3.2.2	Other Iron and Aluminum Salts	573
6.3.2.3	Commercial Polymer Flocculation Agents	573
6.3.2.4	Hydrate of Lime	574
6.4	BIOLOGICAL METHODS OF WASTE WATER TREATMENT	575
6.4.1	General Fundamentals	576
6.4.1.1	The Nature of Microorganisms	576
6.4.1.2	Processes in Metabolism	582
6.4.1.2.1	Factors Influencing the Biological Processes in the Treatment of Commercial Waste Waters	584
6.4.2	Processes of Decomposition	586
6.4.2.1	Aerobic Decomposition of Organic Carbon Compounds	586
6.4.2.2	Nitrification	587
6.4.2.3	Denitrification	591
6.4.2.4	Desulfurization	593
6.4.2.5	Fermentation Processes	594
6.4.2.6	Other Microorganisms Involved in the Purification Processes	595
6.4.3	Requirements for the Use of Biological Processes	596
6.4.3.1	Relationship between BOD_5 and COD	596
6.4.3.2	pH Value	597
6.4.3.3	Temperature	599
6.4.3.4	Necessary Minerals	601
6.4.3.5	Inhibiting Agents and Toxicants	603

6.4.3.5.1	Determination of Threshold Values for Substances	604
6.4.3.5.2	Disturbance of the Methanogenic Fermentation of Sewage Sludge in Municipal Sewage Treatment Plants	606
6.4.3.5.3	Disturbance of the Aerobic Decomposition in the Sludge Activation System in Municipal Sewage Treatment Plants Caused by Chromate	607
6.4.3.5.4	Disturbance in the Biological Purification Processes	607
6.4.3.5.5	Disturbance of the Purification Processes by Heavy Metals	608
6.4.4	Biological Methods of Waste Water Treatment	610
6.4.4.1	Systems for Biological Waste Water Treatment	613
6.4.4.1.1	Utilization of Agricultural Waste Water	614
6.4.4.1.1.1	Fundamental Principles	614
6.4.4.1.1.2	General Conditions	615
6.4.4.1.1.3	Design of Irrigation Systems	621
6.4.4.1.1.4	Methods of Irrigation	623
6.4.4.1.1.4.1	Extensive Land Irrigation Fields	623
6.4.4.1.1.4.2	Sewage Lands	625
6.4.4.1.1.5	Irrigation Facilities	626
6.4.4.1.1.5.1	The Mechanical Sewage Plant	626
6.4.4.1.1.5.2	The Distribution Network	627
6.4.4.1.1.5.3	Pump Stations	627
6.4.4.1.1.5.4	Drainage Network	628
6.4.4.1.1.5.5	Accessory Facilities	629
6.4.4.2	Waste Water Ponds	629
6.4.4.2.1	Sedimentation Ponds	633
6.4.4.2.2	Nonaerated Waste Water Ponds	634
6.4.4.2.3	Aerated Waste Water Ponds	636
6.4.4.2.4	Polishing Ponds	637
6.4.4.2.5	Safety Ponds	638
6.4.4.2.6	Waste Water Fish Ponds	638
6.4.4.3	The Biofiltration Process	638
6.4.4.3.1	General Fundamentals	639
6.4.4.3.2	Components of Biological Filters	646
6.4.4.3.3	Biological Filter Design	650
6.4.4.3.4	Biological Filter Towers	658
6.4.4.4.	Immersed Biological Filters	659
6.4.4.5	Sludge Activation Systems	667
6.4.4.5.1	General Fundamentals	667
6.4.4.5.2	The Activated Sludge	668
6.4.4.5.3	Oxygen	672
6.4.4.5.3.1	Aeration with Atmospheric Air	675
6.4.4.5.3.2	Procedural Methods	676
6.4.4.5.3.3	Aeration with Pure Oxygen (Gassing with Oxygen)	680
6.4.4.5.4	Aeration Equipment	684
6.4.4.5.4.1	Pressurization	684
6.4.4.5.4.2	Surface Aeration	688
6.4.4.5.4.3	Combined and Other Aeration Equipment	693
6.4.4.5.5	Design of Sludge Activation Systems	694
6.4.4.5.6	Two-Stage Sewage Plants	702
6.4.4.5.6.1	Two-Stage Biological Filter Systems	703
6.4.4.5.6.2	Two-Stage Sludge Activation Systems	703
6.4.4.5.6.3	Adsorption Processes in Sludge Activation	705

6.4.4.5.6.4	Biological Filters Combined with Activation Processes	706
6.4.4.5.6.5	Biological Filter Method with Preceding Activation	708
6.4.4.5.6.6	Downstream Polishing Ponds	710
6.4.4.6	Anaerobic Waste Water Treatment	713
6.4.4.7	Final Clarification	716
6.4.4.7.1	Final Clarification Basins Downstream of Biological Filter Systems	716
6.4.4.7.2	Final Clarification Basins in Activation Systems	720
6.4.4.7.2.1	Final Clarification Basins with Horizontal Flow	722
6.4.4.7.2.2	Final Clarification Basins with Vertical Flow	730
6.4.4.7.3	Notes on the Structural Design of Final Clarification Basins	735
6.5	**METHODS OF PHYSICAL AND PHYSICO-CHEMICAL WASTE WATER TREATMENT**	737
6.5.1	Removal of Suspended Matter	739
6.5.1.1	Installations in the Sedimentation Containment	739
6.5.1.2	Microstraining	741
6.5.1.3	Filtration	743
6.5.1.3.1	Sand Filters	745
6.5.1.3.2	Land Filtration	747
6.5.1.3.3	Ultrafiltration and Hyperfiltration	748
6.5.2	Elimination of Dissolved Organic Substances	750
6.5.2.1	Adsorption by Activated Charcoal	751
6.5.2.2	Oxidation by Ozone	756
6.5.2.3	Desorption of Volatile Substances	756
6.5.2.4	Reverse Osmosis (Hyperfiltration)	756
6.5.3	Elimination of Nutrients	758
6.5.3.1	Nitrogen Elimination	759
6.5.3.1.1	Nitrogen Elimination by Microbiological Processes	760
6.5.3.1.1.1	Nitrogen Elimination in Activation Systems	760
6.5.3.1.1.2	Nitrogen Elimination in Fixed-Bed Reactors	760
6.5.3.1.1.3	Nitrogen Elimination in Oxidation Ponds	761
6.5.3.1.2	Nitrogen Elimination by Physicochemical Processes	762
6.5.3.1.2.1	Ammonia Desorption (Stripping)	762
6.5.3.1.2.2	Selective Ion Exchange	763
6.5.3.1.2.3	Breakpoint Chlorination	763
6.5.3.2	Phosphorus Elimination	764
6.5.3.2.1	Phosphorus Elimination by Biological Treatment Processes	765
6.5.3.2.1.1	Conventional Sludge Activation Processes	765
6.5.3.2.1.2	Pond Treatment	765
6.5.3.2.1.3	Phosphate Elimination by Algae	766
6.5.3.2.2	Phosphorus Elimination by Physicochemical Processes	767
6.5.3.2.2.1	Iron Phosphate Precipitation	767
6.5.3.2.2.2	Aluminum Phosphate Precipitation	768
6.5.3.2.2.3	Calcium Phosphate Precipitation	768
6.5.3.2.2.4	Methods of Treatment	768
6.5.4	Elimination of Dissolved Inorganic Substances	769
6.5.4.1	Ion Exchange Methods	770

6.5.4.2	Electrodialysis	770
6.5.4.3	Hyperfiltration	771
6.6	TREATMENT OF COMMERCIAL AND INDUSTRIAL WASTE WATERS	772
6.6.1	Inorganic Industrial Waste Water	773
6.6.1.1	Waste Waters from Mortar, Sandy Limestone, Cement and Porcelain Production Plants	773
6.6.1.2	Waste Waters from Metal-Processing Industries	776
6.6.1.2.1	Waste Waters from Ironworks	776
6.6.1.2.2	Waste Waters from Steelworks and Rolling Mills	778
6.6.1.2.3	Waste Waters from Machine Tool Production and Mechanical Workshops	779
6.6.1.2.4	Waste Waters from Metal-Pickling Plants	779
6.6.1.2.5	Waste Waters from Anodizing Plants	781
6.6.1.2.6	Other Waste Waters from Metal-Processing Industries	782
6.6.1.2.6.1	Chromate Detoxication	784
6.6.1.2.6.2	Cyanide Detoxication	787
6.6.1.2.6.3	Neutralization	794
6.6.1.3	Waste Waters from Mines and Ore-Dressing Plants	798
6.6.1.3.1	Mine Waters	798
6.6.1.3.2	Waters from Pithead Baths	798
6.6.1.3.3	Waste Waters from Coal Washing	799
6.6.1.3.4	Waste Waters from Coal Coking Plants	799
6.6.1.3.5	Waste Waters from Coal Gasification	801
6.6.1.3.6	Waste Waters from Charcoal Production	802
6.6.1.4	Waste Waters from Chemical Industry	802
6.6.1.4.1	Waste Waters from Mineral Acid Production	802
6.6.1.4.2	Waste Waters from Potash Industry	803
6.6.1.4.3	Waste Waters from Soda Works	804
6.6.1.4.4	Waste Waters from Fertilizer Production	804
6.6.1.4.5	Waste Waters from Antimony Red Production	805
6.6.2	Organic Industrial Waste Water	805
6.6.2.1	Waste Waters from Pharmaceutical and Cosmetics Industry	806
6.6.2.1.1	Waste Waters from Pharmaceutical Industry	806
6.6.2.1.2	Waste Waters from Cosmetics Industry	807
6.6.2.2	Waste Waters from Dye and Colorant Production	807
6.6.2.2.1	Inorganic Dye and Colorant Waste Waters	807
6.6.2.2.2	Organic Dye and Colorant Waste Waters	808
6.6.2.3	Waste Waters from Soap and Synthetic Detergent Production	810
6.6.2.3.1	Waste Waters from Soap Production	810
6.6.2.3.2	Waste Waters from Synthetic Detergent Production	811
6.6.2.4	Waste Waters from Plastics Industries	812
6.6.2.4.1	Waste Waters from Plastics Industry Producing Cellulose	812
6.6.2.4.2	Waste Waters from Plastics Industry Producing Condensed Products	812
6.6.2.4.3	Waste Waters from Plastics Industry Producing Polymerized Products	813
6.6.2.5	Waste Waters from Tanneries and Leather-Producing Plants	814
6.6.2.5.1	Waste Waters from Tanneries	814

6.6.2.5.2	Waste Waters from Leather Production	815
6.6.2.6	Waste Waters from Textile Industry	815
6.6.2.6.1	Waste Waters from Spun Fabric Production	816
6.6.2.6.1.1	Waste Waters from Flax Retting and Hemp Roasting	816
6.6.2.6.1.2	Waste Waters from Silk Cooking	817
6.6.2.6.1.3	Waste Waters from Rayon and Staple Fiber Production	817
6.6.2.6.2	Waste Waters from Cotton Bleaching	819
6.6.2.6.3	Waste Waters from Wool Washing	819
6.6.2.6.4	Waste Waters from Cloth Production	820
6.6.2.7	Waste Waters from Cellulose, Paper, and Cardboard Production	821
6.6.2.7.1	Waste Waters from Pulp Factories	821
6.6.2.7.2	Waste Waters from Cellulose Production	822
6.6.2.7.2.1	Waste Waters from Sulfite Cellulose Production	822
6.6.2.7.2.2	Waste Waters from Sulfate Cellulose Production	824
6.6.2.7.2.3	Waste Waters from Chemical Straw Pulp Production	824
6.6.2.7.2.4	Waste Waters from Paper Mills	825
6.6.2.7.2.5	Waste Waters from Straw-Board Production	827
6.6.2.7.2.6	Waste Waters from Wood-Fiber Board Production in a Wet Process	827
6.6.2.8	Waste Waters from Petroleum Processing Industry	828
6.6.2.8.1	Waste Waters from Crude Oil Production in Oil Fields	828
6.6.2.8.2	Waste Waters from Mineral Oil Refineries	829
6.6.2.8.3	Waste Waters from Petrochemical Plants	830
6.6.2.8.4	Waste Waters from Gasoline Stations	831
6.6.2.9	Waste Waters from Foodstuff Production	831
6.6.2.9.1	Waste Waters from Sugar Factories	831
6.6.2.9.2	Waste Waters from Milk Processing Plants	833
6.6.2.9.3	Waste Waters from Production of Nonalcoholic Beverages	834
6.6.2.9.4	Waste Waters from Breweries	834
6.6.2.9.5	Waste Waters from Margarine, Edible Fat, and Oil Production	835
6.6.2.9.6	Waste Waters from Slaughterhouses and Meat Processing Plants	836
6.6.2.9.7	Waste Waters from Fish Curing Plants	837
6.6.2.9.8	Waste Waters from Fruit and Vegetable Canneries	838
6.6.2.9.9	Waste Waters from Starch Production and Potato Chip Processing Plants	838
6.6.2.9.10	Waste Waters from Wine and Champagne Production	839
6.6.2.10	Waste Waters from Rendering Plants	840
6.6.2.11	Waste Waters from Fish Meal Production	841
6.7	HANDLING RESIDUES FROM WASTE WATER TREATMENT	842
6.7.1	Screenings	843
6.7.2	Sand Trappings	845
6.7.3	Matter Collected in the Oil and Fat Separators	846
6.7.4	Sludge Treatment	847
6.7.4.1	Sludge Volume and Composition	848
6.7.4.1.1	Sludge Volume	849

6.7.4.1.2	Sludge Composition	853
6.7.4.2	Crude Sludge Stabilization	859
6.7.4.2.1	Anaerobic Stabilization (Digestion)	863
6.7.4.2.1.1	Unheated Sludge Digesters	863
6.7.4.2.1.2	Heated Sludge Digesters	865
6.7.4.2.2	Aerobic Stabilization	877
6.7.4.3	Sludge Dehydration	883
6.7.4.3.1	Thickening	886
6.7.4.3.2	Natural Dehydration	890
6.7.4.3.3	Sludge Conditioning	893
6.7.4.3.4	Technical Dewatering	897
6.7.4.3.4.1	Dewatering in Centrifuges	898
6.7.4.3.4.2	Dewatering with Band Filters	900
6.7.4.3.4.3	Dewatering with Filter Presses	902
6.7.4.3.4.4	Dewatering by Vacuum Filtration	904
6.7.4.4	Sludge Utilization and Removal	905
6.7.4.4.1	Sewage Sludge Utilization in Agriculture	905
6.7.4.4.1.1	Pasteurization	906
6.7.4.4.1.2	Aerobic Thermophilic Sludge Treatment	907
6.7.4.4.1.3	Sludge Consolidation with Quicklime	908
6.7.4.4.1.4	Silo Composting	908
6.7.4.4.1.5	Composting in Bio-Reactors	909
6.7.4.4.1.6	Problems with Heavy Metals	909
6.7.4.4.2	Sewage Sludge Deposition in Sanitary Landfills	912
6.7.4.4.3	Sewage Sludge Composting	914
6.7.4.4.4	Sludge Discharge in the Open Sea	914
6.7.4.4.5	Sludge Drying and Incineration	914
7	**SAMPLING, ANALYSIS AND CLASSIFICATION OF WASTE WATERS AND SEWAGE SLUDGE**	916
7.1	WASTE WATER AND SLUDGE SAMPLING	916
7.1.1	Site Inspection	917
7.1.2	Methods of Waste Water Sampling	919
7.1.3	Random Sampling	921
7.1.4	Time-Dependent Sampling	922
7.1.5	Volume-Dependent Sampling	924
7.1.5.1	Volumetric Measurement	925
7.1.5.2	Volume-Dependent Manual Sampling	930
7.1.5.3	Volume-Dependent Sampling Using Instruments	931
7.1.6	Waste Water Sampling in Practice	934
7.1.6.1	Sample Preservation	935
7.1.6.2	Waste Water Volumes for Analysis	938
7.1.6.3	Bottles	939
7.1.6.3.1	Glass Bottles	939
7.1.6.3.2	Plastic Flasks	940
7.1.6.3.3	Special Bottling	941
7.1.6.4	Sludge Sampling	942

7.2	TESTING ON THE SITE	943
7.2.1	Temperature	944
7.2.2	Sensory Testing in Situ	945
7.2.3	Density	948
7.2.4	pH Value	950
7.2.5	Electric Conductivity	952
7.2.6	Redox Potential	954
7.2.7	Oxygen	955
7.2.8	Electrometric Determination of Dissolved Oxygen	955
7.2.9	Sedimentary Matter	958
7.3	LABORATORY ANALYSES	959
7.3.1	General	959
7.3.2	Inorganic Substances	961
7.3.2.1	Ammonia	961
7.3.2.2	Nitrite	965
7.3.2.3	Nitrate	968
7.3.2.4	Phosphate	970
7.3.2.5	Sulfide Sulfur	975
7.3.2.6	Sulfite	980
7.3.2.7	Sulfate	982
7.3.2.8	Cyanide	984
7.3.2.8.1	Total Cyanide	986
7.3.2.8.1.1	Spectrophotometric Determination Utilizing Barbituric Acid Pyridine	991
7.3.2.8.1.2	Titrimetric Determination Utilizing Silver Nitrate	994
7.3.2.8.2	Easily Releasable Cyanide	996
7.3.3	Organic Substances	998
7.3.3.1	Oxidizability	1001
7.3.3.2	Chemical Oxygen Demand (COD)	1007
7.3.3.2.1	Method Utilizing Potassium Dichromate	1007
7.3.3.2.2	Accelerated Method	1013
7.3.3.2.3	Removal of Hg and Ag from Residue Solutions for COD Determination	1017
7.3.3.2.4	Total Organically Bound Carbon (TOC) and Dissolved Organic Carbon (DOC)	1018
7.3.3.3	Biochemical Oxygen Demand (BOD)	1022
7.3.3.3.1	BOD_5 Determination after Dilution	1025
7.3.3.3.2	BOD_5 Determination with Dilutiona as a Field Test Method	1030
7.3.3.3.3	General Information and Special Directions	1031
7.3.3.4	Phenols	1036
7.3.3.4.1	Phenols Volatile in Steam	1036
7.3.3.4.2	Total Phenols	1038
7.3.3.5	Oils and Fats	1040
7.3.3.5.1	Gravimetric Determination after Extraction with n-Hexane	1041
7.3.3.5.2	Infrared Analysis	1046
7.3.3.5.3	Gas-Chromatographic Analysis	1047
7.3.3.6	Detergents (Surfactants)	1048
7.3.3.6.1	Quick Test for Detergents	1048
7.3.3.6.2	Determination of Anion-Active Detergents	1049

7.3.3.6.2.1	Photometric Determination of Anion-Active Detergents with Methylene Blue	1050
7.3.3.6.2.2	Titration with Cetyl Trimethyl Ammonium Bromide Against Methylene Blue (EPTON Titration)	1053
7.3.3.6.3	Determination of Non-Ionogenic Detergents (Preliminary Test)	1055
7.3.3.6.3.1	Determination of Non-Ionogenic Detergents (Surfactants), Quantitative Determination	1056
7.3.3.6.4	Separation of Detergents by the Foaming Method	1064
7.3.3.7	Extractable Organic Halogen Compounds (EOX)	1065
7.3.3.8	Fermentability Testing	1069
7.3.3.9	Waste Water Toxicity	1071
7.4	FISH TEST	1076
7.5	ASSESSMENT OF THE ANALYSES	1080
7.5.1	Concentration or Load	1081
7.5.2	General Load of Waste Water	1082
7.5.3	Pollution Load of Waste Water	1082
7.6	SEWAGE SLUDGES	1083
7.6.1	Directions for Sludge Sampling and Decomposition for Heavy Metal Determination	1084
7.6.2	Sludge Volume and Sludge Index	1092
8	REFERENCES	1094
9	SUBJECT INDEX	1105
10	SUPPLEMENTS	1137

Water Supply and Sanitation Projects
in Developing Countries
(Sector Paper of the German Federal
Ministry for Economic Cooperation)

International Drinking Water Supply
and Sanitation Decade
(Prepared under the Interregional
Cooperation Programme between the
World Health Organization and the
German Federal Ministry for Economic
Cooperation)

1 INTRODUCTION

In nature, water undergoes a continuous, natural cycle, owing to evaporation and precipitation, during which it is used by humans for many purposes, for example:
- hydropower plants,
- shipping,
- irrigation,
- drinking and industrial water,
- recreation.

The extension of sewerage systems in communities, the increase in industrialization, the merging of many delocalized small and intermediate-sized companies into large businesses, and the establishment of sanitary landfills and other water disposal measures have led to a situation in which the waste water inflow into natural waters overtaxes their natural self-purification capability.

In the course of time, legislative measures to protect these waters have had to be introduced and strengthened continuously. The contamination and overburden of natural waters with pollutants and harmful substances were to be avoided, and the multiple use of water was to be made possible in densely populated and industrialized regions.

The result was an increased development of waste water treatment facilities with the necessary provisions for waste water collection.

1.1 Classification of Waste Water

Waste waters are classified according to their origin and composition. When designing waste water disposal plants, i.e., waste water collection and treatment, one differentiates among:

- domestic waste water with discharges from households as well as public buildings and other facilities, including water for street cleaning and fire fighting and also waste water from small local industries connected to the same sewerage system,
- commercial waste water coming from commercial businesses, e.g., slaughter houses, small industrial operations, and other public facilities - and usually connected to the common sewerage system,
- industrial waste water produced by large industrial plants of all kinds as well as similar activities,
- agricultural waste water from livestock production and from plant and animal processing operations,
- seepage water (foreign water) from managed drainages and drainage pipelines and the artificial lowering of the groundwater level as well as groundwater that leaks into the sewerage system through pipelines and other installations,
- rainwater, including all forms of precipitation: rain, snow, hail, and fog,
- surface water from those water bodies that feed directly into the sewerage system in question.

These different types of waste water are collectively termed "municipal waste water" and arise in the municipal sewerage system in cities. Thus, municipal waste water consists mainly of domestic sewage and foreign water; larger or smaller percentages of rainwater, depending on the local conditions; and commercial, industrial, and possibly agricultural waste water.

Waste waters contain different substances of natural and artificial origin, which can be more or less harmful when used by humans, animals, or the environment (e.g., the atmosphere or soil). The composition of waste water depends on its origin and treatment before discharge.

1.2 Composition of Waste Waters

Apart from its physical properties, all natural water contains components of active and inactive material. Plants and animals are also found in surface waters.

Even in the form of precipitation, water acquires organic and inorganic particles, gases, microorganisms and trace amounts of ammonia and nitrates when falling through the atmosphere. Up to 50 mg/l of sulfuric acid have been found in rainwater in industrial regions. Near factories, rainwater also contains carbonic acid, hydrochloric acid, nitric and other acids, the oxides of lead, zinc, copper, and other metals; soot; and phenols. With the rainwater runoff, many soluble and insoluble substances found on the ground, including animal excreta, are also washed into surface waters and the groundwater. Criteria that represent the degree of pollution of waste water are the BOD, the COD, the ammonia nitrogen content, the TOC, and the TOD.

The BOD, or biological oxygen demand, is the amount of oxygen required for the decomposition of the organic substances by microorganisms in 20 days at 20 °C. It is designated as the total BOD or as BOD_{20}. The BOD_5 (BOD in 5 days) is also commonly determined. In domestic waste water a certain relation exists between BOD_5 and BOD_{20}. Thus, when BOD_5 is known, BOD_{20} can easily be calculated (Table 1.2.-1).

Tab. 1.2.-1: The value of BOD at time T in days in relation to BOD_5

Days	5°C	10°C	15°C	20°C	25°C	30°C
1	0.11	0.16	0.22	0.30	0.40	0.54
2	0.21	0.30	0.40	0.54	0.71	0.91
3	0.31	0.41	0.56	0.73	0.93	1.17
4	0.38	0.52	0.68	0.88	1.11	1.35
5	0.45	0.60	0.79	1.00	1.23	1.47
6	0.51	0.68	0.88	1.10	1.31	1.56
7	0.57	0.75	0.95	1.17	1.40	1.62
8	0.62	0.80	1.01	1.23	1.45	1.66
9	0.66	0.85	1.06	1.28	1.49	1.69
10	0.70	0.90	1.10	1.32	1.52	1.71
12	0.77	0.97	1.17	1.37	1.56	1.73
14	0.82	1.02	1.21	1.40	1.58	1.74
16	0.85	1.06	1.24	1.43	1.59	1.75
18	0.90	1.08	1.27	1.44	1.60	1.76
20	0.92	1.10	1.28	1.45	1.61	-
25	0.97	1.14	1.30	1.46	-	-
Complete oxygen demand of first phase	$1,02 = 0,7 \cdot 1,46$	$1,17 = 0,8 \cdot 1,46$	$1,32 = 0,9 \cdot 1,46$	$1,46 = \frac{1}{0,684}$	$1,61 = 1,1 \cdot 1,46$	$1,76 = 1,2 \cdot 1,46$

The decomposition, and the accompanying oxygen demand, can be expressed as a first order decay function. This means that a constant incremental amount of the remaining oxygen demand disappears per unit time.

In addition to the oxygen demand for bacterial oxidation (BOD_5), waste water pollution can also be expressed through the COD (chemical oxygen demand). This is the oxygen required for the chemical oxidation of the organic and inorganic constituents.
- Potassium dichlorate ($K_2Cr_2O_7$) or
- potassium permanganate ($KMnO_4$)

is used as the oxidizing agent.

Potassium dichlorate is the stronger oxidizing agent and results in a higher oxygen demand at the same level of pollution.

Table 1.2.-2 from IMHOFF /73/ shows as an example the oxygen demand of various organic substances and its dependence on the method of determination.

Tab. 1.2.-2: Value of the oxygen demand determined by different methods according to IMHOFF/73/

	Theoretical oxygen demand mg/g	Empirical oxygen consumption			Organic carbon
		with $KMnO_4$ mg/g	with $K_2Cr_2O_7$ mg/g	as BOD_5 mg/g	TOC mg/g
Lactic acid	1067	260	970	540	400
Glucose	1067	600	990	580	400
Lactose	1122	3	920	580	421
Dextrin	1185	220	950	520	444
Starch	1185			680	444
Phenol	2383	2360	2340	1700	760
Casein	1410*	150	1150	580	560

* Without the oxidation of nitrogen

The pollution of commercial and industrial waste waters can be compared with one another when related to a common parameter. This is accomplished with a population correction factor (see 2.2.3).

In Table 1.2.-3 the BOD_5 values (from LIEBMANN-VIEHL) for some highly contaminated liquids are given.

Tab. 1.2.-3: Some BOD_5 values of liquids from LIEBMANN and VIEHL /96/

Blood (Blood amount near 70 % of live weight)	160 - 210 g/l
Liquid manure	7 - 18 g/l
Whey	45 g/l
Seepage from beet foliage	80 g/l
1st. wash water by butter production	17 g/l
Condensation from steaming potatoes	4 - 9 g/l
Phenol	1.7 g/l

- The TOC (total organic carbon) is a defined analytically exact quantity. It is determined by the thermal oxidation of organic substances, e.g., through combustion at high temperature, and the subsequent measurement of the amount of CO_2 formed. Theoretically, the amount of organic carbon can vary between approximately 8 % (for CCl_4) and 94 % (for $C_{10}H_8$, naphthalene). The variation is much less for the mixtures found in water and waste water. The organic carbon content in waste water, however, is only a part of the pollution.

- The content of ammonia nitrogen is a further criterion for pollution due to city and commercial waste water. Even relatively small amounts of it are toxic for fish. Apart from domestic or commercial waste waters, ammonia can originate from agricultural and forestry sources (fertilizer).

- The <u>TOD</u>, (total oxygen demand) represents the theoretical amount of oxygen required to oxidize all oxidizable substances present in waste water. These are essentially organic compounds, but also include oxidizable inorganic substances. When the TOD is determined by oxygen consumption, for example by combustion at high temperature in the presence of oxygen, then its value is the sum of the necessary O_2 to form CO_2, H_2O, nitrogen oxides, and sulfur oxides from the organic compounds, as well as the corresponding O_2 present in the bonds of the oxides in some inorganic compounds.

Various substances are contained in surface and ground waters that in small concentrations over a long period of time can have a negative effect on human health.

1.3 Harmful Effects of Waste Water on Natural Waters

Waste water is considered harmful when the normal use of water is impaired or hindered or when reusable "waste products" are fed into natural waters. <u>Direct damage</u> can arise, for example, where
- the water is used for fish production,
- beaches are used by tourists,
- seas or lakes are used for drinking water supply or
- are zoned as recreational areas.

That damage can be quantified and compared to the costs (investment and operation costs) of waste water treatment. Economic advantages that result from the efficiency of sewage plants are demonstrable.

An abundance of plant and animal life exists in and on waters not polluted with waste water, from water bacteria, microbes, crustations, and mussels up to fish, reptiles, mammals (such as beavers and otters), and birds. The flora extends from lower forms of algae to water and swamp plants such as trees, e.g., the alder and willow, that grow in wet areas. The ecological function of the individual forms of life and their interdependencies, for example for the biological purification of water, are not completely known.

Waste waters could destroy an entire ecological system and, as such, eliminate a source of food production and natural resources. Losses in tourism are also a consequence.

Economic losses additionally occur through the valuable substances used in production processes that are discarded in waste water instead of being recovered by recycling. Many of these substances could be recovered and reused. This is the case, for example, in waste products from paper mills and oil refineries, in particular, the sulfite waste liquor, which can be employed in the production of other products.

The related damage is hard to quantify. The costs associated with the onchocerciasis-caused blindness, especially in developing countries, due to contaminated waste water or in Japan due to mercury poisoning (Minamata disease 1968) can only be estimated.

Waste waters having exceptionally toxic effects on humans and animals contain the following components:
- organic solvents,
- organic halogen compounds,
- organic phosphorous compounds,
- substances with proven carcinogenic effects,
- hydrogen sulfite,
- cyanide,
- fluoride,
- heavy metals, especially mercury and cadmium, and
- compounds of these metals,
- disease-producing organisms and/or living parasite eggs.

Tab. 1.3.-1: Some diseases contagious to man by which waste water could be the source of the infection

Disease	Spread	Carrier and living space of the infectant	Mode of communication
Amibiasis: intestinal disease	Spread all over the world; often to be found with 50 % and more of the population in areas with no sanitary facilities, especially in tropical regions.	Entamoeba histolytica: unicellular animal, excreted in human feces	Water, communication of fresh feces from hand to mouth, spoilt vegetables, flies, the dirty hands of the persons handling food.
Ascariasis: intestinal disease	Spread all over the world, especially widespread in humid tropical zones where it occurs in more than 50 % of the population.	Ascaris lumbricoidis: Roundworm excreted in human feces	By direct and indirect communication through larvae from ground to mouth. Communication also by dust.
Cholera: acute general infection	Endemic in India and Bangladesh, from where, at times, it spreads epidemically; the El-Tor group is endemic in the South Pacific, in Asia, and the Near East.	Vibriocholera: bacteria excreted in man's feces and vomit.	Communication mainly by water, but also by contaminated food, by flies and contaminated soil.
Ancylostomiasis: intestinal disease caused by blood-sucking worms	Prevailingly endemic in most tropical and subtropical regions of America, in Mediterranean countries and in Asia.	Necator Americanus and Ancylostoma duodenale: Nematodes excreted in the feces of infected persons.	The larvae existing in the warm and humid soil penetrate the skin, especially that of the foot.
Leptospirosis: violent general infection	Spread all over the world; occupational disease common with farmhands coming into contact with contaminated soil or water.	Leptospira icterohaemorrhagiae and others, excreted in the urine of infected animals, especially cattle, dogs, rodents, pigs.	Contact with water, mud, or soil contaminated by the urine of the infected animals.

Cont. Tab. 1.3.-1

Leptospirosis: violent general infection	Spread all over the world; occupational disease common with farmhands coming into contact with contaminated soil or water.	Leptospira icterohaemorrhagiae and others, excreted in the urine of infected animals, especially cattle, dogs, rodents, pigs.	Contact with water, mud, or soil contaminated by the urine of the infected animals.
Shigellosis: acute intestinal disease	All regions of the world: the arctic, the milder and the tropical zones	Twenty-seven serum types of the order Shingella: bacteria excreted in the feces of infected persons.	By direct contact when fecal matter is communicated to the mouth; but also by contaminated food, by flies and by the soil.
Strongyloidiasis: infectious disease, usually in the bowels	The geographical spread is very much that of the Ancylostomiasis	Strongyloid stercoralis: worm excreted in the feces of infected dogs or humans.	The infectious larvae occurring in infected humid soil penetrate the skin, especially that of the leg.
Tetanus: violent disease, often lethal, caused by the toxin of the tetanus bacillus	Spread all over the world; lethal cases occurring mainly in agrarian areas of tropical zones	Clostridium tetani: bacillus excreted in the feces of animals, especially horses, and of longlasting virulence in the soil.	Spors in the soil, in the dust of roads, and in the feces of man and animals; they penetrate the organism especially when the victim is injured.
Infection with Trichuris: infectious disease of the colon	Spread all over the world especially in regions with hot and humid climates	Trichuris trichura: worm excreted in the feces of the person infected.	Intake of worms' eggs stemming from contaminated soil.
Typhoid fever: intestinal disease	Very common all over the world. Widespread disease in the Far East, the Near East, South America and Africa	Salmonella typhii: bacillus excreted in the feces and the urine of the infected persons.	The most important carriers are water and contaminated food. In some countries other important factors are flies and vegetables grown on contaminated soil.

*) According to AMERICAN PUBLIC HEALTH ASSOCIATION (1965)

Waste waters invariably contain microorganisms, especially the bacteria of intestinal diseases such as typhus, para-typhus, internetus, and dysentery, as well as viruses such as those for polio and infectious jaundice. In addition, domestic and some commercial waste waters contain living eggs from human and animal parasites (mar worm, band worm).

Contagious diseases for which waste waters serve as the source of the infection are shown in Table 1.3.-1, according to the American Public Health Association (1965). The purity of water is influenced by natural and artificial (technical) pollution. In general, artificial pollution is stronger and more permanent than natural pollution.

Artificial pollution occurs when an area surrounding a water body is developed with houses, factories, and roads, or through agricultural and forestry use etc. The influence on the water is greater the more the area is used, i.e., the degree to which the land is developed and used for economic production. Drainage from developed land carries pollutants directly into natural waters or the pollutants are carried by rainwater into the storm drainage. A further consequence of land development is the reduction of seepage, i.e., the precipitation that would otherwise filter into the ground. Thus, the amount of groundwater is reduced, and the groundwater level sinks.

The strongest pollutant of natural waters is the discharge of waste water from cities and industry. As a consequence the natural water properties are changed considerably. Water removed for reuse must be purified beforehand. In some instances, the water treatment is so complex that certain usage become uneconomical.

Further detrimental effects on natural water are caused by:

- Waste waters that greatly reduce the oxygen content due to their constituents, which are chemically or biologically oxidizable and consume the dissolved oxygen in the natural water through the oxidation of
    - organic compounds (carbohydrates, fats, proteins, etc.),
    - ammonium,
    - nitrites,
    - sulfites.
- Waste waters that cause the sedimentation of sludge. Their contents can be mineral (for example, from clay factories) or organic (for example, yeast residues from wineries). Solids that are organically decomposable cause further damage by reacting with sedimented sludge (formation of hydrogen sulfide and polluted organic substances) or by the floatation of sludge.
- Waste waters with fertilizing agents, i.e., with nutrients such as compounds of nitrogen and phosphorous (e.g. ammonium nitrate and ammonium phosphate) which are accessible to algae and higher water plants.

Owing to these reasons, waste water collection and treatment is absolutely necessary.

2 TYPES AND AMOUNT OF WASTE WATER

2.1 Domestic Sewage Water

2.1.1 Origin and Types of Domestic Sewage

The various types of human water use create waste waters that occur alone or in mixtures at different concentrations.

According to PÖPEL /128/, domestic sewage water originates
- in private households from:
    a) cooking, dishwashing, house cleaning, washing clothes, and bathing,
    b) the use of the toilet,
    c) washing outside paved surfaces and automobiles;
- in public buildings from:
    a) cleaning the building, personal hygiene and cooking and washing in the cafeteria, if present
    b) the use of public toilets,
    c) washing outside paved surfaces and automobiles;
- in small commercial businesses from:
    a) cooking, dishwashing, house cleaning, washing clothes, and personal hygiene,
    b) the use of the toilets,
    c) washing outside paved surfaces and automobiles.

Fresh domestic sewage emerges as a turbid, gray or yellowish fluid with a stale odor, in which sludge particles, feces, vegetable remains, strips of paper, and synthetic materials are suspended. The longer and more turbulent the flowline in the sewer is, the finer the sewage particles will be.

2.1.2 Amount and Composition of Sewage Water in Private Households

2.1.2.1 Amount of Domestic Sewage

The amount of domestic waste water is the water consumed from the supply system minus the water used for cooking, drinking, and watering the lawn and garden. The additional feces and other waste products that are added to waste water amount to only ca. 1.4 kg per person per day (kg/(P d)). It can be approximated that the amount of domestic sewage water is ca. 80 % of the water consumption.

Since the water consumption depends essentially on living standards and habits, the amount of sewage rises with increasing standard of living.

The amount of waste water undergoes hourly, daily and yearly variations. According to IMHOFF /73/, at the beginning of the week the increase in sewage due to washing clothes, and at the end of the week due to household cleaning, is quite evident. On Sundays and holidays the amount of waste water is distinctly reduced. Especially apparent is the quantitative change in the occurrence of waste water throughout the day. After a low point during the night, the amount of water and its concentration with wastes increase sharply in the morning hours, usually reaching a maximum at midday and declining steadily thereafter. An example for a city with a population of 50,000 is shown in Fig. 2.1.-1 by IMHOFF /73/.

A certain correlation is apparent between the amount of waste water and its waste load, which is represented here as sedimentary substances. Other load indicators, for example the BOD_5, also display a similar behavior. The coincidence of the waste water peaks with the maximum load during a day explains why the daily waste water load is calculated at 14 or even 16 hours when designing sewage plants. In the example shown in Fig. 2.1.-1, only one main surge of waste material occurs and this is at midday. Occasionally, several inflow peaks appear within one day.

The larger the city, the more uniform the amount and composition of the waste water.

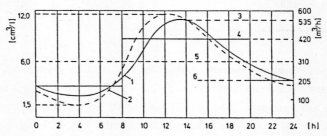

Fig. 2.1.-1: Fluctuations in the amount of waste water and the sedimentary substances contained therein for 50,000 inhabitants in a 24 hour period.

1 - Hourly waste water amount; 2 - Sedimentary substances; 3 - Daily peaks; 4 - Daytime average; 5 - Average over 24 hours; 6 - Nighttime average

In Table 2.1.-1, the factors for calculating the peak hourly load are shown, as well as the proportionate parts of the hourly dry-weather discharge q_h on the total daily discharge Q_H for various community sizes (BÖHNKE /31/).

$$q_h = \frac{Q_H}{t}$$

Tab. 2.1.-1: Hourly discharge q_h in relation to the daily discharge Q_H

Community size in 1000 P	Max. hourly discharge t	Averages	
		Daytime t	Nighttime t
< 5	10 - 12	12 - 14	> 84
5 - 10	12 - 13	14 - 16	84 - 48
10 - 50	13 - 15	16 - 18	48 - 36
50 - 250	15 - 18	18 - 20	36 - 30
> 250	> 18	20 - 22	30 - 27

In Table 2.1.-2, the amount and composition of sewage water arising under normal conditions are given according to PÖPEL /128/ for the individual installations and activities in private households. Such values are required for dimensioning the appropriate sanitary fixtures or house connections as well as for planning sewerage systems.

Tab. 2.1.-2: Amount and composition of sewage water and the water demand in private households per inhabitant and day /128/

Type	Amount in l/(P·d)		Pollution in g/(P·d)						
	Water demand	Water discharge	Total solids	Mineral solids	Organic solids	BOD_5	C_o	N	P
Drinking and cooking	3	–					8	0.2	
Dish-washing	4	4							
Washing clothes	20	19							
Personal hygiene	10	10					7	–	
Baths/Showers	20	20							
House cleaning	3	3							
Excreta –Feces	20	22	27	4	23		17	1.5	0.6
Excreta –Urine			55	15	40		5	12.2	0.8
Total	80	78	(190)*	(80)*	(110)*	(54)*	37	13.9	(2.3)*

* for domestic sewage (IMHOFF)

With rising living standards, the water demand values given in the above table and, consequently, the amount of waste water will increase considerably, especially for bathing, showers, cleaning larger apartments, operation of new household utilities, watering the lawn and garden, and washing automobiles and outside surfaces. The load in waste water will increase, however, only minimally, and thus with increasing water consumption, the concentration of contaminants in the waste water will decrease.

The water demand for various household activities is given in Table 2.1.-3 by PÖPEL /128/.

In areas with livestock production, water supply for the animals must be considered. The water demand for livestock (not including stall cleaning) is estimated as follows:

- for cattle   50 - 200 l/(animal and day)
- for calves   10 - 40 l/(animal and day)
- for poultry   0.3 l/(animal and day)

Tab. 2.1.-3: Water demand for household activities /128/

Activity	l per occurrence
Dishwashing (per meal for 4-6 persons) by hand automatic dishwasher Washing clothes (4 kg) by hand washing machine House cleaning pail of water Personal hygiene washing hands showering shower bath bath medicinal bath child's bath Toilet with raised water tank with integrated tank with direct connection to water supply Garbage disposal per cycle Washing automobiles with pail with water hose	 10 - 25 20 - 45 250 - 300 100 - 180 8 - 10 2 - 5 40 - 80 80 - 140 200 - 250 30 - 50 30 - 40 8 - 12 12 - 15 6 - 14 4 - 5 20 - 40 100 - 200
	l/m²/a
Watering lawn when required	5 - 10

In an African country (1978), the following values for the specific water usage were taken in planning the village water supply:

Inhabitants: First planning phase   75 l/(P·d)
                  Second planning phase   120 l/(P·d)
                  Future demand   150 l/(P·d)
Livestock: Large animals   50 l/(animal and day)
                  Small animals   8-10 l/(animal and day)

2.1.2.2 Composition of Domestic Sewage

Domestic waste water is dangerous owing to its high microorganism content and infectious properties. Essentially harmless escherichia coli bacteria, usually appearing in massive colonies, are particularily characteristic, and originate in the human and animal intestine. They are indicators for extreme pollution. Approximately 10^{11} to 10^{13} coli bacteria per day and per person are introduced into waste water. The total number of bacteria, including groups considered to be relatively harmless, is about 1000 times larger.

Microorganisms are present in waste water partly in the form of viruses and bacteria, e.g., the typhus- or para-typhus-producing salmonella, and partly in the eggs of worm parasites. They come from hospitals, people sick at home, disease carriers, etc. Purified waste water is thus not completely hygienic and, in some cases, must be sterilized in addition to the mechanical-biological treatment.

Apart from physiologically questionable organisms, domestic sewage contains harmless bacteria that decompose waste material through hydrolysis, reduction, and oxidation. Fermenters and enzymes also participate in this decomposition. Finally, domestic sewage contains hormones, stimulants, and vitamins from human and animal excreta.

The waste load values of domestic sewage are seldom used in designing sewage plants. They may be needed only in rare cases for separate buildings or very small sewage plants.

Table 2.1.-4 by IMHOFF /73/ gives the daily pollution caused by one inhabitant. Assuming a daily water consumption per person of 150 l/(P·d) (for German conditions), the resulting concentrations of waste substances are given in Tab. 2.1.-5.

With low water consumption, higher waste concentrations can in general be expected since the waste load per person is nearly constant and depends mainly on living habits.

Tab. 2.1.-4: Average daily waste load per person /73/

	Total g/(P·d)	Organic g/(P·d)	Mineral g/(P·d)
Total amount of waste	190	110	80
Portion of dissolved substances	100	50	50
Portion of undissolved substances	90	60	30
- sedimentary	60	40	20
- nonsedimentary	30	20	10

Tab. 2.1.-5: Waste concentrations in domestic sewage /73/

	Total mg/l	Organic mg/l	Mineral mg/l
Total amount of waste	1260	730	530
Portion of dissolved substances	660	330	330
Portion of undissolved substances	600	400	200
- sedimentary	400	270	130
- nonsedimentary	200	130	70

The typical chemical composition of domestic waste water is shown in Tab. 2.1.-6 as determined by the AMERICAN PUBLIC HEALTH ASSOCIATION.

Tab. 2.1.-6: Analysis of domestic waste water by the AMERICAN PUBLIC HEALTH ASSOCIATION (values in mg/l).

Substances mg/l	Pollution		
	High	Average	Low
Total Solids	1000	500	200
- soluble	700	350	120
- solid	300	150	80
total suspended	600	350	120
- soluble	400	250	70
- solid	200	100	50
total dissolved	500	200	100
- soluble	300	100	50
- solid	200	100	50
sedimentary, ml/l	12	8	4
BOD_5	300	200	100
Oxygen consumption	150	75	30
Dissolved oxygen	0	0	0
Total nitrogen	85	50	25
- organic	35	20	10
- ammonia	50	30	15
- nitrites	0,1	0,05	0
- nitrates	0,4	0,20	0,1
Chloride	175	100	15
Alkalinity (as $CaCO_3$)	200	100	50
Fats	40	20	0

* not including sediments

2.1.3 Amount and Composition of Sewage Water from Public Buildings

The sewage water from public buildings consists mainly of washing, cleaning, and toilet waters and, when present, kitchen and cafeteria discharges. The composition is similar to that in private households. The water demand and corresponding amounts of waste water for individual public facilities are given in Tab. 2.1.-7. In most cases the discharge occurs only during working hours, which leads to large fluctuations.

These values are the basis for designing only the sanitary fixtures in the building and the connection to the sewerage system.

Tab. 2.1.-7: Amount and load of waste water and the water demand in public buildings and facilities /128/

Facility	Average number per 10 000 P.	Units	Water demand	Waste water discharge	Waste water per day	Waste water per week	Waste water per year	Remarks
Kindergarten	300 children	l per child and day	10	10	x			
Schools without showers without swimming pool	1700 pupils	l per pupil and day	10	10	x			a)
with showers without swimming pool		l per pupil and day	20	20	x			b) c)
with showers with swimming pool		l per pupil and day	30-50	30-50	x			b) c)
Nurseries and day care centers	250 children	l per child and day	30	30	x			b) c)
Gymnasium	100 users daily	l per user	10	10	x			a)
Indoor swimming pools	100 users daily	l per user	150-180	150-180	x			b) d)
Outdoor swimming pools	500 users daily in season	l per user	150-180	150-180			x	b)
Public buildings	600 employees	l per employee/day	40-60	40-60	x			a) e) also DIN 4262 (3)
Military barracks	10 men	l per man and day	200-400	200-400	x			a) e)
Prisons	1 prisoner	l per pris. and day	400-600	400-600	x			a) e)
Hospitals	100 beds	l per bed and day	250-600	250-600	x			a) also DIN
Medical clinics	100 beds	l per bed and day	unt.1500	unt.1500	x			b) 19250 (4)
Rail stations Feeder water		m³/lok filling	8-22	-				
Cleaning water		m³ per train car cleaning	1.5	1.5		x		b)
Supermarket buildings		l per m² surface and market day	3-5	3-5	x			b)
Weekly markets		l per m² and market day				x		b)
Slaughter houses		l per animal unit	300-400	300-400		x		see Ind. wastewater
Watering public surfaces gravel surface paved roads	50 ha	l per m²/d	0.1-1.5 0.1-1	- (0,5)			x	
lawns and flowerbeds		l per m²/d	until 3	-				
Street cleaning		l per m²						
Urinals controlled flushing		l per urinal	40-60	40-60	x			a)
continuous flushing		l per flush period/hour	200	200	x			a)
Public faucets controlled ventile wells		m³/d	3	(3)	(x)			
continuous flow ventile wells		m³/d	14-20	(14-20)	(x)			
Fountains		l/sec	1-350	(1-350)		(x)		
Hydrants		l/sec	5-10	(5-10)			(x)	

a) Pollution as with domestic sewage.
b) Pollution less than domestic sewage.
c) When gymnasiums and swimming facilities are used by private sport clubs, the values increase correspondingly.
d) Depending on the individual swimming facility, the values can be much higher.
e) With garages and washing areas, the values are correspondingly higher.

2.1.4 Amount and Composition of Sewage Water from Tourist Facilities

Numerous tourist facilities are built outside of cities away from sewerage networks. There the sewage water must also be collected and treated.

Such facilities include, for example:
- restaurants,
- hotels,
- camping grounds,
- organized beaches.

The separate sewage plants for these facilities are to be designed for the number of persons (P) that are likely to be accommodated.

2.1.4.1 Restaurants

According to the German standard DIN 4261 for restaurants, the following equivalencies can be defined:

- Restaurants with normal use   3 seats = 1 P

  Corrections:
  for restaurants with greater frequency:
  9-10 guests per seat in 24 h   1 seat = 3 P
  11-14 guests per seat in 24 h   1 seat = 4 P
  15-18 guests per seat in 24 h   1 seat = 5 P
  for outdoor restaurants   15 outdoor seats = 1 P
- Club and sport buildings
  without accommodations   10 users = 1 P
- Camping facilities   2 users = 1 P

2.1.4.2 Hotels

Hotels built <u>inside cities</u> are usually connected to the public sewerage system. The waste water can be fed directly into the public system or fed in after pretreatment, depending on the size of the hotel and its installations. Sewage waters arise from toilets, baths, showers, swimming pools, restaurants, etc.

2.1.4.3 Camping Facilities

Camping grounds are classified according to their usage:
- <u>long-term</u> camping grounds, usually far from train connections or main roads and intended for longer stays
- <u>short-term</u> camping grounds
- <u>weekend</u> camping grounds.

The long-term and weekend facilities must have all sanitary facilities and a complete sewage treatment plant installed. Domestic as well as kitchen and garage waste water should be treated.

Short-term camping grounds do not usually have garages or washing facilities for automobiles. Thus settling tanks for sludge and oil separators are not necessary.

2.1.4.4 Organized Beaches

Organized beaches are often equipped with toilets, showers, or even swimming pools.

The water discharges must be treated and may not be fed directly into fresh waters or the ocean. When possible, the waste waters should be collected at pump stations and then pumped into the public sewerage system or to the local sewage plant.

2.2. Municipal Sewage Water

The term "municipal sewage water" refers collectively to the different types of waste water, which are found in the public sewerage systems in cities and municipalities (see also 4.3.3).

2.2.1 Amount of Municipal Sewage Water

The percentage amounts of individual types of waste water in municipal sewage are given in Tab. 2.2.-1 (PÖPEL /128/).

Tab. 2.2.-1: Total and municipal waste water amounts in the Federal Republic of Germany /128/

	%	$10^9 m^3/a$	$m^3/(P \cdot a)$	$l/(P \cdot d)$
Total waste water amount	100	15.4	262	714
Municipal waste water	32	5.0	84	230
Industrial waste water (without recycling)	47	7.2	125	340
Agricultural waste water	1	0.2	3	7
Rainwater runoff in areas with sewerage system	20	3.0	50	137
Municipal waste water	100	5.0	84	230
Domestic sewage	50	2.5	42	115
Seepage (foreign) water	14	0.7	12	32
Industrial and commercial waste water	36	1.8	30	83

The water consumption and municipal waste water occurrence are, however, essentially dependent on the living standard of the population. Thus the specific water consumption in highly industrialized and wealthy countries like the USA and Switzerland is many times greater than that in less developed countries, as shown in Tab. 2.2.-2.

Tab. 2.2.-2: Water consumption of several foreign cities

Country	Location	Year	Consumption		l/(P·d)
Finland	Helsinki	1930	mean consumption		175
		1950			235
		1963			360
France	Rural areas		mean drinking water consumption		126
	Cities				245
Norway	Oslo	1963	mean total consumption		580
			industrial only		230
Austria	Vienna	1969	total average consumption		313
			maximum		416
Sweden	Stockholm	1961	maximum consumption		422
			mean consumption, made up of:		337
			domestic consumption		198
			industrial consumption		102
			other consumers		37
Switzer-	Basel	1968	mean consumption		720
land	Zürich	1968	mean consumption		420
	La Chaux-De-Fonds	1968	mean consumption		270
Spain	Rural capital	1964	public water supply		226
	Cities over 10000 P		public water supply		111
	Cities under 10000 P		public water supply		42
	Average		public water supply, made up of:		138
			domestic consumption		86
			public faucets		14
			city maintenance		8
			other consumers		17
USA	Average	1961	public water supply	sector	mean
			domestic consumption	60-280	140
			industrial consumption	40-400	120
			public buildings	20- 80	40
			other consumers	40-160	100
			total consumption	160-920	400
	San Francisco	1952	mean total consumption		400
	Los Angeles	1950	mean total consumption		630
	Chicago	1950	mean total consumption		875
	Beverly Hills	1950	mean total consumption		2000

The factors that influence the amount of waste water can be seen in Table 2.2.-3. This table can be used as a checklist when determining the actual amount of waste water in a community.

The specific waste water occurrence is additionally dependent on the size and characteristics of the community. Among other determinants, the availability and price of water can also be determining factors for the consumer. Thus the specific value is larger in large cities and municipalities (Table 2.2.-4) and the lowest in rural areas.

Small commercial businesses are included in these waste water amounts. Usually at least 150 l/(P d) is assumed. The values in cities with millions in population are often much higher.

Thus for concrete project planning, the relevant waste water amounts should be taken from measurements or reliable estimates.

Tab. 2.2.-3: Factors which effect the sewage water occurrence /31/

Influential factor	Effect on amount of domestic sewage	
	Decreases	Increases
Rising living standard		x
Greater housing density		x
Larger households	x	
Stronger industrialized areas		x
Estate settlements		x
Welfare housing	x	
Agricultural areas without irrigation	x	
Warm climate		x
Ample water resources		x
Poor water quality	x	
Higher water network pressure		x
Metering water use	x	
Higher water price	x	
Expanded sewerage system		x

Tab. 2.2.-4: Sewage water occurrence in communities of different size (P)

Community size P	Sewage water amount l/(P·d)
< 5 000	150
5 000 - 10 000	180
10 000 - 50 000	220
50 000 - 250 000	250
> 250 000	300

For exact calculations, especially for very large cities, the general equation from the United Nations guideline /163/ can be used to determine the domestic and public water demand and, with it, the waste water occurrence.

The equation is:

$$Q_p = K_s \cdot K_t \cdot K_p \cdot (K_j \cdot q \cdot N + Q_i) \frac{1}{1000} \qquad [m^3/d]$$

with the quantities defined as follows:

Q_p = Water demand in m^3/d.

K_s = Coefficient representing the losses that occur from the water source to the water treatment station. These losses vary according to the material and length of piping, the water pressure, the kind of soils and the type of operation. The normal value of K_s is 1.01 and should not in any case be greater than 1.04 to 1.05.

K_t = Coefficient representing internal consumption and possible losses in the treatment station. This coefficient varies depending on the method used for water treatment, the type and condition of the installations, the quality of treated water, and the water's thermal characteristics. Depending on the particular conditions, the coefficient lies between 1.06 and 1.10.

K_p = Coefficient accounting for the losses from the drinking water treatment facility to the consumer. The water losses in the pipelines depend on the quality of the installation work, the age of the pipeline, the quality of the pipe-laying, and the maintenance work. These losses can be as high as 10 to 40 %, and, generally, the technical losses in a well built and maintained pipe network should not exceed 15 to 20 %. Accordingly, the value of K_p ranges from 1.15 to 1.20. However, one should always try to reduce the losses.

K_j = Coefficient accounting for the deviation of the mean specific drinking water demand per day from the yearly average. This coefficient depends on the size of the site, its characteristics (degree of industrialization, kind of sanitary installations in houses, presence of seasonal activities, etc.) and its climate. The value of this coefficient usually varies between 1.3 and 1.6 and is larger for smaller sites, for areas with large climatic changes, or sites characterized by seasonal activities.

q = The global mean daily specific water consumption for <u>one</u> person in liters $(l/(P \cdot d))$. The value is set according to existing norms or determined through comparison with other sites. The mean daily specific consumption is also different depending on the character of the site (urban or rural), structure of the population (type of predominant activity, permanent or seasonal character, etc.), size of the community, living standard, climate and price of water.

N = Number of inhabitants assumed as a basis for dimensioning the drinking water supply, considering realistic population projections as well as social, technical, and economic factors. This prediction depends on the water supply system, the possibility of expansion, the size of the investment, the accuracy of development projections for the community, etc. The following values serve as guidelines for the planned lifetime of different structures:

- General waterwork systems   40 to 50 yr
- Collection and holding works including structures difficult to expand later   30 to 40 yr
- Water containments, large conduits, underground storage facilities   20 to 25 yr
- Wells, pumps, pumpstations, and treatment and distribution systems (expandable)   10 to 15 yr

Based on these lifetimes for designing water supply systems, the population number <u>N</u> is determined as follows:

If the community has a development plan, the population number is calculated according to this plan. If the community has no plan, the number is estimated with the equation:

$$N = N_a (1 + 0.01 p)^n$$

where

N_a = current population

p = percentage increase of population taken from statistical analyses. When demographic information is not available, p = 2.5 to 3 % is assumed.

n = number of years for which the population number (N) is to be calculated, as given above.

For tourist or resort sites, or where larger events occasionally take place, e.g., sporting or cultural events, the population number must be modified. Accordingly, one should differentiate between temporary and permanent residents and calculate separately for each case.

Q_i = Water demand as the supply reserve for fire-fighting. The amount is calculated with various guidelines, depending on the size of buildings and the fire protection equipment.

It is usually assumed that water from fire-fighting is not fed into the sewerage system. Therefore, it is not considered in calculating waste water volumes.

In dimensioning sewerage systems, the hourly flow rate is calculated with the equation:

$$Q_s = \frac{1}{24} \cdot K_s \cdot K_t \cdot K_p \cdot K_j \cdot K_h \cdot q \cdot N \cdot \frac{1}{1000} \quad [m^3/h]$$

where

Q_s = Sewage flow rate in m^3/h (or in l/s or m^3/s)

K_h = Coefficient accounting for hourly fluctuations in water consumption expressed as the ratio between the peak hourly consumption and the yearly average.

This coefficient varies according to the size of the community, the magnitude of water consumption, and the type of climate.

2.2.2 Fluctuations in the Amount of Sewage Water

Variations in the amounts of sewage water also arise from the living habits of the population, manufacturing processes in industry, commercial and public services, and climate changes. Waste water occurrence depends more on the frequency and amount of precipitation than on the air temperature. In both combined and separate collection systems, a larger amount of water arrives at the treatment plant during wet weather, and in the rainy months, the average amount is higher.

Even in dry weather, however, daily fluctuations also occur that depend on the size of the drainage area (connected residents) and its structure (industrialization). Fig. 2.2.-1 /31/ shows two typical flow curves over 24 h for different sized communities.

The maximum flow rate assumed in dimensioning treatment plants and sewerage systems is different for drainage areas of the same size but with different-sized communities. For the sewerage system, the maximum flow per second serves as the calculatory basis. In dimensioning treatment plants, the 12 hour daytime average (8 am to 8 pm) and also the 24 hour average are considered, apart from the maximum hourly flow. For structures to be dimensioned hydraulically, the maximum hourly flow rate is generally used.

The hourly flow peaks and daily averages in municipal sewage flow depend also on the connection dimensions (community size or population equivalent).

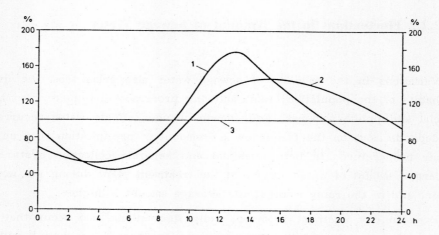

Fig. 2.2.-1: Hourly waste water flow as percentage of the average waste water flow /31/

1 - middle-sized city; 2 - large city; 3 - average daily waste water flow

Tab. 2.2.-5: Fluctuation range of municipal sewage water /1/; $Q_{max} = Q_{day} : Q_h$

Community size in 1000 P	daily peaks	Hourly values Q_h for average over daytime hours	average over nighttime hours
> 300	18	20 — 23	30 — 27
100 — 300	16 — 18	18 — 20	36 — 30
20 — 100	14 — 16	16 — 18	48 — 36
5 — 20	12 — 14	14 — 16	84 — 48
< 5	10 12	12 14	⩾ 84

The calculatory flow rates are set under the assumption that the hourly peaks are distributed over a period of 24 hours or more.

The daily sewage flow as a function of community size is thus to be divided by the hourly values given in Tab. 2.2.-5.

2.2.3 Sewage Water per Inhabitant

The composition of municipal sewage is determined by
- the original composition of the water,
- the added wastes from households, commerce, industry, and agriculture and,
- substances washed in from the ground surface or leaked into the sewer lines.

The variety of substances is condensed into a few general characteristic values with respect to their importance for waste water treatment and sludge disposal plants and for receiving bodies of water.

The division of water contents into suspended (settleable and floatable) and dissolved substances has proven practical.

The organic components are of special importance for the decomposition processes in the sewage plant and in receiving bodies of water. This importance is measured by the chemical and biological oxygen demand caused by the organic compounds.

Plant nutrients in waste water, such as nitrogen, phosphorous, and potassium, stimulate plant growth and thus produce a secondary pollution.

Additional contaminants (metal compounds, carcinogenic substances, insecticides, etc.) found in small amounts will become increasingly important for the future of water protection. General characteristic values for these substances do not yet exist.

Municipal waste water contains an additional commercial waste load, which is accounted for by the population equivalent (PE).
PE values (see also Section 2.4) are given in Tab. 2.2.-6 for different industrial and commercial sectors. The table also contains data on sedimentary substances and waste water volumes per unit of production.

The actual pollution load varies from place to place and depends on the number of connected residents and the population equivalents of the

Tab. 2.2.-6: Waste factors (population equivalents), sedimentary substances, and waste water discharge for various industries

No.	Type of business	Units U	BOD_5 g/U	PE/U	Sediment. substances kg/U	Waste water amount m³/U
0	1	2	3	4	5	6
I	HOUSEHOLDS, ACCOMODATIONS, LODGING					
1.	Dwellings	persons	60 g BOD_5/(P·d)	1	40 g/(P·d)	200 l/(P·d)
2.	Hotels, hospitals	occupancy		1.5		
3.	Barracks, prisons, priv. schools, camp grounds	occupancy		1.0		
4.	Offices and businesses					
4.1	with cafeteria	employees		0.4		
4.2	without cafeteria	employees		0.2		
5.	Schools	pupils, teacher		0.2		
II	FOOD, LUXURIES, ANIMAL FEED					
1.	Bakeries, coffee roasting, food processing	employees		1.5		
2.	Starch production					
2.1	from potatoes	1 t potatoes	3-6 g/l w.w.	500		15-25
2.2	from grain	1 t grain	6-10 g/l w.w.	350-1000	1.5-3	9-11
3.	Potato processing	1 t potatoes 1 l w. water	25	500	6	8
4.	Canned vegetables	1 t product	3000-5000	200-500		2-14
	Canned fruit	1 t fresh fruit 1 l w. water	0.2-1.4	500		1-3
5.	Marmalade, cocoa, chocolate, confectionery production	employees 1 t product		3 60		
6.	Fruit juice production					
6.1	with fruit	1 t fresh fruit		250		
6.2	with extracts	1 m³ product		15	0.75-1.5	

cont. Tab. 2.2.-6

0	1	2	3	4	5	6
7.	Breweries (without malthouse)	1 hl beer 1 l w. water	0.6-1.2	100-350	0.2-0.4	1.7-2.4
8.	Malthouses	1 t grain 1 l w. water	1.62	10-110		10
9.	Distilleries					
9.1	potatoes	1 t potatoes	5-10 l w.w.	1500		25
9.2	grain	1 t grain	5-10 l w.w.	2000-3000		
9.3	molasses	1 l w. water	20	600		
9.4	wood alcohol production	1 hl alcohol (100 %)		700		
9.5	yeast production	1 t yeast 1 t molasses		5000-7000 6000	7.5	15-80
10.	Wineries	1 m³ product		100-140		
11.	Oil mills, margarine production	1 t product		500		
12.	Dairies					
12.1	only milk processing	1000 l milk		25-70	0.4-0.75	2-6
12.2	butter production	1 t butter		+1000		
12.3	cheese production at dairy	1 t cheese 1000 l milk		200 45-230		10
12.4	cheese production separated	1 t cheese		4000		
12.5	condensed milk	1 t milk		100		
12.6	powdered milk	1 t product		0.7		
13.	Fish processing					
13.1	canned fish, smoked fish	1 t fish		500		
13.2	fish meal	1 l w. water 1 t fish	4-8	200-470		1-3 30-50 (cool.w.)
14.	Peanut roasting	1 t peanuts		0.7		
15.	Meat processing					
15.1	slaughter-houses	1 t live weight per slaughtered animal	0.8 l w.w.	130-400 20-200	2-7.5	300-2000

cont. Tab. 2.2.-6

0	1	2	3	4	5	6
15.2	butcher	employees per animal (1 ox = 2,5 swine)		15 90-200		2-6
III	ANIMAL CARCASS UTILIZATION					
1.	Covering material from hides	1 t raw material 1 large animal 1 small animal 1 l w. water	5-9	55-100 300 150		0.5-1 (25-30 cool.w.)
2.	Tannery	1 t skin		1000-4000	2.5-4.5	40-140
3.	Leather processing	employees		1		
4.	Fur	employees 1000 furs		50 400		45
5.	Glue production	1 t glue 1 t raw material		1000-15000		
6.	Animal husbandry, brushes, bristles	1 t raw material		500		
IV	AGRICULTURAL OPERATIONS					
1.	Livestock production					
1.1	poultry	1000 birds		120-300		150-200 l/d
1.2	swine, calves, sheep	1 animal	80-200	3		10-20 l/d
1.3	cattle	1 animal		8		45 l/d
2.	Silage	1 t silo filling		200-650		
V	TEXTILE INDUSTRY					
1.	Wool washing	1 t wool employees		2000-4500 100		20-70
2.	Bleaching facilities	1 t wares employees		1000-3500 50	50-100	
3.	Dying facilities	1 t wares employees		1000-3000 80		
4.	Hemp and flax processing	1 t hemp or flax straw		700-3000		20-90
5.	Synthetic silk and viscose production	1 t wares		700		
6.	Laundry	1 t articles	19	350-900	2-3	20-29
7.	Dry cleaners	employees		4		

cont. Tab. 2.2.-6

0	1	2	3	4	5	6
VI	PAPER AND BOARD INDUSTRY, PULP, CELLULOSE AND WOOD PROCESSING					
1.	Pulp	1 t pulp 1 t wood		45-70 10-30		100-125
2.	Cellulose					
2.1	sulfit pulp	1 t pulp		3500-5600		8-11 (200-400)
2.2	soda pulp and straw pulp	1 t pulp		500		500-1000
3.	Paper production					
3.1	from cellulose and pulp	1 t paper 1 m³ w. water	20-100	200 -	3.5-13.5 0.15-1	125-2000
3.2	from rags	1 t paper		900	20-33	
4.	Straw pulp	1 t pulp 1 l w. water	2.5-3.9	1000-2500		
5.	Plywood and chip boards	1 t boards	4000	1200		12
VII	CHEMICAL INDUSTRY					
1.	Paints, varnish, lacquer	employees		20-31		110/d
2.	Cosmetics, perfume	employees		10		
3.	Plastics	1 t plastic				500
4.	Soap	1 t soap		1000		15
5.	Rubber	1 t finished				100
6.	Potash, saline	1 t carnalite				1
7.	Acids, alkali chloride production	1 t chloride				50
VIII	MINING (COAL AND ORE)					
1.	Coal					
1.1	pit water	1 t coal m³ w. water		5-50	200	2-10
1.2	coal washings	1 t coal		100		
2.	Ore					
2.1	pit water	1 t ore m³ w. water		3	0.6	2
2.2	ore washings	employees m³ w. water		40	1.5	1000 m³/d
3.	Natural gas production	1000 m³ gas		3		

cont. Tab. 2.2.-6

0	1	2	3	4	5	6
4.	Coking	1 t coal		300		
5.	Charcoal making	1 t charcoal		2000		
6.	Carbonization (lignite)	1 t coal		500		
7.	Gas production generators					
7.1	from lignite coal	1 t coal		500		
7.2	from mineral coal	1 t coal		300		
7.3	from anthracite coal	1 t coal		50		
7.4	from coke	1 t coal		5		
8.	Coal briquettes	1 t coal		50		
IX	MINERALS INDUSTRY					
1.	Clay preparation	employees 1 m³ w. water		25	0.5	
2.	Stone, gravel and sand washing	employees		40		
3.	Lime and cement production	employees		3		
4.	Construction material					
4.1	conduits, pressed cement-asbestos	1 t construct. elements		30		
4.2	calcareous sandstone	m³ w. water			1	60 m³/h
5.	Porcelain, glass					
5.1	manufacture	1 t glass, porcelain				3-38
5.2	processing					
5.2.1	optical processing	employees		3		
5.2.2	engraving, cutting, silver plating	employees 1 t product m³ w. water		400 240	0.001	
5.2.3	glasswool production	employees		8		
X	METAL PRODUCTION					
1.	Smelting	1 t raw steel		7.5		65-200
2.	Steel mill with gas scrubber	1 t raw steel		8		20

cont. Tab. 2.2.-6

0	1	2	3	4	5	6
3.	Rolling mill	1 t end product 1 m³ w. water		8-50	8-15	8-50
4.	Cast and tempered iron	1 t cast iron		12-30		3-8
XI	METAL WORKING INDUSTRY					
1.	Steel structures	employees		1		40-200 l/d
2.	Metal working	employees		1		40 l/d
2.1	emulsion broken	1 m³ emulsion		10		
2.2	emulsion not broken	1 m³ emulsion		350		
3.	Iron, steel and sheet metal processing	employees		1-10		50 l/d
4.	Pickling facilities	employees		10-15		0.1-25 7-10
5.	Hot galvanizing	employees		10		
6.	Galvanizing	employees		100		20-30 m³/h
7.	Colored metals and light metals					
7.1	pickling					
7.1.1	copper	employees 1 kg product		10 340		
7.1.2	zinc	1 kg product		100		
7.2	eloxating	employees		10		
8.	Auto repair, rail car repair workshops	employees		5-10		
XII	OIL REFINERIES					
1.	Mineral oil refinement	1 m³ oil		700		3-70

industry. Accordingly, the total waste load B is given by

$$B = B_D + B_I = (P + PE) \cdot 60 \qquad [g \cdot BOD_5/(P \cdot d)],$$

where

- P = number of persons
- PE = population equivalent
- B_D = domestic load
- B_I = industrial load

The waste load per person can be taken from the compilation in Tab. 2.2.-7 from IMHOFF /73/.

Tab. 2.2.-7: Average specific waste load per person and day /73/

	min.	org.	total	BOD_5
Settleable suspensions	10	30	40	20
Nonsettleable suspensions	5	10	15	10
Dissolved substances	75	50	125	30
Combined	90	90	180	60 g/(P·d)

The per capita value can also widely vary. In the USA, higher values have been used for a long time. For German conditions, the value 60 g $BOD_5/(P \cdot d)$ has proven to be reliable.

With these numbers it is possible to estimate the waste load from industrial and commercial waste water through the population equivalent values. Usually, the calculation is made in relation to the BOD_5 value per person. For untreated water the population equivalent (PE) is set at 60 g BOD_5/d, while for settled (pretreated) waste water a PE of 40 g BOD_5/d is taken.

In isolated cases, the waste loads can deviate considerably from these averages, as shown by the examples in Tab. 2.2.-8.

Tab. 2.2.-8: Specific waste loads per person and day /128/

		Solids miner. organ. tot.			Oxygen consumption KMnO$_4$ BOD$_5$ cons.		Plant nutrients org.N tot.N Phospho- Potassium rus				pH-value	Temperature
		g/(P·d)			g/(P·d)		g N/(P·d) g P/(P·d) g K/(P·d)					°C
Mean value for Europe (Imhoff)	settling subst.	20	40	60	19							
	floating subst.	10	20	30	12							
	dissolved subst.	50	50	100	23							
	total	80	110	190	54 / 74*		12.8	2.3	5.8			12
Mean value for the USA (Fair/Geyer)	settling subst.	15	39	54	19							
	floating subst.	10	26	36	23							
	dissolved subst.	80	80	160	12							
	total	105	145	250	54		10					
In the FRG — Golzheim (Kaes)	settl./float. subst.	19	50	69							7.6	14.5–15.0
	dissolved subst.	172	84	256								
	total	191	134	325	214	68	2.6	6.2	1.8	8.0		
In the FRG — Kaiserswerth	settl./float. subst.	19	56	75	118	28					7.8	winter 14 summer 17.8
	dissolved subst.	153	65	218	81	24						
	total	172	121	293	199	52	5.7	13.8	1.9	6.2		
In the FRG — Stuttgart-Büsnau	filterable substances settled	9	21	30	49	55		9.6	3.0			

* For countries with a high living standard, for mixed systems in the USA

For the remaining waste water constituents, the following values can be assumed for the specific waste load per person and day.

Total nitrogen (mainly ammonia and organically bound nitrogen)
in raw waste water   10 – 15 g N/(P·d)
in mechanically preclarified waste water   7 – 10 g N/(P·d)
of which ca. 80 % is in ammonia-nitrogen

Total phosphorus
in raw waste water   3.0 g P/(P·d)
in mechanically preclarified waste water   2.7 g P/(P·d)
in biomechanically purified waste water   2.1 g P/(P·d)

In raw waste water, about 50 % of the phosphorus is present as ortho-phosphate, about 40 % as condensed phosphates (from detergents), and up to 10 % is bound organically. The condensed phosphates are transformed nearly completely to ortho-phosphates through hydrolysis caused by enzymes in aerobic biological treatment methods such as biological filters or activation processes. Thus, after treatment, up to 95 % of the total phosphorus is present as ortho-phosphate.

Total sulfur
in protein and sulfate compounds   1.0 g S/(P·d)

Nitrogen and sulfur can also originate from nitrates and sulfates in industrial waters.

2.2.4 Waste Concentrations

Apart from the daily waste load B, (kg BOD_5/d), the waste concentration in municipal sewage L_{om} (mg BOD_5/l) and the daily flow Q_d (m³/d) are also important in dimensioning and designing treatment plants and associated water containments. The concentration L_{om} is given by

$$Lom = \frac{B}{Q_d} \quad [mg\ BOD_5/l]$$

As can be seen in the equation, this value is largely dependent on the magnitude of the waste water flow. With an increased flow rate, the waste concentration decreases. For normal sewage from a German city without commercial pollution, concentrations are given in Tab. 2.2.-9 as 24 hour averages in g/m³ (= mg/l). The given values hold for a water consumption of 200 l/(P·d), which is average for European cities.

If the daily consumption is greater or less than 200 l/P, the waste water is correspondingly more diluted or more concentrated, since the waste discharge of the inhabitants only slightly increases with increased water consumption. Concerning the dissolved substances, it should be noted that a part of them, especially the mineral component, is already present in the tap water.

Tab. 2.2.-9: Average waste water concentrations in g/m³ /73/

	min.	org.	total	BOD_5	
Sedimentary substances	50	150	200	100	
Nonsedimentary substances	25	50	75	50	200
Dissolved substances	375	250	625	150	
Combined	450	450	900	300	g/(m³)

Table 2.2.-10 gives a survey of low, medium and high waste concentrations for European conditions.

Tab. 2.2.-10: Waste concentrations for European conditions

Concentration	pH-value	Sediment. subst.	Suspended subst.	Dissolved subst.	BOD_5 unfiltered	Potassium permanganate consumpt. unfiltered	COD unfiltered	Ammonia (NH_3)
		ml/l after 2 h	mg/l	mg/l	mg/l	mg/l	mg/l	mg/l
Low	7.0-8.0	2.0	300-500	400-600	100-200	150-250	150-300	15-30
Medium	7.0-8.0	4.5	500-700	600-800	200-400	250-600	300-450	30-40
High	7.0-8.0	5.0	700	800	400	500	450	50

2.2.5 Fluctuations in the Waste Concentration and Waste Load

The hourly and daily variations of waste concentration and waste load in municipal sewage are also determined by living standards, industrial operations (working shifts, etc.), commercial and public service facilities, as well as the flow conditions in the sewer network. During night, the waste concentration (mg BOD_5/l) drops sharply. This is due to reduced waste production, but also largely results from waste retained in the sewer lines due to low nighttime flow rates. With rising water volumes and increasing load in the morning, these deposits are partially washed onward, so that the waste concentration and load rise sharply.

With insufficient dilution and mixture, industrial waste waters discharged in surges and having widely varying compositions during the day can produce unfavorable effects. This is especially the case for industrial waters containing large amounts of sludge or those containing high concentrations of dissolved inorganic or organic components.

These difficulties can be overcome with mixture and equilibration basins, which are integrated into the sewage plant. An example of the variation in waste concentrations is illustrated in Fig. 2.2.-2 for a small community with discharge surges of different magnitude and their equilibration by preclarification and sedimentation basins.

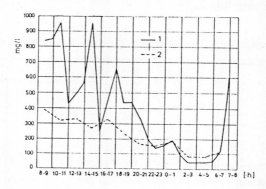

Fig. 2.2.-2: Equilibration of waste water surges through a preclarification basin and a 24-hour equilibration basin (NIERSVERBAND)

1 - Biochemical oxygen demand at inlet of sewage plant (unfiltered); 2 - Biochemical oxygen demand at outlet of preclarification and 24-hour equilibration basin (unfiltered)

2.3 Foreign Water

2.3.1 Types of Foreign Water

Foreign water refers collectively to types of water that, although not polluted, are not diverted to reservoirs due to technical or economic reasons. Thus they are channeled together with waste water into the sewerage system and, consequently, increase the dry weather flow.

Rainwater or snow-melt runoff that is inadvertently or intentionally diverted into a sewage water network instead of the storm drainage may be considered foreign water.

2.3.2 Origin and Amount of Foreign Water

Considering the origin of foreign water as discussed in Section 2.3.1, it is difficult to state exact amounts. As such, the proportionate amount due to groundwater seepage into the system cannot be exactly determined. Tab. 2.3.-1 from FAIR and GEYER /48/ contains data on groundwater seepage for the FRG and USA.

Tab. 2.3.-1: Groundwater seepage into sewer systems /48/

	Groundwater seepage		
	relative to total sewage water in municipal sewerage system	relative to surface area served by sewerage system	relative to sewer line length
	%	l/(sec·ha)	l/(sec·km)
Federal Republic of Germany	17 – 50	0.02 – 0.06	0.17 – 0.5
USA minimum	5 – 40	0.058	0.14
mean	20 – 75	0.23	0.81
maximum	30 – 85	0.58	2.78

An approximate determination of the groundwater seepage contribution is possible by measuring amounts during daytime and nighttime hours or by comparison with the water consumption.

Where possible, no unpolluted spring and stream water should be fed into the public sewerage system, since this would raise water treatment costs and reduce the volume of clear surface water. In urban areas, however, it is sometimes unavoidable, due to economic reasons or space requirements, that small streams and springs must be diverted into the sewerage system as foreign water.

The water volumes of individual spas and thermal baths can be measured quite exactly.

For separated sewerage systems (sewage and storm drainage), rainwater and snow-melt runoff is still considered foreign water. With poor inspection procedures, isolated roof or courtyard drainages may be erroneously connected to the sewage line. Occasionally, rainwater is intentionally fed into the sewerage network at certain points for flushing purposes. If the wet weather flow in a separated sewerage system is more than twice the dry weather flow, the sewage network should be inspected and improper connections corrected.

2.4 Commercial and Industrial Waste Water Amounts

In industry, water is used as a raw material, as a means of production (process water), and for cooling purposes.

The freshwater used comes either from the public water utility or directly from ground or surface waters when the company has its own water supply facilities. The water remaining after use is discharged as waste water.

In addition to being altered quantitatively, the water is also altered qualitatively. Sometimes, superfluous water is separated and diverted during the extraction and processing of raw materials.

The discharged waste water contains a great variety of substances. In order to determine their amounts and composition, local analyses must be carried out.

Water supply requirements and waste water output depend on several factors. The differences in the amount of waste water produced by different industries, which are to some extent subject to considerable daily and hourly fluctuations, have the following causes:

- <u>Differences in the type of industry</u>. Different amounts of water are required for the production of raw materials, the processing of raw materials, and the manufacture of finished products.
- <u>Different basic materials</u>. Contaminated or pure raw material, or

semifinished products in different stages of processing or preparation, lead to varying amounts of waste water.
- Different manufacturing processes. For example, dry or wet processes using the same primary products give different amounts of waste water.
- Size of the plant.
- Mode of operation. The uniformity in the amount of waste water and its composition over a period of a day or a week is affected by whether a single shift or multiple shifts are operated.
- Seasonal activity. Sugar refineries, canning and jam factories, fruit-juice factories, and wine presses have a higher waste water output only during and shortly after the harvest (working season or "campaign").
- Differences in power supply. The need for cooling water varies according to whether a plant has its own power supply or takes power from outside. The type of power is also significant, e.g. electricity, steam power, diesel-engine, or water power.
- Local conditions. Scarcity of water or high water prices make it necessary, for instance, to use water very economically. The condition of the receiving water or official restrictions can at the same time limit the discharge of waste water.
- Use of in-plant recirculation systems. These have a decisive influence on the amount of waste water and the time at which it is discharged.
- Range of production. Very few processing industries manufacture only a single article. It is usually impossible to establish exactly the water requirement per unit of production from the production figures given in industrial statistics.

Many industrial plants have their own water supply. Table 2.4.-1 shows the use of water from a private supply in industry (1965) in West Germany.

Table 2.4.-2 gives the uses of water from own supply by industry in West Germany.

Tab. 2.4.-1: Water from own supply used by industry itself (1965) in West Germany

Industrial group	Water supply	Discharged unused	Given to third party	Own use
	Mill.m³/a	Mill.m³/a	Mill.m³/a	Mill.m³/a
Mining	3 064	1 410	232	1 423
Primary products and capital goods	6 755	82	157	6 515
Investment goods	567	4	9	553
Consumer goods	468	3	4	460
Food and beverages	536	6	4	525
Total	11 390	1 505	406	9 476

Tab. 2.4.-2: Uses of water from own supply by industry (1965) in West Germany

Industrial group	Amount used	Cooling water	Boiler feed-water	Process water	Used by personnel
	Mill.m³/a	%	%	%	%
Mining	1 423	80.4	2.4	14.6	2.6
Primary products and capital goods	6 515	74.2	2.1	22.0	1.6
Investment goods	553	56.5	1.5	27.4	14.6
Consumer goods	460	36.4	5.3	52.0	6.2
Food and beverages	525	52.3	4.1	40.9	2.7
Total	9 476	71.1	2.4	23.7	2.8

Tab. 2.4.-6: Waste water flow in the consumer goods industry /1/

Industry	Unit of measurement	Type of waste water	Waste water amount in m³ min.	max.	average
0	1	2	3	4	5
Wool washing factories	1 t wool	total waste water	8	15	
Cotton bleaching factories	1 t cotton 1 t cotton	boiler solution total waste w.	4 400	5 4000	1000
Artificial silk and viscose staple fibre factories	1 t fibre 1 t triacetate	(viscose process) (acetyl cellulose proc.)	80 10	1000 15	100
Spinning mills	1 t yarn	total waste water	3	250	125
Weaving mills	1000 m² woven cloth	total waste water	10	350	50
Dye mills	1 t cloth	total waste water	30	160	
Tanneries	1000 m² hides	total waste water	0.7	0.5	1.0-1.5
Leather manufacture	1000 m² leather	total waste water	280	450	400
Glue factories	1 t bones 1 t wet glue stock 1 t hide glue	total w.w. total w.w. total waste water	30 300	8 140 1200	
Silk factories	large-scale plant	total waste water	ca. 200.0 m³/h		
Candle factories	medium-sized plant	total waste water	ca. 50.0 m³/h		
Foundry and forge	1 t raw material	total waste water	1	30	10
Flax and hemp rettery	1 t flax straw 1 t flax straw	(tank retting) (canal retting)	40	60	20
Silk boiling plant	1 t raw silk	total waste water	15	70	
Buna manufacture	1 t finished product	total waste water			750
Oil refineries	1 t crude oil	total waste water	3	70	17
Hydrogenation works	1 t benzine	total waste water	1.0	1.25	

Tab. 2.4.-7:   Waste water flow in the food processing industry /1/

Industry	Unit of measurement	Type of waste water	Waste water amount in m³ min.	max.	average
0	1	2	3	4	5
Sugar refineries	1 t beet	flume and washing water	5	7	
	1 t beet	diffusion waste and pulp press water	1.4	2	
	1 t beet	hot well water	4	5	
	1 t beet	total w.water	10	20	
Saccharification of wood	1 t prod. sugar	total waste water	1.7	2.4	
Potato starch manufacture	1 t potatoes	washing water	5	8	
	1 t potatoes	blanching w.		12	
	1 t potatoes	starch washing water	1	3	
	1 t potatoes	pulp press water	0.4	0.6	
	1 t potatoes	total waste water			20
Maize, wheat, rice starch	1 t prod. starch	total waste water	24	28	
Pressed yeast factories	1 t molasses	total waste water	15	80	
Distilleries	1 t potatoes	total w.water	8.5	25	
	1 t grain	total w.water			10
	1 t grapes	total w.water			0.75
Sulfite spirits manufacture	1 t cellulose	total waste water	9	10	
Malt houses	1 t barley	total w.water	5	18	10
Breweries	1 m³ beer	(without malting)	3	17	15
	1 m³ beer	(with malting	15	60	33

DIN 4045 distinguishes between industrial = Q_{sj}, commercial = Q_{sg}, and other sewage discharges.

The meaning of the term commercial is not clearly established. It can, however, be assumed that, in general, all local commercial enterprises are covered, e.g. butcher, baker, laundry, restaurant, and other small businesses. Their water consumption and waste water flow are included under "domestic" or "communal" waste water flow. When these busines-

cont. Tab. 2.4.-7

0	1	2	3	4	5
Alcohol-free beverages	1 m³ beverages	total waste water	1	10	6
Dairies	1 m³ processed milk	process water	0.5	3	
	1 m³ processed milk	cooling water	2	4	
	1 m³ processed milk	total waste water	2	7	5
Margarine and oil factories	1 t vegetable oil	washing water			0.9
	1 t vegetable oil	refuse water			0.17
	1 t edible fat	refinery waste water			0.06
	1 t edible fat	churning waste water			0.02
	1 t edible fat	total waste water	2	3	
Canning factories	1 t fruit or vegetables	total waste water	5	500	35
Fish meal factories	1 t raw material	blood water	0.05	0.1	
	1 t raw material	stick water	0.65	1.0	
	1 t raw material	condensation water	5	10	
	1 t raw material	total waste water		50	30

ses serve more than just the local residents, however, they must be considered as larger commercial enterprises or industrial enterprises, e.g. wholesale butcheries, which also supply outside the town, hotels catering largely to tourists, large-scale laundries, etc.

Where no definite data are available, it is recommended that the following assumptions (from LAUTRICH /92/) for future commercial enterprises without cooling water be taken into consideration:

   with low water consumption   = 0.5 l/(s·ha)
   with medium water consumption   = 1.0 l/(s·ha)
   with high water consumption   = 1.5 l/(s·ha)
   without data as medium   = 1.0 l/(s·ha)

For each larger enterprise Q_{sg} or Q_{sj} must be determined separately. In addition to the daily amount, the distribution over the day and over the week must be taken into account.

Commercial and industrial waste waters contain substances that also differ greatly in their degree of contamination. Even within one branch of industry the values can vary considerably, owing to particular production processes or environmental conditions.

Particularly in the last few years, more industries have been developing water-conserving processes or expanding production so that, through reprocessing of waste water, it is possible to obtain water for industrial use, as well as to recover production materials. These changes have taken place in the industrially developed countries partly as a result of orders from the water-supervising authorities and partly due to considerations of economy in the various industries.

Tab. 2.4.-8: Waste water flow in other industrial and commercial enterprises /1/

Industry	Unit of measurement	Type of waste water	Waste water amount in m^3		
			min.	max.	average
Electric power generation	1000 kWh	total waste water	1	600	200
Gas production	1000 m^3 gas	total waste water	1	23	10
Laundries	1 t laundry	total waste water	10	40	15
Slaughter houses	per slaughtering	total waste water	0.3	4.0	1.0
	1 t slaughter animals	total waste water	2	40	14
Meat processing	1 t meat	total waste water	0.3	7.5	2
Animal utilization	1 t raw material	total waste water	2	30	

Tab. 2.4.-9: Water consumption and waste water flow in various industries according to various data and water consumption per employee per year /158/

Serial no.	Industry	for	Water consumption	Waste water flow	Water consumption per employee in m³/year (KELLER)
0	1	2	3	4	5
1	Brewery	1 l brewed beer	24 l	5.65 l	1000
2	Malt houses				1115
3	Distilleries	1 t potatoes	20 m³	8.5 m³	
4	Distilleries	1 t molasses	17.5 m³	7.5 m³	
5	Brandy distilleries and liquor factories				294
6	Saccharification of wood (BERGINS) (SCHOLLER)	1 kg produced sugar 1 t wood 1 kg produced sugar 1 t wood	2.4 l 700 m³ 1.7 l 500 m³	2.4 l 1.7 l	
7	Confectionery industrie				148
8	Sugar refineries	1 t beet	10-20 m³	10-20 m³	10 000
9	Prepared cereals industry				50
10	Dairies	1 l treated milk	5-6 l	5-6 l	900
11	Margarine factories	per 50 kg margarine		2-3 m³	1100
12	Soap and detergent industry				300
13	Mineral water and lemonade industry				450
14	Slaughter houses	per slaughtering	until 4 m³	1.5-2 m³	180
15	Fish meal factories	per t raw material	80 m³	50 m³	180
16	Fruit and vegetable canning				105
17	Tanneries (small) Bark tanning (uppers) Bark tanning (sole leather) Tanneries (large)	per hide per hide per hide		1.2 m³ 1.5 m³ 2-3 m³	512

cont. Tab. 2.4.-9

0	1	2	3	4	5
18	Shoe and clothing industry				5
19	Starch factories Blanching water Pulp water Starch washing water	per kg potatoes per 5000 kg potatoes	20 l	20 l 50-60 m³ 2-3 m³ 5-15 m³	
20	Paper industry White pulp Unbleached pulp Bleached pulp Bogus paper Printing paper Fine paper Newsprint Pasteboard Printing houses Tar roofing paper, Asphalt board	1 t 1 t 1 t 1 t 1 t 1 t 1 t 1 t	300 m³ 200 m³ 500-550 m³ 350-450 m³ 500 m³ 900-1000 m³ 200 m³ 135 m³	130 m³	20 000 6500 3470 10 300
21	Textile industry Flax (canal retting) Flax (tank retting) Artificial silk and viscose staple Cotton yarns Wool weaving mills Bleaching plants Laundries Customer laundry Worsted yarns	1 t flax 1 t flax 1 t artificial silk or viscose staple 1 t yarn 1 t laundry	200 m³ 20 m³ 150-200 m³	100 m³ 1000-4000 m³ 10-20 m³	4500-7500 120 300 385 450
22	Fine mechanical and optical industry				8
23	Mechanical engineering				13
24	Electrical industry				14
25	Fine ceramics industry				16
26	Pharmaceutical industry Chemical industry Chemicotechnical industry				34 1600 5000

cont. Tab. 2.4.-9

0	1	2	3	4	5
27	Coal industry	1 t coal	10 m³	10 m³	
	Coal washing	1 t coal	10 m³	10 m³	1650
	Coking plant	1 t coal	1.2-1.5 m³		
	Coking plant	1 t coke	5.0 m³		
	Gas cooling		30-35 %		
	NH_3 benzole production		35 %		
	Lignite carbonization water	1 t low-temperature coke		0.5 m³	
28	Coal by-product industry	1 t synthetic fuel	80 m³		2500
29	Paints and fuels				34
30	Plastics processing industry				70
31	Iron industry				20
	Steel, sheet metal and metal goods industry				44
	Motor vehicle construction and repair				72
	Iron, steel and malleable-iron casting				70
	Steel construction				300
	Drawing shops and cold rolling mills				300
	Forging, hammer and shingling works				350
	Iron mining ore				750
	Metal semi-finished products				1750
	Blast furnaces, steel works and warm rolling mills	1 t pig iron	15 m³		
32	Potash industry	1 t carnallite		1 m³ discard solution	350
	Soda factories	1 t soda		1.5 m³ discard sol.	350
	Glauber salt factories	1 t Na_2SO_4		1 m³ discard solution	54
33	Glass industry				150
	Crystal cutting shops and glass etching works				
34	Explosives industries	1 t nitrocellulose		50 m³	
35	Cement industry				1200-2500
36	Synthetic rubber	1 t Buna	500 m³		

Table 2.4.-10 gives general observations concerning the substances contained in the waste waters of some industrial and commercial branches, and Table 2.4.-11 gives their special features (BISCHOFSBERGER /136/).

Tab. 2.4.-10: Observations concerning the substances contained in the waste waters of some industrial and commercial branches /136/

Photographic products.	Used solutions from developing and fixing baths.	Alkalinity, various organic and mineral reducing agents. Toxic elements.
Textile industry, laundries.	Fabric washing.	High proportion of alkali and organic substances. Detergents.
Fiber-manufacture.	Synthetic fibers. Viscose. Polyamide. Polyester. Vinyl products.	Presence of solvents, enzyming products, dye-stuffs, neutral water with BOD load.
Fiber treatment.	Washing, Color fastness test. Bleaching. Printing and finishing. Wool-combing.	High or medium amounts of suspended matter, alkaline or acid water. Very high and fluctuating BOD. Dyestuffs, chemical products, reducing or oxidizing agents, sometimes sulfides. Fat, wool fat.
Paper and pulp industry.	Digestion. Bleaching. Fiber washing. Pulp refining.	High COD and BOD. Dyestuffs, high amounts of suspended matter, colloidals and dissolved substances. Sulfites. Fluctuating pH.
Paper and pasteboard.	Mechanical manufacturing processes. Dosage. Mixing, superfluous water.	White and organic water, fibers, china clay, titanium, kaolin, baryta, pigments, latex, mercury salts.
Other industries, aerospace.	Beryllium. Titanium and tantalum metallurgy. Hydrofluoric acid.	Toxic metals. Surface treatment agents, acid waters containing corrosive products.

Tab. 2.4.-11: Distinctive features and constituents of some industrial waste waters /136/

High temperature	Power stations, all industries, laundries, bottle washing plants in breweries and beverage industries
High amount of suspended matter	Paper mills, pasteboard mills, pulpwood grinding works, cellulose factories, wool-scouring plants, canning factories, coal-washing plants, tanneries, breweries, slaughter-houses
High amount of sediments	Sugar refineries, coal mines, rolling mills, blast furnaces, glass works, gravel washing plants
High amount of organic substances (BOD_5)	Slaughter-houses, meat product factories, rendering plants, glue factories, tanneries, leather factories, sauerkraut factories, canning factories, soap factories, cellulose factories, strawboard factories
High amounts of dissolved substances	Oil industry, coal mines, iron pyrite salt works, potash industry, soda factories, chemical industry, tanneries, softening plants, sauerkraut factories
Acids	Margarine and sauerkraut factories, manufacture of synthetic fatty acids, soap factories, bleaching plants, iron pyrite mines, pickling plants, electroplating plants, powder and explosives factories, chemical industry, candle factories, coal mines, viscose factories, wool-scouring plants (with waste waters treated with acids)
Alkalis	Textile factories, metalworks, chemical industry, buna factories, tanneries, laundries, gas works, wool-scouring plants
Oils and fats	Dairies, margarine factories, slaughter-houses, meat product factories, soap factories, oil industry, tanneries, wool-scouring plants, bleaching plants, dye-works, cloth printing works, laundries, candle factories, metal processing
Toxic substances	Tanneries, leather factories, dye-works, carbonizing plants, gas works, coking plants, electroplating plants, powder and explosives factories, spinning material factories, chemical industry, pesticides
Radioactive substances	Uranium mines, laboratories, hospitals, nuclear power stations
Detergents	Soap factories, textile factories, dye-works, laundries
Color	Paper and pasteboard mills, tanneries, dye-works, paint factories, artificial silk factories, electroplating plants
Infectious properties	Rendering plants, animal carcass disposal, tanneries, glue factories
Odor	Tanneries, yeast factories, distilleries, fish meal factories, slaughter-houses, knackeries, lignite carbonizing plants, coking plants and gasworks
Lack of nutrients	Plants that produce purely inorganic effluents, paper and cellulose industries, coking plants and gasworks

2.4.1 Inorganic Industrial Waste Water

Inorganically polluted waste waters are produced chiefly in the coal and steel industry, in the nonmetallic minerals industry, and in commercial enterprises and industries for the surface processing of metals (iron pickling works, electroplating plants, eloxal works).

These waste waters contain a large proportion of suspended matter, which can be eliminated by sedimentation, often together with chemical flocculation through the addition of iron or aluminum salts and flocculation agents (in this case mostly organic polymers).

The purification of warm and dust-laden waste gases from blast furnaces, converters, cupola furnaces, refuse and sludge incineration plants, and aluminum works results in waste waters containing mineral and inorganic substances in dissolved and undissolved form.

The precooling and subsequent purification of blast-furnace gases requires up to 20 m^3 water per t of pig iron. On its way into the gas cooler the water absorbs fine particles of ore, iron, and coke, which do not easily settle. Gases dissolve in it, especially carbon dioxide and compounds of the alkali and alkaline earth metals, if they are water-soluble or if they are dissolved out of the solid substances by gases washed out along with them.

In the separation of coal from dead rock, the normal means of transport and separation is water, which then contains large amounts of coal and rock particles and is called coal washing water. Coal washing waters are recycled after removal of the coal and rock particles through flotation and sedimentation plants.

Other waste waters, e.g. from rolling mills, contain mineral oil and require additional installations, such as scum boards and skim-off apparatus, for the retention and removal of mineral oils. Residues of emulgated oil remaining in the water also need chemical flocculation.

In many cases, waste waters are produced that in addition to solid substances and oil, also contain extremely harmful solutes. These include blast-furnace gas washing waters containing cyanide, wastes from the metal processing industry containing acids or alkaline solutions (mostly containing nonferrous metals and often cyanide or chromate), wastes from eloxal works and from the waste gas purification of aluminum works, which in both cases contain fluoride. Small and medium-sized nonmetallic minerals plants and metal processing plants are so situated that they discharge their waste water into municipal sewerage systems and have to treat or purify their effluents before discharge, in compliance with local regulations.

There follows a list of some industrial wastes containing inorganic substances, the composition and purification of which show the special problems of such wastes.

2.4.1.1 *Waste Waters from Nonmetallic Minerals Industries*

In the manufacture of mortar binding agents and glass, as well as concrete blocks and ready-mixed concrete, waste waters arise from sand, gravel, and crushed stone washing or from the relevant working processes, such as wet dust removal, washing of the raw materials, or further processing of the materials, such as grinding and polishing.
The manufacture of aluminum and bauxite produces liquid and solid wastes.
In the manufacture of concrete blocks, concrete parts, and ready-mixed concrete, wastes arise from washing of the moulds, machines, and transport vehicles.

In sandy limestone producing plants, bricks are manufactured from the slaked lime and fine-grained silica sand by pressing the moulds and heating for 8 - 10 hours with steam under high pressure. The only wastes that occur in the process are cooling waters with a slight oil content from the condensation pumps of the steam engines and the condensates from the hardening boiler. These hot discharges normally flow continuously with a pH over 10 $[Ca(OH)_2]$ and contain suspended and dissolved substances, sometimes also organic substances and oils up to 50 mg/l.

In cement factories, various types of cement and building materials are manufactured by grinding, mixing, and burning materials containing limestone and clay. Here the waste waters chiefly arise from the wet removal of dust caused by the different working procedures, and also from the removal of the residue left after washing. These effluents normally contain only inorganic substances. Today in cement factories electrofilters are often used to remove dust, which means a considerable reduction in the amount of waste water.

In porcelain factories, kaolin, quartz, and feldspar are used. The wastes from porcelain factories can be divided into four types: precipitation, domestic, the so-called "white" wastes, and wastes containing phenol. The main waste water is the so-called "white" waste. If washbasins, sinks, and similar articles are manufactured, there may be a hardening shop and an electroplating shop joined to the works. The white waste waters contain mainly inorganic suspended solids from the raw materials used. They are cloudy, and the pollutants only settle very slowly. The greater part of these suspended solids is of a size from 10 µm to 1 µm and less (LIEBMANN /96/). In general, the waste waters contain 1000 to 5000 mg/l suspended matter and during the washing of the machines and the floors up to 10 - 50 g/l.

Porcelain factories also need large amounts of heat for firing their products. If gas or oil is available, there is no waste water. In most cases, however, the factories have to produce this gas in generators. In addition to tar, gas water containing phenol, with a phenol content between 4000 and 8000 mg/l, results from the cooling of the producer gas in the pipes, and from gas cooling and washing water.

In other ceramic industries, wastes come from cleaning the mills, transport vehicles, and processing equipment. The wastes are mainly polluted by finely ground minerals. This inorganic turbidity is so stable that it is impossible to clarify the water satisfactorily even after days or weeks of sedimentation.

2.4.1.2 *Waste Waters from the Metal Processing Industry*

In the manufacture of basic materials and finished goods for different industrial branches - such as the motor vehicle and aircraft industries, shipbuilding, the manufacture of jewelery, household goods and tools, and the optical, fine-mechanical and electrotechnical industries - the metals used are treated with a large number of different chemicals in order to cool the metal, remove grease and oxides, prepare the metal for painting and varnishing, and make surfaces smooth and corrosion-resistant. During manufacture, the products pass through chemical baths in various sequences. As they come out of each bath, they are normally rinsed with water in order to remove traces of the bath liquor from the surface of the metal.

The waste waters can be subdivided as follows on the basis of the degree of concentration of harmful substances:
- waste waters from rinsing the products, which contain emulsions (soaps, naphthene, naphthenosulfonates, detergents, etc.).
- concentrates discarded after decomposition of the active substances (acids, alkalis, metals) and enrichment with foreign substances.
- half-concentrates; this refers to the closed rinsing baths used after chemical baths. After certain concentrations are reached, they are no longer effective for holding back harmful substances and they are in general only partly used for topping up the chemical bath liquor, which is reduced by evaporation and dragging out. Therefore, they must also be discarded from time to time.
- Regenerates from the ion exchangers. For economic and technical reasons, ion exchange plants, especially in larger factories, are operated with a water recirculation system for the rinsing waters.

All solutions with more than 100 g of dissolved substances per liter count as concentrates, those with lower concentrations as half-concentrates (HARTINGER /64, 65/). For technical reasons within the production process, their concentration should not exceed c. 10 to 20 % of the concentration of the preceding bath. For rinsing waters, HARTINGER gives approximate ranges:
rinsing water c. 0.5 to 2 times the contents of the runs used per hour, concentration of the substances in the rinsing water c. 0.5 to 3 $^o/o$ of the concentration of the preceding bath.

2.4.1.2.1 Waste Waters from Iron Works

Two types of iron works are distinguished, "wet" and "dry".
For the manufacture of cast iron, large quantities of cooling water are needed, up to 40 - 50 m³ water/t cast iron. The cast iron granulation water contains among other things 700 - 11000 mg/l suspended solids, 0 - 1.6 mg/l cyanide, 18.8 - 1400 mg/l H_2S, 3.0 - 4.8 mg/l ammonia. In addition, there is ore washing water, which is purified and recycled for as long as possible. ROELEN /158/ gives the water demand of an iron works with extensive use of water recirculation as follows:

Gas washing	7 %
Generation of electricity	14 %
Circulation system	19 %
Blast-furnace	14 %
Steel works	21 %
Rolling-mill	22 %
Other uses	3 %

The <u>blast-furnace gases</u> contain 45 - 135 kg dust per t cast iron. The dust is separated by means of a wet or dry washing process. The wet washing process produces over 20 m³ water/t cast iron; however, the introduction of water recycling systems has reduced the water consumption to 5 - 6 m³/t.

Washing waters from the blast-furnace gas recirculating system contain coke, ammonia, cyanide, thiocyanates, naphthalene, and metal traces (Zn, Pb).

2.4.1.2.2 Waste Waters from Steel and Rolling Mills

The pig-iron obtained in the blast-furnace is processed in special furnaces (converters) with various added elements and a variety of methods are used to make cast iron and steel (wrought iron)/forgeable iron.

For this, large quantities of cooling water are needed, as for instance in Thomas-Gilchrist steel works: 3 - 4 m³/t raw steel, in Siemens-Martin steel works: 12 - 18 m³/t steel. The water consumption can be reduced by up to 2 % by means of hot cooling.

The brown vapors from the converter, which contain iron oxides and manganese, are washed in rinsing columns and the substances are separated. The waste water from this process contains substances in the range 8 - 15 g/l.

Washing waters from the converter gas recirculating systems can contain fluoride.

Iron and other metals can be processed mechanically for the manufacture of metal sheets, plates, strips, wires, etc., by <u>rolling</u>; two or more rolls together with the driving engines form a rolling mill train.

Rolling mill trains produce wastes that contain in addition to mill scale, fine dust particles, oil, and grease. The scale has a high specific gravity and can form in terms of weight over 1 g/l of the waste water. In purified discharges from copper rolling mills, metal oxides, e.g. of copper, have been found in amounts up to more than 100 mg/l. Waste waters from an aluminum rolling mill contain small amounts (mostly under 1 mg/l) of aluminum compounds. As these substances are to some extent very finely distributed and coated with traces of grease and oil, their settling is hindered or almost completely prevented.

Discharges from cold-rolling mills contain:
- acid rinsing wastes from sulfuric, hydrochloric, or nitric acid pickling, which are treated by neutralization, air oxidation, and iron precipitation.
- waste waters from chrome passivation, which sometimes occur mixed together with the above-mentioned wastes and are conditioned by ion exchange.
- wastes from electrolytic degreasing, which have high COD values.
- lubricant wastes from rolling mill trains, which form emulsions.
- soluble cooling oils, which from time to time have to be destroyed and replaced.

Other discharges come from dezincing and concentration baths, and these have to be regenerated.

In the cold-rolling of metal sheets, there is a newer process whereby the rolls are cooled with an emulsion instead of water, and this contains about 10 % oil, emulsifying agent, and stabilizing agent against bacterial decomposition.

2.4.1.2.3 Waste Waters from Machinery Production and Mechanical Workshops

In engineering works and metal working plants, coolants, lubricants, or abrasives are used in metal-cutting, especially for turning, milling, drilling, groove-cutting, sawing, and grinding (MEINCK et al. /107/). These additives are mainly emulsifiers such as soaps, naphthenates, naphthenosulfonates, and other anionic, possibly also cationic, detergents, resins, etc., often referred to as soluble oil. In addition, anti-corrosive agents, e.g. nitrite, chromium compounds, or phosphates, are added to the soluble oil.

Commercially available soluble oils are mixed with a certain proportion of water, according to their purpose, and can be recycled for several months. After this time, they are drained off as used emulsion and replaced by fresh emulsion. The used emulsion forms the waste water of these works.

The quantity discharged is generally small, forming only a few m^3, but the biochemical oxygen demand can reach values of over 7500 mg/l (OFFHAUS /122/).

2.4.1.2.4 Waste Waters from Metal Pickling Plants

The iron parts that come from the rolling mills (wires, strips, sheets) have a layer of oxide on their surfaces, which interferes with further processing and therefore has to be removed.

The metal surfaces are cleansed by:
- degreasing and
- pickling.

Degreasing is carried out in three processes: thermal, chemical, and electrochemical.

Thermal and chemical degreasing takes place by means of organic solvents (benzine, benzole, kerosene, trichloroethene) and creates no waste water.

For chemical degreasing by means of alkaline solvents and for electrochemical degreasing, sodium hydroxide, phosphates, silicates, carbonates, and synthetic detergents are used. These degreasing processes are always followed by a washing process in warm and cold water. In general, effluents from degreasing plants present no particular problems, except when cyanide solutions (from Na or K) are used.

Iron pickling takes place in drawing mills, plate and sheet rolling mills, galvanizing and tin-plating plants, press, punching, and enameling works, sheet-metal ware factories, etc. In order to remove the oxide layers (mixture of FeO and Fe_2O_3, also Fe_3O_4), which, according to the processing of the materials, are known variously as hammer scale, iron scale, forge cinder, tinder, or sinter, dips are normally used with dilute sulfuric or hydrochloric acid, nitric acid alone or with admixtures of these acids, also very rarely with hydrofluoric acid or acid salts. They cause a purely chemical dissolution of the oxide layers, together with the formation of the corresponding iron salts. The concentration of the fresh pickling liquors varies according to the type of acids used and the materials to be pickled. In most cases it is between 5 and 20 % by weight; this acid content is consumed except for a remainder of 2 - 7 % by weight. In general, pickling is done with hydrochloric acid at a low temperature (not above 80 degrees).

The process of pickling results mainly in two kinds of waste water: concentrated waste pickling liquors and diluted rinsing waters. Concentrated waste pickling liquors are discharged in batches, when the fully exhausted pickling baths are drained off; the rinsing waters flow continuously. The amount and concentration depend entirely on the type of works, so that generally valid data cannot be given.

According to MEINCK et al. /107/, in a tinplate plant, for 200 t sheet iron about 20 m³ waste hydrochloric pickling liquor (10 % by weight HCl) and about 2000 m³ rinsing water were discharged daily; furthermore, in a transformer sheet rolling mill, for an average of 12 t sheet metal, at least 300 m³ waste pickling liquors and rinsing waters are discharged daily; and in a medium-sized galvanizing plant with an acid consumption of 0.6 to 1 t hydrochloric acid, 7 m³ rinsing water are discharged daily. The consumption of sulfuric acid generally varies according to the type of material to be pickled between 30 and 50 kg per t of iron and can be reduced to 10 - 15 kg per t of iron by the installation of recovery plants. A substantial reduction in acid consumption is achieved by the use of mechanical descaling.

The composition of sulfuric and hydrochloric pickling wastes in the iron and steel industry in the Ruhr district is shown in Tables 2.4.-12 and 2.4.-13 (MÜHLE /115/).

Tab. 2.4.-12: Composition of sulfuric pickling wastes /115/

Firm	Material to be pickled	H_2SO_4 free	$FeSO_4$ 7 H_2O	Corresp. Fe	Total concentration of H_2SO_4	Acid loss (col.3: col.6)
		g/l	g/l	g/l	g/l	in %
1	2	3	4	5	6	7
A	Metal strips	37.7	320.0	64.5	151.1	25
B	Sheet	16.4	449.0	90.7	175.2	9
B	Steel wire	22.6	106.0	21.4	59.5	38
E	Bars	3.2	289.0	58.3	105.6	3
F	Pipes	41.0	231.0	46.6	122.9	33
G	Steel wire	25.3	455.0	91.5	186.2	14
H	Wire rods	22.6	577.0	116.2	225.8	10
R	Metal strips	24.5	632.0	127.3	248.0	10
T	Metal strips	86.7	537.0	108.3	277.1	31
V	Metal strips	15.4	631.0	127.0	237.5	6

Tab. 2.4.-13: Composition of hydrochloric pickling wastes /115/

Firm	Material to be pickled	HCl free g/l	$FeCl_2$ g/l	Corresp. Fe g/l	Total concentration of HCl g/l	Acid loss (col.3: col.6) in %
1	2	3	4	5	6	7
A	Galvanizing plant	69.1	361.9	159.4	276.5	25
B	Galvanizing plant	56.9	185.5	81.8	163.0	37
L	Metal strips	41.3	258.5	114.2	91.0	22
N	Tin sheets	27.6	240.0	105.8	164.4	16
E	Wire	90.5	261.0	115.0	238.2	38

In copper and other non-iron pickling plants, the flow of pickling wastes and rinsing water is usually considerably lower than in iron pickling plants. HUPFER in /115/ gives the following data on wastes from various brass works with pickling baths and chromic acid dips (Table 2.4.-14):

Tab. 2.4.-14: Composition of waste waters from various brass works /115/

Components	Used pickling baths g/l	Used dips g/l	Rinsing water effluents (mg/l)		
			Tube works	Wire and drawing works	Rolling mill
Sulfuric acid	59.7-163.5	5.6-85.8	4-209	192-4942	140-1997
Copper	4.0- 22.6	6.9-44.0	34-147	385-1582	10- 87
Zinc	4.3- 41.4	0.2-37.0	19- 73	350-4300	28- 112
Chromium, hexavalent	-	4.3-19.1	0- 5	0- 67	0- 35
ditto, trivalent	0.0- 0.56	13.5-47.7	3- 78	345-1100	6- 84
Iron	0.1- 0.2	0.03-0.36	1- 5	9- 93	1- 13

Seepage liquors result from leaks in the pickling vats during the whole time they are filled; they drain greater or lesser amounts of acids and salts, according to the composition of each pickling bath.

In addition to pickling wastes, rinsing and seepage waters, there are waste waters containing phenol when generators to produce heating gas are installed. They contain greater or lesser amounts of free mineral acids and metal salts.

2.4.1.2.5 Waste Waters from Eloxal Works

The surface of aluminum oxidizes on exposure to atmospheric oxygen. A flat-white or gray film of oxide adhering firmly to the metal is formed. Since this oxide film is ineffective against more aggressive attacks by corrosion, artifical processes are used to obtain oxide layers of any desired thickness. Most important in this respect is the <u>el</u>ectrolytic <u>ox</u>idation of <u>al</u>uminum, in short, the "eloxal process".
This is achieved by anodic oxidation. These oxide layers can be permanently dyed, normally with organic dyestuffs.

In the eloxal process the following procedures are generally necessary:
- precleaning of the basic product with perchloroethylene (tetrachloroethene),
- pregrinding, fine grinding, and polishing (mechanical processes),
- second degreasing with prechloroethylene (tetrachloroethene),
- final degreasing by short pickling in 10 to 20 % soda lye,
- chemical polishing in baths containing hydrofluoric acid. The polishing baths contain hydrofluoric acid, nitric acid, and ammonium hydrogen fluoride in aqueous solution.
- oxidation in sulfuric acid baths with 10 to 30 % by weight sulfuric acid. The baths are most effective within a range of 6 to 9 g aluminum content per liter. Levels up to 24 g/l are possible, and then the liquor must be replaced. Baths are often kept at the proper concentration level through frequent removal of small amounts of the

liquor and topping up with fresh acid; this means they are never completely replaced.

Used acid concentrates, half concentrates, and rinsing waters are frequently neutralized with calcium hydroxide.

An analysis of the discharge from a neutralization and sedimentation plant in an eloxal works for mixed domestic sewage and industrial effluent, with a very high proportion of effluent containing fluoride, showed a fluoride content of between 7.8 and 11.7 mg/l.

The neutralization of concentrates and half concentrates in the manufacture of eloxal resluts in very large amounts of sludge. The sludge mainly consists of calcium fluoride, calcium sulfate, and aluminum hydroxide. They can be drained efficiently with a chamber filter press. Solid matter between 40 and 45 % by weight is obtained.

2.4.1.2.6 Waste Waters from Electroplating Plants

In the metal processing industry, electrolytic processes are used to obtain metallic deposits for improving surface quality. The metal is first pretreated to remove rust, dirt, and grease. This pretreatment is an electrolytical process, carried out in baths having a cyanide concentration between 5 and 10 g CN^-/l. In addition, the metal is also descaled, i.e. any metallic coating is removed, and passivated.

Galvanization, or the coating of the surface with a metallic layer, involves physical-chemical processes. The following metals are used for this purpose, as shown in Table 2.4.-15 /158/: copper, brass, zinc, chromium, nickel, cadmium, and possibly also precious metals such as silver and gold. The products are hung in galvanizing baths. The bath liquors are aqueous solutions which in general contain chiefly cyanide salts, in addition to metal salts.

The wastes from electroplating plants contain the same substances as were used to prepare the baths. When alkaline baths are used, also alkaline salts are discharged. On the other hand, the effluent from a hard chrome plating plant contains uncombined chromic acid and dichromates, which produce a neutral or faintly acid reaction.

Tab. 2.4.-15: Chief metals used in electroplating baths /158/

	alkaline with cyanogen	acid without cyanogen
Silver	only cyanidic	-
Cadmium	mainly cyanidic	-
Chromium	-	mainly acid
Copper	mainly cyanidic	-
Nickle	-	mainly acid
Zinc	mainly cyanidic	-
Brass	mainly cyanidic	-

The concentration of the waste waters fluctuates considerably and depends on the technical working conditions. For normal plant conditions the proportion of toxic substances characteristic in each case (CN^-, Cr^{3+}, Cd^{2+}, etc.) is rarely below 10 mg/l and can reach 1000 mg/l. Table 2.4.-16 shows the composition of cyanide bath liquors (BORCHERT /34/).

Tab. 2.4.-16: Composition of cyanide bath liquors /34/

Copper electrolyte Copper (Cu) Cyanide (CN) Tartaric acid	14-20 g/l 22-27 g/l 0-15 g/l	Cadmium electrolyte Cadmium (Cd) Cyanide (CN)	10-20 g/l 40-50 g/l
Zinc electrolyte Zinc (Zn) Cyanide (CN) NaOH	20-35 g/l 30-50 g/l 40-90 g/l	Silver electrolyte Silver (Ag) Cyanide (CN) K_2CO_3	3-30 g/l 20-40 g/l 0-100 g/l
Brass electrolyte Copper (Cu) Zinc (Zn) Cyanide (CN) NaOH	10-30 g/l 5-30 g/l 25-80 g/l 0-15 g/l	Gold electrolyte Gold (Au) Cyanide (CN) Na_2HPO_4	0.5-4 g/l 2-10 g/l 70 g/l

A chromium bath can contain up to 400 g chromic acid/l, in addition to sulfuric acid and chromic salt.

Simple cyanogen compounds are lethal when they reach a concentration of about 1 mg CN^- /1 kg body weight.

In the case of chromium and its compounds, levels of about 2 - 3 mg Cr/l are likely to be harmful. Copper salts also begin to show harmful effects in similar concentrations. Zinc compounds are slightly less toxic.

However, when mixed together some substances can have a toxic effect that is higher than the sum of their individual levels of toxicity. Furthermore, toxic effects increase as the water temperature rises and its oxygen content diminishes.

The composition of the waste washing waters from electroplating plants is shown in Table 2.4.-17 (NEMIROV /119/).

Tab. 2.4.-17: Composition of some waste washing waters from electroplating plants

Treatment	Concentration in waste washing waters, mg/l

Copper coating using:
    Copper cyanide +
    Sodium cyanide +
    Sodium carbonate
	2.8 - 14 (Cu) and 58 - 290 (CN)

Cadmium coating using:
    Cadmium oxide +
    Cadmium cyanide +
    Sodium cyanide
	48 - 240 (Cd) und 120 - 600 (CN)

Zinc coating using:
    Zinc cyanide +
    Sodium cyanide +
    Sodium nitrate
	70 - 350 (Zn) und 102 - 510 (CN)

2.4.1.3 Waste Waters from Mines and Ore Dressing Plants

The waste waters from mines contain mainly inorganic substances from ore extraction and dressing, from coal pits, from slate quarries, or from salt mines.

In collieries, there are five different types of waste water, which are very different in quantity and composition (HUSMANN /71/):
- pit waters,
- pithead bath waters,
- wastes from coal washing and dressing,
- coking wastes,
- coke quenching waters.

2.4.1.3.1 Pit Waters

In coal pits, underground water accumulates, either from the groundwater or by seepage from surface water bodies. This waste water must be drained and pumped away.

The quantity depends mainly on the geological character of the area. In the Emscher district, about 1 m^3 groundwater has to be pumped away from every ton of coal extracted. In collieries with no overburden, the amount of groundwater is 7.75 m^3/t coal extracted. In collieries with overburden, on the other hand, it is only 0.96 m^3/t.

The pit waters are acid. They can be contaminated with suspended solids and dissolved substances of various kinds, e.g. mainly sodium chloride, but also calcium, magnesium, and barium chloride, and sometimes the corresponding sulfates. Pit waters also contain various metal salts (iron, nickel, and magnesium sulfate and arsenic compounds). The proportion of sodium chloride sulfate is subject to considerable fluctuations. In the Rhine-Westphalia coal-basin, the pit water often contains 10 to 20 g/l chlorides (SIERP /158/). In rare cases, the level rises to 50 g/l, with individual levels of 200 g/l. HUSMANN /71/ gives chloride levels between 0.5 g/l and 100 g/l.

Table 2.4.-18 shows the composition of coal pit waste waters in the Ruhr district (MEINCK et al. /107/).

Tab. 2.4.-18: Composition of coal pit waste waters in the Ruhr district /107/

	Gneisenau	Engelsburg		Constantin	
	Level 2	554 m level	Level 7	883 m level	1/2/2a	Level 2
Appearance:	clear	–	cloudy	clear	clear	clear
pH:	7.70	6.74	7.34	7.14	7.39	7.83
Density (kg/l) at 20 °C:	1.0015	1.0042	1.0123	1.059	1.0072	1.0110
			mg/l			
Dry residue at 105 °C:	1150	4368	16870	100700	10050	16700
Susp. matter:	18	38	153	–	29	7
organic:	7	14	109	–	12	4
inorganic:	11	24	44	–	17	3
Argillaceous earth + iron oxide:	22	–	5	–	10	–
calcium	200	254	684	4148	616	2598
strontium oxide	–	–	–	–	–	–
barium	–	–	–	1634	–	–
magnesium	45	218	223	1692	208	262
chlorides (Cl)	532	–	9928	53899	4964	8865
sulfates (SO_4)	232	1991	449	–	515	0
Silicic acid (SiO_2)	20	12	18	18	10	20
Hardness (DG)						
total	26.3	55.9	100.6	711.2	100.7	296.2
alkaline earth	20.0	25.4	68.4	474.4	61.6	259.8
magnesia	6.3	30.5	31.2	236.8	39.1	36.4
Uncombined carbon dioxide	–	–	14.7	–	18.4	0.5
Uncombined oxygen	–	–	–	–	–	8.2

2.4.1.3.2 Pithead Bath Waste Waters

Washing waters from the washing and bathrooms during shift changes and water from the toilets flow in batches. They contain soap, dirt, feces, and coal dust. The amount of this water is likely to be between 30 and 100 l/head and day according to water consumption, on average 60 l/head (SIERP /158/), MEINCK /107/).

According to analyses from the Emscher district published by BACH in /107/, this water normally has a slight alkaline reaction and sometimes shows over 1 g/l undissolved substances (mostly organic), several mg/l dissolved substances (including chloride) and a potassium permanganate consumption of over 300 mg/l.

2.4.1.3.3 Waste Waters from Coal Washing and Dressing

After extraction, the coal is classified on the spot according to fineness of grain, separated from the dead rock (discard), and then won by wet or dry dressing. Most often, wet dressing processes are used.

The quantity of washing water in circulation depends on the coal to be dressed and on the degree of purity required by the end product. On average, about 3 to 4 m^3 are needed per t fine coal, and about 5 to 6 m^3 per t nut coal. Additional water can be calculated at a rate of about 30 m^3 per 100 t fine coal, and about 20 m^3 per 100 t nut coal.

The washing waters from coal wet dressing look cloudy, almost black; they contain varying amounts of suspended coal and plastic clay particles in fine-grained to powdery distribution, also dissolved salts and decay-resistant organic substances. The coal washing water from a colliery in the Emscher district that used tap water for washing contained per liter 1.28 g suspended solids (0.747 g of which were inorganic) and 0.575 g dissolved substances (0.515 g of which were inorganic), while the coal washing water from another colliery in the same district that used pit waters in a recirculation system had per liter 18 g suspended solids (of which 3.85 g were inorganic) and 1.38 g dissolved substances (of which 1.3 g were inorganic; 0.56 g were chlorides). The potassium permanganate consumption in both waste waters had a level between 30 and 40 mg/l, and the amount of nitrogen compounds was under 10 mg/l.

2.4.1.3.4 Waste Waters from Hard-Coal Coking Plants

When hard-coal is carbonized, the gas that comes out of the coke furnaces contains water vapor, which condenses when the gas is cooled. The condensate contains waste products from the coal. Quantity, concentration and components depend on the type of coal used, the plant conditions during carbonization - such as temperature and heating time - as well as the processes used for treating the coke gas. The quantity of waste water, consisting of condensate and, according to the process used, washing and steam stripping waters, is about 0.08 to 0.38 m³ per t of coal, varying according to the type of gas treatment process used. GROSS and DENNE* give the following figures for the carbonization of Saarland hard-coal:

medium heating temperature, in °C	1188 —	1317
amount of waste water	0.163 —	0.154 m³/t coal
amount of phenol	0.434 —	0.202 kg/t coal

The effluents contain in dissolved form ammonia, carbolic acid derivatives, phenols, pyridine bases, cyanides, thiocyanates, and sulfides in addition to small quantities of inorganic cations, occurring as salts of mineral acids. Tar and oils are dispersed and are removed in a primary treatment stage by sedimentation (tar) or flotation in gravity separators.

The remaining substances are as follows:
<u>Ammonia</u>. Amounts contained may reach levels up to about 10 g total NH_3 per liter. 65 to 90 % of the ammonia is bound to weak acids. When the aqueous solutions are boiled, the ammonia is separated and driven off with steam. The ammonia that is bound to strong acids, about 10 to 35 % of the total ammonia, remains in the solution during this process and can only be driven off with steam in the same way after being released by means of strong bases (calcium hydroxide or soda lye).

This behavior is significant from an analytical and technical point of view, and is the reason for distinguishing volatile and fixed ammonia compounds.

Volatile: NH_4OH, $(NH_4)_2CO_3$, NH_4HCO_3, $(NH_4)_2S$, NH_4HS, NH_4CN
Fixed: NH_4Cl, $(NH_4)_4Fe(CN)_6$, $(NH_4)_2S_2O_3$, $(NH_4)_2SO_4$, NH_4CNS

---

* GROSS, R. u. DENNE, A., "Betriebserfahrungen mit einer Anlage zur biologischen Behandlung phenolhaltiger Kokereiabwässer", Stahl und Eisen 88, Nr. 6, 1968

In addition there exist NH_4^+ compounds with organic acids and phenols.

Phenols. When river water containing phenol is chlorinated in drinking water treatment plants, chlorophenols can form, which give the water an unpleasant taste. In contrast, the phenol concentrations in coking plant wastes that contain phenol are between about 0.5 to 3.5 g/l.

Cyanides and sulfides are also substances that could have harmful effects.

2.4.1.3.5 Coke Quenching Water

The quality of coke quenching water much depends on the coal that is charged. The quantity of effluent is actually small, since most of the water evaporates. The amount may be considered to be from 0 to 0.5 m³/t quenched coke.

Discharges from hard-coke low-temperature carbonizing plants with coke-furnace temperatures of about 600° C contain more phenol than the discharges from hard-coal coking plants that have phenol levels of ca. 6.5 to 8.5 g total phenol/l, of which ca. 85 to 95 % is distillable with steam.

Lignite carbonization wastes contain up to 30 g total phenol/l, up to 16 g volatile organic acids/l, and a BOD_5 up to 50 g/l.

2.4.1.3.6 Waste Waters from Charcoal Manufacture

The distillation of wood in retorts - the heap charring produces the waste waters - similarly gives, together with products of decomposition, enriched condensates that are acidic. The quantity of condensate is greater in wood distillation than in coke production from hard-coal. With a pH of 3, the condensates contain about 90 g organic acids/l (related to acetic acid), about 12 g methanol/l and about 6 g phenol/l, related to carbolic acid.

amount of final liquor per 100 t raw carnallite processing as:

with 10 %   KCl, corresp. 12.8 % $MgCl_2$, 32.7 m³ final liquor
with 16 %   KCl, corresp. 20.4 % $MgCl_2$, 52.3 m³ final liquor
with 20 %   KCl, corresp. 25.5 % $MgCl_2$, 65.4 m³ final liquor and
with 26.9 %   KCl, corresp. 34.3 % $MgCl_2$, 87.9 m³ final liqour

When untreated water containing these substances is used for preparing drinking water, the taste and wholesomeness of the drinking water are impaired. In this respect, the following taste limits are established:

for magnesium chloride	160 mg/l
for magnesium sulfate	250 mg/l
for calcium chloride	500 mg/l
for calcium sulfate	500 mg/l
for sodium chloride	400 mg/l
for sodium sulfate	400 mg/l

The fishing and cattle industries also suffer from high salt levels. Grazing cattle will not drink salty water.

When water containing salt is used for irrigation purposes, a distinction must be made between the effect of the salt on the soil and on the plants. With regard to the effect on the soil, common salt and magnesium chloride may be considered as roughly equivalent. The effect is due to potassium and lime being taken out of the soil by exchange processes and the humus gel dissolving by exchange of bases. In this way the soil loses its crumb structure and its water and nutrient holding capacity. Sandy soils are particularly affected in this respect.

There is no danger of adverse effects on plant growth in meadows when the sodium chloride content of the irrigation water stays below 0.5 g/l. It should however be noted that the salt concentration may rise through evaporation.

Possible <u>nitrogen fertilizers</u> are ammonium salts, nitrates, cyanamides, and urea compounds. The basic material is usually ammonium nitrate. Ammonia is obtained by catalytic high pressure synthesis at high temperatures from a mixture of 1 part nitrogen and 3 parts hydrogen. Various processes are used according to the raw material available.

Hydrogen from coking gases was previously used for ammonia synthesis. Today hydrogen and nitrogen for this purpose are prepared exclusively by a process in which steam and air are passed alternatively over glowing coke. This gives a mixture of producer gas and water gas with the components H_2, N_2 and CO_2. The CO_2 is then washed out. From this gas production process come gas washing waters (with 7.5 cm³ suspended matter per liter) which give the waste water a blackish color due to ash and coal particles. The greatest quantities of process water are the cooling waters. Contaminated water consists only of washing waters from plants and shops, alkaline solutions from final washes of the carbon dioxide, and condensation waters from evaporation plants in the manufacture of ammonium nitrate.

The gas washing waters are often rich in sulfur. In contrast to the washing waters from producer gas and water gas, the discharge from carbon dioxide washes in the coking gas process contain no solid matter and are, except for the carbon dioxide content, practically free from dissolved foreign substances. They show only a faint acid reaction as a result of the carbon dioxide absorbed, and this can be largely removed in deaeration towers or cascades. For the manufacture of 1 t ammonium nitrate there are about 5 - 15 m³ condensation water with 200 - 250 mg NH_4NO_3/l and 10 to 20 mg NH_3/l.

Fertilizers <u>containing phosphates</u> are used mostly in agriculture as lime superphosphate. Lime superphosphates are obtained by decomposition of the raw phosphates (apatite and phosphorite) through treatment with sulfuric acid.

Part of the fluorine contained in the minerals escapes as hydrogen fluoride and fluorsilicum. These acid gases are very poisonous. They have to be precipitated through water or common salt solutions in closed condensation plants with built-in irrigation or atomizing plants. Insoluble silicic acid (SiO_2) and sodium silicum fluoride (Na_2SiF_6) are precipitated, by reactions of the fluorine compounds with water and salt, from the concentrated salt solution in which the waste gases from the superphosphate factories are absorbed. The latter acid is filtered off, washed with water and dried. If the waste waters after this are still acid due to a low sulfuric acid content, they can easily be neutralized and rendered harmless by adding lime.

About 0.2 m³ waste water/t product are produced in the making of superphosphates.

2.4.1.4.5 Waste Waters from Antimony Pentasulfide Production

In the manufacture of technical (arsenic-containing) antimony pentasulfide (antimony sulfide of varying composition, chiefly Sb_2S_5), which is used for fireworks, for medical application, and especially for vulcanizing and reddening rubber, as well as in the preparation of the ignition compound for matches, about 180 - 200 m³ waste water per t antimony pentasulfide is produced. This water contains 4 - 5 g/l common salt, and 1 - 1.5 g/l calcium chloride, as well as small amounts of arsenic compounds. With the exception of the arsenic compounds, its composition and properties are similar to those of the final liquors from ammonium soda factories.

2.4.2 Organic Industrial Waste Water

Contaminated organic industrial wastes flow chiefly from those chemical industries and large-scale chemical works, which mainly use organic acids for chemico-technical reactions.

The effluents contain organic substances having various origins and properties. These can only be removed by special pretreatment of the waste waters, followed by biological treatment. In this category are effluents from the following industries and plants:

- factories manufacturing pharmaceuticals, cosmetics, organic dyestuffs, glue and adhesives, soaps, synthetic detergents and biocides (pesticides and herbicides),
- wood distillation plants (mentioned in Section 2.4.1.3.6) and refuse incinerator plants,
- tanneries and leather factories,
- textile factories,

- cellulose and paper manufacturing plants,
- plants of the oil refining industry,
- tanker washing plants,
- the metal processing industry,
- hospitals.

2.4.2.1 Waste Waters from the Pharmaceutical and Cosmetic Industries

2.4.2.1.1 Waste Waters from the Pharmaceutical Industry

The quality of the wastes from the production of pharmaceuticals varies a great deal, owing to the variety of basic raw products, working processes, and waste products.

It is a characteristic of the pharmaceutical industry that very many products as well as intermediate products are manufactured in the same plant. Thus different kinds of effluent with widely varying qualities flow from the different production areas.

It is also usual for large-scale chemical industries to manufacture pharmaceutical products together with other chemical products.

Drugs are generally classified according to their therapeutic effects.

Waste substances include extraction residues of natural and synthetic origin, used nutrient solutions, specific poisonous substances, and many other products.

For example, in the making of antibiotics there are the following waste products (PÖPEL /129/):
- used nutrient solutions (fermentation liquids, slops) with a BOD_5 of 4 000 to 8 000 mg/l, contaminated by, among other things, organic acids, proteins and residues of the extraction agents, carbohydrates, and nutrient salts, pH 2 - 3;
- mycelium sludge impregnated with kieselgur (amount of mycelium sludge about 120 kg/m³ nutrient solution) as filter residue with over 100 g/l BOD_5. The sludge contains remains of extraction agents, such as acetone, amyl acetate, dichloroethylene, and it is possible these may also be in the waste water in large amounts. Also, the sludge forms a decomposable mass;

- purification waters with a BOD_5 of 600 to 2 500 mg/l and a high proportion of solid matter (8000 mg/l). The amount of this water is ca. 1.7 m³/m³ processed nutrient solution;
- washing waters contaminated by chemicals (acids, alkalis) and solvents (acetone);
- discharges from injection condensers. They contain volatile substances that make up to 25 % of the BOD of the liquid to be evaporated; its BOD_5 is 60 - 120 mg/l.

The total water amount rises to levels above 2 000 m³/day.

2.4.2.1.2 Waste Waters from the Cosmetic Industry

Raw materials used in the cosmetics industry are vegetable oils, fats, fruits and other extracts, animal fats, components and extracts of chemical compounds such as alcohols, bactericides, drugs, detergents. There are large-scale cosmetic industries, but also very small firms. The quality of the waste waters from these industries is determined by the range of products. Therefore, it is usually significant whether the wastes come from a small enterprise with a limited range of products or whether they are from an integrated production branch of a large-scale chemical works.

Waste waters from the small commercial enterprises are normally discharged, after appropriate treatment, into the public sewerage system. Large-scale cosmetic industries, on the other hand, have their own purification plants.

2.4.2.2 Waste Waters from Dyestuff Manufacturing

There are very many possible processes for the manufacture of dyestuffs. Moreover, production is only rarely limited to the making of only one dye. Normally a variety of products are manufactured in one factory, in some cases 1200 compounds with a great variety of chemical compositions.

For this reason it is impossible to describe the composition and the treatment of the different waste waters.

The amount and composition of these wastes must be measured individually in existing production plants.

2.4.2.2.1 Waste Waters from Inorganic Dyestuffs

Inorganic dyestuffs, which chiefly contain dissolved and undissolved contaminants, do not give rise to decomposable effluents. The chief inorganic dyes and the waste waters they produce are (MEINCK /107/ et al.):

<u>Prussian blue</u> (ferric ferrocyanide), manufactured by precipitation of potassium ferrocyanide (yellow prussiate of potash) with iron sulfate and aeration. Treatment is with dilute sulfuric acid, pressing out, and washing. The waste waters consist of the acid reaction liquid and the washing waters. In addition to free sulfuric acid (pH 1.5) and sulfates, they contain chlorides, iron compounds, and cyanogen compounds, with a high amount of suspended solids.

<u>White lead</u> (basic lead carbonate), made by treating metallic lead with acetic acid and carbonic acid. The resulting waste waters (washing waters) usually contain lead acetates, lead carbonate, and acetic acid residues.

<u>Cadmium yellow</u> (α- cadmium sulfide) made by reaction of cadmium chloride lyes with barium sulfide or sodium sulfide. The resulting waste waters generally contain considerable amounts of barium or sodium sulfide as a product of reaction, excess sulfide and pigment.

<u>Cadmium red</u> (β- cadmium sulfide) made by effect of sodium sulfide solution on precipitated cadmium carbonate. The waste waters contain, in addition to pigment residues, considerable amounts of hydrogen sulfide and up to 30 g/l common salt. Their reaction is primarily alkaline.

<u>Chrome yellow</u> (lead chromate) is made by precipitation of soluble lead salts with potassium dichromate. The waste waters contain potassium acetate and unconverted residual potassium chromate (dry residue of the reaction solution: about 9 g/l, acetic acid content: 4.2 g/l, BOD_5: 3.0 g/l).

<u>Viridian green</u> (chromic oxide) is made by heating to red heat potassium dichromate with crystallized boric acid and decocting with water; this produces waste waters containing borate. This dye is also made by deflagrating alkali chromates with sulfur in rotary hearth furnaces.

Waste waters containing alkali sulfate and small amounts of dichromate come from the washing of the residues in lye and in water.

Artificial colcothar (iron (III) oxide) is made from the waste waters containing iron salts that come from iron pyrite mines, pickling plants, and aniline manufacturing. Depending on the basic material and the processes used, waste waters result which contain calcium sulfate and calcium chloride or sodium sulfate or sodium chloride.

Lithopones (mixture of zinc sulfide and barium sulfate) give quenching and washing waters from the reaction of zinc sulfate with barium sulfide solution. These wastes can contain zinc, barium sulfate, and hydrogen sulfide ions and are always contaminated with residues of the pigment.

Lampblack is made by incomplete burning of oil or naphthalene or organic residues, as well as by clearing acetylene under high pressure and at high temperature. Cooling, washing, and purification waters are produced.

Paris green (copper (II) acetate arsenite) is made by reaction of copper sulfate with arsenite sodium in acetic acid, and gives waste waters containing acetic acid, sodium sulfate, and sodium arsenite.

Titanium white (titanium dioxide) is made from titanium slag or ilmenite (ferrotitanium, $FeO \cdot TiO_2$) by disintegration with 86 % sulfuric acid and subsequent leaching of the disintegrated material. During this process, titanium white separates and is filtered off. This gives a weak acid with about 18 % H_2SO_4 and 12 % $FeSO_4$. The pigment is then washed, which produces large quantities of washing waters containing acid and iron.

Vermilion (mercury sulfide) is made by reaction of metallic mercury with concentrated potassium polysulfide solution. The washing waters contain potassium sulfide residues; the reaction lye is regenerated and reused.

2.4.2.2.2 Waste Waters from Organic Dyestuffs

In most cases, the basic materials for the manufacture of organic dyes are cyclic hydrocarbons such as benzene, toluene, their derivatives, anthracene, or intermediate products obtained from them such as aniline, toluidine, anthraquinone, etc.

The process steps can include: sulfonation (with sulfuric acid), nitration (with sulfuric and nitric acids), reduction of nitro compounds to amino compounds (with iron filings and acid, zinc, ammonium sulfide, sodium sulfide, sulfurous acid, etc.), diazotization (with nitrites and free acids), condensation (with aluminium chloride among other things), oxidation (with chlorine, manganese dioxide, nitric acid, etc.), melting (with caustic alkalis), salting (with common salt), precipitation (with organic solvents), alkalizing (with caustic alkalis), and others.

The waste waters from these processes therefore contain a great variety of organic and inorganic contaminants in dissolved, and sometimes undissolved, form. Particularly frequent constituents are:
- residues of primary and intermediate products (benzene, aniline, cyclic nitro compounds, etc.),
- residues of finished products (dyes),
- methanol and other organic solvents,
- sulfuric acid and its salts,
- glycerine,
- nitric acid and its salts,
- nitrites,
- common salt,
- lime,
- iron salts,
- aluminum chloride,
- acetic acid and its salts,
- and secondary reaction products of these substances.

The reaction can be an acid, alkaline, or neutral. Although the dye content consists only of final residues that are technically impossible to identify, color is a characteristic of these wastes. To the layman, they appear to be highly contaminated, even when the amount of dye is very small and quite harmless.

2.4.2.3 Waste Waters from Soap and Synthetic Detergents Production

In the manufacture of soap, vegetable, or animal fats, oils or artificial or natural fatty acids are boiled in alkaline solutions.

The primary fats used are mainly those that are not suitable for human consumption, e.g. waste fats, fats from waste waters, bone fats, horse fats, mutton tallow, fat from rendering plants, etc. In the manufacture of curd soaps and finer toilet soaps, vegetable oils such as olive oil, sesame oil, corn oil, soya oil, etc., are used.

The process steps in a soap factory are essentially the following:
- purification of the primary fats,
- fat cleavage (or manufacture of fatty acids) and
- saponification.

The following waste waters are produced.
- various wastes from the purification of the raw fats,
- wastes containing glycerol from fat cleavage and from glycerol refining,
- nigres from the saponification process,
- washing waters,
- condensation, rinsing, and cooling waters.

2.4.2.3.1 Waste Waters from Soap Production with Fatty Acids

To make soap (alkaline salts of fatty acids), the fats are treated with adsorbent substances such as charcoal, bone black and activated coal, kieselguhr, fuller's earth, smectite or silica gel at temperatures of 70 to 90 °C.

Bleaching is done by using sodium peroxides, percarbonates, hypochlorites, sodium hydrogen carbonate, potassium permanganate, sulfuric acid, and other reducing agents.

Fat cleavage is carried out in order to separate the fatty acids needed for soap-making from their bonds with glycerol. The quality of the waste waters depends on the processes used for fat cleavage.

In all listed fat cleavage processes, a glycerol water containing 10 - 14 % glycerol is obtained as a by-product. This is processed in the factory to give an 80% concentration of pure glycerol.

Saponification, mostly in the form of carbonate saponification, gives a "soap glue" from which the "soap curd" is separated by adding common salt and collected as a half-liquid product above the "nigre" (lower aqueous layer). The waste waters usually have a temperature of 90 - 100° C, and as they cool, layered soaps and gluey components are separated.

The amount of effluent can vary considerably according to the type of factory and the processes used, so that generally valid data cannot be given.

The following figures from a large-scale soap factory may serve as a rough guide. The total amount of water in 8 hours was around 1500 m³ (MEINCK /107/).

This is accounted for by:

cooling and condensation water	1360 m³
discharges from raw fat refining	58 m³
washing water from fat cleavage facility	20 m³
nigre and soap washing water	20 m³
purification water	25 m³

The rest arises from other waste waters, e.g. rinsing water from water purification plants, which are not restricted to soap factories.

Effluents from soap factories have several characteristic qualities that have to be taken into account when treating the water. They have an unpleasant smell. They still contain various fats and greasy, soaplike substances, which are detrimental when the water is discharged into the public sewage systems, since they cause greasy deposits on the sides of the sewers.

Discharges from raw fat refining and injection condensors in the glycerol evaporation plant have a strong odor. This can be almost wholly removed by chlorination when treating condensation waters, and to a large extent removed in the case of the washing waters. The refining of raw fats by treating with absorbent agents such as fuller's earth, activated coal, silica gel, etc. actually produces no waste waters, but there are wastes from the purification of the adsorbent agents. This is done by means of alkaline solutions. These waste waters show a distinct alkaline reaction, contain much suspended matter and small amounts of fat or oil.

In the purification of raw fats by washing, weakly acid waste waters are produced, containing an amount of oxidizable organic substance many times greater than that of normal municipal sewage. Data showing the analysis of waste water of this type are summarized in Table 2.4.-20.

If the raw fat refining process includes bleaching, the waste waters contain residues of excess bleaching agent or its conversion products. Bleaching agents used in the soap industry are: peroxysulfate, hypochlorite, sodium peroxide and other oxidizing agents.

Tab. 2.4.-20: Composition of soap factory waste waters /107/

Name of sample	pH-value	Evaporation residue total	Evaporation residue loss on ignition	chlorides	$KMnO_4$-consumption	sulfuric acid free	sulfuric acid bound	zinc
		mg/l	mg/l	mg/l	mg/l	mg/l	mg/l	mg/l
Discharge from refining	4.3	3 764	1 854	120	1 966	-	853	-
Washing water from fat cleavage (SIERP)	below 1.0	199 800	158 400	-	124 800	6 300	49 200	28 400
Nigre	8.5	317 840	94 220	124 000	18 720	-	-	-
Nigre (SIERP)	-	190 000	58 000	80 000	28 990	-	-	-

2.4.2.3.2 Waste Waters from Synthetic Detergent Production

The worldwide use of fat and improvements in manufacturing techniques in oil factories led to a shortage of lower fatty acids and a rise in their price.

An attempt was therefore made to synthesize molecules similar to soap molecules that need no natural fats for their manufacture. Such substances are detergents.

The surface active detergents belong to various groups, which can be divided into three different types according to their electric charge (ionization):
- anionic detergents, e.g. alkyl sulfates and alkyl aryl sulfonates, which are most commonly found in household use. Only "soft" alkyl aryl sulfonates should be used; they have a straight side chain and are easily biodegradable.
- non-ionized detergents, which, for example, are based on polyoxyethylenes and are used in combination with fatty acids and alkyl phenols. These detergents are chiefly used in industry.
- cationic detergents, such as pyridine derivates and quaternary ammonium bases, which because of their germicidal properties are mainly used for sanitary purposes, such as washing baby's diapers, as a softening agent in the laundries of restaurants and hospitals, etc.

Heavy-duty detergents have other added ingredients, such as bleaching agents, stabilizers, dirt-absorbing colloids, hydrotropic substances, skin protecting agents, brighteners, etc. In addition, they contain alkalis and phosphates. They are aptly called "builders" and are substances that have no surface active properties themselves, but which increase the effectiveness of the active substances. In the past, such builders were stale urine, ashes, potash, and, later, soda. More recently, however, certain phosphates for stabilizing the hardness (calcium and magnesium compounds) of the water (tetrasodium phosphate, pyrophosphate, trisodium polyphosphate) have become much more important.

Table 2.4.-21 shows the composition of detergents (KLING, LIEBMANN /95/).

Tab. 2.4.-21: Composition of detergents /95/

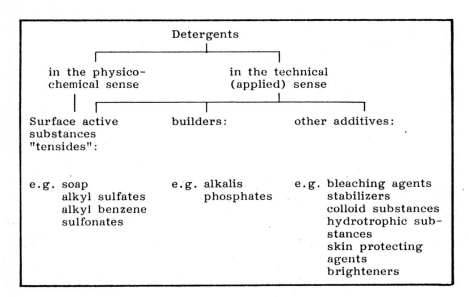

From the manufacturing process, various waste waters are produced with the following qualities (MEINCK /107/):
- marked tendency to form foam (due to reduction of the surface tension),
- contain emulgated substances, e.g. fats and oils,
- sudden changes in composition and reaction due to batch discharges of wastes from individual phases of the manufacturing process.

Table 2.4.-22 shows the composition of waste waters from the manufacture of detergents, and Table 2.4.-23 shows the same for wastes from the purification of detergents.

Tab. 2.4.-22: Composition of waste waters from the manufacture of detergents /95/

Value	pH	COD	Active substances	Residues		
				oil	total	on ignition
		mg/l	mg/l	mg/l	mg/l	mg/l
Min.	0	14 000	117	660	10 300	2 200
Max.	11	90 800	10 000	78 700	56 900	20 500
Av.	-	50 000	3 200	22 400	33 000	8 300

Tab. 2.4.-23: Composition of waste waters from the purification of detergents /95/

Value	pH	COD	BOD_5	Active substance	Residues			SO_4
					oil	total	on ignition	
		mg/l	mg/l	mg/l	mg/l	mg/l	mg/l	mg/l
Min.	5	200	33	10	30	500	250	38
Max.	8.7	24 000	160	730	36 100	20 100	1 600	575
Av.	-	5 200	72	130	6 300	3 000	680	260

Fig. 2.4.-1 shows a diagram of detergent manufacture from HEINZ in /100/.

Table 2.4.-24 shows the physiological behavior of some types of detergents (MEINCK /107/).

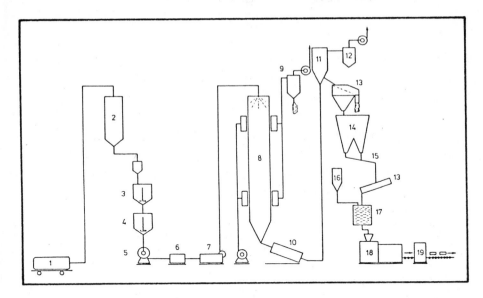

Fig. 2.4.-1: Diagram of detergent manufacture using the "WELTER tower process" /100/

1 - tank vehicle; 2 - pre-storage; 3 - mixing tank; 4 - feeder tank; 5 - centrifugal pump; 6 - pulverizer; 7 - press pump; 8 - tower; 9 - separator; 10 - BOOSTER drum; 11 - air-lift; 12 - separator; 13 - sifter; 14 - bunker; 15 - perfuming; 16 - perborate; 17 - mixing cascade; 18 - filling machine; 19 - packaging machine

Tab. 2.4.-24: Physiological behavior of some types of detergents /107/

Chemical structure	physiochemical type	physiological behavior
alkyl sulfates	anion active	easily biodegradable
alkyl aryl sulfonates (with straight side chain)	anion active	not easily biodegradable
alkyl aryl sulfonates (side chain branched with quaternary carbon atom)	anion active	hardly biodegradable
tertiary ammonium or pyridinium salts	cation active	toxic, bactericidal
polyglycol ether	non-ionogenous	not easily biodegradable

2.4.2.4 Waste Waters from Plastics Industries

Plastics can be manufactured through the action of chemicals on high-molecular-weight natural substances, or produced synthetically from low-molecular-weight basic materials by condensation, polymerization, and other processes. Waste waters are discharged both from factories producing primary and semifinished products and from factories for finished products.

The successive processes may vary to some extent, may or may not use water, and give liquid and solid waste substances with organic and inorganic components.

Of the many different types of plastics, the following groups are discussed here:
- plastics based on cellulose,
- plastics as products of condensation,
- plastics as products of polymerization.

2.4.2.4.1 Waste Waters from Cellulose-based Plastics Industry

Table 2.4.-25 shows a summary of these products from MEINCK /107/.

Tab. 2.4.-25: Summary of Cellulose-based plastics /107/

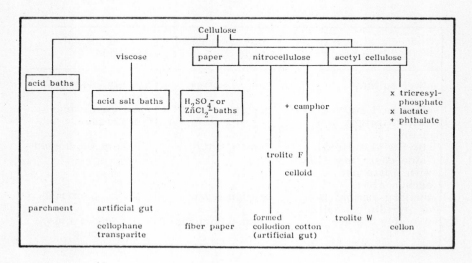

Plastics based on cellulose, parchment, and formed collodion cotton are manufactured by the effect of concentrated acids on cellulose. The waste washing waters from this process are acid, the waste acids can be evaporated or reused after adding fresh acid.

In the production of artificial gut, there are diluted alkaline waste liquors that can contain disturbing sulfur compounds. However, these can be converted by exhaust gas or acid waters containing carbon dioxide arising from manufacturing processes. Other waste waters to be considered are spent acid precipitation baths, and weakly acid rinsing and washing waters containing sodium sulfate or hypochlorite.

They are similar in composition to wastes from viscose synthetic fiber factories. This also applies to wastes arising from the manufacture of transparent viscose films. The manufacturing processes for fiber paper give leaching or rinsing waters.

2.4.2.4.2 Waste Waters from Plastics Production (Condensation Products)

Among the products of condensation (combination of various molecule groups accompanied by elimination of water or simple molecules), the most important are phenoplasts and aminoplastics. Phenoplasts (phenolic resins, e.g. Bakelite) are made by catalytic condensation of formaldehyde and phenols. Phenol synthesis (Fig. 2.4.-2) gives cooling, condensing, and rinsing waters, which react neutrally or weakly alkaline. The condensing processes produce, in addition to quite large amounts of cooling water (over 100 m^3/t product), small amounts of alkaline purification waters, and so-called reaction waters (about 60 - 80 l per t product). These waters are partly alkaline (red water), partly acid (acid water). Table 2.4.-26 shows their composition.

Aminoplastics (aminoplast resins, e.g. Pollopas) are made by condensation of formaldehyde with urea or thiourea, using catalysts, while maintaining certain conditions. The manufacture of aminoplastics and melamine resins gives waste waters containing formaldehyde, methanol, and suspended resin.

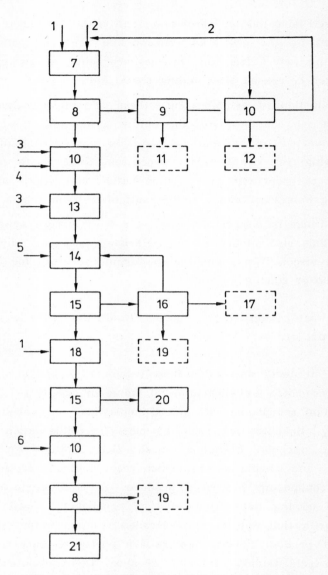

Fig. 2.4.-2: Diagram of phenol synthesis

1 - H_2SO_4; 2 - benzene; 3 - NaOH; 4 - Na_2CO_3; 5 - Water; 6 - Na_2CO_3; 7 - sulfonation; 8 - distillation; 9 - separation; 10 - neutralization; 11 - acid waste waters; 12 - alkaline waste waters; 13 - alkaline melting; 14 - quenching of the melts; 15 - sedimentation; 16 - washing; 17 - sulfite liquor; 18 - degradation; 19 - phenolic wastes; 20 - sulfatic wastes; 21 - phenol

Tab. 2.4.-26: Composition of waste waters from the condensation of artificial resins /107/

	red water	acid water
Bromine consumption, related to phenol (phenolic bodies)	6 100 mg/l	24 000 mg/l
Volatile phenols	1 700 mg/l	15 000 mg/l
Free formaldehyde	1 200 mg/l	8 100 mg/l
$KMnO_4$ consumption	60 000 mg/l	210 000 mg/l
pH value	11.3	1.7

2.4.2.4.3. Waste Waters from Plastics Production (Polymerization Products)

The lower, unsaturated hydrocarbons obtained in petrochemical works have attained a greater importance as raw material for the technically more important polymerization products. The processing of the raw material to a semifinished product, and the processing of the semifinished product to the finished polymer are often carried out in separate plants.

In the presence of catalysts acetylene forms vinyl chloride, with fatty acids it forms vinyl esters; and with hydrogen cyanide it forms acrylonitrile. These can be polymerized or processed to polyvinyl chloride (Igolite, Mipolam, Vinidur, PVC fiber, Vestolit, etc.), polyvinyl esters (Vinapas, Vinoflex, etc.) and polyacryl products (Acronal, Plexigum, Dralon, and Orlon fiber, etc.).

Ethylene is the primary material for polyethylene (Lupolen, Vestolen, etc.) and polystyrene (Vestyron, Trolitul, Styroflex, etc.), and ethylene oxide and its derivates, e.g. polyglycol.

Polypropylene and propylene glycol, which is used extensively in the manufacture of polyester resins, are obtained from propylene.

Oxidation of butene gives butadiene, which polymerizes to buna rubber when mixed with other equally polymerizable components, e.g. styrene or acrylonitrile.

Polyacrylonitrile fibers (Dralon, Orlon, PAN fiber, Zefran fiber) are made by dissolving acrylonitrile in dimethyl formamide. The liquid solution or paste is spun hot through spinning nozzles. Waste waters result from both the polymerization of the acrylonitrile and the subsequent treatment of the polyacrylonitrile fibers. The former are evaporated before being discharged in order to recover the monomeric acrylonitrile they still contain; the waste waters then contain only relatively small amounts of acrylonitrile in addition to larger amounts of other nitriles and aldehydes, as well as polyacrylonitrile not involved in the process. The latter appears as a white, crumbly suspension. The discharge from the distillation column has a BOD of approximately 3 000 mg/l. Wastes from subsequent treatment or the waste gas washes are noted for their dimethyl formamide content.

Polyamide fiber (Perlon) is made from caprolactam. The polymerization of caprolactam to Perlon is carried out under pressure with the exclusion of air and at a high temperature. Waste products from this process are: exhaust gases, column sump residues, trickle waters, concentrate filter rinsing waters, leakage waters, and cleaning waters.
The quality of these wastes is shown in Table 2.4.-27 (KAEDING /107/).

Tab. 2.4.-27: Composition of waste waters from Perlon manufacturing

	Amount mg/l	BOD_5 mg/l	COD mg/l
Hot grid spinning			
Lactam and oligomers	763	870	1023
Preparations	669	543	1025
Sapal (surface active substances)	25	1	31
Direct spinning			
Lactam and oligomers	1737	1983	2328
Preparations	877	667	1298
Sapal (surface active substances)	42	1	54

The synthesis of rubber from isoprene produces waste waters containing isoprene and intermediate products. Among these, dioxans and dioxan alcohol deserve attention from a health viewpoint since they are toxic. The polymerization of butadiene to a synthetic rubber can be carried out according to various processes; emulsion polymerization and solution polymerization are most frequently used.

With respect to waste water technology, the solvent process is worthy of attention, because with it only small discharges are produced from the purification plant for the butadiene and the solvents used, i.e. only vapor condensate and cooling waters arise.

2.4.2.5 Waste Waters from Tanneries and Leather Producing Plants

In the leather industry, the specific water demand per unit finished product is very high. Tanneries are among the most water intensive plants. The quantity of water depends only to a slight degree on the type of hides and the mechanical and chemical methods used in tanning. The literature gives 100 m^3 water per ton of rawhide. The actual water demand in the factories is usually higher and also includes the water used to clean the machines and workshops, as well as in the staff welfare rooms.

Many processes have been developed for the manufacture of leather, so that exact knowledge of the processes used in individual plants is necessary before the waste water problem can be solved.

2.4.2.5.1 Waste Waters from Tanneries

In tanneries, appropriate animal hides and skins are made into leather.

The hides and skins which are dressed in tanneries have three separate layers: the epidermis - outer epidermal layer, the corium - middle keratin layer, which is the true skin, and the subcutis - inner flesh layer consisting of fatty tissues. In tanning, the outer and inner layers are removed, in order to obtain the middle layer as leather or true skin. The true skin consists of protein, collagen (about 65 %), and elastin.

Leather is treated in two phases: a preliminary treatment in the <u>water workshop</u>, where the hides are prepared for the real tanning process, and in the <u>tanneries</u>.

The <u>preliminary treatment</u> includes:
- soaking and washing the hides, which are often very salty. For dried hides, caustic soda, wetting agents, or lactic acid is added to the soaking water. The washing and soaking waters are very salty and enriched with dirt, blood, dung, soluble proteins (albumins, globulins), and other organic contaminants.
- liming, in which the hides are treated with lime milk in piles that are turned over from time to time. This loosens the true skin from the upper and lower layers, so that it is easy later on to separate them mechanically. In the manufacture of chrome leather, sodium sulfide or sodium arsenite is added. After liming, the hides are rinsed with water and then dehaired and degreased.
- deliming and plumping, which are carried out in order to remove the lime, which is partially combined with fatty acids, from the pelt.

<u>Tanning</u> is done by various processes:
- vegetable tanning process: The tanning agents used here are mainly quebracho extract and the bark of oak, spruce, and other trees. Recently, artificial tanning agents have increasingly been used. These are obtained by condensation of formaldehyde with phenol, cresols, anthracene, and their sulfonic acids. Sulfite waste liquor from cellulose factories, and the lignin obtained from them, are also used to a limited extent to dilute the tanning liquor.
- acid pit tannage: At one time, this was the only tanning process used. The plumped pelts were layered alternately with ground bark and the pit was filled with water. The pelts had to be repiled and the process was repeated several times, so that the whole tanning

process sometimes lasted several years. Pit tannage (still widely used today in developing countries) has now largely been replaced by quick tanning in tanning tumblers. Here the pelts do not lie still in the liquors but are kept moving constantly. To prevent mold formation, very often p-nitrophenol is added to the tanning liquor as a preserving agent. The spent liquors are discharged as waste water.

- Chrome tanning with chromic salts is mainly done by the single bath process. For this process, a solution of chrome alum, aluminum chromate, or other chromic salts, with added common salt, is used. The solution is usually somewhat neutralized with soda. The tanning process lasts from 6 hours to 2 days. Excess salt and soda are washed out of the leather after tanning.
- chamois dressing: This process, in which the pelt is made into leather by kneading with fatty oils, is used for treating sheep and lamb hides, to make so-called "chamois leather".
- combined processes: In these processes, various metallic salts are used together with vegetable or synthetic tanning agents. For example, vegetable tanning often follows chrome tanning. There are chemical works that specialize in producing whole tanning programs with the appropriate chemicals. The mixtures of chemicals for each stage of treatment are then sold only with a brand name, and no description of the composition. This is not very helpful for the analysis and judgment of the relevant waste waters and for decisions as to their treatment.

Fig. 2.4.-3 shows a diagram of the processes used in a tannery and the sources of waste water (WHO /179f/).

The amount of waste water in tanneries is 0.7 to 5.0 m³ for every larger hide, depending on the size and equipment of the plant and the type of tanning used. On average, 1.0 to 1.5. m³ water are used. In the processing of cowhides SCHOLZ in /107/ gives Ia waste water flow of 140 l/kg rawhide for the manufacture of chrome upper leather, 80 l/kg for sole leather, and 90 l/kg for vegetable tanned flat leather.

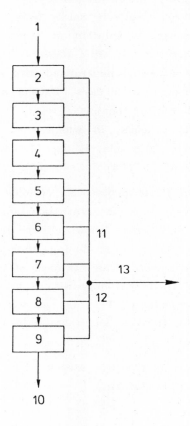

Fig. 2.4.-3: Diagram of processes used in a tannery and the sources of waste water /179f/

1 - salted hides; 2 - washing; 3 - soaking; 4 - liming; 5 - dehairing; 6 - lime pit; 7 - deliming; 8 - tanning; 9 - finishing; 10 - leather; 11 - workshop wastes; 12 - tannery wastes; 13 - to purification plant

In a tannery with chrome and bark tanning, the waste water amounts from the different processes are as follows:

soaking and washing 22.5 %
liming 17.5 %
rinsing 5.5 %
plumping and bating 19 %
chrome tanning 2 %
bark tanning 2 %
washing and drumming 31.5 %

There can be considerable variations from these figures in plants using different methods. Thus for example the daily waste water flow from a tannery, according to the U.S. Department of the Interior, is given in Table 2.4.-28.

Tab. 2.4.-28: Daily waste water flow from a tannery according to U.S. Department of the Interior, Federal Water Pollution Control Administration /179f/

Main sub-processes	% treated hides	Waste water flow from about 6 factories processing about 500 hides/day			Waste water
		BOD_5	susp. solids	total solids	m^3/day
Washing and short soaking	100	160	590	1360	360
Dehairing	40	320	790	790	170
Fleshing	60	540	1520	1810	250
Bating	100	90	20	180	110
Tanning (veg.) (chrome)	20 80	20 50	20 70	110 290	20 60
Final processes	100	20	20	50	60
Total		1200	3030	4770	1090

The waste water flow is irregular. During two hours in the morning there is normally a peak flow of 250 % of the hourly average. During this time, the analytical ratings of the different wastes also reach their highest values. During the other hours of the day, the flow amounts to only 50 % of the hourly average.

Table 2.4.-29 gives the waste water flow and the BOD_5 load of tannery wastes /179f/.

The waste water from a tannery (including preparation of the hides) has a fairly acid pH, a high chloride content (5 g Cl/l), a high $KMnO_4$ consumption (750 - 1250 mg/l), a high amount of settleable substances (10 - 12 mg/l), and contains emulgated fat and tends to form foam.

The dichromate content can reach peak values of 2000 mg/l. The population equivalent of tannery waste water is given as 500 P/t hides (WAGNER /107/).

Tab. 2.4.-29: Waste water flow and BOD_5 load in a tannery /179f/

Type of waste water	% proportion of total waste water	% proportion of total BOD_5 load
A. **Preparation wastes**		
Washing and soaking	7.5	4.2
Fleshing	5.6	4.9
Liming	4.4	13.1
Dehairing and washing	34.5	19.1
Washing in container	37.3	0.3
B. **Tanning wastes**		
Tanning	6.1	46.3
Bleaching and finishing	2.6	9.4
Floor washing wastes	2.0	2.5

Tannery wastes carry anthrax infection, and this is the decisive factor when considering them from a hygienic viewpoint. These wastes are therefore not only dangerous for the transport and tannery workers who come into contact with them, but they also infect the waste waters that arise from their processing in tanneries and leather factories.

2.4.2.5.2 Waste Waters from Leather Producing Plants

In leather factories, the tanned leather is made ready for marketing by finishing processes.

Different types of processing may produce no waste water.

If in the leather factories there is no dyeing or processing of the fats in glueing, for example, then the wastes are not particularly harmful.

In plants where the leather is dyed after finishing, the following substances are used: acid, basic, and substantive aniline dyes, rare sulfur dyes (for chamois leather), and dyeing woods, such as logwood, redwood, or fustic (for glacé leather). In this way, dye residues also get into the waste water, usually in small amounts, however, as the dyes are put on with a brush. As well as the dye liquors, there are many times the amount of rinsing waters. The pH is 5 - 7. The $KMnO_4$ consumption is about 500 mg/l.

2.4.2.6 Waste Waters from the Textile Industry

In the manufacture of textiles, vegetable, animal-mineral, and artificial fibrous materials are processed. The most important raw materials are hemp and flax, raw sheep's wool, cotton, raw silk, artificial silk, rayon staple, asbestos and synthetic fibers. Textile fibers cannot normally be used in their original (natural or synthetic) form. They first have to be finished in various ways according to the type and quality of the fibers and the purpose they are required for.

In general, two types of textile works are distinguished: those that produce raw materials and those that process raw materials. Plants that produce raw materials are those which process and produce natural or synthetic raw materials; these plants include flax retteries, wool scouring plants, cotton bleaching plants, silk boiling plants, and artificial silk and rayon factories. Processing and finishing plants, which

turn semifinished products into finished products, include spinning and weaving mills, bleaching plants, dyeworks, finishing plants, laundries and cloth mills.
Organic or inorganic waste waters result from the various processes.

In these processing plants, a fairly large proportion of the waste water constituents are dissolved organic substances, or, as for example with wool, organic substances such as wool fat and wool dirt. On the other hand, wastes from finishing plants often contain substances that are toxic or do not decompose easily. Thus for example, viscose wastes can contain hydrogen sulfide, carbon disulfide, free mineral acids, and zinc salts.

In Table 2.4.-30 KEHREN /158/ gives the following waste water amounts in m³ per t finished product, including all rinsing waters.

Tab. 2.4.-30: Waste water amounts in the textile industry in m^3/t product /158/

Wool scouring plant	20 - 70 m^3/t product
Dye-works	20 - 50 m^3/t product
Bleaching plants	50 - 100 m^3/t product
Cloth factories	600 - 1000 m^3/t product
Viscose, reprocessed wool, or silk factories	50 - 100 m^3/t product
Rayon staple factories	350 - 1000 m^3/t product

Waste water flows mainly in two phases of textile finishing, i.e. treatment and washing process (wool scouring, cotton bleaching, flax retting, silk degumming), and the bleaching and dyeing processes.

2.4.2.6.1 Waste Waters from Spinning Mills

2.4.2.6.1.1 Waste Waters from Flax and Hemp Retteries

Flax and hemp stems consist of an epidermis (outer layer), containing the fibers, and a woody core, the cambium. In order to release the fibers, the link between the fibers and the woody core must be decomposed. This link consists of pectic substances, i.e. polymolecular carbohydrate mixtures.

Flax and hemp retting takes place in two stages:
- in the first stage, the fibers are released from the stems by "retting",
- in the second stage, the semifinished product is washed, centrifuged, and dried, and the fibers removed from the retted stems.

Land or dew retting and water retting are distinguished.

<u>Water retting</u> is carried out by a suitable method in large-scale plants in tank or stand retteries and in channel or running water retteries.

An anaerobic or an aerobic process may be used. Soluble sugars and coloring matter, as well as starches, proteins, and resins contained in the plants go into solution and are thus removed.

When air is excluded in an aerobic pectin fermentation, gases (carbon dioxide, hydrogen, nitrogen, hydrogen sulfide) and, depending on the age and type of the stems, varying amounts of volatile organic acids are formed, which mixed together give the retting liquor its typical odor. The formation of acid in the retting liquor increases in this case as the fermentation progresses; if the retting goes on too long, cellulose fermentation, purification, etc. begin, and attack the fibers themselves.

When air is supplied, in a more modern process, the possibly harmful formation of acids is impeded or prevented through the partial reduction of the organic acids by oxidation right down to carbon dioxide and water.

Chemical additives (carbonates and hydrogen carbonates of alkalis or of lime) to combine the organic acids in biological retting, and purely chemical disintegration processes with water containing sulfuric acid, which are supposed to release the fibers more quickly, prevent unpleasant odors, and make the waste waters harmless, have in the long run not proved to be useful (MEINCK /107/).

In tank or stand retting with no freshwater inflow, about 20 m³ water/t flax straw (or hemp straw) are considered necessary.

In channel retting, there are several channels side by side, through which water is constantly passed. Wooden crates containing the bundles of straw are moved against the current by a lever apparatus. At the end of the channels, water is discharged, corresponding in quantity to the water inflow (from 40 to 60 m³ water/t flax straw).

The waste water consists of this discharged retting liquor, enriched with products of decomposition and washed out nitrogen, phosphoric acid, potash, lime and other similar compounds. Added to these, there are also rinsing waters at certain times (from washing of workshops and machinery) and toilet water.

Waste waters from modern retteries are light yellow to dark brown in color, cloudy to some degree due to colloids, and have a typical slightly sour odor. Tank rettery liquors usually smell distinctly of hydrogen sulfide as soon as the flax is taken out. The temperature of the waste waters (25 - 30° C) is approximately the same as that of the retting process. Wastes from both tank and channel retting generally contain only a small amount of suspended solids (mostly below 100 mg/l), these being chiefly of an organic nature. More details concerning their composition are given in Table 2.4.-31.

Tab. 2.4.-31: Properties of retting waters /107/

Total residue	1000 - 6500 mg/l
$KMnO_4$ consumption	1000 - 6000 mg/l
Volatile organic acids	1500 - 6000 mg/l
BOD_5	1300 - 3600 mg/l

The degree of contamination can be higher, however.
A population equivalent for flax retting is 300 P per 100 kg flax straw (WAGNER /107/).

The volatile organic acids consist almost entirely of acetic and butyric acid. In addition, formic acid, propionic acid, and valeric acid occur

(Table 2.4.-32). The proportion of these components in relation to the total amount of volatile organic acids is (CARLSON AND LUNDIN /107/):

Tab. 2.4.-32: Organic acids in retting waters /107/

Formic acid	0.7 %
Acetic acid	61.2 %
Propionic acid	6.5 %
Butyric acid	31.7 %
Valeric acid	0.6 %

Comparative data on the amount of vegetable nutrients in these waste waters are given in Table 2.4.-33 (MEYER /107/).

Tab. 2.4.-33: Amount of vegetable nutrients in flax retting wastes in comparison with domestic sewage and starch factory wastes /107/

Type of waste	Total N	P_2O_5	K_2O	CaO	BOD_5	$N : P_2O_5 : BOD_5$
	mg/l	mg/l	mg/l	mg/l	mg/l	
Domestic sewage	109	42	68	116	460	2.5 : 1 : 11
Flax retting wastes	40	60	350	160	2500	0.7 : 1 : 42
Starch factory wastes	319	175	604	–	3000	1.8 : 1 : 17

As shown in Table 2.4.-34, the values fluctuate very considerably (STOOF /170/).

Tab. 2.4.-34: Composition of flax retting wastes /107/

Total suspended solids	7 - 159 mg/l
Dry residue	1050 - 6571 mg/l
Loss at red heat of dry residue	551 - 3465 mg/l
Total nitrogen (N)	7 - 109 mg/l
Ammonia (NH_3)	3 - 19 mg/l
Phosphates (P_2O_5)	5 - 235 mg/l
Potash (K_2O)	101 - 918 mg/l
Permanganate consumption	1122 - 6045 mg/l

In many retteries, to render flax or hemp cottonlike, the decorticated flax or hemp straw is made into harl by a dry (mechanical) process through chemical treatment.

There are two processes for this:
- The KORTE process: Preliminary treatment of the green tow with a dilute hydrochloric acid solution, disintegration with hypochlorite and dilute hydrochloric acids, rinsing with freshwater, removal of the final acid residues in a soda bath, further rinsing and boiling in diluted soda lye, several rinses, washing, and glazing.
- The BASTFASER A.G. leaching process: Green tow is soaked in water, boiled without pressure in a very diluted soda lye, disintegrated under pressure, disintegration liquors are removed by rinsing baths, remaining alkali is neutralized in an acid bath, removal of acid residues in a soda bath, washing, and glazing.

The KORTE process yields acid and the leaching process yields alkaline wastes. The wastes differ also in the amount of inorganic salts they contain.

2.4.2.6.1.2 Waste Waters from Silk Boiling

Natural silk is obtained from threads spun by the silkworm into cocoons.

fiber. There can be considerable differences in specific water consumption. By far the greater part of the various viscose wastes shows a clearly acid reaction, while a small part shows an alkaline reaction. Acid predominates in the mixture. The effluent normally smells of hydrogen sulfide and carbon disulfide, which can be present in amounts from a few mg to over 100 mg/l. The total amount of suspended solids (mainly organic) may be less than 100 mg/l, but can be considerably more or even up to nearly 1 g/l.

In the manufacture of artificial silk or rayon staple by the cuprammonium process, the dissolution of the cellulose (which may be previously refined with soda lye) is achieved by the effect of a mixture of basic copper sulfate (20 %) and ammonia (25 %). After the addition of organic hydroxyl compounds (tartaric acid, sugars) this solution is filtered and deaerated under pressure, and then forced out of wide spinning nozzles into slowly coagulating baths (first warm, softened, deaerated water, then dilute sulfuric acid). Drawing then produces fine threads, which are hardened and freed from copper in dilute sulfuric acid; they are then treated with a weak soda solution, rinsed several times with cold water, treated with a soap bath, and then dried. The liquid wastes consist of the spent precipitants, containing copper, washing and rinsing waters, spent soap solution, and waste water containing ammonia waste (spinning water). The water consumption is about 1300 m^3/t fiber.

The wastes contain the substances given in Table 2.4.-37 (GERSTNER /107/).

Tab. 2.4.-37: Composition of waste waters in artificial silk manufacture /107/

	Waste spinning acid	Waste spinning water
Copper (Cu), mg/l	8 000 - 16 000	80 - 200
Sulfuric acid (H_2SO_4), mg/l	12 000 - 65 000	750 - 1 200
Ammonia (NH_3), mg/l	-	over 800
pH value	-	10

In the acetylated cellulose process, cellulose is transformed into cellulose triacetate by the effect of acetic anhydride or glacial acetic acid in the presence of catalysts such as sulfuric acid, perchloric acid, zinc chloride, etc. The triacetate is washed, dissolved in acetone, and spun dry, i.e. in hot air. The free inorganic acids, zinc chloride, etc., are drained off as wastes with the washing waters, while the valuable organic chemicals (acetone, acetic acid, alcohol) are for the most part recovered. The quantity of waste water is less than in other processes used for the manufacture of artificial silk; it amounts to about 10 to 15 m³ per t triacetate. The wastes are strongly acid with few suspended solids (below 100 mg/l), but very many dissolved constituents.

2.4.2.6.2 Waste Waters from Cotton Bleaching

Cotton is a very economical vegetable fiber, containing mainly cellulose. The non-cellulose constituents have to be removed before the fiber is used. Cotton is normally made into fibers by dry processes; here there are no waste waters.

The cotton, but mainly the yarns and fabrics made from it, is first boiled in strongly alkaline solutions made with soda and sodium hydroxide. This removes the pectins contained in the cotton. The process is followed by bleaching with thin solutions of chloride of lime or other chlorine preparations, and the treatment with dilute acids. After each process, the treated materials are rinsed with plenty of water.

KNIESEL /158/ gives the amount of waste water in cotton bleaching plants as about 50 - 100 m³ water/t product, with a population equivalent of 250 to 300 P/t product and the following types of waste water:

desizing baths,

boiling solutions (alkaline),

spent chlorine solutions (alkaline),

spent acid solutions (acid),

spent mercerizing solutions (alkaline),

rinsing waters from each stage of treatment.

The total amount of contaminants contained in cotton bleaching wastes is 196.6 kg per t material (GEYER /107/). About 50 % of the total organic contaminants is accounted for by the boiling solutions and desizing baths.

2.4.2.6.3   Waste Waters from Wool Washing

Wool is delivered to cloth factories either uncleaned or partially cleaned. The substances shown in Table 2.4.-38 were found by SCHULE, MAERKER, KRÜGER /158/.

Tab. 2.4.-38: Substances found in sheep's wool /158/

	SCHULE and MAERKER	ULRICH G. KRÜGER
Water	10 - 23 %	2 - 24 %
Wool fat	7 - 14 %	12 - 47 %
Suint	20 - 22 %	3 - 24 %
Dirt	2 - 23 %	
Pure wool fiber	20 - 50 %	15 - 72 %

In order to remove the contaminants adhering to the wool (suint, wool fat, vegetable contaminants, dust, and excrement), it is normally given a superficial cleaning on the animal before being shorn. Afterwards, the wool is thoroughly cleaned in wool washing machines.

Four to six wash bowls in succession are generally used in wool washing processes. This enables the wastes to be treated separately.

The washing process gives highly concentrated washing and rinsing waters, containing non-wool amounting to 25 to 60 % (depending on the origin of the raw wools, the degree to which the sheep were washed before shearing, etc.) and possibly even more of the weight of the unfinished wool. This non-wool consists of wool fats, suint, organic and inorganic components, in suspended, colloidal and emulgated form. In addition, there are detergents, such as soda, soap, and also oils, used to soften the wool, which are rinsed away again at the end of the washing process.

There are usually up to around 100 m^3 waste water per 1 t wool, and their composition can vary considerably according to the amount of water used, as shown in Table 2.4.-39 (FABER /107/).

Tab. 2.4.-39: Composition of wool washing water /107/

		Max.	Min.	Average
Fat	mg/l	25 800	3 000	8 650
Suspended solids	mg/l	30 300	2 400	11 520
Alkalinity	mg/l	29 400	3 430	6 780
Permanganate value 4 h at 28° C		7 400	398	1 830
BOD$_5$	mg/l	22 000	1 200	5 500

350 to 400 kg total contaminants per t product (of which up to 200 kg/t are organic substances) are eliminated in the waste waters.

2.4.2.6.4 Waste Waters from Cloth Production

In cloth factories, fibrous tissue is made into cloth by "fulling" in weak alkaline solutions. In this process, lint and loose threads are rubbed off and rinsed away by mechanical processes. Before fulling, disturbing foreign substances are removed, such as threads, wood and straw particles, etc.

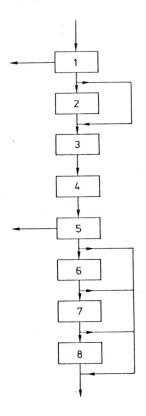

Fig. 2.4.-5: Flow chart of a wool washing plant
1 - cleaning; 2 - dyeing of raw material; 3 - combing; 4 - spinning; 5 - washing; 6 - carbonizing; 7 - parts dyeing; 8 - bleaching

When wastes containing fibrous material are drained, fibers are deposited at the sides of canals, ditches and rivers and form felt layers and ridges. After some time, depending on the season, these begin to rot, and usually cause very unpleasant odors.
In dye-works, materials are first washed with soap solutions or wetting agents, so that later the dyes go on more easily and more evenly. This produces washing solutions and rinsing waters. For lighter colors, the

materials also have to be bleached.

In general, sulfur, aniline, azo, and other dyes are used. Waste waters arise from preparation of the cloth, making up the dyes, removal of the spent dyes, and rinsing of the vats, machinery, and dyehouse.

Table 2.4.-41 gives the composition of textile wastes (JUNG /158/). However, the amount and composition of these wastes differ from plant to plant. The fluctuating quality of textile wastes can also be assessed from the minimum and maximum values of a textile plant (about 70 % cotton and 30 % wool) given in Table 2.4.-40 (KEHREN in /158/).

Tab. 2.4.-40: Minimum and maximum values for the waste water quality of a textile plant /158/

		Minimum	Maximum
Temperature		10° C	30° C
pH		7.6	11.5
Transparency	cm	0.5	17
Suspended solids	mg/l	17	2713
Suspended solids lost on ignition	mg/l	5	1374
Dry residue	mg/l	706	4376
Residue from glowing	mg/l	314	3886
Soda lye	mg/l	136	896
Oxygen consumption	mg/l	21	603
Total sulfate	mg/l	102	979
Hydrogen sulfide	mg/l*	3.47	30

* found in eight samples

Tab. 2.4.-41: Composition of textile wastes /158/

Type of plant	Dye works	Dye works	Bleachery, dye-works
Processing	Wool	Cotton	Cotton
Appearance	reddish cloudy	dark blue opaque cloudy	brown opaque cloudy
pH	6.8	9.1	11.5
Dry residue (filtered)	2068 mg/l	1240 mg/l	2327 mg/l
organic	460 mg/l	437 mg/l	838 mg/l
inorganic	1608 mg/l	803 mg/l	1489 mg/l
Permanganate consumption	312 mg/l	733 mg/l	534 mg/l
BOD_5	93 mg/l	188 mg/l	255 mg/l
Chlorides (Cl)	114 mg/l	118 mg/l	255 mg/l
Organic nitrogen	4 mg/l	16 mg/l	22 mg/l
Ammonia (NH_3)	6 mg/l	trace	trace

The population equivalent for a cotton dye-works can be seen in Table 2.4.-42 (KUISEL /158/).

Tab. 2.4.-42: Waste water amount and population equivalent of a cotton dye-works /158/

	Waste water amount in m³ per t product	Population equivalent per t product
Direct dyes	about 70	160
Sulfur dyes	about 60	700
Indigo	about 100	60

In cloth printing works the dyes are applied by pressure only to certain areas. Thickening agents used for the dye pastes are starches of various origins, dextrine, alginates, tragacanth, and other natural rubbers, also albumin, casein, and, in many cases, acetylated cellulose. Apart from the pigments and the thickening agents, printing dye contains other additives necessary for fixing the pigment, such as tannin, acetic acid, metallic salts, reduction agents, alkali, etc. Printing is followed by rinsing and soap baths.

The wastes contain pigment residues and the chemicals listed above or their conversion products. They are thus similar to the wastes from dye-works.

In finishing plants, the cotton materials and artificial silk blended fabrics are given stiffness, density, and weight, as well as a more attractive appearance. Starches, dextrine, glue, wax, clay or kaolin, heavy spar, and, more recently, synthetic resins, are normally used as finishing agents; they are applied to the materials as a paste. Waste waters arise from cleaning the workshops, containers, and apparatus. They contain varying amounts of the finishing agents in collodial solution or suspension.

2.4.2.7 *Waste Water from Wood Processing, Pulp, Paper, and Cardboard Mills*

The paper industry is divided into the half stuff industry, such as the cellulose industry, and the actual industry for manufacturing paper from these half stuffs.

In all these manufacturing processes, there are waste products which ought to be recovered or reused.

In the production of half stuffs, there are various types of waste water according to different processes, e.g. from white and brown mechanical pulp mills, from sulfite pulp mills, from rag carbonizing, from board and straw mills, and from paper production proper.

In recent years, it has been possible to make considerable progress with regard to economizing freshwater in paper mills, with the result

as much as the actual boiling solutions). The washed cellulose is then further processed in paper, rayon staple or artificial silk mills.

The spent solutions, together with the washing waters, form the sulfite waste liquors. Other waste waters come from the bleaching plants. They contain chlorine or hypochlorite. Fig. 2.4.-7 shows a diagram of sulfite pulp manufacture (SCHEPP /146/).

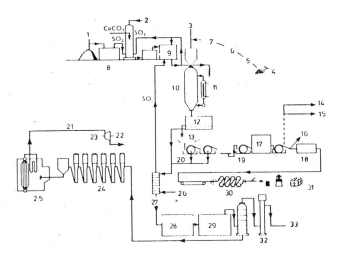

Fig. 2.4.-7: Diagram of sulfite pulp manufacture /146/

1 - iron pyrites; 2 - water; 3 - chip storage bin; 4 - spruce; 5 - barking; 6 - chipping; 7 - screening; 8 - pyrite burner; 9 - cooking liquor; 10 - digester; 11 - cymene; 12 - washing tank; 13 - rejects; 14 - flash dryer; 15 - flact dryer; 16 - to paper mill; 17 - bleaching plant; 18 - draining chest; 19 - table sorter; 20 - vacuum washer; 21 - high pressure steam; 22 - steam of varying tension; 23 - turbine; 24 - evaporation plant; 25 - combustion boiler; 26 - steam; 27 - to fermentation; 28 - neutralization, purification; 29 - fermentation; 30 - paper-making machine; 31 - dispatch; 32 - rectification; 33 - alcohol

For the production of 1 t bleached pulp, the total waste water flow is about 1000 m^3 (MANGOLD /104/). This high amount can be reduced by half if the water is recycled.

The washing wastes are similar to the sulfite waste liquors, only about 50 - 100 times diluted. In European sulfite pulp mills, the population equivalent rises to 5000 and over, with an average of 3500 per t pulp from spruce wood.

In the manufacture of in-plant sewage systems, strict separation of the different kinds of waste water should be observed.

2.4.2.7.2.2 Waste Waters from Sulfate Pulp Production

The sulfate pulp process also makes it possible to process high-resin woods, especially pine. The process is used for the manufacture of very strong paper, e.g. insulating papers, bag paper, packing paper.

Preliminary treatment of the wood is the same as in the sulfite process. Disintegration of the wood is achieved with the help of caustic soda and added sodium sulfate (1/3 of the caustic soda) to obtain kraft or sulfate pulp. The sulfate cooking process makes it easier to dissolve the lignin out of the wood, with a yield that is higher and better in quality. Further processing of the pulp is the same as in sulfite pulp mills.

In the sulfate process, there are the following wastes: digester condensates (135 l/t pulp), condensates from the boiling vats and diffusers (270 l/t pulp), rinsing waters from the diffusers, to a large extent dependent on the wash-out effect, vapor condensates from the evaporation of the black liquor (13 to 15 m^3/t pulp).

2.4.2.7.3 Waste Waters from Straw Pulp Production

Straw is often used as a raw material for the production of cellulose. However, for 1 t pulp, around 4.4 t chopped straw have to be processed. The straw is cooked with caustic soda, of which about 0.65 t per t pulp is needed. During this process, the cooking liquor absorbs 3.4 t straw constituents per t pulp; these are mainly of organic composition, and are distributed in about 11.4 m^3.

2.4.2.7.4 Waste Waters from Paper Mills

For paper production, various raw materials are needed e.g.

- papers and boards from chemical pulp, with or without mechanical pulp,
- papers and boards with a high proportion of fillers and used paper,
- papers and boards from chemical pulping of rags, straw, and wood, and from processing of synthetic substances.

The raw materials, like chemical pulp, rags, used paper, etc., are broken up in a pulper. After this, heavy dirt, like sand, glass, and metal scraps, and light dirt, like plastic scraps, are removed from the fibrous material in several stages, which then goes into large chests via refiner hollanders (beating machines). In the stuff chests, fillers and other mineral matter are generally added to the stuff. This highly liquid pulp is passed through round or long sieves onto the paper machine, where a sheet is formed by removing water through filtration. It is then thermally dried and in some cases further processed. Fig. 2.4.-8 shows the flow chart of a paper mill, including the waste water flow points (WHO /179 d/).

Paper and board factories have a high water consumption. In the course of paper manufacture, the fibrous material is diluted to such an extent that there are between about 125 and 2000 m³ waste water per ton paper (MEINCK /107/). The water demand, e.g. for

>   packing paper   is   125   m³/t
>   newsprint   is   200   m³/t
>   fine paper   is   400   m³/t .

For the manufacture of fine paper, e.g. cigarette paper, even larger amounts of water are needed.

The wastes from loading, sizing, and other processes contain, in addition to organic, more or less easily putrescible constituents, considerable amounts of inorganic substances, such as barium and aluminum sulfate, argillaceous earth, and loaders (china clay). The BOD_5 of these wastes can be very high, up to 3000 mg/l, especially when used material (textiles, paper, etc.) is included in processing. The waste water flow and the population equivalent per t product are shown in Table 2.4.-46 (BUCKSTEEG /38/).

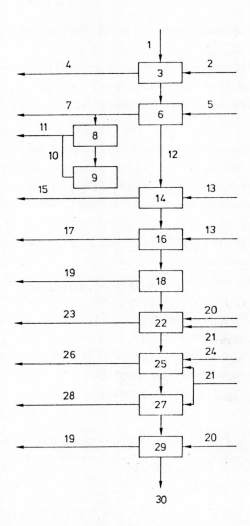

Fig. 2.4.-8: Diagram of paper manufacture /179d/

1 - tree trunks; 2 - fresh or recycled water; 3 - wood preparation; 4 - wood wastes; 5 - sulfite solution, alkali sulfate; 6 - wood paste; 7 - exhausted sulfite solution; 8 - evaporation; 9 - concentrated feed stuff; 10 - condensate; 11 - acid neutralization plant; 12 - paste; 13 - white or freshwater; 14 - sorting; 15 - diluted solution; 16 - washing; 17 - washing water; 18 - drying; 19 - waste water; 20 - chemicals; 21 - raw or recirculated water; 22 - bleaching; 23 - bleaching wastes; 24 - precipitants and pigments; 25 - stock preparation; 26 - cleaning waste water; 27 - paper machine; 28 - white water; 29 - finishing; 30 - finished paper

detergents, solvents, pesticides, and many other products or intermediate products.

Fig. 2.4.-10 gives the composition of some oil refinery products (RUGGLES).

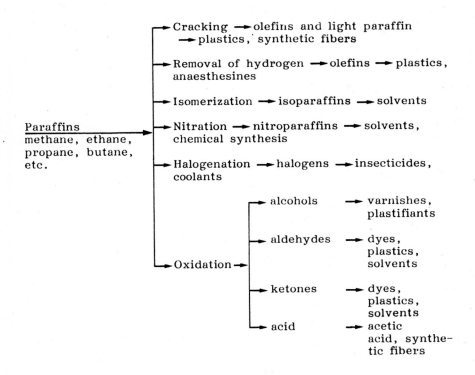

Fig. 2.4.-10: Composition of some oil refinery products

The most important components of cracked gases are olefins (ethylene, propylene, butylene, butadiene) and the aromatic substances (e.g. benzole). The cracked products are cooled quickly and condensed. In the process, steam is also condensed, in which various organic compounds, phenols, and carbon disulfide compounds are dissolved. These pollute the waste water. Wastes also come from purification of the basic materials and in the production of pure products. By far the largest amount of waste water comes from cooling waters.

Separate drainage systems are needed to carry out the treatment of wastes economically and effectively. Water recirculation systems should also be introduced, as well as the use of oil or other chemicals instead of water for cooling the cracked gases. Other production processes giving no wastes could be used, water recirculation systems in the cooling plants, or air instead of water cooling should be introduced, and finally waste products should be collected before they come into contact with water, and disposed of by other means.

It is very rare that only one product is manufactured in petrochemical plants. Normally one finds large plants with a great variety of processes, where the wastes generally flow to one place and are purified together. Small, highly concentrated process discharges should in this case be suitably pretreated where they are produced, before being mixed with the wastes intended for biological purification.

2.4.2.8.4 Waste Waters containing Oil from Oil Storage, Gasoline Stations and Workshops

Wastes containing oil come from oil storage tanks, gasoline stations, workshops, distribution centers, and similar places. They can contain petroleum and other fuels.

These wastes may not be discharged untreated into the public sewerage system, because they are very harmful for the sewerage workers, the sewers and the receiving water. They contain hydrocarbons, which can easily be separated and used in the same plant or in special oil recovery and utilization plants.

There is always a danger that harmful amounts of oil might get into surface and underground water bodies through accidents, leaks in the sewage installations and through carelessness. If these waters are used for preparing drinking water, it is possible that the chlorine used to treat the drinking water may combine with the organic substances (e.g.

humins) to form chlorine compounds (chlorophenols, chlorocresoles), which even in concentrations of 0.5 to 1.0 ppm are detectable as a medicinal odor. This makes the water unusable and further treatment of the drinking water is necessary.

Wastes containing oil should therefore be treated as well as possible before discharge.

2.4.2.9 Waste Waters from the Food Processing Industry

Waste waters from the food processing industry are by nature rich in organic substances.

The waste water often flows irregularly during only a few hours of the day; there may also be great fluctuations in the amount and in the concentration of the constituents. Further, in the course of the year, there may be considerable variations in the waste water conditions, due to seasonally irregular utilization of capacity. This applies for example to milk processing plants, beverage factories, slaughterhouses, and meat and sausage factories.

The biggest fluctuations, however, are in those plants that produce only during a few weeks or months of the year, e.g. sugar refineries, wine and sweet must producers, canning factories, and fruit and yeast distilleries. These plants are known as campaign or seasonal plants. As long as there are not too many of them, a good solution from an economical and technical point of view is to purify these wastes together with domestic or municipal sewage in purification plants using mechanical-biological processes. However, these plants are often situated in relatively thinly populated areas and produce large quantities of highly contaminated waste water.

Wastes from food and beverage plants mostly contain energy-rich compounds, such as proteins, peptides, amino acids, sugar and carbohydrates, animal and vegetable fats, lower organic acids, alcohols, aldehydes, and ketones in their original form or as conversion products from fermentation.

In many branches, there are high degrees of waste water contamination from product losses that could be avoided.

Improvements in the retention of solid and liquid substances have a twofold advantage here: there is a rise in yield, and the cost of waste water purification is reduced.

- The wastes often have a very one-sided composition, which can, for example, when there is a lack of nitrogen, lead to insufficient formation of biologically active sludge or to degenerate, unsettleable biological sludge.
- The wastes decompose very quickly as a result of their good degradability, which can lead to the formation of acid and bad odors, and hinders aerobic decomposition.
- Depending on the production processes used, the wastes may have a high salt or acid content, which similarly hinders aerobic biological decomposition.

The quantity and amount of wastes from the food processing industry vary according to the manufacturing branch, production processe and the nature of the raw materials.

Table 2.4.-50 describes wastes from various food processing industries. Table 2.4.-51 shows the amount and composition of wastes from some plants in the food processing industry. Table 2.4.-52 gives examples of reduction in freshwater consumption in the food processing industry. Table 2.4.-53 gives examples of reduction of substances in wastes from the food processing industry.

Tab. 2.4.-50: Wastes from various food processing industries /1/

Industry Manufacturing branch	Process from which waste water originates	Type of waste water	Waste water constituents
1	2	3	4
Food processing industry			
Fresh milk dairies	Milk reception, bottling unit	Cooling and condensation waters	Lactose, protein
	Cleaning of: tankers, sterilizers, evaporators, other machines	Rinsing and cleaning waters	
Cheese dairies	Skimming Churning	Cooling and condensation waters	
	Cheese manufacture	Rinsing and purification waters	
	Whey processing	Butter washing water	
	Manufacture of: condensed milk, dried milk, other by-products		
Slaughterhouses	Slaughtering, bleeding skinning cutting up viscera processing emptying of pauch stall cleaning disinfection of vehicles	Discharges from: slaughtering shed viscera shed dung shed stalls etc.	Blood, plasma content, gut contents, gut and skin scraps, disinfectant, urine, dung
Fish processing industry	Preliminary treatment of fish washing, scaling, filleting, etc.	Washing waters	Blood, flesh scraps, disinfectant
	Marinade production: curing process cold marinades boiled marinades	spent bleeding baths (curing baths) draining water spraying water washing water	
Juice industry	Fruit preparation	Washing water rinsing water transport water	Fruit scraps, dust, fine suspended solids
	Bottle washing	Bottle washing waters	Organic substances, alkalis
Mineral water industry	Bottle washing	Bottle washing waters	Broken glass, labels

cont. Tab. 2.4.-50

1	2	3	4
Breweries	Brewing house fermenting cellar storage cellar barrel washing barrel filling	Rinsing and puri- fication waters	Carbohydrates, proteins, organic nitrogen compounds, Solid constituents: grain and hop residues, cooler sludge containing protein, yeast
	Bac cooler beer cooler ice machine	Cooling waters	
Malthouses	Steeping	Steeping tank liquors washing waters Cleaning and rinsing waters from workshops	Sugar, floating substances (coarse and fine) suspended and settled sub- stances
Yeast factories, distilleries	Yeast and/or alcohol manufacture		Slops, yeast
Potato distillery	Malting Potato washing Mashing Fermentation of mash and yeast propagation Rectification of raw spirit Cleaning of distilling apparatus and workshops	Washing and steeping waters Washing waters First fractions from steaming Discharges from fermentation, cooling waters Singlings water Cleaning waters	Yeast slops, vinasses, butanol, grain slops, wine slops
Starch factories	Potato washing	Transport and washing waters	Soil, protein
Potato:	Production of pulp Starch washing Pulp pressing	Potato juices Starch washing waters Pulp pressing waters	
Potato flakes	Prewash	Transport and washing waters	
Potato slice factories	Peeling: mechanical steam process lye process Cleaning	Rinsing waters	
Vegetable and fruit canning factories	Washing Cutting up Boiling	Washing waters Fruit juices or blanching waters Waters from clean- ing workshops and machines	Mainly carbohydrates

cont. Tab. 2.4.-50

1	2	3	4
Sauerkraut factories	Cleaning	Cleaning waters	free lactic acid, common salt, carbohydrates, protein compound
	Lactic fermentation	Vegetable juices	
Margarine, edible fat, and edible oil factories	Purification of raw oils and fats Vacuum distillation	Spent acids and/or lyes, washing waters, condensation waters	
	Recovery of fatty acids		
	Fat hardening	Hot gas washing waters	
Margarine	Milk collection point	Rinsing and cleaning waters	
	Spraying off of milk fat emulsion	Cold rinsing water	
	Rolling and kneading	Press water	
Sugar industry Cane sugar refineries	Beet washing Root removal	Fluming and washing water	Leaves, roots, other substances
	Slicing		Soil
	Diffusion process	Diffusion wastes	Fine beet scraps
	Pressing	Press water	
	Purification of raw juice	Lime cake water	
	Evaporation of raw juice	Condenser water	
Sugar refineries	Recovery of activated carbon	Spent lyes and washing waters	
	Injection condensers	Condensates	
Extraction of sugar from molasses	Evaporation and distillation of vinasses	Washing waters	
Confectionery industry	Marzipan manufacture	Wastes containing cyanide	
In all cases:	Cleaning, rinsing of workshops, machines, apparatus	Cleaning and rinsing waters, cooling waters	

Tab. 2.4.-51: Amount and composition of wastes from some plants in the food processing industry

Type of plant	Amount of waste water	BOD$_5$	Population equivalent
1	2	3	4
Dairy (without cooling water)	0.5-3 m³/t milk	0.5-3 kg/t milk	9-55/t milk
Milk drying spraying tower roll drying	0.2 m³/t milk -	0.4 kg/t milk 1.0 t/kg milk	8/t milk 19/t milk
Manufacture of condensed milk	3-5 m³/t milk	1.6-4.4 kg/t milk	30-80/t milk
Manufacture of ice-cream	4-6 m³/t ice-cream	8-10 kg/t ice-cream	148-185/t ice-cream
Slaughterhouse cattle (400 kg live weight) pigs (100 kg live weight) poultry	0.6-9.6 m³/animal 0.3-0.4 m³/animal 20-40 l/slaught.	2.4-10.4 kg/animal 0.43-2.1 kg/animal 6-30 g/slaught.	44-192/animal 8-39/animal 0.1-0.5/slaught.
Fish processing	24 m³/t	41 kg/t	760/t
Carcass disposal	0.9-1.1 m³/t	1.8-4.6 kg/t	33-85/t
Brewery (without malthouse)	0.4-1.2 m³/hl beer	0.4-2.7 kg/hl beer	7-50/hl beer
Fruit distillery (independent distillery)	0.5-0.8 m³/d	6-35 kg/d	110-650/d
Wine growing small and medium vineyards large-scale vineyards	- -	0.2-0.4 kg/ha area 0.7-1.4 kg/ha area	3.8-7.6/ha area 1.3-2.5/ha area

cont. Tab. 2.4.-51

1	2	3	4
Sweet must production	—	0.45-0.9 kg/100 kg fruit	8-17/100 kg fruit
Beverage industry lemonade, carbonated beverages	1.9 m³/1000 bottles	1.24 kg/1000 bottles	23/1000 bottles
Canning industry peas processing carrots processing green beans proc.	10-25 m³ per 1000 kg 7-23 m³ 3.7-5.3 m³ (cans)	7-18 kg per 1000 kg 6-32 kg (cans) 7-11.6 kg	130-330 per 1000 kg 110-590 (cans) 130-215
Sauerkraut production	5-9 m³/t cabbage	4.2-9.2 kg/t cabbage	80-170/t cabbage
Yeast industry	21 m³/t molasses	150 kg/t molasses	2800/t molasses
Potato industry transport and washing water (without recirculation)	5-8 m³/t	1-3 kg/t	19-56/t
steaming plant (without transport and washing water peeling water	0.2 m³/t —	0.9-1.8 kg/t 3-5 kg/t	17-34/t 56-93/t
Flakes and farina Potato crisps	13-20 m³/t 16.6 m³/t	20-39 kg/t 11.3 kg/t	17-34/t 210/t
Starch industry corn starch without recirculation with recirculation potato starch	9-11 m³/t 0.4-3 m³/t 4-10 m³/t	13.5-15.1 kg/t 0.8-10.8 kg/t 19-35 kg/t	250-280/t 15-200/t 350-650/t
wheat starch without recirculation with partial recirc. rice starch	19-20 m³/t 8-12 m³/t 8-12 m³/t	86-92 kg/t 48.5-57 kg/t 5.4-9.2 kg/t	1600-1700/t 900-1060/t 100-170/t
Sugar refinery	0.5-1.0 m³/t beet	0.8-1.6 kg/t beet	15-30/t beet

Tab. 2.4.-52: Examples of reductions in water consumption in the food processing industry /186/

Fresh water demand for	Reference unit	Fresh water consumption		Reduction %	Ways of achieving reduction in fresh water consumption
		now min-max av=average	reduced min-max av=average		
Malthouses - total	m³/t barley	3.1-9.0 av = 4.5	av = 0.8	82.3	Complete recycling of total water used
for first steeping water	m³/t barley	1.1-7.0 av = 2.4	av = 0.4	83.3	Recirculation, treatment with kieselgur and asbestos, addition of activated carbon
Beer production tot.	hl/hl beer	5-33 av = 13.3	av_1 = 7.6÷8.1 av_2 = 6.6÷7.1 av_3 = 5.6	42.8 50.4 57.9	
cooling water	hl/hl beer	0.3-13.8 av = 4.8	av = 0.6	87.5	Better technical installations for cooling and water recycling
process water	hl/hl beer	4-14.5 av = 8.5	av_1 = 7.0÷7.5 av_2 = 6.0÷6.5 av_3 = 5.0	17.7 29.4 41.2	Economical use of water Use of non-returnable bottles Optimal conditions

Tab. 2.4.-53: Reduction of substances in the waste water in the food processing industry

Waste water	BOD_5 mg/l*	Recoverable substances	% dry solid matter*
1. Dairy			
1.1 Whey	32000	Carbohydrates, protein, fat	6
1.2 Buttermilk	40000	Carbohydrates, protein, fat	8.5
1.3 Butter washing water	4000	Protein	0.5
2. Brewery/malthouses			
2.1 Waste yeast	180000	Protein, carbohydrates, ethanol, vitamins	13
2.2 Barley steeping water	2000	Protein, carbohydrates	0.2
3. Distillery			
3.1 Potato juice	6000	Protein, carbohydrates	1
3.2 Potato slops	12000	Protein, carbohydrates	6
3.3 Grain slops	16000	Protein, carbohydrates	4
3.4 Beet vinasses	10000	Protein, carbohydrates	4
3.5 Sugar cane vinasses	10000	Protein, carbohydrates	4
3.6 Pomaceous fruit slops	15000	Protein, carbohydrates	5
3.7 Small fruit slops	15000	Protein, carbohydrates	5
3.8 Grape slops	20000	Protein, carbohydrates	2
4. Starch factories			
4.1 Potato juice	6000	Protein, carbohydrates	1
4.2 Wheat starch waste w.	8000	Protein, carbohydrates	1
4.3 Maize steeping water	10000	Protein, carbohydrates	2.5
4.4 Maize starch waste w.	8000	Protein, carbohydrates	1
5. Yeast factories			
5.1 Yeast washing water	8000	Protein, carbohydrates	2.5
5.2 Molasses waste water	10000	Protein, inorganic salts	2.5
6. Fish meal factories			
6.1 Fish press water	30000	Protein, fat	8
7. Slaughterhouses			
7.1 Cattle blood	170000	Protein	16
7.2 Pig blood	200000	Protein	16
8. Textile industry			
8.1 Wool scouring water	6000	Fat	2
9. Juice and must industry			
9.1 Fruit scraps		Pectin, carbohydrates, vitamins, essences,	
9.2 Pectin washing water		Pectin	
10. Wine industry			
10.1 Waste yeast	180000	Protein, carbohydrates, fat, ethanol, vitamins	
11. Fruit and vegetable canning industry			
11.1 Fruit and vegetable scraps		Pectin, carbohydrates, vitamins, essences, colouring matter	

* The data for BOD_5 and % dry solid matter are average guide values, obtained from a large number of measurements and published data. No data are given where the fluctuations are too great.

2.4.2.9.1 Waste Waters from Sugar Refineries

Sugar is manufactured from sugar beet and sugar cane.

2.4.2.9.1.1 Beet Sugar

Sugar refineries normally process the beet they receive within 7 to 12 weeks, working 24 hours a day. This short production period is necessary in order to keep the loss of beet or beet sugar as low as possible, and it means that waste water only flows during this short time. There are three areas of sugar production:
- production of raw sugar,
- sugar refining,
- extraction of sugar from molasses.

<u>Raw sugar production</u> begins (Fig. 2.4.-11) with washing the sugar beet received, which gives fluming and washing water. This is contaminated by sand, clay, leaves, and juice from damaged beet.

The tips are then removed from the washed beet, they are cut into strips in slicing machines, and the juice is extracted in diffusion batteries with hot water. The spent slices are then pressed. The resulting process water from diffusion and pressing contains a considerable amount of sugar, which constitutes a heavy pollution load. More modern processes, however, enable it to be returned to the diffusion battery.

In settling tanks, which normally work in several stages, contaminants are precipitated out of the raw juice by addition of lime. The resulting lime cake water can be used in agriculture as a valuable fertilizer. The purified raw juice (clear juice) is then evaporated to syrup in vacuum evaporators, and crystallized to raw sugar by cooling. The condenser water is drained off as waste water, or it can be recycled. The semiliquid mass produced in the crystallizers is then separated in centrifuges into raw sugar and molasses.

Tab. 2.4.-59: Amount and composition of dairy wastes /138/

	Unit	Fresh milk	Further processing Average	Range
Cooling water	m³/1000 l milk			2.0 - 4.0
Contaminated water	m³/1000 l milk	1.0	2	0.5 - 5.0
BOD_5	kg/1000 l milk	0.1 - 2.5	2.5	0.3 - 5.0
BOD_5	mg/l	100 - 2500	1250	1 - 50 000
BOD_5/COD	-		0.69	0.35 - 0.9

- use of high pressure cleaning systems, which make it possible to clean effectively while at the same time saving water,
- use of non-returnable bottles, so that no bottle washing is necessary,
- batch cleaning,
- use of warm water and excess vapors for rinsing or for setting up suitable water recirculation systems,
- blowing product residues out of pipelines with air,
- water economizing measures in cooling processes, such as replacing fresh water with ice water, heat recovery by large exchange surfaces, use of cooling water as warm water, use of evaporator-condensers with cooling towers,
- introduction of fixed, non-removable pipelines,
- regular inspections,
- collection of leakage drips (if a leak drips once every second, organic material of about 200 g BOD_5 collects in 10 hours),
- setting up of washing water recirculation systems with an increased concentration of products in the wash water and the subsequent possibility of re-utilization, then,
- development of maximum utilization of residual substances, including marketing of them.

2.4.2.9.3 Waste Waters from Non-alcoholic Beverage Production

Non-alcoholic beverages include
- fruit juices (sweet musts), which are delivered to the consumer in concentrated form, or, in the case of juices from berries, stone fruit or wild fruit, with added water. After the addition of water, the acid content must not be below 5 to 10 g/l.
- fruit juice drinks with a juice content of
  at least 30 % for pomaceous fruit juices or grape juice,
  at least 6 % for citrus juices or citrus juice mixtures,
  at least 10 % for other juices or juice mixture,
- lemonades, cold and hot drinks,
- mineral waters, spring waters, and possibly table waters.

Wastes from plants manufacturing beverages containing fruit juice are distinctly different according to whether the plants obtain the juice from fruit supplied, or whether they prepare supplied juices and concentrates to make them ready for drinking. In the first case, similar conditions prevail in the plants at the time of the fruit delivery to conditions in wine-producing plants up to the product "cleared must"; similar measures are also required for the retention of fruit scraps, sediments and juice. Fig. 2.4.-14 shows a diagram of fruit juice manufacture (WHO /179j/).

DITTRICH* gives production-specific values, as shown in Table 2.4.-60.

(In all processes, with the exception of the bottling of naturally cloudy juices, cooling water is included in the waste water amount.)

The composition of wastes from bottle cleaning is given in Table 2.4.-61.

A production-specific amount of:
- with bottle-washing machines   107 - 123 l/100 bottles
- total   166 - 223 l/100 bottles

can be expected for bottle washing.

---

* Dittrich, V.: Abwasserbeseitigung in der Fruchtsaftindustrie. Lecture

Tab. 2.4.-60: Amount and quality of wastes in the manufacture of non-alcoholic beverages

	Waste water amount m³/t or m³/1000 l	COD kg O_2/t	BOD_5 kg O_2/t
24 000 l orange juice from syrup plus fruit pulp 2 t (bilberry, strawberry)	4.2-5.8	9.9-12.5	7.2-9.1
20 000 to 40 000 l per day bottling various naturally cloudy juices from concentrates	2.6-13.5	5.2-9.9	2.1-6.8
Pressing of 42 to 84 t cherries per day	5.0-8.5	3.9-4.5	2.8-4.5
41 t apricots/d	4.6	5.3	4.4
92 to 97 t plums/d	4.0	3.4	2.4-2.9
319 t apples/d	1.2	2.0	1.4
325 t pears/d plus 4 t apples/d	1.2	4.0	1.9

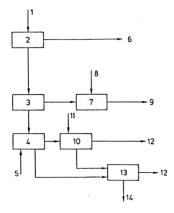

Fig. 2.4.-14: Flow diagram of fruit juice manufacture /179j/

1 - water; 2 - water treatment; 3 - cooler; 4 - carbonator; 5 - carbon dioxide; 6 - sludges, washing water, spent salt solutions; 7 - bottle washing; 8 - detergents; 9 - washing water with detergents; 10 - syrup mixer, store; 11 - sugar, aroma, color, acid, etc.; 12 - waste water; 13 - bottling; 14 - fruit juice drink

The amount of waste water in beverage plants can only be reduced a little. Water can only be saved by bottle washing in stages.

However, organic contamination can be reduced by up to 70 % by retention and filtering of sediment and yeast.

Tab. 2.4.-61: Quality of wastes from bottle washing for various beverages /186/

Type of bottles to be washed	BOD_5 mg/l	$KMnO_4$ consumption mg/l	pH	settleable substances mg/l	population equivalent 40 per 1000 bottles
Wine	4.5-15	12-16	8.1-8.4	0-0.15	0.20-0.75
Beer	185-705	290-1930	9.5-11.9	0.2	8.4-42
Milk	6.3-25	9.5-46	8.6-10.1	0.1	0.95-1.3
Sweetened clear lemonade	295-600	928-1196	8.4-8.5	1.5-3.8	17-29
Sweetened cloudy lemonade	660	2170	8.1	0.1	27
Cola drinks	340	1370	10.1	0.2	2.6
Fruit juice drinks	450	1320	9.4	2.5	26
Vinegar	15	27	7.7	0.2	0.8
Sweetened clear lemonade and mineral water in ratio of 1:1	64	237	10.0	0.7	3.4

2.4.2.9.4 Waste Waters from Breweries

Barley is the most important grain used for brewing beer, with the addition of rice, oats, rye, wheat and millet. Sometimes, syrup and caramel are also added. Fermentation takes place by means of a yeast culture /179g/.

The manufacture of beer consists of three processes: preparation of malt from barley, preparation of beer wort, and fermentation.

Large breweries manufacture malt from grain in their own plant; smaller breweries usually get their malt from special malthouses.

For the production of malt, barley (or wheat, in the case of white beer) has to be freed from dust, washed and sorted. This leaves dust, awns, and rejected grains, which are used as cattle fodder.

In the steeping tanks, the grain is allowed to swell for 2 - 4 days in water of 15° C to 20° C with aeration. The water is changed twice a day. The steeped grain is then kept for 7 - 9 days on the malting floor with moist air until it germinates. This changes the starches into maltose. The malt is then dried at 105° C until it contains about 3 % water. After drying, the rootlets are removed in the degerminating machine, and the malt is stored. The main pollution load comes from batch discharging of the steeping tank waters.

In the brewery, the malt is crushed and mixed with water in mash tuns. The insoluble residues are separated as spent grain in the lauter tun and subsequent filter presses. This spent grain is marketed as livestock feed. If it cannot be used fresh, it is dried; this gives only condensation water as waste, which is only slightly contaminated. The hot work is then strained in the hop sieve to remove the hop residues, which are washed several times. The spent hop residues (spent hops are utilized as fertilizer, or as additional raw material in the paper industry) are in many plants finely ground and discharged with the waste water. In addition to the wastes mentioned above, there are above all washing, cleaning, and rinsing waters from the cleaning of the machines, containers, filter cloths, and especially bottles and barrels.

Fig. 2.4.-15 shows a diagram of beer manufacture (WHO /179g/).

In malthouses, there are wastes from washing, germination and wet transport of the brewing barley. The waste water amounts fluctuate considerably, due to the batch discharge of the contents of large containers. A peak amount can be 30 % of the daily total in one hour. The following amounts can be given as examples:

waste water amount:	2.1 to 3.9 m³/t barley
settleable substances:	3 to 13 ml/l after 2 hrs. settling time
BOD_5 overlaying water	1.4 to 4.1 kg O_2/t barley
COD overlaying water	2.4 to 7.0 kg O_2/t barley

In breweries, the retention of excess fermenting cellar yeast, storage cellar dregs, total cooler sludge and kieselgur-yeast residues is of decisive significance in reducing waste water contamination. In this case, the amount of waste water is 0.5 to 2.08 m³/100 liter consumer beer (KÜHBECK /179g/). The higher values include cooling water discharges.

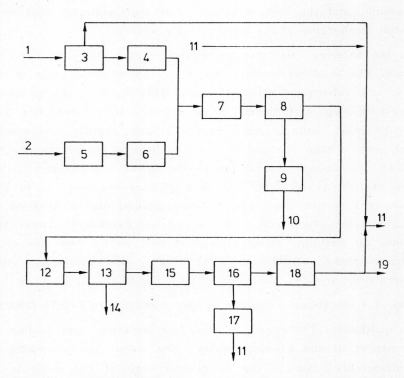

Fig. 2.4.-15: Simplified flow chart of beer manufacture with waste water discharge points /179g/

1 - barley; 2 - malt; 3 - malting; 4 - feed grinder; 5 - grain mill; 6 - oven; 7 - mash tun; 8 - lauter tun; 9 - filter press; 10 - livestock feed; 11 - rinsing and washing wastes; 12 - brew kettle; 13 - hop sieve; 14 - spent hops; 15 - bac cooler; 16 - fermentation; 17 - yeast turbidity; 18 - beer barrels store; 19 - beer

With retention of the above-mentioned production wastes
the BOD_5, unsettled, is: 0.35 to 0.97 kg O_2/100 liter beer and
the BOD_5, settled, is: 0.33 to 0.89 kg O_2/100 liter beer.

Without retention of the production wastes

the BOD_5, unsettled, is: 0.69 to 1.06 kg O_2/100 liter beer and
the BOD_5, settled, is: 0.65 to 1.30 kg O_2/100 liter beer.

With careful plant management, it is possible to maintain waste water levels of 0.4 - 0.6 m³/hl beer and pollution levels of 0.5 - 0.6 kg BOD_5 (sed)/hl beer with a ratio of COD/BOD_5 (sed) of 1.5. The BOD_5 of beer itself is about 80 g O_2/l, and the COD about 120 g O_2/l.

Examples of the composition of various waste water types in a brewery are given in Table 2.4.-62.

Tab. 2.4.-62: Amount and composition of various waste water types in a brewery /107/

Type of waste water	pH	dry residue	susp. solids		BOD_5	amount of w.w.
			total	ash		
		mg/l	mg/l	mg/l	mg/l	m³/100 hl
Barrel cleaning						
a) steel barrels	7.1	980	250	-	21	1
b) wooden barrels	7.3	-	-	-	62	3
Bottle cleaning						
a) washing solution (beer)	11.5	71700	310	-	870	3
b) rinsing water (beer)	7.2	940	95	-	16	32
c) washing solution (lemonade)	11.4	7900	1010	-	854	1
d) rinsing water (lemonade)	8.1	1050	34	-	44	10
Filter cloth washing						
a) mash filter	6.7	1070	1846	96	325	9.5
b) cooler sludge filter	6.7	1290	456	32	694	4.2
Cleaning waters						
a) fermenting vat without yeast	5.3	2060	3944	332	3550	1.7
b) fermenting vat with yeast	5.0	-	-	-	70250	-
c) storage vat without yeast	6.8	1010	164	28	502	1.4
d) storage vat with yeast	5.2	-	10900	-	84500	-
e) beer filter	5.9	1940	37835	36400	2000	2

2.4.2.9.5 Waste Waters from Margarine, Edible Fat, and Edible Oil Production

Natural fats and oils are triglycerides of higher fatty acids.

Table 2.4.-63 shows world fat production.

<u>Oilseeds</u> or those plant parts containing oil are processed in three steps:
- oil extraction,
- purification of the oil,
- in some cases, conditioning of the oil.

Tab. 2.4.-63: World fat production (in 1000 t) /107/

Type of fat	1935/39	1950	1954
Vegetable fats			
Babassu oil	27	45	23
Cottonseed oil	1560	1361	1742
Coconut oil	1932	1805	1973
Groundnut oil	1506	1760	1882
Wood oil	136	113	93
Linseed oil	1039	1040	1012
Olive oil	871	1064	712
Palm oil	962	1102	1229
Palm kernel oil	372	435	404
Rapeseed oil	1207	1524	1497
Castor oil	181	209	218
Sesame oil	653	703	690
Soybean oil	1229	1769	1978
Sunflower oil	562	758	803
Animal fats			
Butter	3611	3021	3293
Fish and fish liver oil	454	340	445
Lard	2495	2494	3620
Suet	1442	2055	2450
Sperm oil	494	386	413

The raw materials from which oil is obtained are solid fats and fatty oils.

Dry oilseeds are sieved and air classified to remove dust, sand, wood, etc., while moist raw material, such as olives, is cleaned by washing and sieving. Oil is then obtained by melting out, pressing, or extracting, and all husks, skins, and stones are removed.

Fig. 2.4.-16 shows a simplified diagram of the production of olive oil.

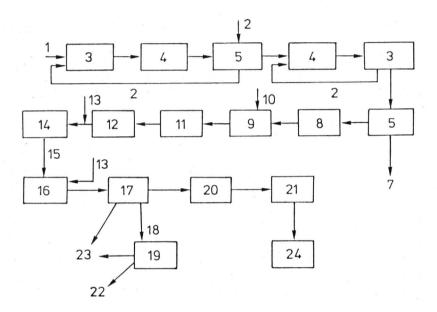

Fig. 2.4.-16: Simplified diagram of the production of olive oil /107/

1 - olives; 2 - washing and transport water; 3 - sieve; 4 - washing pit; 5 - stone remover; 6 - stone drier; 7 - dried stones; 8 - hammermill; 9 - intermediate container; 10 - salt; 11 - mixer; 12 - must pit; 13 - water; 14 - revolving screen; 15 - olive must; 16 - must container; 17 - separator; 18 - separator sludge; 19 - drainage; 20 - pure oil scales; 21 - oil intermediate container; 22 - livestock feed; 23 - fruit juices; 24 - oil storage tanks

The amount and composition of oil factory wastes is largely determined by production capacity, the raw material to be processed, and the processes used.

In addition to the normal rinsing and cleaning waters, the following types of waste water can be expected:
- extraction water containing benzine residues (0.4 % of the oil flow),
- distillate with fatty acids, neutral oil, and unsaponifiable fat from steam treatment of the oil (25 kg steam per 100 kg oil),
- cooling water from distillation and extraction.

In margarine factories, wastes come from the milk souring plant, and from cooling or quenching of the emulsion with cold water. The quenching water can be clarified and reused in the oil refinery before being finally discharged. In margarine factories there are also rinsing and cleaning waters from the milk reception point, which are similar to wastes from fresh milk dairies.

No generally valid data can be given for the amount of the various wastes (MEINCK /107/). The washing of 1 t vegetable oil gives about 0.9 m^3 washing water and about 0.17 m^3 waste water containing sulfuric acid. The manufacture of 1 t edible fat (margarine) gives refinery wastes of about 0.06 m^3 and churnery wastes of about 0.02 m^3.

The waste water is normally acid and contains considerable amounts of easily putrescible organic substances containing nitrogen and a high amount of residual oil.

For example, the waste water of a medium-sized margarine factory showed on analysis the following compositions:

undissolved substances	230 mg/l
dissolved substances	6400 mg/l
chlorides	1500 mg/l
permanganate value	1800 mg/l
total fat (ether extract)	130 mg/l
nitrogen	7 mg/l
pH	6 - 7

In order to reduce pollution, care should be taken that all solids separated in centrifuges or presses are not washed away with the waste water but are utilized as a valuable livestock feed. The amount of waste water can be reduced by recirculation systems and multiple use of cooling and distillation waters. For this, it is necessary to separate the waste water streams.

2.4.2.9.6 Waste Waters from Slaughterhouses and Meat Processing Plants

A distinction must be made between large plants which do all the slaughtering for a whole municipality, and small private businesses in municipalities that have no slaughterhouse. Here, slaughtering may take place only once a week, or even less.

In large slaughterhouses, the animals to be slaughtered are kept in stalls which have to be cleaned. This gives solid and liquid waste substances comparable to those from agricultural operations. Before slaughtering, the animals are cleaned to remove excrement and dirt. The blood should be collected and not discharged with the waste water. After bleeding, the animal is scalded for 4 to 6 minutes at a temperature of about 60 degrees centigrade, and the hair removed. This gives scalding water, containing bristles, hair, and sometimes small amounts of fat.

In the viscera processing section, the viscera are cleaned. This gives cleaning waters contaminated with mucous, excrement and detergent residues. The first stomach or paunch is emptied in another section and also cleaned, giving wastes.

Fig. 2.4.-17 shows a flow diagram of a slaughterhouse including the waste water points.

The following data were obtained in a slaughterhouse where 100 small animals and 75 head of cattle were slaughtered daily:

		Small animals	Cattle
Waste water amount	m³/animal	0.26	0.98
Sedimentary substances after 2 h	l/animal	6	13.5
Dry solid matter	kg/animal	0.19	0.42
BOD_5 in overlaying	kg O_2/animal	0.43	2.39

The blood was collected.

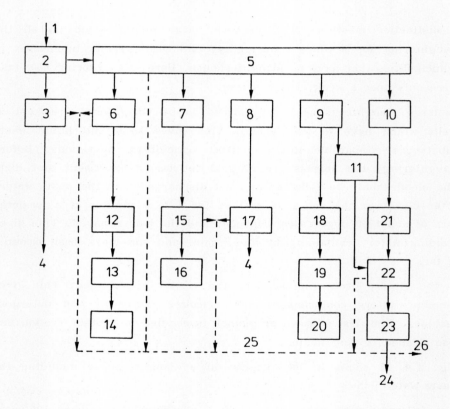

Fig. 2.4.-17: Flow diagram of a slaughterhouse /179k/

1 - animals; 2 - holding pens; 3 - settling tank; 4 - compost (fertilizer); 5 - slaughtering section; 6 - blood; 7 - skins; 8 - viscera contents; 9 - by-products; 10 - skeleton (bones); 11 - liver, kidneys, etc.; 12 - coagulates; 13 - screening filter; 14 - drying of blood; 15 - drying and desalting; 16 - tanning; 17 - pressing; 18 - discharge; 19 - degreasing; 20 - fat and tallow; 21 - dressing; 22 - washing; 23 - cold store; 24 - market; 25 - waste water; 26 - purification plant

2.4.2.9.8 Waste Waters from Fruit and Vegetable Canning

Fruit and vegetable canning factories are often seasonal plants. However, they may also use fruit brought from overseas, and in this case they operate throughout the year. Because different fruits ripen and are harvested at different times of the year, the raw materials and finished products change during this period, and as a result the condition of the wastes from such plants may often vary considerably.

The raw material is washed after delivery, using fresh or recirculated water. Vegetables grown in the ground have to be washed intensively.

In peeling, the surface layer of the fruits or vegetables is slightly parboiled by steam or corroded in lye, and removed by subsequent washing. The vegetable and fruit pieces are blanched in special containers, or blanchers, with hot water, steam, or, for products which are afterwards dried, hot air.

The products can be preserved by heat treatment; cold treatment; addition of vinegar, salt, or sugar; drying; or fermenting. Wastes in any amount worth mentioning come only from fermenting in the form of fresh brine and acid brine.

The different types of contaminated wastes from canning factories are shown in Table 2.4.-66.

Generally, large amounts of water are needed for cleaning the raw products. Data given in the literature for the amounts of waste water vary. In general, however, these amounts can be expected: 1 to 2 m^3 per t raw product, and in special cases even up to 3.6 m^3 per t raw product. ELDRIDGE /107/ gives the following specific total waste water amounts for 1000 cans (containing 570 g):

green beans	about 6.6 m^3
beetroots	about 3.9 m^3
carrots (RYAN)	about 3.6 m^3

peas   about 3.9 m³
spinach   about 3.1 m³
tomatoes   about 2.4 m³
(whole fruit)
tomato pulp   about 1.2 m³
cherries   about 6.2 m³
apricots   about 8.6 m³
pears   about 8.0 m³

With respect to its composition of carbohydrates, protein, and fat, the waste water is only slightly different from domestic sewage; however, the carbohydrates predominate.

Tab. 2.4.-66: Types of waste water from a fruit and vegetable canning factory, and substances they contain /186/

Stage of processing	Type of waste water	Substances contained
Delivery of raw products	Rainwater	Soil, fruit, and vegetable scraps
Washing, sorting and transport of raw products	Washing water transport water	Soil, other surface dirt, vegetable scraps, vegetable juice
Preparation and cutting up	-	Vegetable juices, additives, vegetable scraps
Steam, acid, lye, and mechanical peeling	Peeling solution condensed steam rinsing water	Peel, rubbed off vegetable scraps, lye, organic acids
Blanching	Blanching water vapours rinsing waters	Vegetable juices, vegetable scraps, blanching additives
Pressing and other extraction of liquors	Rinsing water	Vegetable juices, brine, common salt
Conditioning (preparation, packing, canning, bottling)	Rinsing water	Vegetable scraps, vegetable juice, residues, and constituents of infusion
Preserving	Cooling water condensates	Product constituents
Cleaning of plant and containers	Cleaning water	Product losses detergents
Water treatment	Rinsing water	Sludges

The possible range of fluctuation and pollution load can be seen in Table 2.4.-67.

The composition of the wastes is subject to considerable change according to the kind of raw products processed and the mode of operation. The following list gives a survey of the more important raw products. The figures are from a study by RYAN /186/ (Table 2.4.-68).

Data for the composition of wastes from the fruit and canning industry are given in Table 2.4.-69 (GREGORY and KIMBALL /107/).

Tab. 2.4.-67: Amount and composition of wastes from various product areas of the canning industry /186/

Product	Waste water amount m^3/t	Pollution amount	
		kg BOD_5/t	mg BOD_5/l
Beans	14 - 23	4 - 8	140 - 600
Peas	30 - 60	8 - 15	300 - 4700
Sweet corn	16 - 27	20 - 30	
Cucumbers	15 - 30	4 - 6	
Sweet peppers	35 - 45	25 - 35	
Tomatoes	1.8 - 3.6	2.1 - 5.5	180 - 4000
Celeriac	10 - 15	15 - 30	
Carrot	20 - 40	20 - 40	520 - 3030
Beetroot	4.5 - 35	15 - 20	2500 - 4000
Ready-to-serve food	30 - 50	14 - 25	
Mixed salads	50 - 100	8 - 30	
Baby food	80 - 160	30 - 69	
Sauerkraut	1.5 - 2.0	3.5 - 10	
Strawberries	30 - 40	50 - 60	
Apples	34	24	1658 - 5530
Pears	8 - 24	6 - 36	450 - 2600
Cherries	4 - 10	7	400 - 2600
Apricots	13	6	200 - 1020

Tab. 2.4.-68: Composition of total waste waters from canning factories /186/

	Suspend. solids mg/l	Ignition residue mg/l	Dissolved constituents mg/l	Ignition residue mg/l	pH	$KMnO_4$ consumption mg/l	BOD_5 mg/l
Tomatoes	450	80	2500	580	4.9	1100	1150
Peas	300	25	6000	3360	4.7	2150	2710
Wax beans	60	10	1670	970	7.6	-	240
Spinach	580	150	1700	950	7.0	40	280
Carrots	1830	170	5800	1900	7.1	-	1110
Beetroots	1600	220	5000	800	6.0	2700	1500
Sauerkraut	60	5	3300	1600	5.6	800	1400
Cherries	20	4	4100	1700	6.2	-	750

Tab. 2.4.-69: Composition of wastes from the fruit canning industry /107/

Type of fruit	Suspended solids mg/l	BOD_5 mg/l	$KMnO_4$-consumption mg/l	pH	Temperature °C
Peaches	600	1400	2000	7.6	31
Apricots	260	200	700	7.6	30

Waste waters from the canning industry contain large amounts of degradable substances, and can therefore cause foul odors, corrosion, etc. in municipal sewer systems.

The amounts of waste water and pollution load in the fruit and vegetable canning industry can be reduced in the following ways:
- improved processes to avoid production losses, e.g. changing to steam peeling, or replacement of hot water blanchers by steam blanchers,
- reduction of freshwater demand by means of recirculation or multiple use of water, e.g. use of cooling water as washing water,

- separation of waste water streams with different pollution levels, so that highly concentrated streams can be evaporated and utilized for making yeast or livestock feeds, or sent directly to sludge fermentation plants,
- filtering out of fruit and vegetable scraps, which can then be used as livestock feed.

2.4.2.9.9 Waste Waters from Starch Production and Potato Processing

Starch is manufactured from two main groups of raw vegetable materials:
- roots or tubers; potatoes, cassava,
- cereals: maize, wheat, rice, rye, barley.

Sago palms or chestnuts can also be used.

2.4.2.9.9.1 Waste Waters from Potato Starch Industry

Only about 1/5 (9 to 35 %) of their weight in starch is obtained from potatoes. For this, 15 to 25 m^3 iron-free water per t potatoes are required.

To obtain starch, the potatoes are first cleaned by dry or wet processes and conveyed into the factory in a hydraulic carrier. They are then put through saw teeth and at the same time rinsed with large quantities of water. The ground potato is then further crushed in brush machines or pulp mills. Water is added, and the starch is separated from the pulp by sieving. The starch milk is filtered and the starch granules, which have a high specific gravity, are collected in settling tanks. The starch is thoroughly washed so that it is completely clean, dried in centrifuges or hydrocyclones, and made into powder.

The pulp is dehydrated and used as livestock feed.

Fig. 2.4.-19 shows a flow chart of potato processing (SEYFRIED, MEINCK /107/).

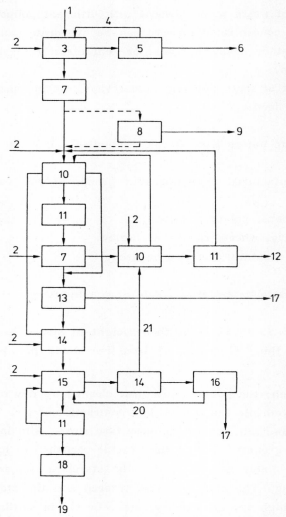

Fig. 2.4-19: Flow diagram showing the processing of potatoes to starch /107/

1 - potatoes; 2 - freshwater; 3 - transport, washing; 4 - recycling; 5 - purification; 6 - waste water; 7 - grinding; 8 - rotary sieves; 9 - pulp water; 10 - washing; 11 - dewatering; 12 - further processing; 13 - separating; 14 - filtering; 15 - starch washing; 16 - thickening; 17 - waste water; 18 - drying; 19 - starch; 20 - very fine starch granules; 21 - overflow

In potato starch factories there are the following wastes:
- potato washing water, about 6 - 8 m³ per t potatoes. In addition to sand and soil, which can amount to 5 - 20 % of the weight of the

potatoes, these contain potato scraps and dissolved potato extracts.
- protein water, about 7 - 12 m³ per t processed potatoes. These contain large amounts of organic, fermentable, and putrescible, dissolved and undissolved potato solids, as well as inorganic salts, especially potassium and phosphorus compounds. Characteristic of these wastes is their marked propensity to start acid fermentation, giving lactic and butyric acid, which often cause pollution due to their foul odors.
- starch washing waters, about 1 - 3 m³ per t potatoes. These contain fine pulp particles and starch granules.
- pulp press waters, amounting to about 0.4 - 0.6 m³ per t potatoes, and of a similar quality to the protein waters.

The population equivalent is given by MEINCK as about 500 PE per t potatoes processed and by SEYFRIED as 350 - 650 PE per t.

Table 2.4.-70 shows the composition of protein water from potato flour production.

Tab. 2.4.-70: Composition of protein water from potato flour production /158/

	Undissolved mg/l	Dissolved mg/l	Total mg/l
Dry residue	1290	7055	8345
inorganic	185	1910	2095
organic	1105	5145	6250
Permanganate consumption			9449
Free ammonia			36
Nitrogen as NH_3			30
Organic nitrogen			528
Total nitrogen			558
Potassium (K_2O)			114
Phosphorus pentoxide (P_2O_5)			322
Calcium (CaO)			96

2.4.2.9.9.2 Waste Waters from Wheat Starch Production

The manufacture of this type of starch is fundamentally different from that of potato starch. In this case, the valuable (vegetable) gluten has to be separated from the starch as completely as possible.

Wheat flour is washed in an "extractor", with added water. The gluten masses are washed, producing gluten washing waters, and are then further processed. The starch milk is concentrated in thickening vats, water is added to it, and the mixture is then separated in a centrifuge.

Fig. 2.4.-20 shows a flow diagram of wheat flour processing (SEYFRIED /107/).

The population equivalent is given as about 500 PE/t grain processed and by SEYFRIED as about 350 - 650 PE/t.
The amount of waste water in wheat starch factories can be as much as 20 m³ per t wheat, with a population equivalent of 1600 to 1700 per t.

Table 2.4.-71 gives the composition of the wastes (SEYFRIED /107/).

Tab. 2.4.-71: Composition of wastes from wheat starch production /107/

	Without recirculation	With partial recirculation
pH	3.43	3.45 - 4.70
Suspended solids, mg/l	216 - 283	100 - 2630
Sedimentary substances, ml/l	0.3 - 1.2	1.2 - 106
$KMnO_4$ consumption, mg/l	6350 - 6696	8400 - 10200
BOD_5, mg/l	4529 - 4620	5024 - 6900
Total nitrogen, mg/l	334 - 375	222 - 293
Dry residue, mg/l	4750 - 5720	5370 - 6200
ignition loss	82 - 84	90 - 95
Volatile organic acids, mg/l	550 - 687	670
Methylene blue test, h	2 - 6	3.5

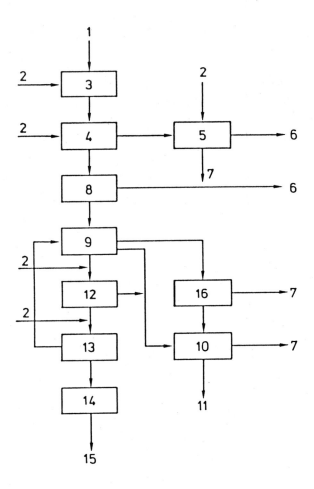

Fig. 2.4.-20: Flow diagram of the processing of wheat flour to starch /107/

1 - wheat flour; 2 - fresh or process water; 3 - dough kneader; 4 - washing; 5 - gluten washing; 6 - further processing; 7 - waste waters; 8 - filters; 9 - thickening centrifuge; 10 - concentration; 11 - secondary starch; 12 - fine milk centrifuge; 13 - elimination of water; 14 - drying; 15 - primary starch; 16 - separator

2.4.2.9.9.3 Waste Waters from Maize Starch Production

The maize is shelled and then mixed with milk and sulfurous acid to help it disintegrate.

Fig. 2.4.-21 shows a flow diagram of the processing of maize to starch (SEYFRIED /107/).

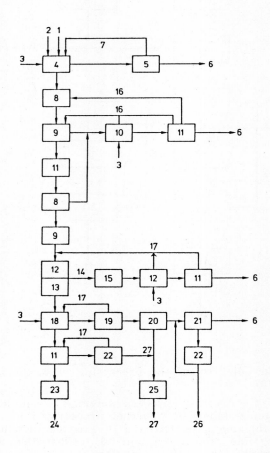

Fig. 2.4.-21: Flow diagram of the processing of maize to starch /107/

1 - maize; 2 - SO_2; 3 - fresh or process water; 4 - steeping tanks; 5 - evaporator; 6 - further processing; 7 - vapours; 8 - grinder; 9 - degermination; 10 - germ washing; 11 - elimination of water; 12 - washing; 13 - sieving; 14 - hulls; 15 - grinder; 16 - washing milk; 17 - starch milk; 18 - separating; 19 - starch recovery; 20 - thickening; 21 - filter presses; 22 - separator; 23 - drying; 24 - starch; 25 - purifying; 26 - waste water; 27 - process water

2.4.2.9.9.4 Waste Waters from Rice Starch Production

The starch in rice grains is very intimately bonded, and in order to extract it, the rice has to be soaked in successive soda solutions. Then the rice is washed and ground in double rice mills.

Table 2.4.-72 shows the composition of wastes from rice starch production (SEYFRIED /107/).

Tab. 2.4.-72: Composition of waste waters from rice starch production /107/

pH	4.16	BOD_5, mg/l	1012
Suspended solids, mg/l	162	Starch, mg/l	120
Sedimentary substances, ml/l	2.8	Total nitrogen, mg/l	96
Dry residue, mg/l	2930	Ammonia nitrogen, mg/l	15
ignition loss	49	Phosphates (PO_4), mg/l	12
$KMnO_4$ consumption, mg/l	1321	Methylene blue test, h	21

2.4.2.9.9.5 Waste Waters from Starch Sugar and Syrup Production

These products are mainly manufactured in connection with maize or potato starch extraction. By inversion with diluted sulfuric or hydrochloric acid, the starch is first changed into dextrin and later into starch sugar. Color is removed from the sugar solution with activated carbon and then the solution is evaporated.

In this process, wastes come from washing the activated carbon with hydrochloric acid or soda, and from the precipitation in injector condensers of vapors from the evaporation of sugar juices.

2.4.2.9.9.6 Waste Waters from Dried Potato Products

Potatoes are prepared for drying by careful washing, followed by steaming at pressure in a cooker. From this process come small amounts of highly concentrated condensation and cooking waters.

The main wastes from these factories are washing waters.

2.4.2.9.9.7 Waste Waters from Potato Chip Production

In potato chip factories, potato processing comprises pre-washing, peeling, washing, cleaning and slicing of the potatoes, washing and rinsing of the slices, and further processing of these to the finished product (drying, frying, salting, packaging). Potato substance is lost during peeling, and this goes into the waste water. For chips, the yield is about 250 kg per t potatoes. The pollution load is 25 kg BOD_5/t processed potatoes. This corresponds to a population equivalent of about 430 per t potatoes.

2.4.2.9.10 Waste Waters from Wine Production

Wine production from the harvested grapes to wine that is ready for bottling comprises the following processes:
- receiving the grapes,
- destalking of grapes, if required,
- crushing and macerating,
- preliminary juice extraction and pressing,
- treatment of musts,
- alcoholic fermentation and maturing.

Experience shows that the highest levels of pollution occur in wastes from the following processes:
- clarification or separation of the must,
- first racking (after fermentation),
- second racking (after fining).

The purpose of these three processes is to separate the must or wine from the solids. This is usually done first by sedimentation, giving pomace in the case of must, and yeast and fining residues later on in the racking processes. The wine is generally not clear enough after sedimentation and is therefore also clarified mechanically. For this, separators, filters, and finally degerminating filters are used, according to the degree of maturity, and with the aim of achieving the best

the amount of time spent by the animals in the community and out at grass.

Table 2.4.-75 shows the amount and composition of silage wastes from PÖPEL /128/.

Tab. 2.4.-75: Amount and composition of silage wastes /128/

Type of silage Dried matter in %	Fodder	Silage liquid			
		% of charge after 20 days	total	Amount in m³ per ha cultivated area	Pollutants in g BOD, per ha cultivated area
Wet silage up to 20 %	sugar beet leaf, catch crops	22.5	30	10	648 000
				4	22 000
	green maize, grass, grass-clover ley	8	10	4	22 000
				3	17 000
Slightly wilted silage 20 - 35 %	grass, grass-clover ley, lucern, silage maize, steamed potatoes	4	5		
Thoroughly pre-wilted silage over 35 %	grass, grass-clover ley, lucern, silage maize, wet grain	-	-	-	-

Both agricultural and industrial livestock production lead to the concentrated production of excrement. Hatcheries, combined slaughtering and dairy plants, and the processing of animal raw materials, all produce waste materials that can cause pollution of the waste waters.

In the U.S.A., the Environmental Protection Agency (EPA) has found it necessary to demand that wastes from large-scale farms (i.e. exceeding a certain size: 750 dairy cows; 10,000 sheep; 55,000 turkeys; 50,000 ducks; 100,000 hens or broilers) should be purified where liquid dung is produced. Even in the fish production industry, limits are set, since too many fish contaminate the water.

Up to this point, various production processes have been described. We now present some additional special processes, needed for the manufacture of certain important products.

2.4.2.10.1.1 Waste Waters from Pectin Production

Pectins or pectic raw materials are substances that become transparent and form jellies when they are thoroughly moistened. Pectins are used a great deal in the food processsing industry, for the manufacture of jams, jellies, sweets, etc., and also in the textile industry.

Pectin occurs in many plants, e.g. fruit or beets. There are soluble and insoluble pectins; however, the latter can be dissolved by boiling in slightly acid water.

In many countries, pectins are chiefly obtained from apple pomace, citrus peel, and dried beet pulp. Apple pomace contains about 8 to 16 % pectin, dried beet pulp about 27 % pectin, in relation to dry solid matter. Pectin extraction and jam making are frequently carried out in the same factory.

Pectin is dissolved out of the raw material by hydrolysis in hot dilute acid. First, however, the raw material is washed in warm water (34 - 40° C), which causes soluble solids such as sugar, salts, tanning and coloring substances, as well as pectins that have already decomposed and are no longer able to form jelly to go into solution. This solution is discharged as waste water or - in the case of apples - it may be evaporated and used in the production of apple syrup. The material is dehydrated in filter presses and it is then absorbed by about 10 times the amount of warm water, acid (lactic, citric, or sulfurous acid) is added, and it is hydrolyzed at 50 to 100° C. The decoction obtained is deaerated and centrifuged; color is removed by means of activated carbon or silica gel; it is then treated with starch-decomposing enzymes, and finally filtered.

The waste waters from these processes in pectin production are: washing waters, condensates, and pectin-free decoction from the precipitation process.

The volume of these wastes is relatively small, amounting only to a few m³ per day, not counting the condensates, in medium-sized factories. Fig. 2.4.-22 shows a diagram of pectin extraction.

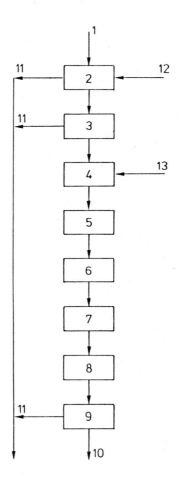

Fig. 2.4.-22: Diagram of pectin extraction

1 - raw materials; 2 - washing; 3 - filtering; 4 - hydrolysis; 5 - centrifuge; 6 - removal of color with activated carbon; 7 - making starch soluble; 8 - filtering; 9 - drying; 10 - pectin; 11 - waste waters; 12 - warm water; 13 - acid water

2.4.2.10.1.2 Waste Waters from Gut String and Gut Processing Plants

The guts of sheep, dogs, cats, and other small animals are used for manufacturing gut strings. For this purpose, the guts are soaked for a day in water with added caustic sodas (NaOH, KOH), soda (Na_2CO_3), or potash (K_2CO_3). They are then rolled several times to separate the outer tissues, and are twisted together into strings. This is normally followed by steaming with sulfur dioxide, which sterilizes and bleaches, and the final process is smoothing with pumice stone and oil. When very fine sheep or goat guts are made into catgut (surgical thread), then they are additionally soaked in disinfecting solutions. Guts of larger animals (e.g. calves, etc.), which are used among other things for sausages, are cleaned with water in a similar way.

The volume of wastes from these processes are usually small. In composition and characteristics, they are similar to slaughterhouse wastes.

Animal hair (except for sheep's wool), bristles, and feathers are used for making furs, upholstery, mattresses, felt, carpets, hats, brushes, and feather bedding. Before processing they are prepared by washing, boiling, and sometimes dyeing and other processes involving water. Raw feathers are often contaminated with excrement, blood and bloody quills, scraps of skin, etc., and they are delivered to feather bedding factories without any preliminary treatment. In cleaning them, it is not normal to add any special disinfectant to the washing water. Therefore, it is possible that wastes from these factories contain germs when, in cases of fowl pest and fowl cholera, feathers are collected from infected farms after drying and are only packed in airtight sacks.

The amount, composition and properties of these wastes vary a great deal according to the type of factory, so that it is not possible to give generally valid data. It should be noted, however, that wastes could be infected with anthrax spores if material from overseas is processed; for this reason, the same measures concerning public hygiene apply here as in the tanning industry.

2.4.2.10.1.3 Utilization of Animal Excreta

The following possibilities exist for using excrement, regardless of the kind of animal:
- fertilizer, including compost,
- livestock feed, either directly (fresh or dried) or after treatment by biological, chemical and physical processes,
- utilization of the solids in industry,
- production of fuel,
- combustion.

The use of industrial methods in livestock production means that very large amounts of excrement are produced. This makes its use as a fertilizer difficult. There is a limit to the amount of semi-liquid manure that can be spread on cultivated land. Various authors give the normal amount as 40 to 50 t per ha; too high amounts have already caused damage. In spite of this, a large part of the excrement will be used as fertilizer also in the future, whether treated or untreated.

As an example of the amount of excrement and deep litter from poultry production, the values given in Tables 2.4.-76 and 2.4-77 (HENNIG and POPPE /68/) can be taken as representative.

Tab. 2.4.-76: Excrement from battery management in poultry production /68/

	Weeks	Amount in kg
Pullet rearing	0 to 8 0 to 18	1.3 5.8
Broiler fattening	0 to 8	1.8
Turkey fattening	0 to 16 0 to 24	23 35
Duck fattening	0 to 8	17
Goose fattening	0 to 9	43

Tab. 2.4.-77: Deep litter from floor management in poultry production /68/

	Weeks	Amount in kg
Pullet rearing (9 birds/m²)	0 to 18	3.5
Broiler fattening (18 birds/m²)	0 to 8	1.2
Turkey fattening (13 birds/m²) (8 birds/m²)	0 to 8 9 to 15 16 to 22	4.5 8.5 10.0

2.4.2.10.2 Waste Waters from Carcass Disposal Plants

The raw materials of carcass disposal plants are animal carcasses, parts of carcasses, and animal products, which due to special circumstances (slaughtering wastes, disease, epidemics, accidents, putrefaction) have to be disposed of. The products obtained from these raw materials are meat meal and fat.

For reasons of hygiene, the raw material must be heated and stirred constantly until it disintegrates. It must then be kept for at least 20 minutes at a temperature of at least 133° C and a pressure of $3 \cdot 10^5$ Pa. This is done in batches in pressure-resistant cookers (autoclaves). The time between filling and emptying is 1.5 to 4 hours, depending on the material. The shorter times are for bones, the longer ones for loads with a large liquid component, especially blood.

After the cooking process, the cooked material is emptied into containers that function as buffers and equalizers between the driers, which operate intermittently. After drying, the dried material is degreased. This process gives a qualitatively better feeding meal and a higher yield as it utilizes the collagenous solutions, these being discharged as waste water in the wet process. In "wet systems", the boiled mass is first degreased and then dried. Degreasing can be done mechanically, using centrifuges, hydraulic presses, or worm extruders, or it can be done chemically by extraction with solvents (e.g. benzine or tetrachloroethane).

In some plants, the raw material is moved through a hot fat bath in a continuous process, during which it is heated, dehydrated, and then degreased.

Steam (vapors) from cooking, or from release of pressure in the cooker, as well as from the subsequent drying process, forms condensed vapors, which in turn form highly polluted and evil-smelling carcass disposal wastes. The volume of the condensates fluctuates between 0.43 and 0.7 m^3/t raw material. The lower values are for material that consists largely of bones. The condensates contain large amounts of ammonia, hydrogen sulfide, organic acids, amines, aldehydes, ketones, and mercaptans.

The following values have been reported for diluted condensates:

pH	7.1 - 9.6
COD	4 480 - 39 210 mg O_2/l
BOD_5	3 342 - 31 926 mg O_2/l
ammonia-nitrogen	250 - 4 300 mg N/l

There are also contaminated wastes from washing of transport vehicles, from the slaughtering shed and sanitary installations, and from cleaning of workshops, machines and traffic areas.

In carcass disposal plants, it is possible to reduce waste water contamination in the following ways:
- immediate collection of raw material from its source and fast processing while it is still as fresh as possible. Especially in warm weather, the material begins to decompose very quickly, and this causes a rapid rise in waste water contamination, measured in terms of COD, BOD_5 and nitrogen compounds.
- keeping a stock of spare parts, so that repairs can be effected

quickly whenever necessary. Even complete aggregates can be kept in reserve, ready for immediate use, so that raw material can always be processed in the freshest possible condition.
- preventing blood and other highly contaminated liquids from reaching the waste water without passing through the cooking and drying plants.

The amounts of waste water and pollutants produced per t raw material are subject to considerable fluctuations. Long-term analyses carried out at the outlet of a carcass disposal plant processing about 200 t raw material/day give the following average values per t raw material for the total waste water (vapors plus cleaning and sanitary wastes):

Waste water amount	3.21	m^3/t
Sedimentary substances, after 2 hrs.	48.2	l/t
COD after settling	21.98	kg O_2/t
BOD_5 after settling	14.95	kg O_2/t
BOD_5 - total	15.25	kg O_2/t
Ammonium nitrogen	2.46	kg N/t
Total phosphorus	0.055	kg P/t

For the BOD_5 total, product-specific values of 3.3 to 21.8 kg O_2/t raw material were determined.

Carcass disposal plants are normally situated near receiving waters with a small water flow and, because of the environmental pollution they produce, far from any other buildings; for this reason, they have to purify their wastes to a large extent themselves, with no other wastes.

2.4.2.10.3 Waste Waters from Fish Meal Production

In fish meal production, fish and fish scraps are processed to an easily digestible livestock feed (fish meat) and train oil. The raw material for fish meal comes chiefly from fish that cannot be marketed for human consumption; these are mostly remainders from auctions, or fish that have been damaged or begun to decompose during transport, i.e. whole fish. A further source of raw material is wastes from fish smoking or frying plants, and fish marinade factories.

In many developing countries, fish meal factories are situated almost exclusively in the lower reaches or near the estuaries of rivers, or at the edge of lakes or the sea. Great difficulties in disposal of the waste waters are caused by the presence of beaches and recreation areas.

The raw material is processed continuously: collection in bunkers, mechanical crushing, disintegration by boiling in cooking drums, then water extraction by pressing or centifuging, and finally drying to finished fish meal. In addition, there is the recovery of oil from the liquid portion of the press water, or by extraction from the dried fish meal.

The amount and qualitative chemical composition of the various waste waters is normally subject only to slight fluctuations. However, there are substantial differences in the concentration, due to the various processes used.

The amount of waste water can reach 30 to 50 m³ per t processed raw material, of which carcass and stickwater, as well as vapor condensates (excluding cooling water), make up only a little more than 1 m³ per t raw material. Thus when the wastes are discharged, they are 30 - 50 times diluted and therefore only slightly contaminated.

Table 2.4.-78 shows the composition of the total waste waters from two fish meal plants from MEINCK /107/. In Plant 1, the press waters were treated with old-fashioned recovery processes, while in Plant 2 they were treated with thickening devices.

Tab. 2.4.-78: Composition of total waste waters from fish meal plants /107/

	1	2
Total suspended solids	1718	617
ignition loss, mg/l	1626	575
Fat, mg/l	2540	1253
Dry residue, mg/l	3692	900
ignition loss, mg/l	1982	470
Chlorides (Cl), mg/l	880	212
Total nitrogen, mg/l	689	126
ammonia nitrogen, mg/l	568	78
organic nitrogen, mg/l	121	48
Permanganate consumption, mg/l	957	369
pH	6.3	6.1

2.5. Storm Water

Storm water (water from all forms of precipitation) is one of the most important components of waste waters that have to be drained by the sewerage system.

Its flow rate is often 10 times that of domestic sewage and in a combined sewerage system it is therefore a decisive factor in determining the dimensions of the sewers. It can also be a reason for choosing a separate sewerage system, the so-called storm drainage.

The amount of rainwater and its degree of contamination must be measured exactly, so that suitable installations are available for its drainage.

2.5.1 Types of Precipitation

The water from precipitation that has to be drained can occur in various forms, as rain, snow, dew, hail, mist, and frost. The form is influenced by meteorological and also topographical circumstances.

Precipitation varies a great deal in intensity, duration and frequency, according to geographical situation and time of year. For example, there might be rain of very great intensity and short duration, or very slight rainfall that continues for days. In the same way, there are countries where it normally rains nearly every day (equatorial region) and other countries, or regions, where not a drop of rain is registered for months on end.

Of all types of precipitation, it is rain which is decisive in the design of sewerage systems. Many observations confirm that a sewerage system that is large enough to contain the rainwater will also be able to cope with water from melted snow.

Furthermore, of all types of rain, only that from cumulonimbus (cb or storm) clouds, or nimbostratus (ns or layer) clouds, is significant. Heavy rainfall originates chiefly in cumulonimbus clouds, which may contain one or more thermals.

Rainfall can be of varying intensity within the same locality. However, in some cases, it is of very extreme intensity and duration. This is known as "historic" maximum rainfall, or "catastrophic" rain. Normally, such rain is not used as a basis for designing sewerage systems. Only that rainfall is taken into account which does not exceed certain standardized parameters; this rain is known as "design storm".

It has generally been established that rain falling in one place is not always the same. It can fluctuate quantitatively and qualitatively. Quantitative fluctuations due to the surface on which the rain falls occur in the amount of rainfall in the specified time and drainage area unit, the total duration of rainfall, etc. Qualitative fluctuations occur in relation to the size of the raindrops, distribution of the rain over the drainage area, etc.

In order to be able to compare different types of rainfall, and to have a basis for design storm or for the corresponding sewer dimensions, the characteristic properties of rain have been defined. For purposes of sewer design, these are: intensity, duration and frequency (average frequency).

The intensity or rate of rainfall is the depth of precipitation on a horizontal surface during a specified time interval;

$$i = \frac{D}{T} \quad [mm/min]$$

Key:
  i = intensity of rainfall, established by observations, in mm/min
  D = depth of precipitation, in mm
  T = duration of rainfall, in minutes

However, for sewerage system design, this definition is impracticable as it would have to be converted to its equivalent in l/s.ha. Therefore, the rainfall per second per area is used:

$$r = \frac{D(mm) \cdot 10000 \ (ha/m)}{T(mm) \cdot 60(s/min)} = 166.7 \cdot \frac{D}{T} = 166.7 \cdot i \quad [l/(s \cdot ha)]$$

Duration is the time between the beginning and the end of the rainfall. The frequency n is the factor that tells how often in the year, as an average over many years, rainfall of intensity i is reached or exceeded. The reciprocal value Z = 1/n gives the number of years in which rainfall of the given intensity i will probably occur once on the average.

2.5.2 Contamination of Rainwater

It used to be thought that rainwater was clean, or that it contained practically no contaminants. Therefore, it was considered as a dilutant for the contaminated water in the sewerage system. This meant that rainwater could normally be discharged into the receiving water without treatment. A great many analyses have shown, however, that this was mistaken and that rainwater during the first few minutes of precipitation is more contaminated than domestic sewage.

Rainwater, or storm water, is contaminated by:
- atmospheric impurities and
- ground surface impurities.

2.5.2.1 Atmospheric Impurities

Atmospheric impurities are dust, combustion emissions, animal excrement, bacteria, etc.

Nothing has been proved concerning the relationship between noxious substances in the atmosphere and the resulting pollution of rainwater. Not even drawing up a balance on the basis of the individual sources of emission gives a reliable result. Depending on the method of measuring, values between 35 and 600 kg/(ha·a) descending particulates can be calculated.

Table 2.5.-1 shows a summary of pollutant values in rainwater from KRAUTH /88/.

Tab. 2.5.-1: Pollutant values in rainwater /88/

Pollutants	Concentration in rainwater, in mg/l		
	1^*	2^*	3^*
Filterable substances	7.0	12.5	3.5
COD	18.7	25.6	–
NH_3	1.26	1.58	0.60
Total P	0.10	0.26	0.031
Cadmium	0.0008	0.001	0.003
Chromium	0.0019	0.002	–
Copper	0.021	0.012	0.007
Lead	0.031	0.110	0.067
Zinc	0.12	0.08	0.06

1^* nach KRAUTH
2^* nach GÖTTLE
3^* nach DAUBER

In addition to dust particles, gas emissions from combustion, especially from coal-fired power stations, affect the atmosphere to a large degree. Sulfur dioxide, SO_2, makes the rain so acid, due to the formation of sulfurous or sulfuric acid, that it has in many countries already caused damage to buildings, made soils and surfaces too acid, and become a health hazard to man.

2.5.2.2 Ground Surface Impurities

Rainwater pollution by the ground surface can have the following origins:
- causes related to traffic, such as refuse, spreading salt in winter, abrasion from road sufaces or vehicle tires, emissions and oil losses from motor vehicles, abrasion from brake linings and metallic components,
- causes not related to traffic, such as vegetation, animal excrement, dust from soil.

Refuse, as produced by pedestrians, cardrivers, and other persons in city and rural traffic, comprises packaging material, fruit and vegetable scraps, litter dropped during household refuse disposal, building site refuse, refuse left over from roadworks, depending on the construction and design of the roads, etc.

Spreading streets in winter consists of either salt between 15 and 40 g/m², or preferably roughening material, where amounts between 70 to 300 g/m² are spread at a time. This is equivalent to an amount of 600 g/m² salt per winter. In order to reduce environmental pollution, the streets in many towns are cleared of snow without using salt.

Abrasion from road surfaces due to the flow of traffic creates large amounts of very fine dust. Annual abrasion on motorway surfaces is about 1.0 mm (SARTOR/BOYD). This is equivalent to about 17 t/(ha·a). City streets of 7.5 m width produce an estimated 0.66 kg per m² in summer.

Table 2.5.-2 shows the amount of abrasion on main roads, using the example of roads with bitumen concrete surfaces in southern and northern Germany, the roads having different traffic volumes (BEECKEN /88/).

Tab. 2.5.-2: Abrasion from bitumen concrete road surfaces on main roads in southern and northern Germany, expressed in millimeters /88/

	Traffic volume cars/24 hrs.	Abrasion in millimeters	
		Summer 70	Winter 70/71
Federal road, northern Germany	7 500	0.3 - 0.8	0.5 - 1.0
Federal road, northern Germany	4 500	0.1 - 0.3	0.5 - 0.6
Federal road, southern Germany	5 000	0.5 - 0.7	1.5 - 1.7
Expressway, southern Germany	9 000	0.5 - 1.0	1.7 - 6.7
Expressway, southern Germany	18 000	0.4 - 0.6	5.9 - 6.7

Abrasion from vehicle tires is equivalent to a loss of 0.03 g/km tire-kilometers driven, or 0.12 g/km auto-kilometers driven. With a traffic volume of 1000 vehicles/day, the daily amount of abrasion from tires is

0.12 kg per kilometer road distance. The BOD_5 caused by this material is small, but the COD is high.

Emissions and oil losses from motor vehicles' enter the environment in the form of particulates and gas via the exhaust. The average annual emission of particulates (mainly soot, tar, and tar derivatives) from a motor vehicle is 2 kg. The gases, i.e. nitric oxides, NO, NO_2, amount annually to 10 kg, and uncombusted hydrocarbon amounts annually to 38 kg per vehicle. Nitrous gases are oxidized in the atmosphere and partially dissolved by rainwater. Gaseous organic hydrocarbons combine with the particulates and are partially absorbed by rainwater. However, no reliable long-term analyses have yet been made which would show the extent to which emissions from motor vehicles contribute to the pollution of storm water flow in residential areas.

Abrasion from brake linings and metallic components are typical elements which deteriorate on the roads. This is mainly inorganic material, containing fair amounts of heavy metals such as copper, nickel, chromium and lead.

Vegetation is an important factor affecting the quality of storm water. Depending on the location, time of year, and atmospheric conditions, such things as leaves, flowers, fruits, fir needles, and other organic material fall and are removed with the rainwater.

Animal excrement contributes to the bacteriological and organic pollution of storm water. Good examples are excrement from pigeons and other birds, dogs, horses, rats, etc. As Table 2.5.-3 (GELDREICH /88/) shows, substantial bacteriological pollution is caused by both animals and man.

Soil dust is also directly blown or washed away by wind and rainwater. The smaller the particles washed away, the greater the pollution, and this also explains why particularly high concentrations occur at the beginning of flow.

Tab. 2.5.-3: Bacteria per g in excreta from man, dog and rat (50 % value) /88/

	Man	Dog	Rat
Fecal coliform bacteria (FC)	13 000 000	23 000 000	330 000
Fecal strepto-cocci (FS)	3 000 000	980 000 000	7 700 000
$\frac{FC}{FS}$	4.3	0.02	0.04

Table 2.5.-4 gives a summary of average concentrations of contaminants in street deposits according to American studies /88/.

Tab. 2.5.-4: Average concentrations of contaminants in street deposits /88/

Parameter	Concentration in mg/g dry solid matter
Organic solids	51
BOD_5	2.3
COD	54
Fats and oils	6.4
Petroleum	3.6
Total phosphorus	0.6
Total nitrogen	0.25
Rubber	2.5
Chromium	0.08
Copper	0.12
Nickel	0.19

2.5.3 Contamination of Wet Weather Flow in the Sewerage System

Storm water is altered in the sewerage system, whether it is a combined or a separate sewerage system.

Storm water contains both the contaminants it collected before ever reaching the ground, and the dust particles that settle directly on paved surfaces in dry weather and that are washed away by the rain runoff.

Water from rain or snow that falls on contaminated ground forms a considerable pollution load for the sewerage system, the sewage plant, or the receiving water, if untreated. Contaminated grounds include filling stations, railway stations where oil products are loaded or unloaded, various stores belonging to chemical works processing organic and inorganic harmful or toxic substances, etc.

However, waste water is affected differently according to whether it is in a combined or a separate sewerage system. Accordingly, there are various advantages and disadvantages with regard to the storm runoff.

2.5.3.1 Contamination of Effluent in Combined Sewerage Systems

In a combined sewerage system, the pollution of the effluent is diluted to some extent by rainwater; on the other hand, the contamination level as a whole rises when rainwater is collected, depending on local and especially structural characteristics of the sewerage system. The following can be considered as sources of contamination:

- contamination of the rain itself before it reaches the ground;
- contamination from dust deposited in dry weather on paved or unpaved ground;
- contaminants produced in the drainage area (e.g. by industrial plants), which are washed into the sewers along with rainwater;

- street spreading material washed into the sewerage system in winter;
- carrying along of material from sewer deposits and from soil erosion in the drainage area. Deposits occur in dry weather as soon as the rate of flow in the sewer falls below 0.5 m/s. A very substantial effect also comes from the foreign water component, and from the component at the beginning of the sewer where very little water flows;
- holding of water in the sewer, in stretches serving as a rain catchment basin, due to pipe subsidence, or similar reasons, and especially in shallow sewers with an insufficient gradient or a gradient less than intended due to a construction error.

2.5.3.2 Contamination of Effluent in Separate Storm Drainage Systems

The storm sewer in a separate drainage system does not only carry the rainwater pollution load. Its effluent is often much more heavily contaminated than the corresponding rainwater. The extra contamination comes from the following sources:
- polluted water which runs directly into the storm sewer in dry weather; e.g. from washing cars on the street;
- polluted water from washing house steps, shops, restaurants, footpaths, etc., which also runs into the street inlets;
- overflow water from public drinking fountains. The result of this is that even in dry weather there is a constant flow of well-oxygenated water in the storm sewer, and a thin slippery layer of bacteria forms on the wet sewer bottom. This thin sewer slime is separated by the tractive force of the water in wet weather and is recorded in analyses of the rainwater quality as organic pollution;
- improper connections which occur when service pipes are installed and which lead to domestic sewage being discharged into the storm sewer, or, in reverse, roof gutters being connected to the sewage network. Improper connections are a great disadvantage in separate sewerage systems;
- washing away of material from sewer deposits and from soil erosion

in contaminated industrial drainage areas;
- unauthorized discharge of highly contaminated wastes into street inlets;
- industrial wastes that can be considered "conventionally clean", e.g. cooling or transport waters, but which may cause deposits;
- contaminated water from wet street cleaning which enters the storm sewers through street inlets;
- contaminated foreign water that reaches the storm sewers;
- drainage waters, although on the whole these can be considered clean.

2.6. Other Industrial Waters

From the point of waste water disposal, other industrial waters include, for example, sea water and reutilized water.

2.6.1 Sea Water

In coastal and bathing resorts, or on camping sites close to the sea, sea water may occur as groundwater even at very shallow depths. In such a situation, concrete, reinforced concrete, and asbestos cement pipes may be heavily corroded by dissolved magnesium salts, ammonium salts, and sulfides. However, there are also other cases where risks caused by sea water must be taken into account, e.g.
- discharge lines laid on the sea bed and stretching several kilometers out into the open sea;
- outlet lines from swimming-pools filled with sea water;
- showers, toilets, and other installations on beaches, which are supplied by sea water;
- municipal sewerage systems which discharge directly into the sea, or river estuaries; in both cases, flooding of the sewer can occur with very high tides, which exposes these sewer pipes to sea water.

2.6.2 Reutilized Water

The reuse of commercial and industrial wastes depends on their composition and the quality of industrial water required. In the Federal Republic of Germany, industrial wastes are divided into four classes according to the concentration of their constituents:

Class 1 - Wastes which can be added to domestic sewage without any further treatment, since they differ from it only slightly in composition.

Class 2 - Organically contaminated wastes which can be added to domestic sewage. When batch discharged, however, these can heavily affect the domestic sewage, so that an equalizing tank has to be used in order to spread the discharge over a longer time period.

Class 3 - Waste waters which can be discharged only after pretreatment, e.g. after removing from it all bulky and settleable suspended matter, fat, and oil. Acid or alkaline waste waters have to be neutralized before discharge.

Class 4 - Wastes which should not be discharged into the sewerage system at all, e.g. liquids such as petroleum and oils, foul-smelling wastes, etc.

In some industrial areas, mainly older ones, storm water is also collected in the industrial effluent sewers. This can be allowed so long as the effluent in the sewer is of the same quality as the storm water. Otherwise, storm water must be collected separately to prevent its being contaminated by the industrial effluent and needing the same sophisticated treatment.

The reutilization of waste water, or the partial recovery of useful material, can be carried out <u>within the same factory</u>, or the waste water can be reutilized directly or indirectly <u>in other commercial branches</u>.

<u>Within the same factory</u>, water can be circulated according to four different flow schemes, as shown in Fig. 2.6.-1, these being /164/:

- open system (Fig. 2.6.-1, 1),
- step system (Fig. 2.6.-1, 2),
- partial recirculation (Fig. 2.6.-1, 3),
- closed recirculation (Fig. 2.6.-1, 4).

- In an open system the water is clarified and discharged into the receiving water. This scheme is normally the most economical for the company, but out-of-date with regard to environmental protection and not to be recommended. In developing countries, it is still the method most commonly used, sometimes even without a treatment plant.
- When water is used in a step system, the clarified water from Section I of the factory is reutilized in Section II or in other factories before being discharged. This pattern may be appropriate, on condition that the quality of the treated waste water from Section I is such that it may be used as process water in other sections or factories. This pattern also presupposes a classification of the water quality requirements of the different sections of the factory.
- The pattern with partial recirculation is the one most commonly used in order to solve water management problems. This pattern makes it possible to dispose of the waste water in a way that normally satisfies the local health and environmental conditions. It may also be the most economical method for the factory.
- Closed recirculation. This model can be useful where there are toxic wastes and special local conditions. Usually it is the most expensive for the factory, but acceptable from the point of view of water resources management. In many European countries where rates are charged for water intake and for discharge of wastes into a receiving water, such methods can reduce the costs for freshwater intake.

In other commercial branches, industrial waste waters may be reutilized directly or indirectly.

- They may be used directly, for example, to irrigate large areas of land, thus utilizing the raw materials contained in the water. This

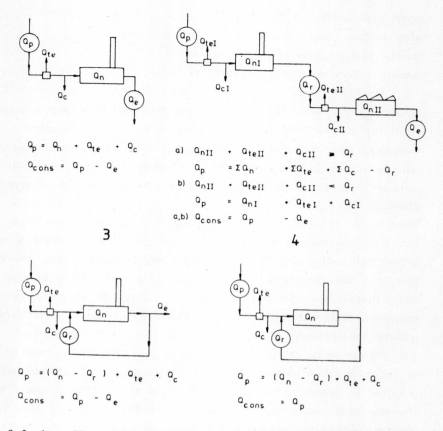

Fig. 2.6.-1: Flow schematics for the utilization of industrial wastes /164/

1 - open system; 2 - step system; 3 - partial recirculation; 4 - closed recirculation

Q_p	= amount of water intake
Q_n, Q_{nI}, Q_{nII}	= water demand of consumer
Q_r	= reutilized and recirculated water
Q_{te}, Q_{teI}, Q_{teII}	= amount used for own purposes by water treatment plant
Q_c, Q_{cI}, Q_{cII}	= water losses in distribution system and other losses
Q_e	= discharge
Q_{cons}	= water consumption (not returned to source)

applies especially to wastes from the food processing industry. It is desirable to return to the soil substances removed from it by food production, such as water, calcium, nitrogen, and phosphorus.
- Industrial wastes can also be discharged into a receiving water and from there recovered <u>indirectly</u> as raw water. The receiving water is

normally a surface water body. In some cases, waste water is stored and treated in underground geological formations, even including drinking water preparation (Namibia, Israel). Naturally, this is not the same as underground storage of drinking water, which is sometimes practiced. Most industrial wastes are reutilized by the indirect methods. In fact, the whole water cycle, via human activity, is in the last resort nothing but the reutilization of waste waters, whether they have been purified naturally or artificially (technically).

2.6.2.1 *Multiple Use and Recirculation of Industrial Waste Waters*

Water authorities often demand that discharged wastes satisfy high standards in quality in order to protect water bodies from pollution. Reutilization of waste waters in industrial plants is especially advantageous when these conditions can only be fulfilled at high cost for the company. Sometimes it is impossible or too costly to obtain good process water for the factory, whether for natural reasons or due to water resources policy. In this case, reutilization may also be appropriate, or even necessary. Especially in areas where water is in short supply and in arid lands, wastes are frequently utilized for this reason. Many plants also find it economical to reutilize their own wastes because they can recover valuable substances from them.

Industrial wastes can be reutilized in the following ways:
- recirculation of wastes within the plant;
- utilization in an industrial plant of municipal and other wastes from outside.

2.6.2.1.1 Reutilization of Waste Waters within the Plant

The reutilization of wastes within a factory is an effective method of reducing the amount of waste water discharged from the factory into the public sewerage system or a water body. This method is used in many factories. Multiple use of process water is especially economical in cases where water is required for transport of raw materials and for cooling and washing processes, providing treatment of the wastes to make them reusable is not too costly. It is also an advantage to use the same water several times when the obtaining water from surface water bodies is only possible at high cost. This is especially the case where the used process water is not too heavily contaminated or where it can be satisfactorily purified by the usual methods. Cooling waters are especially suitable, since they are hardly contaminated and have only been heated during the cooling process. When the temperature of the water is too high, it can easily be reduced by allowing the water to cool down. However, slime and other biological overgrowths can form in cooling waters, and where this is the case, measures are necessary to prevent it.

In other cases, e.g. in foundries, hot "cooling" water is used for cooling processes where higher temperatures are needed.

In many factories, cooling water that has been used once or several times can finally be used as process water. Some examples demonstrate this (MÜLLER in /107/):

In collieries, pit waters are normally used for washing the coal. Cheaply obtained river water, or water from the public supply, is only used when the pit water is too salty. The washing waters are recirculated for as long as possible. To prevent the salt from becoming too concentrated, part of the water is constantly replaced by freshwater with a lower salt content. The amount of solids, i.e. fine coal particles, in the water should not be allowed to exceed 60 mg/l, since from this level onward the water becomes less effective for washing and rinsing. It is therefore necessary to have intermediate clarification installations.

In coking plants, water is used to quench the red hot coke. On average, about 5 m³ quenching water per 10 t coke are needed for this. Coke quenching waters mainly contain large quantities of undissolved substances, chiefly coke fines.

In contrast, the dissolved contaminants, such as phenols, cyanide compounds, and hydrogen sulfide are only present in small amounts. The dissolved substances can therefore be separated from the quenching water in simple sedimentation tanks, and it can be recirculated in the factory.

In iron and steel works, water is used for cooling the blast furnaces, for washing the blast furnace gases, and in the granulation of slag. Blast furnace cooling waters are only slightly contaminated and can be recirculated. Wastes from the granulation plant, however, contain a large amount of solids, such as slag sand and dust. Before reutilization, the water has to be clarified in sufficiently large clarification basins. Even the particularly noxious blast furnace gas washing wastes, which contain cyanides, sulfides, and phenol, can successfully be recirculated. In order to recover the washing waters for recirculation, the solids are separated in sedimentation tanks. The settling of fine dust particles is in many cases speeded up by adding chemical precipitation agents to the water. Part of the water in circulation is regularly replaced by freshwater to prevent the concentration of dissolved substances from becoming too great.

In rolling mills, large quantities of water are used for cooling the rolls. The wastes are heavily contaminated with scale and also frequently with oil. Before being recirculated, they have to be treated in clarification basins to separate the oil and the solids.

Wastes from plating plants consist mainly of spent bath liquors and large quantities of rinsing water. The latter can be recircualted. It is purified by ion exchange processes within the circuit and in this way the freshwater demand is reduced to 0.1 % of the amount of rinsing waters in circulation.

In <u>mechanical pulp mills</u>, the wastes from <u>white pulp grinding</u> mainly contain undissolved wood pulp residues and few organic substances. In contrast, in the production of <u>brown</u>, or steamed, <u>pulp</u>, considerable amounts of organic leaching substances pass into the waste water.

In white pulp mills, therefore, the water can be reutilized after separation of the undissolved substances, and the recirculation system can be so tightly closed that the amount of waste water is very little or none at all.

In brown pulp mills, it is the dissolved substances remaining in the wood after steaming that heavily contaminate the waste water. Wastes from the grinding of brown pulp contain, among other things, resins, pentoses, humates, vanillin, methanol, acetic acid, and formic acid. When the recirculation system is arranged in the same way as in white pulp mills, the concentration of noxious substances in the water becomes so high that the production process runs into difficulties. Therefore, the water in the recirculation system should be cooled and suitable flocculation agents added, in order to separate substances disturbing to the production process and to make constant reuse of the water possible.

<u>Paper and board mills</u> need large quantities of water, mainly for washing and bleaching, according to the quality of the product. Water consumption fluctuates within a wide range in individual mills. On average, it is about 125 m^3/t product for the manufacture of packing paper, about 200 m^3/t finished product for newsprint and about 400 m^3/t finished product for fine paper. By recycling the water or waste water, significant reductions can be achieved in freshwater consumption and the amount of wastes to be discharged. For this purpose, traps are used in the in-plant circulation system to retain fibrous material. This makes it possible, on the one hand, to recover the fibrous material in as fresh a condition as possible and return it to the production process, while, on the other hand, clearing the water of fibers.

The chemical industry needs a great deal of water and produces as a result large amounts of waste water. Cooling water is increasingly being recirculated. Condensation water from indirect heating of process vessels is used as boiler feed. Process wastes are recirculated, among other reasons, particularly in order to recover valuable material from them.

In addition to cooling, water is used in the various sections of the plant for many washing processes, e.g. cleaning of the raw material and of the products, washing soluble substances out of gas streams, purification of particulate-laden gases, and washing of process vessels, especially in intermittent production. In many large-scale chemical works, the cooling water demand is 5 - 10 times as great as the water demand for all other purposes. Instead of using river water only once as cooling water, recirculation methods are increasingly being introduced, to economize water use and to keep the amount of waste water discharge as small as possible, in view of the growing need to protect water bodies from pollution.

Beet sugar refineries operate seasonally. They start at the beginning of the beet harvest (usually early October) and process the beets day and night, during the so-called campaign, within 7 to 12 weeks. Waste water is essentially discharged only during this time. The water demand is very high, giving a correspondingly high volume of waste water. Efforts are being made here, too, to use water several times and to make the recirculation system as complete as possible. Most sugar factories today reutilize their wastes by means of more or less closed recirculation systems. The demand for fresh process water can in this way be considerably reduced, e.g. to about 0.5 to 1.0 m^3/t beet, where the total water usage is 10 - 14 m^3/t beet. Practice has shown that in this case, too, a certain amount of fresh water has to be introduced constantly into the system. The corresponding amount of waste water can be discharged after suitable purification, but it is preferable to collect it in storage ponds and treat it when the campaign is over. It can then be discharged at a rate adapted to the capacity of he receiving water.

2.6.2.2 Utilization of Waste Waters for Other Purposes

There are various ways of using industrial wastes outside industry:
- for drinking and bathing purposes,
- in dual purpose plants,
- for recharging groundwater,
- in recreation areas,
- for agricultural purposes
- for the recovery of raw materials.

2.6.2.2.1 Reutilization of Waste Waters for Drinking and Bathing

As a general rule, water for drinking and bathing purposes should not be obtained directly from waste waters, independently of the degree to which the wastes have been treated by the normal processes. This is because the danger to human health is too great. Even in wastes that have been treated and sterilized to a reasonably high degree, stable organic and other residues from human excreta and from industrial discharges have been found; these could constitute a danger if the treated waste water is used. There is not enough current information available concerning the possible effects on health of long -term use of water containing such substances.

At the moment, drinking water that is temporarily obtained from treated waste water may only be used in emergencies. Such an emergency occurred in Chanute, Kansas, U.S.A., in summer 1956 during a long period of drought, when the River Neosho dried up and no other water was available. Sewage water from the town of Chanute (12000 inhabitants) was biologically purified in trickling filters, run through a shallow river basin, where it was kept for 17 days and finally prepared for the municipal drinking water supply. This direct use went on for 5 months.

In the U.S.A., South Africa and Israel, studies (WHO /166/) have been carried out that show that it is possible to obtain perfectly acceptable

Tab. 2.6.-1: Suggested methods of waste water treatment to satisfy criteria for safeguarding of health /170/

	Irrigation			Recreation purposes		Industrial reutilization	Municipal reutilization	
	Crops, not for direct consumption	Crops to be cooked; fish rearing	Crops to be consumed raw	with contact	without contact		non-drinking water	drinking water
Health criteria (key to symbols below)	A + F	B + F / D + F	D + F	B	D + G	C or D	C	E
Preliminary clarification	●●●	●●●	●●●	●●●	●●●	●●●	●●●	●●●
Biological treatment		●●●	●●●	●●●	●●●	●●●	●●●	●●●
Sand filtration or equivalent process		●	●		●●●	●	●●●	●●●
Nitrification						●		●●●
Denitrification							●	●●
Chemical clarification						●	●●	●●
Activated carbon process							●	●●
Ion exchange or equivalent process						●		●●
Disinfection		●	●●●	●●●	●●	●●●	●●●	●●●[a]

Health criteria:
A No coarse matter; parasite eggs clearly eliminated.
B As A, plus complete elimination of germs.
C As A, plus elimination of most germs and some viruses.
D Maximum of 100 coliform bacteria per 100 ml in 80 % of samples.
E No fecal coliform bacteria per 100 ml, no toxic effects on man, other criteria for drinking water.
F No chemicals that leave residues in crops or fish.
G No chemicals that cause irritation of mucous membranes or skin.

Methods of treatment marked ●●● are essential in order to satisfy the criteria for safeguarding of health.
Methods marked ●● are also important, while those marked ● are only sometimes required.
[a] free chlorine after 1 hour.

- these conditions can only be fulfilled by maintaining sufficient distances or by planting hedges as protection.
- The use of waste water for irrigation is not allowed on land which serves or neighbors on the water supply.
- Irrigation must be timed so as to ensure that the spraying of grasslands and green fodder is stopped at least 14 days before cutting or grazing.
- Beets for fodder or for sugar production, potatoes for industry, oilseeds and fiber plants may only be irrigated up to 4 weeks before harvesting; potatoes for human consumption and cereals may be irrigated until flowering. Vegetable plants may only be irrigated with pure water.
- Waste waters containing pathogenic organisms from isolated institutions such as hospitals for infectious diseases, sanatoriums, carcass disposal plants, epidemic slaughter-houses, etc., must not be used for irrigation.

Suitable for irrigation are, as a rule, domestic and municipal sewage, and other wastes contaminated mainly with organic substances, such as wastes from dairies, canning factories, starch factories, distilleries, breweries, slaughterhouses, sugar refineries, etc. Generally not suited are wastes from the chemical industry, which contain considerably fluctuating amounts of free acids and alkalis, wastes from the potash industry, which have a high salt content, wastes from metal processing plants, which contain noxious metallic salts, acids, or alkalis, wastes from coal, ore, and salt mines, etc.

Suitable waste waters can also be used satisfactorily for irrigating forestry land.

2.6.2.2.6 Recovery of Valuable Residues from Industrial Waste Waters

Many industrial waste waters contain valuable residues from the raw materials and other substances used in the production process. It is frequently in the company's interest to recover these materials; on the one hand they, can be reutilized or sold, and on the other hand, it makes purification of the wastes easier and cheaper.

Preconditions for the recovery of materials are that it can be done economically and that there is a market for them. Industry has lately shown an increasing interest in recovering materials from waste water because many raw materials have become scarce and therefore relatively expensive. Also, it is becoming more difficult to despose of special wastes in view of the ever stricter requirements for waste purification. In many cases, production methods have to be changed in order to meet these requirements economically.

Recovery is most profitable when there is a large amount of valuable material in the waste water, and very few unutilizable substances.

A variety of processes are used for recovering materials from individual wastes: sedimentation, precipitation, fat and oil traps, save-alls, vacuum and filter presses, cyclones, evaporating plants, ion exchange, filtration, etc. Sometimes recovery is part of the in-plant water circulation system, or it may be a part of the in-plant water circulation system, or it may be a prerequisite for economical reutilization of wastes, as for example in electroplating plants, paper mills and chemical industry. These are some examples:

In coal mining, the coal is normally washed after extraction and the coal sludge recovered from the washing water by sedimentation, flotation, or other processes; the coal contained in the sludge is then utilized as far as possible. Wastes from coking plants, gas works, low-temperature carbonization plants, hydrogeneration plants, and similar works, contain mainly phenols, ammonia, sulfates, cyanides, chlorides, and coal tar components. Dephenolization of these wastes and recovery of the phenols is being increasingly carried out. Phenols can be recovered with the aid of special processes, such as extraction with benzene, toluene, and similar solvents, adsorption by activated carbon or steam stripping. The Emscher River Association and the Lippe River Association operate

20 dephenolization plants in coking works, in which at present about 7500 t total phenol are recovered annually. 17 of these plants use the benzene/lye extraction process. 2 plants use the PHENOSOLVAN extraction process (Fig. 2.6.-2), in which phenol is extracted by means of an organic solvent.

The pickling plants of iron and other metal processing works produce highly concentrated waste pickling liquors and diluted rinsing waters. These wastes are treated in order to recover the acids, and iron and metal salts, such as sulfuric acid, iron sulfate, copper sulfate, etc.

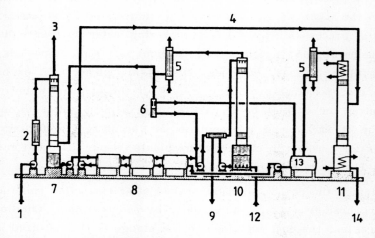

Fig. 2.6.-2: PHENOSOLVAN extraction process (LURGI - Frankfurt/M.)

1 - raw water containing phenol; 2 - cooler; 3 - waste gas; 4 - concentrated phenosolvan; 5 - condenser; 6 - separator; 7 - fumigation tower; 8 - stage extraction; 9 - dephenolized waste water; 10 - stripping column; 11 - distilling apparatus; 12 - direct steam; 13 - phenosolvan tank; 14 - raw phenol

In galvanizing plants, the wastes are treated with ion exchanges in order to recover valuable metallic salts that would otherwise be discharged with the water. Complete desalting of rinsing waters containing chromates with the aid of an ion exchanger makes complete recovery of chromic acid possible.

In artificial silk and rayon staple mills within the synthetic fiber industry, significant amounts of chemicals are used in the processing of sulfite cellulose to artificial silk by the viscose method, these amounts being for 1 t cellulose 800 kg caustic soda, 1100 kg sulfuric acid, 290 kg carbon disulfide and 49 kg zinc sulfate. These chemicals are extracted again from the final product and discharged with the waste water in a different form. In order to make production more economical, the chemicals used must be recovered as far as possible. For the manufacture of artificial silk or rayon staple by the cuproammonium process, considerable amounts of chemicals are also used: for 1 t finished product 600 kg ammonia, 258 kg copper sulfate, 100 kg soda lye and 1000 kg sulfuric acid. The profitability of this process is similarly greater if the chemicals are recovered as far as possible, especially the copper and ammonia.

Wastes from chemical pulp, paper, and textile mills contain considerable amounts of fibrous material; recovery of this material is a necessary part of the production process. It is achieved with the aid of screening fiber traps and flotation systems.

Wastes from vegetable and fruit canning factories are clarified by means of screens and sieves while still as fresh as possible, in order to recover most of the undissolved constituents. These are used as livestock feed. Solid residues from brewery wastes are used for the same purpose.

Fats and oils from many industrial wastes are recovered for utilization with the aid of oil and fat traps. For example, fats and oils from rolling mill wastes, which are recovered in separators, are treated in larger works in special plants and then reused. Oil refineries recover oil from their wastes in oil collectors, which must then be purified before reuse. The recovery of wool fat from the wastes of wool scouring plants has proved to be particularly profitable. Similarly, large amounts of fat are recovered by appropriate treatment of slaughterhouse wastes, which can then be used to manufacture industrial fats, etc.

In <u>slaughterhouses</u>, blood, hair, skin, fats, viscera, bones, and glands are recovered, which are commercially valuable. Of particular interest are the hormone preparations that can be obtained from glands and other waste products (SIERP /158/):

- from the pancreas of mammals and fishes; insulin (regulation of sugar metabolism),
- from the thyroid; thyroxin (regulation of heart activity and metabolism),
- from the parathyroid (epithelial bodies); parathyroid hormone (regulation of the lime metabolism and nerve excitability),
- from the adrenal glands of cattle and from marrow; adrenalin (raising of blood pressure) and adrenal gland extracts (carbohydrate and mineral metabolism),
- from the prehypophysis; hormones promoting lactation (prolactin), increase activity of sexual glands (prolan A and B), and regulation of the thyroid,
- from the posthypophysis; oxytocin, a hormone for inducing onset of labor, and hormones for raising blood pressure and for regulating concentration activity of the kidneys,
- testes of bulls, ovaries of cows and pigs (sex hormones).

3 WASTE WATER DISPOSAL IN RURAL AREAS AND SMALL VILLAGES

The disposal of waste water in small villages and rural areas does not fall within the scope of this technological study. At the World Water Conference in Mar del Plata, it was decided that during the Water Decade the supply and disposal of water in rural areas should be considered of primary importance. Despite current decisions concerning the aims and objectives of the Water Decade, it must be realized that the biggest problems of water supply and disposal are in the peripheral areas (slums) of large cities.

While the problems of rural water supply and disposal can be solved by low cost sanitation, soft technology, or self-help projects, the problems of urban peripheral areas have to be solved with appropriate technological methods.

There are significant differences between rural areas, on the one hand, and urban peripheral areas and cities, on the other hand, as regards water supply and sewerage systems. In view of this fact, the question of water supply and disposal in rural areas and small villages is not included in this study.

Information on this subject can be found in the many publications of the World Health Organization (WHO) and the World Bank mentioned in the list of references, as well as in publications of other international organizations.

4 COLLECTION AND DRAINAGE OF WASTE WATER

In areas with more than a given population density, the various kinds of water and waste water which are found wherever there is human habitation have to be collected and removed. As well as being a sanitary and hygienic necessity, this is done to achieve a reasonable quality of life in these areas.

Problems occur when the erection of buildings leads to large areas of land being paved, making it necessary to drain rainwater. Over a period of years, this can cause a lowering of the groundwater table. Often this effect is exacerbated by the fact that rainwater, which could previously run off slowly and uncontrolled, after building development often leaves the drainage area very quickly in sewers or at least in leveled and possibly lined open watercourses. Vegetation, of which the great importance for human wellbeing is increasingly being recognized, inevitably suffers as a result. The delayed effects of almost completely paving over the ground's surface can be seen today in nearly all large cities. Changes in the flow pattern of the groundwater can occur when drainage systems are installed, either because the groundwater tends to flow along the lines of the system or because it seeps into the system.

A very important point is that the collected water must normally not only be drained somewhere, but must also be purified.

The collection, drainage, and purification of waste waters represent a cost factor that is not matched by short-term benefits. It is understandable that the "uneconomic" drainage and purification of wastes in developing countries can only be pursued at a reasonable, i.e. low specific, cost. However, sewage disposal is a precondition for fulfilling

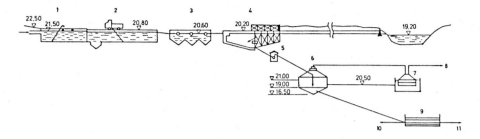

Fig. 4.1.-2: Longitudinal section through a sewage treatment plant

A drainage scheme can be developed in different ways, depending on various circumstances. Fig. 4.1.-3 shows various layout schemes, using a combined sewerage system as an example.

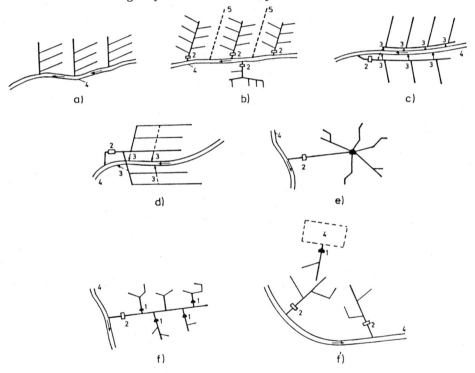

Fig. 4.1.-3: Layout schemes

a - transverse system; b - partial system; c - intercepting system; d - parallel system; e - arterial system f and f' - district systems (especially suitable for very large areas); 1 - pumping station; 2 - treatment plant; 3 - storm-overflow; 4 - receiving water; 5 - limit of drainage areas

These schemes are very important for general designing of waste water disposal systems. They are the basis for planning the individual structures.

The layout of the constituent structures of a sewerage system depends on the location of the receiving water, the volume of sewage, special characteristics of the drainage areas, the relief and type of soil, the position of the sewage treatment plant, etc.

4.2 Types of Drainage Systems and Drainage Methods

4.2.1 Drainage or Waste Water Disposal Systems

A waste water disposal system comprises the whole combination of structures needed for a particular method of collecting and removing waste waters. These structures make up the system of sewers, the treatment plant, and the outfall installations.

For one town there may be a single drainage system or several independent systems. The construction, operation, and cost of the total drainage arrangements depend to a large extent on which solution is chosen.

Fig. 4.2.-1 shows the drainage arrangements for one town using a single system (Fig. 4.2.-1 a) and two independent systems (Fig. 4.2.-1 b).

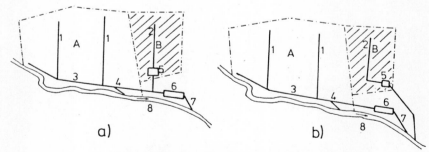

Fig. 4.2.-1: Drainage systems

a - single drainage system; b - two drainage systems; 1 - collectors; 2 - yard collectors; 3 - main collector; 4 - storm-overflows; 5 - pretreatment plant for industrial wastes; 6 - sewage treatment plant; 7 - outfall structures; 8 - receiving water; 9 - treatment plant for industrial wastes; A - town; B - industrial area

4.2.2 Drainage Methods

4.2.2.1 Combined and Separate Sewage Systems

An area can be drained by a combined or a separate sewerage system. In some cases, e.g. when it is more economical, both methods may be used within the same drainage (sewerage) system. Part of the district may be drained by a combined system and the other part by a separate system. This is often the case where an older sewerage system already exists, or when there are several receiving waters.

A combined sewerage system collects and removes all waste waters in one line (domestic sewage, industrial wastes, storm water, seepage and foreign water).

A separate sewerage system collects and removes contaminated (waste) water and storm water in separate lines.

Both methods are successfully used. They both have various advantages and disadvantages which can be decisive in planning a drainage system (Table 4.2.-1).

Two lines are sufficient for a municipal sewerage system: one for foul water, seepage water, and foreign water, and one for storm water. In industrial areas, the number of lines is determined by the number of different types of waste water. Fig. 4.2.-2 shows for example a waste water disposal scheme with 7 different lines in a chemical works /26/.

A combined sewerage system normally costs less to construct than a separate system, since only one line has to be laid in the streets, and it also takes up less space. This is a significant advantage where the streets are narrow. Another advantage is that the pipes need less flushing, since they are cleaned by the storm water. On the other hand, larger treatment plants may be required, and heavy rainfall could lead to basement flooding if no countermeasures are taken.

A separate sewerage system eliminates the disadvantages of the combin-

Tab. 4.2.-1: Comparison of separate and combined sewerage systems (HÖRLER /69/)

Criterion	Separate system	Combined system
a) Economic considerations		
Number of lines	2 lines in every street	Single line
Operation, flushing, cleaning	More expensive	More economical
Cost of sewers	More economical in low lying areas, if storm water can be removed in sewers laid at a shallower depth. The storm sewer can have a gradient.	Costs often substantially lower
Many small water bodies in the drainage	Advantageous, since storm water can be discharged into these water bodies through shorter and shallower sewers.	The maximum water level of these water bodies is often too high for the storm overflows, so that storm water has to be conveyed over long distances.
Treatment plants	Industrial cooling waters can be connected to the storm sewer; this relieves the treatment plant.	These allow a certain degree of overloading of the water body in wet weather, so that capital costs are not much higher than in a separate sysystem, but only if rainfall is low.
Service pipes	More expensive; 2 lines have to be laid: one for storm water and one for foul water	Simpler and cheaper
b) Technical considerations		
Backflow during heavy rain	Backflow during heavy rain does not cause basement flodding. Exception: backflow through house sewer system	Basement flooding possible

cont. Tab. 4.2.-1

Criterion	Separate system	Combined system
Backflow from receiving water flood	As above	Possible, depending on height of basement and flood water level of receiving water
Dry weather flow	Small water depth and low tractive force	Good conditions for cleaning the sewers
Presence of aggressive waste waters	Sewers must be made of resistant material	The sewers are very wide and this provides limited protection against aggressivity
Pumping stations	The pumps operate constantly	In addition to pumps for the foul water, there also has to be pumps for storm water; these only operate for a few hours every year, which makes the stations expensive.
Accommodation of sewers in streets	It is frequently difficult to accommodate two sewers	Less problematic
c) Hygienic considerations		
Storm overflows	None for foul water	Heavy pollution of receiving water (but can be reduced by provision of rainwater clarification basins)
Light rainfall of intensity below about 15 l/(s ha) (approx. 90 % of annual rainfall)	Street refuse is conveyed into the receiving water via the storm sewer	The storm-overflows do not start running and therefore do not pollute the receiving water

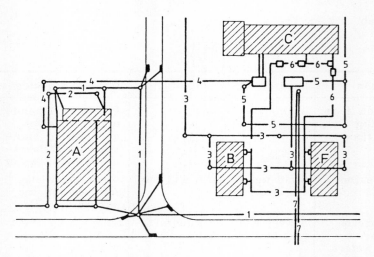

Fig. 4.2.2: Diagram of sewer systems in a chemical works /26/

1 - storm and cooling waters; 2 - domestic sewage; 3 - water-carriage of solid matter; 4 - industrial wastes; 5 - wastes to sedimentation tank, 6 - industrial wastes for neutralization and reutilization; 7 - sludge pipe line

ed system. It can also be more easily enlarged. The foul water sewers can be laid at a shallower depth than the storm sewers. It is also possible to do with less pumping facilities. However, the construction costs are higher, and it is more expensive to operate because of the two lines.

Economic considerations are important when deciding which method to use. In many cases, a combined system is more economical for large towns on level ground, and for villages where supervision is very difficult. In districts with enclosed industrial areas, the separate system may be advantageous. However, the main factors determining costs, and therefore the choice of system, are the type of terrain and the nature of the receiving waters.

4.2.2.2 Pressure Drainage and Vacuum Drainage

In recent years, new drainage methods have been developed, whereby the effluents are conveyed by pressure or by vacuum to the clarification plant and not, or not exclusively, by gravity.

Comparison of the pressure and the vacuum methods shows that the pressure method is especially appropriate for areas where houses or small communities are widely scattered, and for long sewers with a very small gradient.

The vacuum method is efficient for closed communities with more than 30 houses. An area with a radius of about 2 km can be drained from a vacuum station.

However, no general rule can be laid down for the use of pressure and vacuum drainage systems. This depends on local circumstances and needs.

4.2.2.2.1 Pressure Drainage Systems

In a pressure drainage system, the waste water flows from each house into a collecting well or directly into the receiving tank of a pneumatic pressure plant.

Fig. 4.2-3 shows a diagram of a pressure drainage system (COUSIN /41/).

According to COUSIN, pressure drainage could in certain cases be more economical than drainage by gravity, e.g.:
- in rural districts where buildings are scattered, especially if there are difficult subsoil conditions, a high groundwater table, and flat terrain,
- in areas where there are many isolated houses,
- where service pipes are deeper than the bottom of the public sewers,
- when new residential areas are created in the peripheral zones of towns, especially if there are long transport distances and different directions of gravitational flow,
- on camping and other tourist sites,
- when new industrial sites are developed.

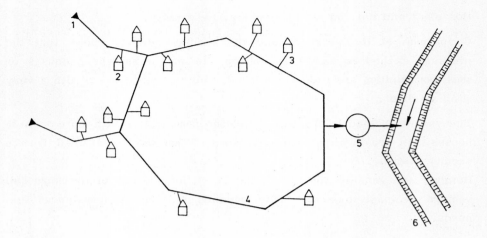

Fig. 4.2.-3: Diagram of a pressure drainage system /41/

1 - flushing station; 2 - connected house with sewage pump; 3 - service connection with pressure pipe; 4 - pressure sewer; 5 - clarification plant; 6 - receiving water

The actual sewer can be laid at a very shallow depth and with an extremely small gradient. The depth of cover only needs to be enough to ensure that the traffic load can cause no damage, and the gradient should be sufficient to ensure that the sewer is emptied, even if this only happens very slowly. The effluent is mainly conveyed by pressure created by the pumps. This method not only ensures a sufficient velocity of flow, it also makes it possible to work with higher velocities.

Depending on the number of pumps operating, a situation can arise in such systems whereby the sewer is under pressure, which does not occur in drainage by gravity. The pressure head can be above street level, making it impossible to connect street inlets. Pressure systems are a variety of the separate sewerage system.

4.2.2.2.2 Vacuum Drainage Systems

This type of system was tested by LILJENDAL in Sweden in 1960. The fecal matter to be removed is diluted with water and "sucked in", i.e. drawn forcibly towards the vacuum pump in the form of droplets. In

economic terms.

The drawings consist of: general plans, site plans, longitudinal sections, plans of structures, and special drawings.

The hydraulic and static calculations comprise all data needed for dimensioning the sewers and other structures. The calculations should be presented in such a way that they can be checked.

The computation of dimensions should be presented in as much detail as possible.

Estimated costs of construction and operation can be of decisive importance and must therefore be worked out accurately.

4.3.2 Criteria for the Design of a Waste Water Disposal System

The design of a sewerage system must be prepared in close connection with the zoning codes for the area.

Generally, the whole designated area is served by one sewerage system. It is also possible for one area to be drained by several systems (see Fig. 4.2.-1), e.g. industrial areas can have their own drainage systems. The important thing is that a proper solution should be found for the disposal of waste water in every part of the area.

In the first place, the design must show the type of drainage system chosen, the receiving water, the treatment plant, and the method of sludge disposal.

4.3.2.1 Choice of Drainage System

The type of drainage system selected will depend on several technical and economic principles. These criteria are interconnected. Every technical solution involves an economic cost and must therefore be considered in terms of an economic evaluation (money, materials, labor). The economical solutions should be examined more thoroughly.

The most important aspects to consider are:
- whether to drain sewage from the city and from neighboring industrial areas together or separately. At this stage, it could also be considered whether neighboring towns should be connected to the system.
- the method of drainage: combined or separate sewerage systems, or a combination of the two.
- various possible alignments of the main sewers.
- various possible receiving waters, of which the nearest and most economical will be chosen.
- various possible sites for the sewage treatment plant.
- outfalls into the receiving water.
- sludge disposal (dumping grounds or utilization in agriculture).

Technical and economic calculations should be made for each possible alternative, containing the following elements:
- total capital expenditure,
- possible graduation of capital expenditure,
- minimum expenditure for maintenance of the system,
- cost (capital and operating costs) of sewage collection and treatment, and sludge disposal.

In order to define the alternatives, the following data are required:
- volume of sewage,
- quality of sewage,
- parameters of sewage treatment process,
- existing water use,
- geotechnical and topographical studies,
- geological, hydrological, and meteorological studies,
- quality of receiving water.

All studies should be carried out taking full account of public regulations and standards.

The land usage in the drainage area should be considered when planning a sewerage system. As such, areas which will have been abandoned in 15 - 20 years, or which are not yet developed, must be planned for.

The sewerage system scheme should give appropriate consideration to possible expansion of the city (Fig. 4.3.-1), so that the system can be expanded without difficulty when necessary. The existence of older sewerage systems should also be taken into account. In such cases, a parallel system or an intercepting sewer system (Fig. 4.3.-2) could be the best solution. When a combined sewerage system is expanded, it is possible to change to a separate system and use the existing sewers as foul water sewers. Such solutions are often difficult to realize.

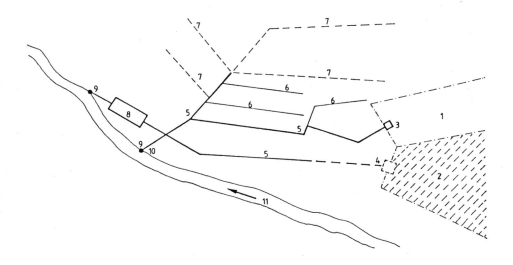

Fig. 4.3.-1 Sewerage system scheme for a developing city
1 - industrial area (first stage); 2 - industrial area (in future); 3 - effluent pretreatment plant (first stage); 4 - effluent treatment plant; 5 - main sewers; 6 - branch sewers; 7 - planned sewers; 8 - sewage treatment plant; 9 - outfall structure; 10 - storm-overflow; 11 - receiving water

A sewerage system should therefore be designed in such a way that future expansion of the city, as far as it is foreseeable, causes no problem. A famous example is the main sewer in Rome (cloaca maxima) which has been in use for 2000 years.

The cost of sewage disposal, whether expressed in national or international currency, is a good indicator and is decisive for the economy of the chosen system. These special costs of both construction and operation must be determined.

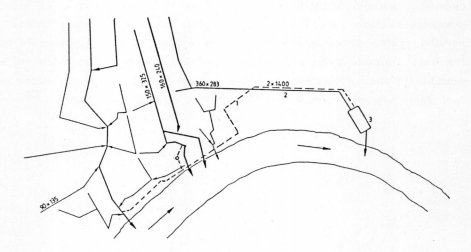

Fig. 4.3.-2: Plan for expanding the capacity of a sewerage system by means of an intercepting sewer

1 - planned intercepting sewer; 2 - existing sewer; 3 - sewage treatment plant

To summarize, the optimal design for a sewerage system will be chosen on the basis of an analysis of the above criteria. From this it will become clear whether one or several systems are more economical. The social, technical, and economic aspects of the various possibilities must be taken into account when choosing the best (not necessarily cheapest!) solution.

4.3.2.2 Choice of Drainage Method

Like the system itself, the optimal method is to be chosen on the basis of comparative studies.

Some solutions are simple, e.g. for towns, where a maximum of two pipe lines can be considered, or villages, where a combined system is usually best. In exceptional cases, where there are very limited financial means, a separate sewer for domestic sewage should be built, and rainwater allowed to run off the surface naturally.

4.3.2.6 Choice of Scheme for the Sewer Network

The first step in working out his choice is to procure all the relevant maps.

An inspection of the area is the next essential step. The planner must be familiar with the topographic conditions and should if possible test the current validity of the available maps by random checks. Experience shows that new buildings or structures do not appear on maps until after several years. To ascertain the general direction of flow, main contours and watersheds should be marked on the maps.

The profile of the alignment must take into account the relief of the area, the planned location of the treatment plant, subsoil conditions, width of the streets and traffic intensity (because of road works), the major producers of waste water, historical monuments, as well as the fact that it might be necessary to acquire buildings or land by compulsory purchase.

The main sewers are generally laid in the lowest part of the town, so that they can drain as large (and as economical) an area as possible. If possible, the water should flow by gravity alone, without the aid of pumps. However, if pumping is necessary, the main sewer should also run near the receiving water, so that the storm-overflow pipes can be kept short.

The sewer alignments should be on ground with a sufficient load-carrying capacity. This holds down the cost of supplementary structures which are needed when the ground has a low load-carrying capacity, including filled ground, old sanitary landfills, and ground where the water table has been lowered. With loess soil, attention should be paid to the distance of the sewers from existing or future buildings.

The special requirements of surface water drainage must also be taken into account. If surface water is drained in open channels, these should follow natural gradients. If they do not, use of the land may be seriously affected where there are broad slopes. It may also be necessary to bank up the slopes against landslides, which means high costs for earth moving and other essential works.

The <u>sewers</u> are normally laid in main roads. During construction, care must be taken to ensure continuous traffic flow. It may sometimes be necessary to reroute the sewers, in spite of the high costs involved, or to build a tunnel under the traffic areas (throughpress method or shield advancing).

This problem commonly occurs in developing countries, where not many wide roads are available. The arrangement of sewers in these streets should therefore be well-planned in advance.

For roads with a grassed or unpaved lane, this is where the sewer should be laid, so that construction and repairs do not interfere with traffic flow. To avoid piercing main sewers with too many connections, another sewer may be laid alongside the main, and even two if the street is more than 30 - 50 m wide. Sewers for municipal waste water should always be laid on public property. Constant access is required for maintenance and repairs or making new connections, and this is much easier if private landowners are not involved.

<u>Branch sewers</u> are laid according to the stage of housing development. Where it is normal in city planning to build houses along public streets and therefore also along the public sewer line, it is a basic principle that property owners should each have their own house connection. As a result of increasingly sophisticated ways of housing development and the need to keep costs low, these housing connections are also made according to given circumstances.

Systems have been built using combined house connections, which reduce capital expenditure by up to 20 %, using lateral sewers and the normal branch sewers (Fig. 4.3.-4 a-c).

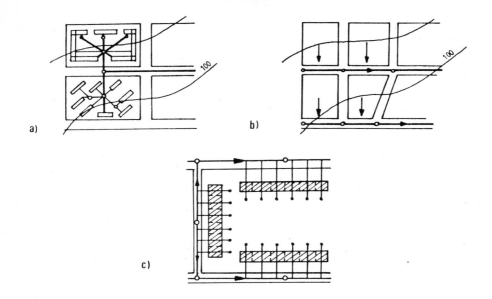

Fig. 4.3.-4: Alignment of service connections and branch sewers
a - with combined house connections; b - with lateral sewers; c - normal branch sewers

The sewers should be at a shallow depth. Table 4.3.-1 shows the relationship between depth and cost (BÖHNKE and DOETSCH /33/).

A separate system for storm water has special characteristics, since within cities this water can be removed on the surface or underground. A storm water collection system can be:
- open,
- closed,
- partly open, partly closed.

The open method (in gutters) has certain advantages: in small villages, in communities on level ground, or as an initial method in developing countries. It is safe to convey storm water openly even within a city in the initial stretches of the sewer system. Here it can also be conveyed in the street gutters. When the volume of water exceeds the capacity of

Tab. 4.3.-1: Cost data for excavation and backfilling of pipe trench /33/

DN mm	Excavation depth (m)	SC 2* DM/m	SC 3 - 5 DM/m	SC 6 DM/m	SC 7 DM/m
100 - 200	1.50 1.75 2.00	56 - 92 64 - 101 121 - 153	42 - 56 48 - 70 106 - 136	62 - 95 70 - 116 129 - 181	95 - 177 110 - 215 174 - 298
250 - 400	1.50 1.75 2.00	71 - 104 73 - 119 142 - 188	53 - 71 60 - 90 121 - 155	79 - 124 90 - 151 152 - 221	124 - 185 143 - 223 212 - 304
500 - 600	1.50 1.75 2.00	87 - 116 100 - 150 162 - 220	64 - 85 73 - 114 137 - 178	96 - 151 110 - 191 174 - 264	152 - 228 166 - 227 250 - 369

* SC 2 = Soil Class 2

the gutters, it is introduced through special inlets into the closed sewer system.

Large cities usually have a separate sewerage system, with the storm water being conveyed in closed sewers.

The same rules generally apply for storm water as for waste water. The most economical solution is often to have several outlets to the receiving water, since this makes it possible to reduce the cross section of the sewers after the water runoff. However, if the storm-overflow lines are long, this may be more expensive than having a large cross section in the main line.

It can be advantageous, for example, to build the sewer in a first stage with its final dimensions but with less storm-overflows, and then to add more storm-overflows as they are needed. A similar procedure can be followed for rain catchment basins. In the first stage, they can be dispensed with, and as the connected drainage area increases, retention tanks can be constructed to help distribute discharge over a longer time period. Storm water from areas outside the community should be conveyed directly to the receiving water in by-pass lines.

4.3.2.7 Allocation and Delimitation of Drainage Area for Each Sewer

The whole area to be drained by a sewerage system is called the drainage area (A_E). On hilly ground, the drainage area normally coincides with the hydrological area limited by watersheds. On level ground, the drainage areas should be delimited in such a way that the natural inclination of the ground can be used for draining storm water.

To determine the drainage area, the angle of the blocks is bisected, and the points of intersection of the bisectors are joined (Fig. 4.3.-5). On sloping ground, the drainage areas should be delimited according to local conditions (Fig. 4.3.-6).

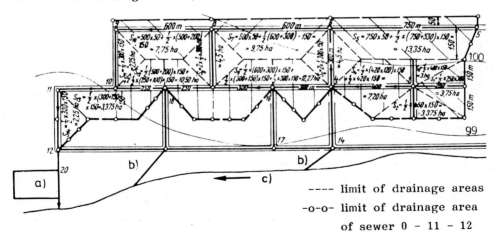

---- limit of drainage areas
-o-o- limit of drainage area of sewer 0 - 11 - 12

Fig. 4.3.-5: Allocation and delimitation of drainage areas
a - treatment plant; b - storm-overflow; c - receiving water

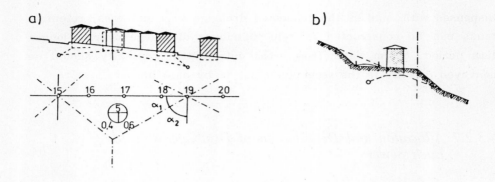

Fig. 4.3.-6: Limits of drainage area
a - on sloping ground; b - with cutting and embankment

Owing to local conditions, in a separate sewerage system on level ground, the drainage areas for domestic sewage are the same as those for storm water. On sloping ground, however, waste water and storm water have different drainage areas.

4.3.3 Calculation of Sewage Volume

The volume of sewage fluctuates, owing to human activity and natural phenomena (precipitation, groundwater infiltration). When designing a sewerage system, it is necessary to know the maximum and minimum values that occur. However, drainage systems are not designed for the maximum possible flow of storm and surface water, because this would be too expensive.

For designing a sewerage system, all types of waste water must be taken into account, but the rate of flow of each type should be determined separately. The calculations are normally done using tables.

The volume of sewage is determined by adding up different types of waste water on the basis of the following calculation principles:
- for domestic sewage, the amount is calculated on the basis of the water demand, taking into account the unfavorable conditions for

the sewerage system;
- for seepage water and other types of water that enter the system unintentionally (also called foreign water), the amount is calculated by projection or from experience (literature);
- for commercial and industrial wastes, calculation is based on data supplied by the industries and other technical sources;
- for precipitation, design storm is used;
- for surface water, the flow is determined by a hydrological study.

4.3.3.1 Calculation of Domestic Sewage Volume

The basis for sewerage system design is the specific water demand per capita. It is assumed that no losses occur through water use, treatment, and distribution. (For exact calculations, the amount of water used by the water works and for firefighting can be deducted).

The sewers and structures that make up the sewerage system are designed on the basis of the daily peak flow rate as shown in Table 4.3.-2 (ATV A 118 /4/). If no other data are available, a peak flow rate of 5 l/(s per 1000 inhabitants) should be used as a basis.

Tab. 4.3.-2: Specific peak flow rate /4/

Size of community P	Daily sewage flow w_s l/person·day	Specific peak flow $\frac{1}{x} \cdot w_s$	q_h l/(s·1000 pers.)
1	2	3	4
< 5,000	150	1/8	
5,000 - 10,000	180	1/10	
10,000 - 50,000	220	1/12	about 5.0
50,000 - 250,000	260	1/14	
> 250,000	300	1/16	

For drainage areas with uniform population density and constant specific water demand, maximum flow is calculated by the following equation:

$$Q_h = A_{E_i} \cdot q_{h_i} \quad [l/s]$$

Key to symbols:

Q_h = maximum sewage flow rate, in m³/s or l/s
A_{E_i} = drainage area with uniform population density, in ha
q_{h_i} = specific sewage flow rate, in l/s; the value q_{h_i} is calculated from the water demand:

$$q_{h_i} = \frac{Q_i}{A_i} \quad \text{with } Q_i = \text{water demand, in m}^3/\text{s or l/s in drainage area } A_i$$

Normally, areas where the sewage flow rate is higher than 10 l/s are regarded as independent and taken into account in calculations as concentrated flow.

Personnel waste waters from industries are calculated as sanitary wastes if they are discharged separately into the public sewerage system. However, it is not normally necessary to include the sanitary water demand when calculating the total water demand of a factory per production unit or per employee, because it is very small in relation to the total waste water from the factory.

4.3.3.2 Calculation of Foreign Water

According to the ATV worksheet A 118 /4/, an allowance should always be made for foreign water as a basic principle when planning the cross section of the foul water sewer.

The flow of foreign water should be taken as 100 % of the volume of sewage. Where necessary, safety margins can be allowed, or the amount can be calculated on a different basis, e.g. according to the size of the area. If no exact data are available, an average flow of groundwater, via leaks in these pipes or possibly via connected drainage systems, of about 2 to 6 l/(s·km²) can be expected.

This additional water leads to a noticeable increase in the waste water flow. Therefore, it is always worthwhile to obtain reliable data by carrying out measurements.

In a combined sewerage system, it is not so important to allow for foreign water when planning the dimensions of the sewer cross sections. However, it should be taken into account when dimensioning special structures and, consequently, the capacity of main sewers. Foreign water is a particularly important factor for these structures and for treatment plants.

4.3.3.3 Calculation Of Commercial and Industrial Waste Waters

Where there are already large commercial and industrial enterprises, the amount of waste water should be determined by means of data collection, questionnaires, and, if necessary, measurements. This also applies to theatres, stadiums, hotels, sanatoriums, barracks, camping sites, etc. Water consumption from the public supply and from private sources should similarly be taken into account, as well as making sufficient allowance for future development.

Where commercial and industrial areas are planned, it is normally impossible to establish exact data concerning the type and size of the future enterprises. Therefore, the following commercial and industrial sewage yield values (l/s·ha) are recommended as a basis for dimensioning:

Plants with low water consumption $\quad q_g = 0.5 \quad [l/(s \cdot ha)]$
Plants with medium water consumption $\quad q_g = 1.0 \quad [l/(s \cdot ha)]$
Plants with high water consumption $\quad q_g = 1.5 \quad [l/(s \cdot ha)]$

If no data are available, experience shows that the following commercial and industrial sewage flow (l/(s·ha)) can be recommended as a basis (ATV A 128 /4/):

$$q_g = 1.0 \quad [l/(s \cdot ha)]$$

In the above figures, cooling water is not taken into account.

4.3.3.4 Calculation of Dry Weather Flow

The dry weather flow Q_t is composed of
- domestic sewage flow Q_h

- commercial and industrial sewage flow Q_g
- foreign water flow Q_f

$$Q_t = Q_h + Q_g + Q_f \qquad [l/s]$$

The following formula is used to calculate the flow rate of domestic sewage Q_h (l/s) for the drainage area A_{E1} (ha) with the given peak flow q_h (l/(s·1000 persons)) and the population density D (P/ha):

$$Q_h = \frac{q_h \cdot D \cdot A_{E1}}{1000} \qquad [l/s]$$

with the definition of A_{E1} in Section 4.3.3.1

The flow rate of commercial and industrial sewage Q_g (l/s) is calculated, using data from the industrial branches, as the product of the drainage area A_{E2} (ha) and the yield value q_g (l/(s·ha)):

$$Q_g = q_g \cdot A_{E2} \qquad [l/s]$$

using the definition of A_{E2} in Section 4.3.3.3

The flow rate of foreign water is calculated as the multiple m of the amount of domestic, commercial and industrial sewage:

$$Q_f = m\,(Q_h + Q_g) = m \left(\frac{q_h \cdot D \cdot A_{E1}}{1000} + q_g \cdot A_{E2} \right) \qquad [l/s]$$

using the definition of m in Section 4.3.3.2

The volume of flow of a design section is calculated according to the following equation (Fig. 4.3.-7):

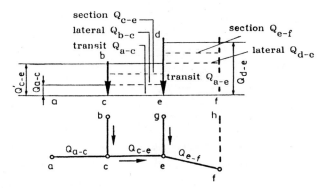

Fig. 4.3.-7: Scheme for calculation of flow volume

$$Q'_{c-e} = Q_{a-c} + Q_{b-c} + Q_{c-e} \qquad [l/s]$$

Here,

Q'_{c-e} = Volume of flow at end of sewer section c-e, in l/s or m³/s
Q_{c-e} = Volume of flow from sewage collected along this section, in l/s or m³/s
Q_{a-c} = Transit flow from sewers upstream of c, in l/s or m³/s
Q_{b-c} = Volume of flow from a lateral sewer, in l/s or m³/s

Example (from ATV A 128)

Q_h = 5.0 l/(s·1000 persons)
D = 100 persons/ha
A_E = 10.0 ha (residential area)
q_g = 0.5 l/(s·ha)
A_{E2} = 5.0 ha (commercial, industrial area)
m = 1.0
Q_h = $\frac{5.0 \cdot 100 \cdot 10.0}{1000}$ = 5.0 l/s
Q_g = 0.5 · 5.0 = 2.5 l/s
Q_f = 1.0 · (5.0 + 2.5) = 7.5 l/s
Q_t = 5.0 + 2.5 + 7.5 = 15.0 l/s

4.3.3.5 Calculation Of Storm Water

The volume of storm water is calculated by various methods. Calculation is based on long-term (at least 20 years) meteorological observation of rainfall and the analysis of these data. Such data can be used to set up national standard procedures, with tables, curves, and graphs, which make the task easier. If no such data are available, then data from neighboring countries or countries with a similar climate must be used.
Many methods of calculation have been developed and various formula worked out. Nearly all methods use the same basic relationships. They differ only in the determination of the calculation factors.
The amount of rainfall is determined as follows:

$$Q_r = A \cdot \psi_s \cdot r \cdot m \qquad [l/s]$$

where,

Q_r = rainwater flow, in l/s
A = drainage area, in ha
ψ_s = runoff coefficient
r = rainfall per second per area, in l/(s·ha)
m = runoff delay coefficient.

The <u>drainage area</u> is determined as described in Section 4.3.2. To repeat, in a combined drainage system, with only one line for all effluents, there is a single drainage area for each sewer and for all effluents. In a separate system, with at least two sewers to drain a given area, each sewer has its own drainage area, making it easier to take account of the relief (i.e. surface rainwater runoff). Drainage area are planimetered and calculated in ha.

The <u>runoff coefficient</u> ψ_s is the ratio between the runoff per second per area q_s (l/(s·ha)) and rainfall per second per area r (l/(s·ha)).

The runoff coefficient allows for the fact that not all storm water runoff reaches the sewers. Part of it evaporates, part seeps into the ground, and only the remainder enters the sewer through street inlets.

$$\psi_s = \frac{q_s}{r} \qquad \begin{array}{l}[1/(s \cdot ha)]\\[1/(s \cdot ha)]\end{array}$$

The runoff coefficient depends on (ATV A 128) /4/:
- The proportion of paved surfaces (roofs, streets, paved footpaths, entrance ways, yards, etc.),
- the incline of the ground (the steeper the incline, the greater the runoff coefficient,
- intensity and duration of rainfall (covered by rainfall frequency).

To make the calculation easier, subdividing the incline of the drainage area into four groups is recommended.

Group	Average incline J_g
1	$J_g < 1\%$
2	$1\% \leq J_g \leq 4\%$
3	$4\% < J_g \leq 10\%$
4	$J_g > 10\%$

Depending on the proportion of paved surfaces, peak runoff coefficients ψ_s are given in Tab. 4.3.-3, obtained from rainfall values of r_{15} = 100 $1/(s \cdot ha)$ and r_{15} = 130 $1/(s \cdot ha)$. Intermediate values can be linearly interpolated.

Rainfall frequency is particularly important for calculating storm water flow. Up to now, a rainfall frequency of n = 1.0 (1/a) has normally been assumed.

For planning new systems or rehabilitating old ones in developing countries, the following base values for rainfall intensity are recommended for dimensioning storm and combined system sewers, depending on the economic importance of the drainage area:

General housing development areas $n = 1.0 - 0.5$ (1/a)

City centers, important commercial and
industrial areas $n = 1.0 - 0.2$ (1/a)

Roads outside developed areas $n = 1.0$ (1/a)

Street and motorway underpasses, underground
railway systems, etc., including drainage $n = 0.2 - 0.05$ (1/a)

Tab. 4.3.-3: Peak runoff coefficients ψ_s for rainfall per second per area of about 100 to 130 l/(s·ha) with a rainfall duration of 15 min (r_{15}) depending on average incline J_g and percentage of paved surface /4/

Percentage of paved surface	Group 1 $J_g < 1\%$		Group 2 $1\% \leq J_g \leq 4\%$		Group 3 $4\% < J_g \leq 10\%$		Group 4 $J_g > 10\%$	
	for r_{15} l/(s·ha)							
%	100	130	100	130	100	130	100	130
0	0.00	0.00	0.10	0.15	0.15	0.20	0.20	0.30
10	0.009	0.09	0.18	0.23	0.23	0.28	0.28	0.37
20	0.18	0.18	0.27	0.31	0.31	0.35	0.35	0.43
30	0.28	0.28	0.35	0.39	0.39	0.42	0.42	0.50
40	0.37	0.37	0.44	0.47	0.47	0.50	0.50	0.56
50	0.46	0.46	0.52	0.55	0.55	0.58	0.58	0.63
60	0.55	0.55	0.60	0.63	0.62	0.65	0.65	0.70
70	0.64	0.64	0.68	0.71	0.70	0.72	0.72	0.76
80	0.74	0.74	0.77	0.79	0.78	0.80	0.80	0.83
90	0.83	0.83	0.86	0.87	0.86	0.88	0.88	0.89
100	0.92	0.92	0.94	0.95	0.94	0.95	0.95	0.96

The rainfall per second per area r is an important factor. It can be determined by various methods. If a rainfall graph is available (dependence on n, i, t), i is easily calculated when n and t are given. Fig. 4.3.-9 shows such a graph. It is assumed that maximum flow occurs in a sewer section when the rainfall duration T_r corresponds to the flow time t_c in that section.

$$T_r = t_c \quad \text{[min]}$$

with

$$t_c = \frac{L}{v} \quad \text{[min]}$$

where,

L = length of sewer, from beginning to calculatory point in m.
In some countries, when the calculatory section is served by two sewers (Fig. 4.3.-8), the longer sewer L_2 may be considered for calculation purposes (e.g. $L_2 > L_1$).

v = velocity of flow of storm water in the sewer, in m/s. This is selected according to the conduit material and other conditions.

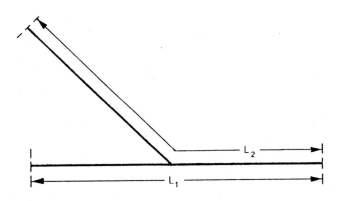

Fig. 4.3.-8: Length of sewer sections

However, T_r should have a minimum value (as shortest duration of rain) depending on incline of drainage area:
- group 1 with 50 % paved surface $\qquad T_r = 15$ min
- group 1 with 50 % paved surface, as well as group 2 and group 3, and group 4 with 50 % paved surface $\qquad T_r = 10$ min
- group 4 with 50 % paved surface $\qquad T_r = 5$ min

The <u>runoff delay coefficient</u> allows for the fact that at the beginning of rainfall Q is not flowing in the sewer, as the raindrops need some time before reaching the sewer. This effect is known as initial delay; it is set on the basis of the rainfall duration. The following values are recommended:

m = 0.8 where $t \leq 40$ min
m = 0.9 where $t > 40$ min

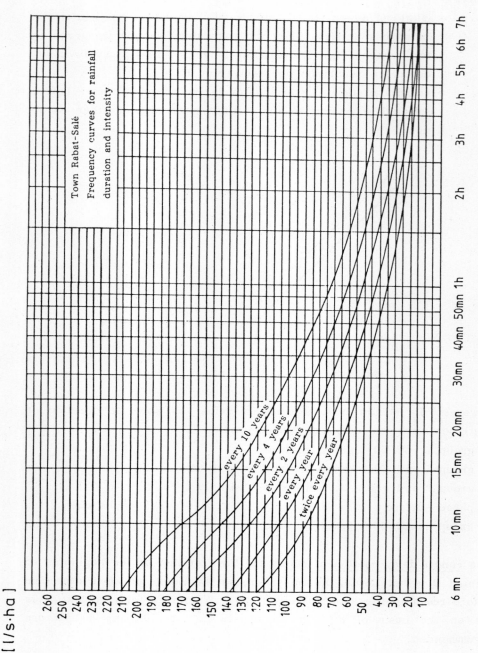

Fig. 4.3.-9: Frequency curve for rainfall duration and intensity

Surface runoff from outlying areas. Outlying areas are unpaved areas outside the developable drainage area whose natural runoff flows over the developed area. This runoff (Fig. 4.3.-10) is admitted into the sewerage if it cannot be diverted directly to the receiving water.

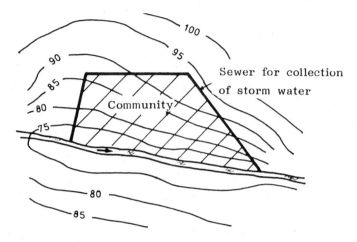

Fig. 4.3.-10: Sewer for storm water from outlying areas

On level ground, maximum storm runoff up to 1 - 5 l/(s·ha) can be expected. A reduction factor should be included to allow for unequal rainfall in the area.
In the literature, for example, the following alpha values are given:

$A(km^2)$	5	5-10	10-15	25-50
α	0.92	0.87	0.80	0.76

4.3.3.6 Calculation of Flow from Surface Waters

Surface water bodies are seldom channeled through waste water sewers. However, this can happen, e.g. in small communities, where the community sewerage system can be used to drain marsh land or for channeling small streams, etc.

The level of a water body can be measured and the flow conditions can be described by hydraulic formulas.

For example, the HEUSSER-formula /69/ can be used:

$$Q_{max} = C \cdot A^{2/3} \qquad [m^3/s]$$

where,

Q_{max} = flood water flow of a water body, in m^3/s
A = size of drainage area, in km^2
C = coefficient:
- on level ground surrounded by hills 2.8 - 4
- mixed relief (level ground crossed by hills) 4 - 6
- mountainous drainage areas with steep gradients 9 - 12
- rare flood water, originating from small basin-shaped valleys 20 - 25

4.3.3.7 Completing the Calculations

The volume of flow for a sewerage system is calculated using tables. Table 4.3.-4 shows an example of such a table for calculating the flow volume.

Tab. 4.3.-4: Calculation of a sewer system

No.	Unit	Symbol	Description	Group
1	-		Sewer number	
2	-		Street name	
3	-		Sewer section number	
4	m	-	Separately	Length
5	m	$\underline{M}$	Together	Length
6	-		No.	
7	ha		35	Area A_E — Percentage of paved surface in %
8	ha		40	
9	ha		45	
10	ha		50	
11	ha		55	
12	ha			
13	-		$J_g < 1\ \%$	Average incline — Peak runoff coefficient c_s
14	-		$1\ \% \leq J_g \leq 4\ \%$	
15	-		$4\ \% < J_g \leq 10\ \%$	
16	-		$J_g > 10\ \%$	
17	E/ha	D	Density	Inhabitants — Number
18	E		Separately	
19	E		Together	
20	-		Inflow from sewer no.	
21	l/s	Q_h	Separately	Waste water flow — Domestic
22	l/s	ΣQ_h	Together	
23	l/s	Q_g	Separately	commercial
24	l/s	ΣQ_g	Together	
25	l/s	Q_f	Foreign water flow	
26	l/s	Q_t	Dry weather flow 22 + 24 + 25	
27	-	c	Time correction value	
28	m/s	v	Intended velocity of flow	
29	s	t_f	Separately	Flow time
30	s	Σt_f	Together	
31	l/s ha	r	Rainfall	
32	ha	A_{E_r}	Drainage area 7+8+9+10+11+12	
33	l/s	Q_r	Storm water flow $Q_r = (\Sigma A_E \times \varphi_s) \times r \times \varphi$	
34	l/s	Q_{tot}	Combined sewage flow 26 + 35	
35	-	J_s	Gradient	
36	-		Shape	Cross section
37	mm	h/b	Size	
38	mm	K_b	Roughness	
39	l/s	Q_v	Capacity	Capacity filling
40	m/s	v_v	Velocity	
41	-	Q/Q_v		Dry weather inspection
42	-	h/h_v		
43	mm	h_t	Dry weather level 42 + 37(42×h)	
44	-	v/v_v		
45	m/s	v_t	Dry weather velocity	
46	m/s	v_m	Velocity	Wet weather
47	mm	h_m	Level	
48	m		Width	
49	m		Ground surface quota	Trench works
50	m		Ground depth	
51	m³		Excavated earth 49-50	
52	-		Soil class	
53	m	h_s	Sand bedding	
54	m	h_k	Gravel packing	
55				Remarks

4.3.4 Hydraulic Calculation of the Sewer Network

The different parts of the sewer system are calculated according to the same hydraulic principles as for drinking or industrial water.

4.3.4.1 General Characteristic Parameters

The physical properties of pure water are shown in Table 4.3.-5.

Tab. 4.3.-5: Physical properties of pure water

Temperature θ	Density ρ	Specific Gravity γ	Compressa- bility β	Elasticity ϵ	Kinematic viscosity ν	Dynamic viscosity η	Surface ten- sion to air σ
°C	kg/m³	kN/m³	10^{-10} Pa^{-1}	10^9 N/m²	10^{-5} m²/s	10^{-5} kg/s·m	10^{-6} N/m
0	999.9	101.9	5.02	1.99	1.794	182.40	7.71
4	1000.0	101.9	4.94	2.02	1.567	-	7.60
10	999.4	101.9	4.82	2.07	1.310	133.19	7.65
20	998.2	101.7	4.65	2.15	1.011	102.44	7.41
30	995.7	101.4	4.56	2.19	0.804	81.60	7.25
40	992.2	101.1	4.27	2.34	0.660	66.62	7.08
60	983.2	100.2	4.08	2.45	0.477	47.94	6.74
80	971.8	99.1	4.15	2.41	0.368	36.36	6.38
100	958.4	97.8	4.30	2.32	0.296	28.77	6.00

Hydrostatic presssure may be taken as an absolute and as a relative value.

Absolute pressure at a point under water is

$$p = p_o + \gamma \cdot h \qquad [kN/m^2]$$

where,

- p = absolute pressure, in kg/m²
- p_o = pressure at water surface; if the water surface is exposed, p_o is equal to atmospheric pressure (Table 4.3-6)
- γ = specific gravity of the water, in kN/m³

Tab. 4.3.-6: Air pressure in relation to height above sea-level

Height above sea-level, in m		0	100	200	300	400	500	600	700	800	900	1000	1500	2000
Air pressure	in torr*	760	751	742	733	724	716	707	698	690	682	674	655	598
	in m H_2O	10.33	10.22	10.11	10.00	9.89	9.77	9.67	9.56	9.45	9.33	9.23	8.60	8.10

Relative pressure is the difference between absolute and atmospheric pressure.

$$p_r = p - p_o = \gamma \cdot h \qquad [kN/m^2]$$

In engineering practice, it is assumed that atmospheric pressure p_o also acts inside, so that only water pressure and earth pressure need be considered for designing the structural components.

The <u>specific gravity</u> of a homogeneous body is the ratio of the weight $\underline{G}$ to the volume $\underline{V}$ of the body;

$$\gamma = \frac{G}{V} \qquad [kN/m^3]$$

The <u>density</u> of a homogeneous body is the ratio of the mass $\underline{m}$ to the volume $\underline{V}$.

$$\rho = \frac{m}{V} \qquad [kg/m^3]$$

Between the specific gravity and the density of a body the following equivalence holds:

$$\gamma = \rho \cdot g \quad \text{or} \quad \rho = \frac{\gamma}{g}$$

where g = gravitational constant, g = 9.81 m/s²

The <u>compressibility</u> of liquids is very low and can be calculated as follows:

$$\frac{\Delta V}{V} = - \beta \cdot \Delta p$$

---

*) 1 torr = 0.0133 10^5 Pa; 10^5 Pa = 750 torr

where ΔV and Δp = changes of volume and pressure, and β = compressibility coefficient.

$$\beta = \frac{1}{\epsilon}$$

where ϵ is the elasticity coefficient.

Viscosity is the property of a fluid whereby it resists forced changes of form. The viscosity of a fluid is characterized by the coefficient of dynamic viscosity.

The ratio of η (coefficient of dynamic viscosity) to ρ (density of the fluid) is denoted ν (coefficient of kinematic viscosity).

$$\nu = \frac{\eta}{\rho} \qquad [\text{cm}^2/\text{s} = \text{Stokes (St)}]$$

ν depends on the water temperature (θ) as shown by Poiseuille's equation:

$$\nu = \nu_0 (1 + 0.0337 \cdot \theta + 0.000222 \cdot \theta^2)^{-1} \qquad [\text{St}]$$

where ν_0 is the kinematic viscosity at 0° C.

Table 4.3.-7 shows kinematic viscosity values for clear water at various temperatures (ATV A 110).

Tab. 4.3.-7: Kinematic viscosity values for clear water at various temperatures /4/

T (°C)	5	10	15	20	25	30
(10^{-6} m²/s)	1.52	1.31	1.15	1.01	0.90	0.80

For practical purposes, it suffices to use a kinematic viscosity value for waste water of $\nu = 1.31 \cdot 10^{-6}$ cm²/s.

The solubility of various gases in water depends on temperature and pressure. The degree of solubility in water at a pressure of $1.013 \cdot 10^5$ Pa can be seen in Table 4.3.-8

Tab. 4.3.-8: Solubility of gases in water, in ml/l

T in °C	0	5	10	15	20	30	50	80
Air	28.64	25.21	22.37	20.11	18.26	15.39	11.40	6.00
O_2	10.19	8.91	7.87	7.04	6.35	5.24	3.85	1.97
N_2	18.45	16.30	14.50	13.07	11.91	10.15	7.55	4.03

4.3.4.2 Theoretical Foundations

The movement of water is not normally uniform and constitutes a three-dimensional problem. For practical purposes, the situation is simplified. This movement is considered as a one- or two-dimensional flow, so that changes in the flow characteristic over the cross-sectional area can be ignored.

Basic principles of one-dimensional flow are:

- the continuity equation for noncompressible fluids:

$$Q = A \cdot v_1 = A \cdot v_2 = \ldots = A \cdot v \quad [m^3/s]$$

where,

Q = flow rate, in m^3/s

A = cross sectional area, in m^2 and

v = velocity, in m/s

- the momentum principle, which to a close approximation can be written as:

$$\Sigma F_x = \frac{\lambda}{g} \cdot Q \cdot (v_2 - v_1)$$

where,

ΣF_x = the sum of x components of the forces acting on the fluid,

v_1, v_2 = the velocities at cross section 1 and 2, respectively.

- the energy equation, whereby a unit weight of the flow has an energy of

$$H_o = z + \frac{p}{\gamma} + \lambda \cdot \frac{v^2}{2g} + h_v \qquad [m]$$

where
- H_o = specific energy head, in m,
- z = elevation of point above horizontal level, in m
- λ = coefficient of nonuniform distribution of velocities:
   - for turbulent flow in smooth circular pipes:
     $\lambda = 1.03 - 1.05$
   - for turbulent flow in rough circular pipes:
     $\lambda = 1.05 - 1.10$
   - for turbulent flow in open channels:
     $\lambda = 1.10 - 1.20$
- h_v = friction losses, in m.

The flow in channels and pipes, whether with free surface or under pressure, can be <u>laminar</u> or <u>turbulent</u>. The distribution of velocity for laminar and turbulent flow is shown in Fig. 4.3.-11.

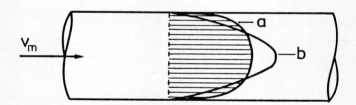

Fig. 4.3.-11: Distribution of velocity in a pipe
a) with turbulent flow, b) with laminar flow

A criterion for the liquid flowing through a pipe is the Reynolds number:

$$Re = \frac{v \cdot D}{\nu}$$

If the REYNOLDS number Re is $\leq$ 2320, the flow is laminar. If Re is > 2320, the flow is turbulent.

In sewers, the flow is normally turbulent.

Three different ranges of turbulent flow are distinguished, in which the friction number may be determined only by the REYNOLDS number, only by the relative roughness, or by both together. Depending on which is the stronger influence, the three ranges of turbulent flow are known as hydraulically smooth, rough, or transitional.

The REYNOLDS number equation can also be used for noncircular pipes with sufficient accuracy. In this case, however, the hydraulic radius is used, and not the diameter.

$$R = \frac{A}{C} \qquad [m]$$

where
- R = hydraulic radius, in m
- A = cross section of flow, in m^2
- C = wetted circumference, in m

As long as the cross section is not much different from a circular shape, the above formulas can also be used for egg, horseshoe, and U-shaped cross sections. In this case, D should be taken as $4 \cdot R$ [m], so that the REYNOLDS number becomes

$$Re = \frac{4 \cdot v \cdot R}{\nu} \qquad [m]$$

4.3.4.3 Calculation of Pipelines

Calculating the pipelines involves
- calculation of frictional losses in the pipes or structures and
- determining the shape and size of the sewer or channel cross section.

4.3.4.3.1 Calculation of Pipes and Channels

4.3.4.3.1.1 Determining Frictional Losses

<u>Frictional losses</u> in the pipes are calculated as friction head (h, in m) or loss of head (in m). Frictional losses may be distributed linearly and irregularly. Total frictional losses are calculated as the sum of the losses

$$h_v = \Sigma h_r + \Sigma h_o \qquad [m]$$

where,

h_v = total frictional losses, in m
Σh_r = the sum of linear frictional losses, in m
Σh_o = the sum of irregular friction losses, in m

<u>Linear</u> friction losses are calculated with the following equation:

$$h_r = J_r \cdot L \qquad [m]$$

where,

h_r = friction head, in m
J_r = dimensionless gradient (gradient of friction)
L = length of pipe line, in m

The general formula for friction head in straight, full circular pipes is, according to DARCY:

$$h_r = \lambda \cdot \frac{L \cdot v^2}{D \cdot 2g} \quad [m]$$

or alternatively

$$J_r = \frac{h_r}{L} = \lambda \cdot \frac{1 \cdot v^2}{D \cdot 2g}$$

λ = friction number [-]

Since sewers are generally laid with only a small gradient, this calculation can usually be simplified by replacing the actual sewer length with the projected length. This simplification is normally valid up to a bed gradient of 1 : 10.

Fig. 4.3.-12 shows the MOODY Diagram, i.e. the friction factor for straight, full circulation pipes. In the hydraulically smooth range, the friction number λ is only a function of the REYNOLDS number and independent of the internal roughness of the pipe (formula by PRANDTL and KARMAN)

$$\frac{1}{\sqrt{\lambda}} = 2 \cdot \lg \frac{Re\sqrt{\lambda}}{2.51} \quad [-]$$

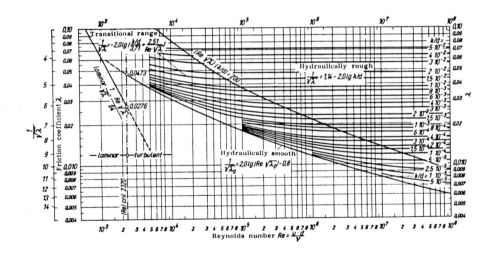

Fig. 4.3.-12: Friction number for straight, full circular pipes (MOODY Diagram)

In the hydraulically smooth conditions, the friction number $\lambda = f(Re)$ is only a function of the REYNOLDS number.

$$\frac{1}{\sqrt{\lambda}} = 2 \cdot \lg \frac{Re\sqrt{\lambda}}{2.51} \qquad [-]$$

In the hydraulically rough range, $\lambda = f(k/D)$ the friction number is only a function of the relative roughness and is independent of the REYNOLDS number (formula by PRANDTL and v. KARMAN)

$$\frac{1}{\sqrt{\lambda}} = 2 \cdot \lg \left(\frac{3.71 \cdot D}{k} \right) \qquad [-]$$

In the transitional range between hydraulically smooth and hydraulically rough, the friction number depends on both the REYNOLDS number and the relative roughness (formula by PRANDTL and COLEBROOK).

$$\frac{1}{\sqrt{\lambda}} = -2 \cdot \lg \left(\frac{2.51}{Re\sqrt{\lambda}} + \frac{k}{3.71 \cdot D} \right) \qquad [-]$$

Flow conditions in sewers nearly always fall within the transitional range (see. Fig. 4.3.-12). Therefore, the PRANDTL-COLEBROOK formula is normally sufficient to calculate the frictional loss.

To calculate velocity, the following formula is derived from the above-mentioned formula (J_r):

$$v = \sqrt{2 \cdot g \cdot J_r \cdot D} \cdot \frac{1}{\sqrt{\lambda}} \qquad [m/s]$$

and according to the PRANDTL-COLEBROOK formula for sewers where $Re = \frac{v \cdot D}{\nu}$, the velocity formula is as follows:

$$v = \left[-2 \cdot \lg \left(\frac{2.51 \cdot \nu}{D\sqrt{2 \cdot g \cdot J_r \cdot D}} + \frac{k}{3.71 \cdot D} \right) \right] \sqrt{2 \cdot g \cdot J_r \cdot D} \qquad [m/s]$$

FULL FLOW: The sewers in combined systems, and the storm sewers in separate systems, are normally dimensioned for flowing full. For circular pipes, these full sewers are calculated according to the continuity equation,

$$Q = F \cdot v = \frac{D^2 \cdot \pi \cdot v}{4} \quad [m^3/s]$$

When the PRANDTL-COLEBROOK formula is applied, this becomes:

$$Q = \frac{\pi \cdot D^2}{4} \left[-2 \lg \left(\frac{2.51 \cdot \nu}{D\sqrt{2 \cdot g \cdot J_r \cdot D}} + \frac{k}{3.71 \cdot D} \right) \right] \sqrt{2 \cdot g \cdot J_r \cdot D} \quad [m^3/s]$$

For oval, horseshoe and U-shaped sections, a similar formula is used, where Re is expressed as R - (hydraulic radius) (D =4R), so that the above formula for Q becomes

$$Q = F \cdot \left(-2 \lg \frac{0.63 \cdot \nu}{R\sqrt{8 \cdot g \cdot J_r \cdot R}} + \frac{k}{14.84 \cdot R} \right) \cdot \sqrt{8 \cdot g \cdot J_r \cdot R} \quad [m^3/s]$$

where k is the natural roughness in mm.

In this formula, average values for the natural internal roughness k are assumed. The term "natural roughness" cannot be defined exactly. It is to a great extent determined not only by the measurable size but also by the shape and distribution of the internal roughnesses.

In sewers, however, the natural roughness, which depends mainly on the building material, has less influence on the flow capacity than the structural or operational roughnesses. The latter result from, among other things, the butt-joints, departures from the designed cross-section, and other influences connected with the production and laying of the pipes, as well as influences from lateral inflows and also deposits.

After considering these influences, which in practice have considerable importance, it is recommended to divide the conduits to be used in the drainage system into four groups according to their type and construction /4/.

According to ATV A 110 /4/, the following should be distinguished:
- Type of conduit:
a) normal sewers, i.e. sewers with manholes, lateral inflows, such as house connections, street inlets and delivery channels, and with

curved sections,
b) straight sewers, i.e. sewers with no manholes, no lateral inflows and no house connections, as for example, constrictions, or presssure pipes.
- Type of construction:
a) normal construction, i.e. sewers laid according to the standards of building construction, especially for pipe laying, and whose joints are caulked, casted or jointed together using some other certified sealing material (construction group I).
b) special construction, i.e. sewers of the above types, but where the individual conduits have a particularly small natural internal roughness. There should be evidence that the velocity when the pipe is flowing full is v > 1 m/s (calculated for a permitted overloading of the sewer once a year). In addition, a high standard of workmanship in laying and jointing of the sewers must be guaranteed by special construction methods, and meticulous care taken during construction (construction group II).

Suggested values for operational roughness k_b are given in the following Table 4.3.-9. /ATV A 110/

Tab. 4.3.-9: Operational roughness k_b (mm) for sewers /4/

Sewer type	Construction	
	Construction group I	Construction group II
Normal sewers	1.50	0.40
Straight sewer sections such as constrictions, pressure pipes	1.00	0.25

Fig. 4.3.-13 shows the standard cross sectional shapes specified in DIN 4263 for closed profiles, together with their geometrical parameters. Figs. 4.3.-14 and 4.3.-15 show the calculation diagrams specified in DIN 19540 for full circular profiles where k = 1.50 mm or k = 1.00 mm.

Figs. 4.3.-16 and 4.3.-17 show the calculation diagrams for normal egg-shaped cross sections or normal horseshoe-shaped cross sections, where k = 1.50 mm.

Figs. 4.3.-18 and 4.3.-19 show the calculation diagrams for full U-shaped cross sections with a step on one or on both sides.

Table 4.3.-10 gives the measurements and hydraulic values of normal cross section shapes.

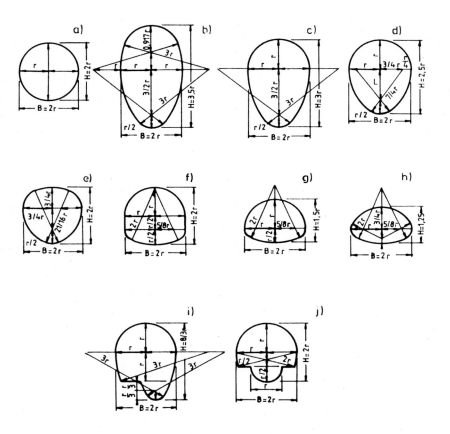

Fig. 4.3.-13: Cross sectional shapes for closed profiles and their geometrical parameters

290

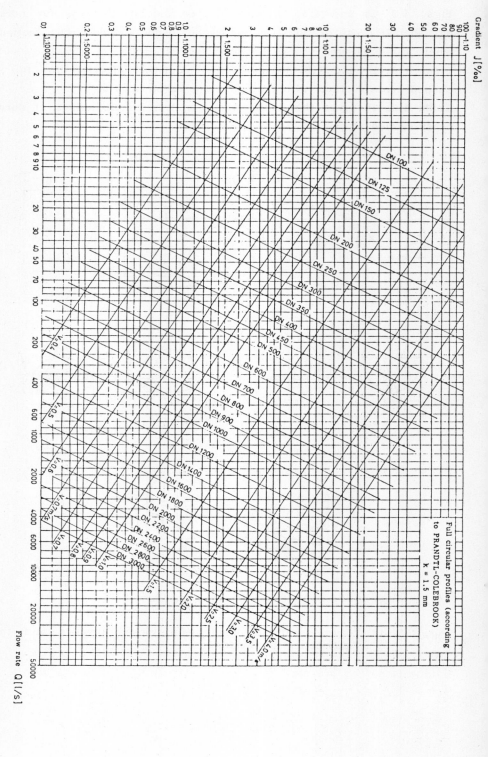

Fig. 4.3.-14: Calculation program for full circular profiles where k = 1.50 mm

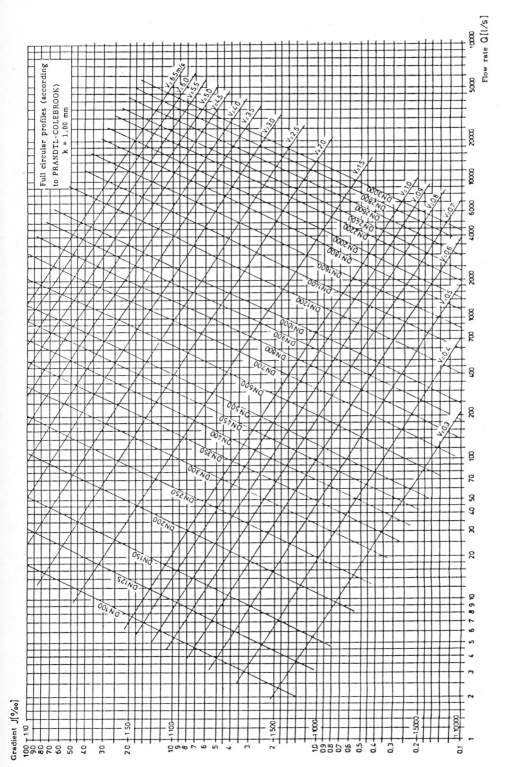

Fig. 4.3.-15: Calculation diagram for full circular profiles where k = 1.00 mm

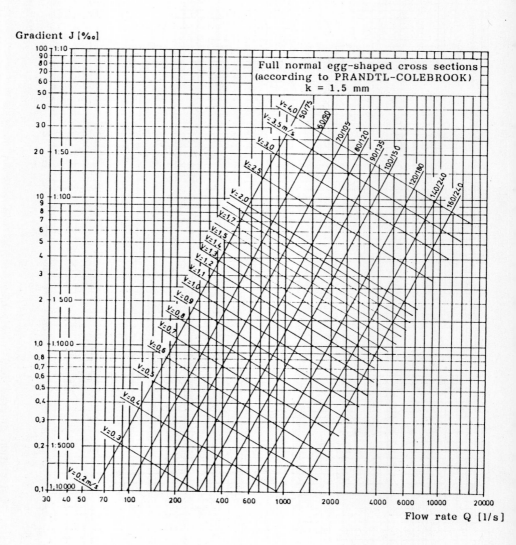

Fig. 4.3.-16: Calculation diagram for normal egg-shaped cross sections where k = 1.50 mm

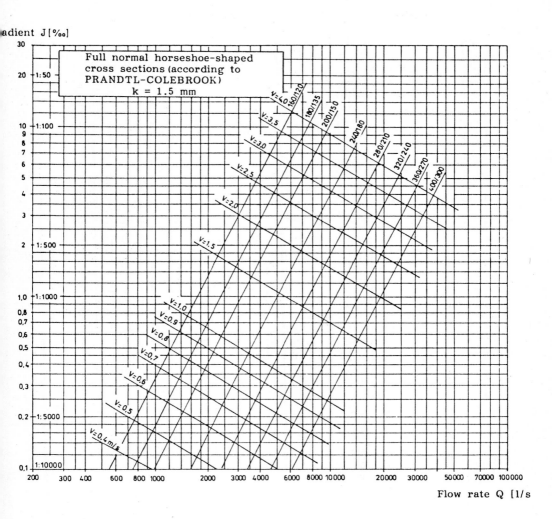

Fig. 4.3.-17: Calculation diagram for normal horseshoe-shaped cross sections where k = 1.50 mm

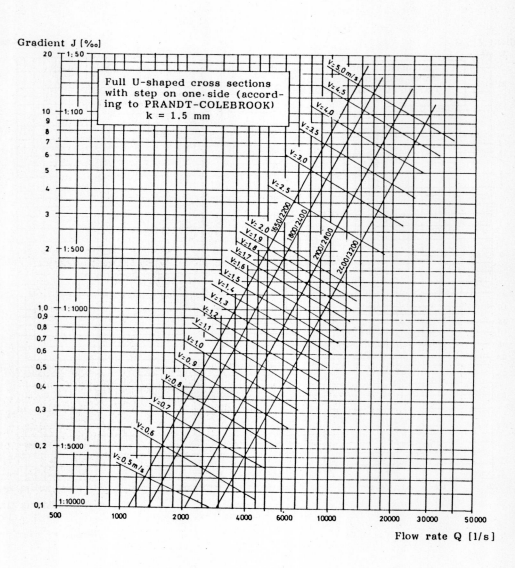

Fig. 4.3.-18: Calculation diagram for full U-shaped cross-sections with step on one side

295

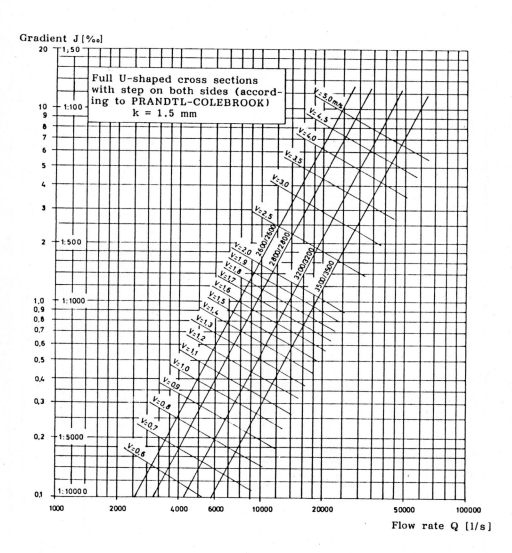

Fig. 4.3.-19: Calculation diagram for full U-shaped cross sections with step on both sides

Tab. 4.3.-10: Measurements and hydraulic data of normal cross-section shapes (k = 1.50 mm)

Circular shape	Nominal diameter = ⌀ cm	r m	F m²	U m	R m
	15 *)	0,075	0,018	0,471	0,038
	20	0,10	0,031	0,628	0,050
	25	0,125	0,049	0,785	0,063
	30	0,15	0,071	0,942	0,075
	35	0,175	0,096	1,100	0,088
	40	0,20	0,126	1,257	0,100
	45	0,225	0,159	1,414	0,113
	50	0,25	0,196	1,571	0,125
	60	0,30	0,283	1,885	0,150
	70	0,35	0,385	2,199	0,175
	80	0,40	0,503	2,513	0,200
	90	0,45	0,636	2,827	0,225
	100	0,50	0,785	3,142	0,250
	120	0,60	1,131	3,770	0,300
	140	0,70	1,539	4,398	0,350
	160	0,80	2,011	5,026	0,400
$F = 3,142 \times r^2$	180	0,90	2,545	5,655	0,450
$U = 6,283 \times r$	200	1,00	3,142	6,283	0,500
$R = 0,500 \times r$	220	1,10	3,801	6,912	0,550
	240	1,20	4,524	7,540	0,600

Egg shape $b:h = 2:3$	Nominal diameter b x h cm	r m	F m²	U m	R m
	50 × 75	0,25	0,287	1,982	0,145
	60 × 90	0,30	0,413	2,379	0,174
	70 × 105	0,35	0,563	2,775	0,203
	80 × 120	0,40	0,735	3,172	0,232
	90 × 135	0,45	0,930	3,568	0,261
	100 × 150	0,50	1,149	3,965	0,290
	120 × 180	0,60	1,654	4,758	0,348
	140 × 210	0,70	2,251	5,551	0,405
	160 × 240	0,80	2,940	6,344	0,463
$F = 4,594 \times r^2$ $U = 7,930 \times r$ $R = 0,579 \times r$					

Horseshoe shape $b:h = 2:1,5$	Nominal diameter b x h cm	r m	F m²	U m	R m
	160 × 120	0,80	1,522	4,482	0,340
	180 × 135	0,90	1,926	5,043	0,382
	200 × 150	1,00	2,378	5,603	0,424
	240 × 180	1,20	3,424	6,723	0,509
	280 × 210	1,40	4,661	7,844	0,594
	320 × 240	1,60	6,087	8,964	0,679
$F = 2,378 \times r^2$ $U = 5,603 \times r$ $R = 0,424 \times r$					

PARTIAL FLOW: In a separate sewerage system, the waste water sewers seldom flow full. In a combined system it is also uncommon for the sewers to flow full, so that they are normally only partially full, with free surface.

The calculation of flow patterns for partial flow is therefore difficult, since the depth of water in a partially full sewer will initially not be known. The values for partial flow have been related to the values for flowing full, and so-called "partial-flow" curves have been set up.

Partial-flow curves depend on the diameter, gradient, and internal roughness. The resulting influence, however, is less than that of the friction of air moving above the free surface. In practice it suffices therefore to use only one curve which takes the decisive influences into account.

Using the velocity formula

$$v = \sqrt{\frac{8g}{\lambda}} \cdot \sqrt{R \cdot J_r} \qquad [m/s]$$

as a base, the following formula is obtained:

$$\frac{v_T}{v_V} = \sqrt{\frac{\lambda_V}{\lambda_T}} \cdot \sqrt{\frac{R_T}{R_V}} \qquad [-]$$

The indices here refer to full flow (V) and partial flow (T). Drawing a partial-flow curve amounts to determining the ratio $\sqrt{\lambda_V/\lambda_T}$ in relation to the relative depth of flow.

However, the expression for this ratio of the friction numbers becomes very complicated if the PRANDTL-COLEBROOK formula is taken as a basis, which logically it should.

FRANCKE in /1/ has shown through experiments that the following is a good approximation:

$$\sqrt{\frac{\lambda_V}{\lambda_T}} = \left(\frac{R_T}{R_V}\right)^{1/8} = \left(\frac{R_T}{R_V}\right)^{0.125} \qquad [-]$$

Fig. 4.3.-20 shows partial-flow curves according to the ATV guidelines for some standard cross sections, in which the influence of air friction on the water surface is also allowed for.

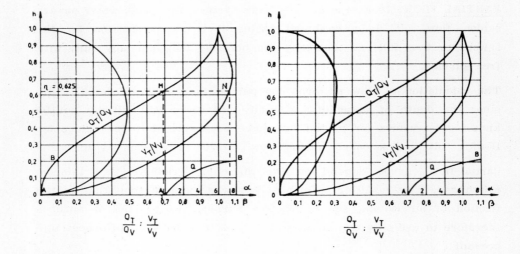

Fig. 4.3.-20: Partial-flow curves for the volume and velocity of flow in some standard pipes (PRANDTL-FRANCKE- THORMANN) for: a - circular pipes; b - egg-shaped pipes

OPEN ARTIFICIAL CHANNELS. For open artificial channels, there is still no scientifically proved flow formula according to the PRANDTL-COLEBROOK equation. Therefore empirical formulas must be used provisionally.

Very many (perhaps more than 100) different empirical formulas for flow in open artificial channels have been suggested. Of these, the velocity formula for artificial channels by GAUCKLER-MANNING-STRICKLER is given here, as it has generally proved to be suitable in waste water engineering. This formula is:

$$v = k_{st} \cdot I^{1/2} \cdot R^{2/3} \qquad [m/s]$$

where k_{st} represents the coefficient of velocity according to STRICKLER, which is shown in Table 4.3.-11.

Tab. 4.3.-11: Coefficients of velocity k_{st}, according to STRICKLER

Type of channel	Condition of walls	$k_{st}/\ [m^{1/3}/s]$
Earth channel	Firm, fine material	50
	Coarse, lumpy material	30
Masonry channel	Well jointed clinker- or brick masonry	75
	Normal brick or rubble masonry	60
	Coarse rubble masonry and paving	50
Concrete channel	Smoothed plaster or steel moulded	95
	Wood moulded	70
	Irregular and coarse surfaces	50

<u>Local or special head losses</u> may occur in mountings and pipe fittings, and through changes in the rate of flow.

In general, the PRANDTL-COLEBROOK formula includes local head losses in the sewer system from manholes (distance more than 50 m), normal bends, changes in the cross section, and main and branch sewers. Large head losses, however, for example from slide and throttle valves, overflows or entry and exit openings, must not be ignored.

The general formula for head losses is

$$h_v = \zeta \cdot \frac{v^2}{2g} \qquad [m]$$

where

h_v = head loss, in m

ζ = a head loss coefficient, which depends on the geometrical shape of the obstacle causing the loss, and on the Reynolds number of the flow. The coefficients for various obstacle shapes should be taken from specialist handbooks.

v = the velocity of flow in the conduit before the obstacle, in m/s

g = gravitational constant.

4.3.4.3.1.2 Determining the Total Discharge

The total discharge Q_{tot} is calculated as follows:
In <u>combined systems</u>, the total discharge is

$$Q_{tot} = Q_t + Q_r \qquad [l/s]$$

where
Q_{tot} = total discharge, in l/s
Q_t = dry weather flow, or the sum of domestic, commercial and industrial waste waters and foreign water, in l/s
Q_r = wet weather flow, in l/s.

In <u>separate systems</u>, dry weather flow is conveyed in the foul water sewer, and storm water in a separate storm sewer. The total discharge is:

in the foul water sewer: $Q_{tot} = Q_t$ [l/s]
in the storm water sewer: $Q_{tot} = Q_r$ [l/s]

4.3.4.3.1.3 Determining the Closed Sewer Cross Sections with Natural Water Head

Measurement is based on the energy gradient.
In a partially full pipe cross section with natural water head and steady flow with no backwater, the energy gradient is equivalent to the bed gradient, so that this can be used for the calculation in these cases.

In pipes <u>flowing full</u>, the calculation is carried out as follows:
- determination of Q_{tot}
- determination of operational roughness k_b (mm) or $k_b = 1.00$ or $k_b = 1.50$, depending on the kind of conduit (see Table 4.3.-9). For developing countries, it is recommended to choose the type of construction common in the area.
- choice of cross-section profile:
  <u>Circular</u> cross sections are used for all sewer sizes. For operational reasons, it is recommended that, independently of the estimated total discharge in the sewers, circular cross sections should generally not

be below the following sizes:
    foul water sewer   DN 250 mm
    storm and combined sewer DN 300 mm

In justified cases, the following minimum cross sections can be used:
    foul water sewer   DN 200 mm
    storm and combined sewers DN 250 mm

Egg-shaped cross sections are normally used for small and medium discharges, mainly in combined sewerage systems (the hydraulic radius and therefore also the velocity of flow increases with the water level more rapidly than with a circular profile), also where only a narrow space is available and where the water table is low; the cross sections should be bigger than 40 x 60.

Horseshoe-shaped cross sections are normally used only for large discharges, mainly as storm and combined sewers, and where the water level is high.

- Selection of the flow velocity

The basic principle for determining the rate of flow is as follows:
- It should be larger than critical velocity, so that no appreciable deposits form in the sewers.

  It is normally recommended not to go below a minimum velocity of 0.5 m/s for the established peak dry weather flow in a foul water or combined sewer, and for the largest storm discharge in the storm sewer.
- The maximum permissible velocity with regard to the construction materials should not be exceeded. However, this maximum permissible velocity should not be set too low (ATV A 118 /4/). Velocities of 6 to 8 m/s can be permitted with the corresponding choice of pipe material. For higher velocities, calculations should be made for the energy dissipation and of static and hydrodynamic load.
- As the pipe cross section increases, the velocity can decrease. However, it should decrease gradually, to avoid the formation of deposits. To achieve this, the available gradient on the individual

sections of a sewer should be distributed in such a way that in the longitudinal section a concave shape is obtained (Fig. 4.3.-21), the upper sections having a greater gradient that gradually decreases as the cross section becomes wider.

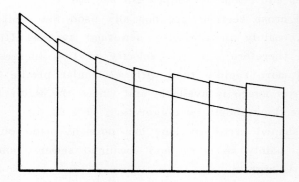

Fig. 4.3.-21: Longitudinal profile of a sewer

- Since gravity flow is normally used in sewers, the hydraulic gradient (J) is assumed to be equal to the bed gradient (J_s). According to CHEZY's law $v = k \cdot \sqrt{R \cdot J}$, the selected velocity should be as low as possible, in order to make use of a small bed gradient and therefore an economical sewer depth. However, since for a given flow, the area increases in proportion to the decrease in velocity (according to the continuity formula $Q = F \cdot v$), several calculations normally have to be made in order to obtain the most advantageous velocity, or the sewer gradient and depth, and the cross section size.

Experience shows that the following gradients can be recommended for the initial calculation:

DN, mm	250	300	400	500	600	800	1000	1200
J_s %	0.35	0.28	0.25	0.2	0.17	0.15	0.08	0.05

Experience further shows that the ratio

$$J_s = \frac{1}{D}$$

can also be used, where D is the pipe diameter (mm).
Gradients below 0.05 % should not be planned, since implementation then becomes very difficult.
After selecting the cross section shape, the cross sectional area is determined on the basis of the graphs or by calculation.

For pipes <u>flowing full</u>, the graphs in Figs. 4.3.-14 to 4.3.19 can be used. For calculating the cross section, only standard sizes, or sizes shown in the graphs, may be used. If for a given Q and a selected J or v a nonstandard size is obtained, it is recommended to choose the next largest cross section.

For <u>partially full</u> pipes, the cross section is determined by progressive approximations, as the water level is not yet known, and it is therefore impossible to determine the hydraulic radius.

For this purpose, dimensionless partial-flow curves are used, as shown in Fig. 4.3.-20 (THORMANN). Here, the effective values v_{ef} and Q_{ef} for partial flow in relation to the value for pipes flowing full are shown as a function of the relative depth of flow.

The relative depth of flow <u>a</u> means the ratio of the effective height <u>h</u> for partial flow to the height of the cross section <u>H</u>:

$$a = \frac{h}{H}$$

According to various standards, <u>all</u> sewer cross sections should be regarded as partially full. When planning new sewers, the flow capacity Q_v should not be fully utilized. If the estimated total discharge Q_{tot} is about 90 % of the flow capacity Q_v, it is recommended to choose the next largest cross section.

Calculation of the cross section of partially full pipes is carried out as follows:
- selection of cross sectional shape or partial-flow curve
- selection of permitted relative depth of flow (a < 0.9)
- selection of approximate cross section size on the basis of the

graphs in Figs. 4.3.-14 to 4.3-17. It should be noted that the flow capacity for complete filling of the selected cross section should be about 10 % greater than the necessary (effective) flow Q_{ef}.
- calculation of the ratio

$$\alpha = \frac{Q_{ef}}{Q_v} ,$$

where Q_{ef} is the effective (necessary) flow and Q_v is the cross section of flow when the pipe is flowing full.
- In Fig. 4.3.-20, for the estimated α, a vertical line up to the Q curve gives a point M, from which a <u>horizontal</u> line towards the left on the y-axis determines the relative depth of flow <u>a</u>, and towards the <u>right</u> on the v line a point N, from which a vertical line downwards gives

$$\beta = \frac{v_{ef}}{v_v}$$

- the value $v_{ef} = \beta \cdot v_v$ is calculated, where v_{ef} is the effective waste water velocity, in m/s.
- comparison of the values <u>a</u> and v_{ef} with permitted values, and, if necessary (if a > 0.9 or v_{ef} < 0.5 m/s), repeat the dimensioning with another cross sectional area.

The hydraulic characteristics Q_{ef}/Q_v and v_{ef}/v_v can also be seen in Table 4.3.-12.

Tab. 4.3.-12: Hydraulic parameters for circular, partially full, closed sewers

$\dfrac{Q_{ef}}{Q_v}$	$\dfrac{h}{D}$	$\dfrac{v_{ef}}{v_v}$	$\dfrac{Q_{ef}}{Q_v}$	$\dfrac{h}{D}$	$\dfrac{v_{ef}}{v_v}$	$\dfrac{Q_{ef}}{Q_v}$	$\dfrac{h}{D}$	$\dfrac{v_{ef}}{v_v}$
0.001	0.02	0.17	0.210	0.31	0.80	0.610	0.57	1.04
0.002	0.03	0.21	0.220	0.32	0.82	0.620	0.57	1.04
0.004	0.04	0.26	0.230	0.32	0.82	0.630	0.58	1.05
0.006	0.05	0.26	0.240	0.33	0.83	0.640	0.59	1.05
0.008	0.06	0.32	0.250	0.34	0.84	0.650	0.59	1.05
0.010	0.07	0.34	0.260	0.35	0.85	0.660	0.60	1.05
0.012	0.07	0.36	0.270	0.35	0.86	0.670	0.61	1.06
0.014	0.08	0.37	0.280	0.36	0.86	0.680	0.61	1.06
0.016	0.09	0.39	0.290	0.37	0.87	0.690	0.62	1.06
0.018	0.09	0.40	0.300	0.37	0.88	0.700	0.63	1.06
0.020	0.10	0.41	0.310	0.38	0.89	0.710	0.63	1.06
0.022	0.10	0.42	0.320	0.39	0.89	0.720	0.64	1.07
0.024	0.10	0.43	0.330	0.39	0.90	0.730	0.65	1.07
0.026	0.11	0.45	0.340	0.40	0.91	0.740	0.65	1.07
0.028	0.11	0.45	0.350	0.41	0.92	0.750	0.66	1.07
0.030	0.12	0.46	0.360	0.41	0.92	0.760	0.67	1.07
0.035	0.13	0.48	0.370	0.42	0.93	0.770	0.67	1.07
0.040	0.13	0.50	0.380	0.43	0.93	0.780	0.68	1.07
0.045	0.14	0.52	0.390	0.43	0.94	0.790	0.69	1.07
0.050	0.15	0.54	0.400	0.44	0.95	0.800	0.70	1.07
0.055	0.16	0.55	0.410	0.45	0.95	0.810	0.70	1.08
0.060	0.16	0.57	0.420	0.45	0.96	0.820	0.71	1.08
0.065	0.17	0.58	0.430	0.46	0.96	0.830	0.72	1.08
0.070	0.18	0.59	0.440	0.46	0.97	0.840	0.73	1.07
0.075	0.18	0.60	0.450	0.47	0.97	0.850	0.74	1.07
0.080	0.19	0.61	0.460	0.48	0.98	0.860	0.75	1.07
0.085	0.19	0.62	0.470	0.48	0.99	0.870	0.76	1.07
0.090	0.20	0.63	0.480	0.49	0.99	0.880	0.77	1.07
0.095	0.21	0.64	0.490	0.49	1.00	0.890	0.78	1.07
0.100	0.21	0.65	0.500	0.50	1.00	0.900	0.79	1.07
0.110	0.22	0.67	0.510	0.51	1.00	0.910	0.80	1.07
0.120	0.23	0.69	0.520	0.51	1.01	0.920	0.81	1.06
0.130	0.24	0.70	0.530	0.52	1.01	0.930	0.82	1.06
0.140	0.25	0.72	0.540	0.52	1.02	0.940	0.83	1.05
0.150	0.26	0.73	0.550	0.53	1.02	0.950	0.85	1.05
0.160	0.27	0.74	0.560	0.54	1.02	0.960	0.86	1.04
0.170	0.28	0.76	0.570	0.54	1.03	0.970	0.88	1.04
0.180	0.28	0.77	0.580	0.55	1.03	0.980	0.91	1.03
0.190	0.29	0.78	0.590	0.56	1.03	0.990	0.93	1.02
0.200	0.30	0.79	0.600	0.56	1.04	1.000	1.00	1.00

4.3.4.3.1.4 Determining the Open Sewer Cross Sections

For uniform flow in open channels, normal discharge is equal to the flow in partially full pipes. In calculating the flow pattern in open channels there are several problems. Trapezoidal channels are shown as an example.

For symmetrical cross sections (Fig. 4.3.-22 a) the following equations are used:

$$F = (b + m \cdot h) \cdot h \quad [m^2]$$

or

$$h = \frac{\sqrt{b^2 + 4 \cdot F \cdot m} - b}{2 \cdot m} \quad ; \quad b = \frac{F - m \cdot h^2}{h} \quad [m]$$

For asymmetrical cross sections (Fig. 4.3.-22 b), the hydraulic dimensions are calculated using the following equations:

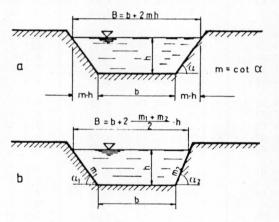

Fig. 4.3.-22: Open channel cross sections:
a - symmetrical; b - asymmetrical

$$F = b \cdot h + \frac{h^2}{2} \cdot (m_1 + m_2) = b \cdot h + h^2 m_m \quad [m^2]$$

$$U = b + h \left(\sqrt{1 + m_1^2} + \sqrt{1 + m_2^2} \right) = b + 2 M \cdot h \quad [m]$$

The relative bottom width ($\frac{b}{h}$) for an hydraulically optimal profile can be calculated with the equation

$$\beta = \frac{b}{h} = 2(\sqrt{1 + m_m^2} - m_m) = 2 (m' - m_m)$$

where $m_m = \dfrac{m_1 + m_2}{2}$; $R = \dfrac{h}{2}$

The values m' in relation to m or m_m can be assumed as follows:

m	0.10	0.25	0.33	0.50	1.00	1.50	2.00
$m' = \sqrt{1+m^2}$	1.005	1.031	1.058	1.118	1.414	1.803	2.236

The relative values $\beta = \dfrac{b}{h}$, $\dfrac{b}{\sqrt{F}}$, $\dfrac{h}{\sqrt{F}}$ in relation to m can be seen in Table 4.3.-13. For cases where b and h are not known while Q, J, m and k_{st} are known, the hydraulically optimal height is

$$h_{1/2} = \left[\dfrac{2^{2/3} n \cdot Q}{(\beta + m) \cdot J^{1/2}}\right]^{\dfrac{2}{1/3 + 5}} \quad ; \quad b = \beta \cdot h \qquad [m]$$

Tab. 4.3.-13: Hydraulic elements for the calculation of optimal trapezoidal channels

m = cotg	0.0	1.0	1.5	2.0	3.0
$\beta = \dfrac{b}{h}$	2.000	0.828	0.606	0.472	0.317
$\dfrac{b}{\sqrt{F}}$	1.4222	0.612	0.417	0.300	0.174
$\dfrac{h}{\sqrt{F}}$	0.711	0.739	0.689	0.636	0.549

4.3.4.3.1.5 Determining the Cross Sections for Pressure Pipes

In sewerage systems, pressure pipes are used as:
- pressure mains and
- dive culverts.

Dive culverts are calculated as culverts where additional bend losses must be allowed for.

There are three basic types of dive culvert:
- straight dive culvert,
- double dive culvert,
- dive culvert with shaft at upper gates.

When calculating dive culverts, the following points should be considered in each case (LAUTRICH /91/):

For a <u>straight dive culvert</u>, it is important for the hydraulic calculation that when the pipe is flowing full the approach velocity in and behind the culvert only changes slightly. For rough estimates, the total local frictional losses can be taken as 1.0.

In a <u>double dive culvert</u>, the dry weather flow (Q_{TW}) and a slight storm-water inflow pass through the straight culvert, so that here, too, about the same flow conditions exist. The inflow to the second culvert normally passes over a sill. Depending on the construction of the overflow, the approach velocity and the corresponding amount of energy is converted before the double culvert. Depending on the height of the sill and of the double culvert, the h_v loss should be set at about 0.5 for partially full and 1.0 for full pipes.

<u>Dive culvert with shaft at upper gates</u>: The inlet shaft usually has a distinctly larger cross section than the culvert or culverts. Therefore, the velocity is very low. In addition, the direction of flow is diverted downward at right angles. In most cases, the frictional loss in the shaft can for practical purposes be set at zero.

The various loss heights (Fig. 4.3.-23) add up to a total loss (h_E) given by,

$$h_E = h_v + h_e + h_k + h_a + h_r \qquad [m]$$

where

h_e = entrance loss, in m
h_k = loss in bend, in m
h_a = exit loss, in m
h_v = velocity head at outlet, in m
h_r = frictional loss of culvert according to the formula for conduit dimensioning corresponding to the length = J L, in m
L = length of culvert, in m

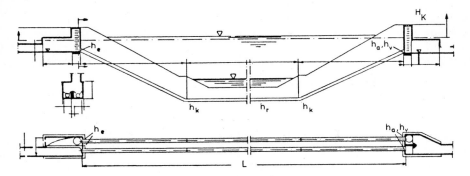

Fig. 4.3.-23: Dive culvert

By introducing coefficients of resistance (sometimes also represented by λ), the following formula is obtained:

$$h_E = (\lambda_v + \lambda_e + \lambda_k + \lambda_a) \frac{v^2}{2g} + h_r \qquad [m]$$

The velocity v relates to the flow cross section of the culvert at the outlet, in m.

In general, the sum of local frictional losses gives approximately the following values:
- culvert longer than 60 m, about 5 % of linear friction losses,
- culvert shorter than 60 m, about 10 %,

As such, the total height of loss can be roughly calculated as,

$$h_E = (1.05 \text{ to } 1.10) \cdot J \cdot L \qquad [m]$$

As a rule of thumb, h_E must always (for all design discharges) be smaller than the height difference (H_K) between maximum water level in the inlet and outlet conduits.

4.3.4.3.1.6 Calculation of Overflows

Overflows are used a great deal in sewerage systems, e.g. in combined systems for discharging storm water into the receiving water, and in sewer systems and treatment plants for discharging waste waters out of various basins, when they rise above a certain predetermined level.

For calculating overflows (Fig. 4.3.-24), the following structures must be dimensioned:
- the overflow structure,
- the overflow conduit,
- the outflow line.

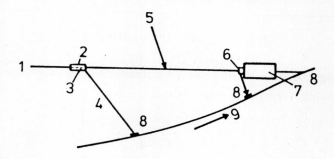

Fig. 4.3.-24: Diagram showing layout of overflows

1 - sewer C_1; 2 - overflow chamber D_1; 3 - overflow structure; 4 - overflow conduit; 5 - sewer B_1; 6 - overflow chamber D_2; 7 - treatment plant; 8 - outfall structure; 9 - receiving water

The <u>overflow structure</u> (Fig. 4.3.-25) is designed and built according to local conditions and economic realities.

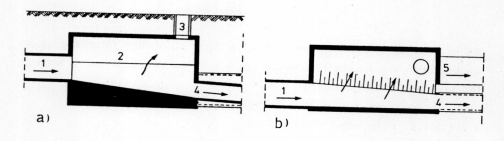

Fig. 4.3.-25: Overflow structure

a - longitudinal section; b - ground plan; 1 - inflow; 2 - weir crest; 3 - manhole; 4 - discharge to receiving water; 5 - discharge to receiving water

In municipal sewerage systems, sills are normally used as <u>side-weirs</u>. The calculation of side-weirs is carried out in the following steps:

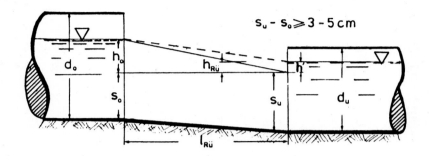

Fig. 4.3.-26: Calculation of side-weirs

- calculation of max. weir head,
- calculation of backwater level up to contact with pipe summit,
- height of weir,
- length of weir.

The weir head is calculated as the difference between the water level in the inlet pipe and the amount of energy in the discharge pipe. For the water level in the inlet pipe, the effective water level is assumed. Dimensioning is done in such a way that the backwater level does not reach or even exceed the pipe summit because of the weir.

For side-weirs, the length can be determined using POLENI's empirical formulas:

$$Q_{Rü} = l_{Rü} \cdot \frac{2}{3} \cdot \mu \cdot \sqrt{2 \cdot g} \cdot c \cdot h_{Rü}^{3/2} \qquad [m^3/s]$$

and further

$$l_{Rü} = \frac{Q_{Rü}}{\frac{2}{3} \cdot \mu \cdot \sqrt{2g} \cdot c \cdot h_{Rü}^{3/2}} \qquad [m]$$

where

$Q_{Rü}$ = overflow discharge, in m³/s. $Q_{Rü}$ can be assumed with reasonable accuracy as the differences between the inflow and the outflow in the interest of a large retention chamber (high sill) with a weir crest, at 0.6 d_o, where d_o is the height of the inflow pipe. If the inflow pipe is sufficiently flushed, it can be laid as high as the permitted backwater level allows. For dry weather flow, the velocity in the inflow pipe should be $\geq$ 0.5 m/s.

$l_{Rü}$ = hydraulic length of weir crest, in m,

μ = the coefficient of overfall. For overfall on one side only, μ is between 0.6 and 0.65, and for a weir with perpendicular approach and rounded crest it is 0.6. For two-sided side-weir, the overfall coefficient for each side should be set at $μ_1$ = 0.50.

g = gravitational constant = 9.81 m/s²

c = reduction value for incomplete overfall: according to investigations by BÜSS, c has the values given in Fig. 4.3.-27, with c = f ($h'/h_{Rü}$).

$h_{Rü}$ = the height of overflow of the weir. The height of overflow occurring with maximum inflow Q_{max} is an estimated average value, where for the sake of simplicity, it is assumed that the water surface above the weir crest is horizontal. In addition to flow conditions, it depends largely on the spatial geometry of the weir and the length of the overflow $l_{Rü}$. The value $h_{Rü}$ should be greater than 0.10 to 0.20 m, as otherwise the weir would be unnecessarily long and involve expensive structures.

$h'/h_{Rü}$	0.0	0.1	0.2	0.3	0.4	0.5	0.6	0.7	0.8	0.9	1.0
c	1.00	0.99	0.98	0.97	0.96	0.94	0.91	0.86	0.78	0.62	0.00

For <u>free surface</u> overflows, the average value for the weir head $h_{Rü}$ is given (using the labels from Fig. 4.3.-26) by means of the equation

$$h_{Rü} = 1/4 \, (h_o - h') \qquad [m]$$

For free surface outlets, the average value $h_{Rü}$ is

$$h_{Rü} = 1/4 \cdot h_o \qquad [m]$$

The formula is purely empirical and should be applicable to constricted and unconstricted side-weirs, disregarding the type of flow (streaming - shooting).

In various countries it is assumed that for small communities or where inexpensive construction is required, the value $h_{Rü}$

$$h_{Rü} = 1/2 \cdot (h_o - h') \qquad [m]$$

or for free surface outlets,

$$h_{Rü} = 1/2 \cdot h_o \qquad [m]$$

can be applied.

These values can be recommended for an initial phase in the construction of a sewerage system in developing countries. Using the c values given in the above Table, the following equation is obtained for a weir with perpendicular approach and for a single sided side-weir:

$$Q_{Rü} = 1,77 \cdot l_{Rü} \cdot c \cdot h_{Rü}^{3/2} \qquad [m^3/s]$$

Calculation is carried out in steps. As a first approximation

$$l_{Rü} = \frac{4}{1000} \cdot \frac{Q_{max}}{d_o} \qquad [m]$$

may be selected, whenever the overfall is complete. With noncircular profiles the diameter to be assumed for d_o is that of a circular profile of equal surface. $H_{Rü}$ is then calculated by the relation

$$h_{Rü} = \left(\frac{3 \cdot Q_{Rü}}{2 \cdot c \cdot 1000 \cdot l_{Rü} \cdot \mu \cdot \sqrt{2g}}\right)^{3/2} \qquad [m]$$

In cases where the overfall is complete (Fig. 4.3.-27 a) the value c = 1.0 is to be selected. An incomplete overfall (Fig. 4.3.-27 b,c) should be avoided by ensuring that the weir crest is high enough.

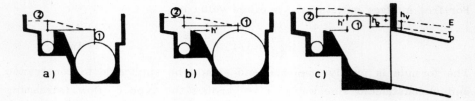

Fig. 4.3.-27: Overfall before the exit channel
1 = $h_v + h_e$, 2 = $h_{Rü}$, a) complete, b) incomplete without back-water, c) incomplete with flood backwater

Fig. 4.3.-28 shows the ATV diagram for the calculation of overflow-weirs or sill length and overflow discharge $Q_{Rü}$ with $h_{Rü} = 1/4\ (h_o - h')$.

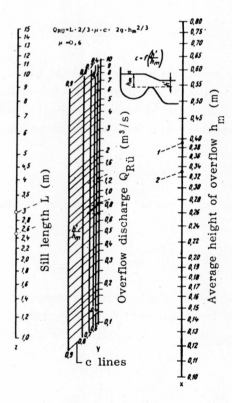

Fig. 4.3.-28: Diagram for calculation of overflow weirs with
$l_{Rü} = 1/4\ (h_o - h')$

The overall length of the overflow structure must allow for the hydraulic contraction at the top of the weir, so that the actual length should be

$$l_U = (1.05 \text{ to } 1.1) \, l_{R\ddot{u}} \qquad [m]$$

For the calculation, first the type of overflow weir should be selected (side-weir or perpendicular weir) and the discharge flow regime (gravity or constricted), for which the relevant $h_{R\ddot{u}}$ and h' are calculated.
If the diagram is used, for gravity flow, a straight line from the h_m value on the x axis and from the Q value on the y axis gives the sill length $l_{R\ddot{u}}$ (m) on the z axis.
For incomplete discharge, first the equivalent Q for gravity flow is determined. For this purpose, a horizontal line to the $h'/h_{R\ddot{u}}$ value on the y axis gives a point from which a parallel line to the c lines on the y axis gives the new Q'. A line from $h_{R\ddot{u}}$ on the x axis and the Q' value on the y axis give the sill length $l_{R\ddot{u}}$ (m) on the z axis.

The <u>length of the constriction</u> (Fig. 4.3.-29) is calculated on the basis of the discharge Q_{crit} to be conveyed through the constriction.

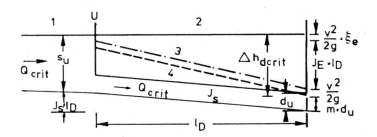

Fig. 4.3.-29: Hydraulic conditions when the overflow sets into operation

1 - weir crest; 2 - constriction; 3 - energy gradient; 4 - hydraulic gradient

The first step is to calculate

$$v = \frac{Q_{crit}}{F} \quad [m/s]$$

J_E is obtained from the nomograms for pipelines (PRANDTL-COLEBROOK) and with

$$h_E = \frac{v^2}{2g}(1 + \zeta_E) \quad [m]$$

the following can be determined:

$$\Delta h_{d\ crit} = \Delta h_E + J_E \cdot l_D \quad [m]$$

where

ζ_E = the coefficient of inflow according to Fig. 4.3.-30. If the constriction mouth is sharp-edged, ζ_E can be set at about 0.35.

At the same time, h_{crit} becomes

$$h_{crit} = S_u + J_s \cdot l_D - m \cdot d_u \quad [m]$$

and since the coefficient $\underline{m}$ for determining the pressure gradient at the end of the constriction can be taken as roughly 1.0 (ATV), l_D becomes

$$l_D = \frac{S_u - d_u - \Delta h_E}{J_E - J_s} \quad [m]$$

The coefficient of inflow ζ_E for sharp-edged constriction mouth and with a water level the same height as the weir crest can be seen in Fig. 4.3.-30.

Fig. 4.3.-31 shows various kinds of overflow.

Fig. 4.3.-32 shows a curve for the calculation of overflows with trapezoidal notches and openings /26/ which are normally used for outlets from basins.

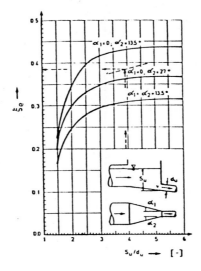

Fig. 4.3.-30: Coefficient ξ_e of inflow for sharp-edged constriction mouth and water level the same height as the weir crest

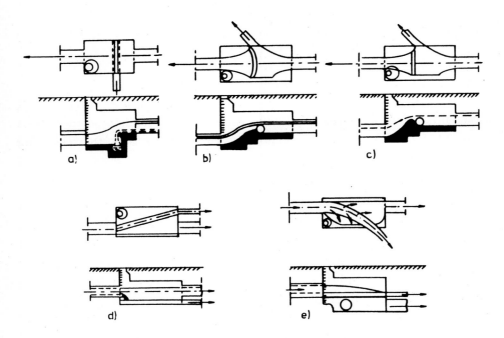

Fig. 4.3.-31: Overflows
a - perpendicular above the outlet; b - perpendicular, curved; c - perpendicular, straight; d - one-sided, straight; e - two-sided, curved

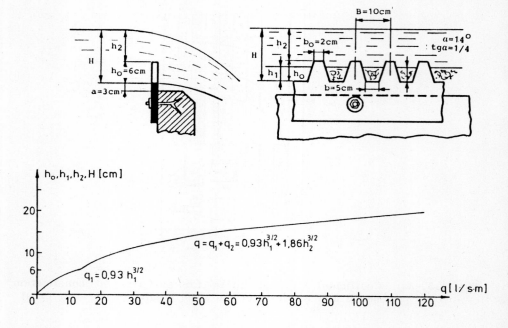

Fig. 4.3.-32: Nomogram for calculation of overflows with trapezoidal notches and openings /26/

4.3.4.3.1.7 Hydraulic Waterhammer

Waterhammer is a series of varying water pressures in pipelines. It is created by a sudden change in the flow volume and the resulting change in velocity that occurs when shutoff devices are opened and closed, or when pumps are switched on and off.

The magnitude of the increase or decrease in pressure in relation to the normal pressure depends on the time the sluice valve is closed, the geometrical parameters (length, diameter, wall thickness) of the pipe, the water density, the elastic properties of the water, the piping material, and the original velocity.

This phenomenon can have a detrimental effect on the pipes. They can be damaged if the pressure fluctuations reach a certain level. Statistics show that over 80 % of damage to pipes in pressure mains is caused by waterhammer.

In pumping stations, waterhammer may be caused by a sudden power cut off, or faulty operation by personnel.

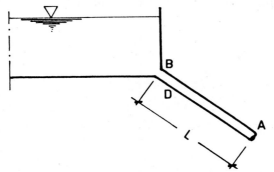

Fig. 4.3.-33: Diagram of waterhammer

The duration t_r of the pressure wave from its starting point to the point of reflection (e.g. to B - end of line) and back to the starting point is (Fig. 4.3.-33) given by

$$t_r = \frac{2 \cdot L}{a} \qquad [s]$$

Here

t_r = duration wave, in s
L = length of the line, in m
a = waterhammer wave velocity, in m/s

$$a = \frac{1425}{\sqrt{1 + \frac{D \cdot E_w}{e \cdot E}}} \qquad [m/s]$$

with

D = diameter of pipe line, in m
e = wall thickness, in m
E_w = elastic modulus of the water = 22000 kN/cm²
E = elastic modulus of the piping material, e.g. for cast iron: 1050000, steel: 2100000, asbestos content: 250000 kg/cm²

The velocity a lies between approximately 1000 and 2000 m/s. Full waterhammmer H occurs when the change of velocity, e.g. due to closing of a sluice valve, occurs in the time t_1 within the run t_r of the pressure wave $(t_1 < t_r)$. The magnitude of the pressure head is:

$$\max \Delta h = \pm \frac{a}{g}(v_1 - v_2) \qquad [m]$$

where

v_1 = flow rate at beginning of change in velocity, in m/s
v_2 = flow rate at end of change in velocity, in m/s

If a sluice valve is completely closed within the duration of the pressure wave, then

$$\Delta H = \pm \frac{a}{g} v_{max} \qquad [m]$$

where

ΔH = rise in pressure in the pipe, in m
v_{max} = velocity of the water in the pipe at the onset of waterhammer, in m/s

The pressure rise in the pipe is very great. For example, it has been observed that even where v = 1 m/s, a = 1000 m/s, and g = 9.81 m/s²,

$$\Delta H = \pm \frac{1000}{9.81} \cdot 1 = 100 \text{ mWS} = 10 \cdot 10^5 \text{ Pa (!)}$$

4.3.5 Static Calculation of Sewer Lines

Static calculation of the sewer lines is carried out in order to determine the stresses on the construction elements that are important for the choice of shape and size of the cross sections of these elements.

There is no difference between the structures in a sewerage system and other types of engineering construction, so that the static calculations are normally carried out using the same methods. However, sewers do have certain features that are of little significance for other types of engineering construction. It is important here to avoid both underdimensioning and overdimensioning, as these affect both costs and operation in an undesirable way.

As a basic principle, just as for other construction elements, finished pipes do not need to be calculated if the prescribed conditions, e.g. earth fill, foundation, width of trench, etc., are fulfilled.

However, due to local conditions, it is not always possible to fulfill these conditions, so that calculations have to be carried out for sewers and special conduits.

In the Federal Republic of Germany, DIN 4032 lists five types of concrete pipe:

    K   - circular
    KW   - circular with reinforced walls
    KF   - circular with base
    KFW - circular with base and reinforced walls
    EF   - egg-shaped with base

According to DIN 4032, the pipes must have a vertical load stability as shown in Table 4.3.-14. Many investigations and methods of calculation have been carried out for determining the stresses arising on conduits.

Tab. 4.3.-14: Vertical load stability of conduits according to DIN 4032

DN in mm Circular	Vertical load kN/m min. K and KF	Vertical load kN/m min. KW and KFW	DN in mm Egg-shaped	Vertical load kN/m min. EF
100	24	-	400 x 600	50
150	26	-	500 x 750	61
200	27	-	600 x 900	69
250	28	-	700 x 1.050	75
300	30	50	800 x 1.200	77
400	32	63	900 x 1.350	80
500	35	80	1.000 x 1.500	83
600	38	98	1.200 x 1.800	86
700	41	111		
800	43	125		
900	The loads must be determined according to static requirements	138		
1.000		152		
1.100		166		
1.200		181		
1.300		194		
1.400		207		
1.500		220		

4.3.6 Layout of Sewers, Gradients and Depths

4.3.6.1 Routing of Sewers

The routing of the sewers is determined by
- local topography,
- location of the treatment plant,
- subsurface conditions,
- width of streets and traffic intensity (because of road works),
- the main sources of waste water,
- historical monuments.

Main sewers are generally laid in a low-lying part of the town, so that an area as large as possible can be drained using the natural gradient. The main sewers should follow the shortest route to the treatment plant.

Sewers for storm drainage should run near the receiving water, in order to keep storm overflow conduits short.

Sewers should be laid in ground with a sufficient load-carrying capacity at a shallow depth, so that the cost of earthworks is kept as low as possible. Former sanitary landfills, wet ground, or ground with an insufficient load-carrying capacity should be by-passed. In loess soil, the sewers must be laid at a suitable distance from high buildings.

Sewers are normally constructed in main roads. If the road has a green strip (alley), the sewer can be laid below this area, so that construction or repairs can be carried out without interfering with traffic. In order to avoid disturbing main sewers by too many connections, extra sewers may be laid alongside them, and two if the road is over 30 - 50 m wide. These sewers can be laid within the street profile in a variety of ways (fig. 4.3.-34).

Storm water drainage should be planned in the same way as domestic sewage disposal. Whichever method provides the most storm overflows appears to be the most economical. However, this might turn out to be uneconomical if it involves very long overflow conduits and a large

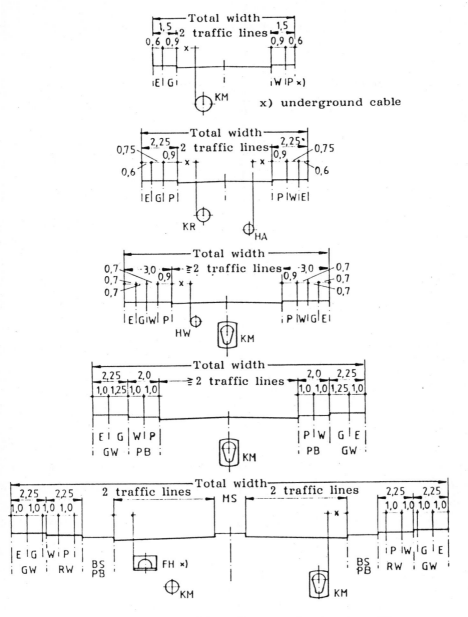

Fig. 4.3.-34: Position of sewers in street profile

E = electricity; G = gas; P = Post Office; HW = water mains; KM = combined sewer; FH = long-distance heat conduction; BS = row of trees; PB = parking bays; GW = footpath; RW = cycle path

number of storm-water tanks. It is also possible to build storm overflows which come into operation at staggered intervals.

On hilly ground, a storm sewer must be laid surrounding the town (see fig. 4.3.-10), and the gradient of this sewer must be adjusted to that of the ground (fig. 4.3.-35).

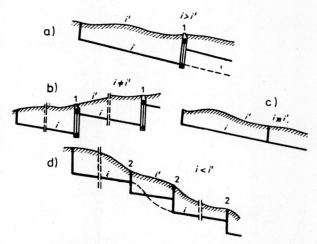

Fig. 4.3.-35: Adjusting the gradient of a sewer to the terrain
1 - pumping station; 2 - drop-manhole

a) sewer gradient bigger than natural gradient; b) sewer gradient in opposite direction; c) sewer gradient equal to natural gradient; d) steep natural gradient

4.3.6.2 Gradient

The gradient must ensure that the waste water flows properly. For this reason, it should be fairly steep; this means greater velocities (because: $v = c \cdot \sqrt{R \cdot I}$), and therefore smaller pipe diameters and cheaper pipes. On the other hand, a steep gradient means greater amounts of earth to be excavated and therefore more expensive earthworks. Between these factors, the right solution has to be found.

The depth of the sewers is determined in relation to the buildings which have to be drained, as well as existing and planned sewers in the street profile, and the nature of the subsurface. The depth should be as shallow as possible, but adequate to ensure proper drainage of buildings and obligatory points.

The depth at the beginning of the sewer is very important. This depth is determined by the buildings to be connected or by their distance A from the street (fig. 4.3.-36).

The minimum depth for branch and main sewers depends on the house connections, depth of obligatory points, freedom from frost, hydraulic conditions for ensuring proper drainage, static conditions, and the kind of foundation soil.

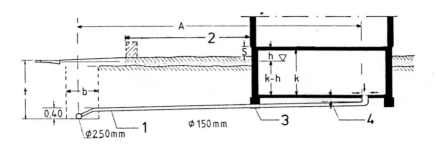

Fig. 4.3.-36: Depth of house connections

1 - house connection; 2 - depth of front garden; 3 - casing; 4 - 0.2 m for vitrified clay or cement pipes; A - distance to furthest drain

Equally important is the depth of the watertable. If the watertable is high, a separate sewerage system is more economical, since the big storm sewers can be laid at a shallow depth, and only the smaller foul water sewers have to be laid in the groundwater.

Obligatory points are normally all points that have to be drained. In waste water engineering, they are those buildings that, owing to their location, require special structures for sewage removal, or obstacles that have to be crossed. Examples of obligatory points are underpasses; sports grounds; main, administrative, and cultural buildings; deeply laid sewers; river beds; or watercourses to be diverted into the sewer system. Such points should be marked on the longitudinal profile, together with appropriate arrangements for draining them.

Sudden large changes of gradient should be avoided (fig. 4.3.-37). For this purpose, transitional gradients can be used.

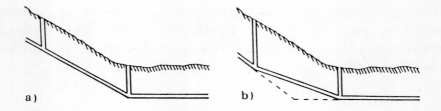

Fig. 4.3.-37: Sudden large changes of gradient

a - avoid sudden large changes of gradient; b - preferable use of transitional gradient for case a

Sections with a very small gradient and increased diameter should also be avoided, since they can cause backwater (fig. 4.3.-38).

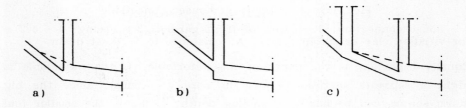

Fig. 4.3.-38: Rapid decrease in gradient with increase in diameter
a - backwater; b,c - acceptable solutions

After a rapid increase in gradient, a smaller pipe diameter is often sufficient. However, the diameter should not become less than DN = 500 mm (fig. 4.3.-39).

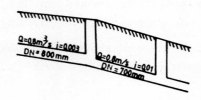

Fig. 4.3.-39: Decrease in pipe diameter

4.3.6.3 Hydraulic Conditions

The velocity should increase in the direction of flow, so that the hydraulic gradient (see fig. 4.3.-21) resembles the descending branch of a Neil's parabola. It is therefore usual to have gradients of e.g. 1 : 10, 1 : 200, and 1 : 300 in the upper reaches of the sewers, and gradients of 1 : 400 and 1 : 500 in the middle stretches.

Open street gutters for storm water should have a gradient of at least 1 : 200. Sewers should be laid with a gradient which ensures that the material being carried is transported even when the volume and velocity of flow is low. This means the velocity of flow must be at least 0.5 m/s, and preferably 0.6 m/s. The upper limit for velocity of flow is considered to be 6 to 8 m/s.

When planning a new sewer system, the flow capacity Q_v should not be fully utilized. If the computed total flow is about 90 % of the flow capacity, it is recommended that the next biggest pipe size be selected.

A change of diameter without a change of gradient can theoretically be effected in one of three ways: keeping a straight invert (fig. 4.3.-40 b), keeping a straight crown (fig. 4.3.-40 a), and according to the water level (fig. 4.3.-40 c).

The diameter change with "straight crown" is preferred, since construction is simpler and cheaper, and it is usually more neatly done than a "straight invert" change. In any case, fitting the pipes together according to the water level is practically impossible, owing to fluctuations in flow.

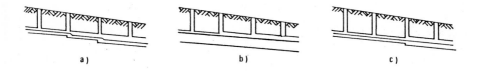

Fig. 4.3.-40: Possible position of diameter change
a - straight crown; b - straight invert; c - straight water level

When an inflow makes it necessary to change the diameter or, alternatively, to change the gradient while keeping the same diameter, it is recommended that a bed drop or an invert threshold with no backwater be constructed, as shown in fig. 4.3.-41.

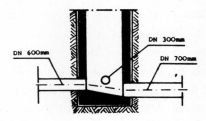

Fig. 4.3.-41: Lateral inflow with manhole and threshold, no backwater

There are two construction methods for changing the diameter: for small diameters, manholes are used (as in fig. 4.3.-42 a), and for large diameters, a direct connection is used, with a reducer and a length of 3 to 3.5 m (as in fig. 4.3.-42 b).

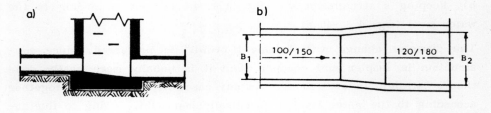

Fig. 4.3.-42: Construction methods for changing the diameter
a - with manhole, for small diameters; b - with direct connection, for larger diameters

Changes in the direction of flow are effected as manholes (fig. 4.3.-43 a), or through intermediate shafts (fig. 4.3.-43 b,c). For larger diameters (B 1000 m), a special bend structure must be provided. In small pipes, a radius of curvature of 1.5 B (B = inside width of sewer) is enough. For larger pipes, the radius of curvature should be $(3 \div 3.5) \cdot B$ or more.

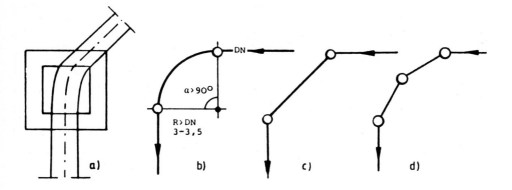

Fig. 4.3.-43: Changes in direction of flow
a - within a manhole; b - through intermediate shafts; c, d - within several shafts with changes in direction of flow

Drop structures have to be provided when the ground has a steep gradient, to avoid exceeding the velocity of flow allowed for the pipe material. Various kinds can be used, depending on the height of the drop and the flow of waste water (see fig. 4.4.-14).

Static conditions

Static conditions can influence the profile of the alignment. The depth must be selected so as to ensure that the perpendicular forces acting on the pipes, such as earth fill and traffic, are as small as possible. The resistance of the pipes normally corresponds to an earth fill of at least 0.80 m.

Longitudinal profiles of the sewers

Longitudinal profiles are made for all sewers. From these, depth, gradient and different underground conditions can be ascertained.

The longitudinal profile should be planned with several possible variants, taking into account the economy of the sewerage system with regard to capital expenditure, operation costs, effect of the sewer construction on existing buildings, etc. Economical solutions are those where the foundation depth of the sewers is shallow, and the route taken by the main sewers to the treatment plant is short.

4.3.6.4 Graphic Representation

Public standard specification should be taken into account when preparing plans and drawings. Graphic representation of the plan can be seen from the following examples:

Tab. 4.3.-15: ATV (Abwassertechnische Vereinigung) - Key to symbols in diagram of public sewers

Fig. 4.3.-44: Layout for hydraulic calculations, to DIN 2425

Fig. 4.3.-46: Detail from a diagram of a separate sewerage system, DIN 2425

Fig. 4.3.-47: Longitudinal section, to DIN 2425

Tab. 4.3.-15: ATV - Key to symbols in diagram of public sewers

Symbol	Description	Symbol	Description
	Connecting sewer (filled in triangle)		Pressure main (existing)
	Branch without connecting sewer (open triangle)		Pressure main (to be constructed)
	Storm-water inlet - street draining (filled in triangle) — Draw triangles in direction of flow		Sewer to be closed and filled in
		• /	Drill hole, numbered
	Branch for street drainage (open triangle)	• ∎	Drill hole, numbered, constructed as groundwater observation well
	Kerbstone storm-water inlet		
	Kerbstone storm-water inlet ("three" part)		NOTATION SYSTEM FOR HEIGHTS
	Storm-water sewer manhole (brick-built)	D +	Height of cover (for foul, combined, and storm sewer manholes)
	Storm-water sewer manhole (precast concrete parts)	S +	Bed height (for foul-water sewers)
	Foul-water sewer manhole (brick-built)	M +	Bed height (for combined sewers)
		Iw +	Bed height (for storm sewers)
	Foul-water sewer manhole (precast concrete parts)	Str +	Bed height (for pressure mains)
		G +	Height of street
	Storm-water sewer manhole (with drop)		Height of ground
			NOTATION SYSTEM FOR SURVEY STATIONS AND FIGURES
	Foul-water sewer manhole (with drop chamber)	⊠	S.S (survey station, disk)
	Foul-water sewer manhole (with chamber and drop chamber)	○	I.P. (iron pipe)
		●	D. and H. (drain and hollow pole)
	Foul water sewage manhole (with two chambers)	⊡	Boundary stone
	Foul-water sewage manhole with side entry (measure distance from sewer center to center of manhole cover)		Right angle (90°)
			Right angle with diagonal
		- - -	Prolongation (of a building, survey line, etc.)
	Outlet (brick-built)	———	Survey line
	Pipe junctions	• •	Starting point of survey line
	Storm sewer (to be constructed)	10.00	Tie line measurement, underline once (for lines and prolongations)
	Storm sewer (existing)		
	Foul or combined sewer (to be constructed)	29.98	End measurement, underline twice (end point of survey line)
	Foul or combined sewer (existing)	- 5.00 -	Offset measurements (measurements between two points)

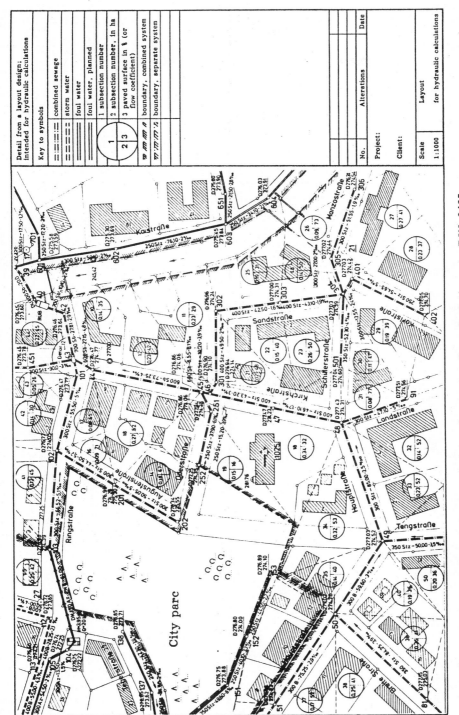

Fig. 4.3.-44: Layout for hydraulic calculations to DIN 2425

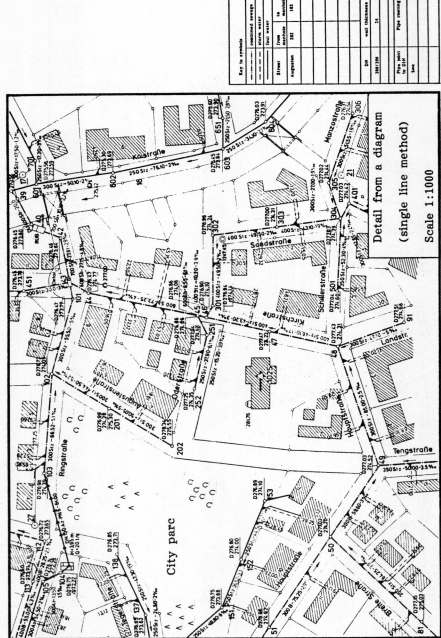

Fig. 4.3.-45: Detail from a diagram (continuous line method), to DIN 2425

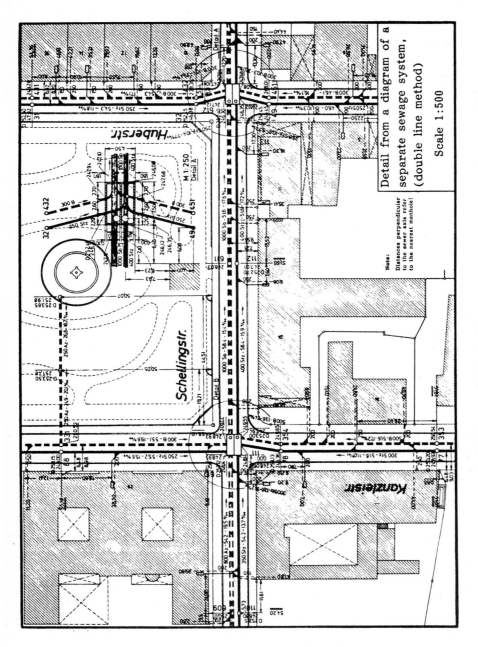

Fig. 4.3.-46: Detail from a diagram of a separate sewage system, to DIN 2425

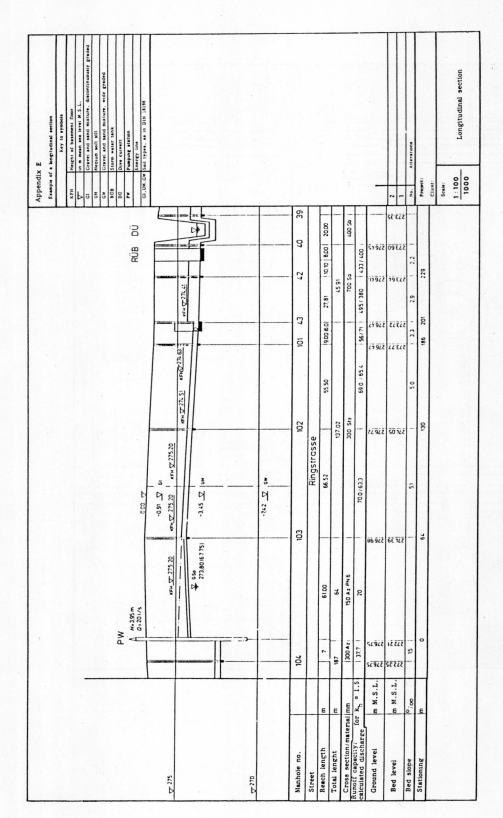

Fig. 4.3.-47: Longitudinal section, to DIN 2425

4.4 Building Components of Drainage Systems

The whole sewer system consists essentially of

- a network of conduits (open and closed) and
- structures such as: intake structures, drop manholes, sewer crossings, junctions, storm-overflow structures, storm-water retention tanks, pumping stations.

All these pipes and structures are made of materials that fulfill certain static, mechanical, and chemical conditions. These conditions are determined on the basis of the quality of the waste water, hydraulic flow, types of soil, and costs.

The quality of the waste water is the most important factor in choosing materials for the pipes and structures. The quality of the waste water is determined by the kinds of waste water. Domestic sewage contains various inorganic salts (e.g. sulfates) and organic salts (e.g. lactates, acetates), vegetable and animal fats, etc. Fresh sewage does not normally attack concrete.

Chemical agressivity is graded according to the hydrogen ion concentration or the pH value as follows:

$$\text{pH } 0 - 4 = \text{strongly acid}$$
$$\text{pH } 4 - 7 = \text{slightly acid}$$
$$\text{pH } 7 = \text{neutral}$$
$$\text{pH } 7 - 10 = \text{slightly alkaline}$$
$$\text{pH } 10 - 14 = \text{strongly alkaline}$$

The fats form a slimy, soaplike, or oily sewer film full of microorganisms, similar to the biological slime in trickling filters. Sewer slime forms on concrete parts. Below the sewer slime, there is a smaller pH, within the acid range, and therefore a greater tendency to attack concrete.

Storm water may be more or less aggressive, depending on where the rain fell and was collected. Groundwater may also be aggressive, owing to the CO_2, sulfates, and other salts it contains, and the same is true of sea water, marsh water, or seepage water from sanitary landfills.

Hydraulic conditions, such as for example gravity or pressure flow, continuous or intermittent flow, and size and fluctuations of velocities, can determine the characteristics of the sewers and other structures.

The type of soil influences the choice of building materials with regard to resistance, corrosion, and stability.

The costs can also affect the choice of construction methods.

To summarize, the materials must satisfy the following conditions:
- adequate mechanical resistance to permanent and variable external loads;
- watertightness, to prevent sewage escaping into the soil, and groundwater seeping into the sewers, in cases where the sewers are laid in the groundwater;
- resistance to aggressive sewage and groundwater;
- resistance to temperature changes;
- resistance to abrasion from solid material carried by the waste water (sand, slag, etc.);
- smooth inner walls, so that the flow causes no additional hydraulic friction losses, except for those allowed for in the design;
- they should permit fast, modern construction of the sewerage system.

These conditions are fulfilled by various materials as follows:
- for domestic sewage in gravity lines: concrete, ceramic, reinforced concrete, brick, cast iron, or plastics. Ceramic pipes are very good, but expensive.
- for slightly alkaline sewage (pH = 7 - 10): ceramic, concrete, and asbestos cement pipes.
- for strongly alkaline sewage (pH > 10): acid-resistant ceramic or brick with special mortar.
- for domestic sewage in pressure lines: reinforced concrete, asbestos cement, cast iron, steel, plastic.
- for aggressive sewage in pressure lines: ceramic, plastic, or coated steel pipes.

Pipes and structures may be built with prefabricated parts or in situ. Normally, the piping tracts are put together from separate, prefabricated pipes, but for large cross sections it is easily possible to build them in situ.

4.4.1 Conduits

For waste water removal
- open conduits and
- closed conduits, in the form of gravity or pressure pipe lines,
are built.

4.4.1.1 Open Channels

Open channels are normally only built for storm water and for clarified waste waters (i.e. treated in a sewage plant). Depending on their size, open channels are known as gutters, ditches, or channels.

<u>Gutters</u> are built to collect storm water from paved surfaces. In streets, gutters should only be built parallel to the direction of traffic. The water should not exceed a depth of maximum 0.10 - 0.12 m, and a water surface width of maximum 1.00 m.

<u>Ditches</u> are small channels, with very little or no hard surfacing, which collect storm water from country roads or unpaved roads. On level ground, they are made with a gradient of at least 2 $^o/oo$. The sides and bottom are protected with grass, stone walls with or without mortar, concrete, or reinforced concrete, depending on the importance of the road.

<u>Channels</u> are open conduits that convey large volumes of waste water. The bottom width is at least 0.4 - 0.6 m and the depth is between 0.6 - 1.50 m. The gradient should be > 0.5 $^o/oo$ for a large flow (>1 m^3/s), 1 $^o/oo$ for a medium flow, and 2 - 3 $^o/oo$ for small chan-

nels. The lining of the sides and the bottom depends on the aggressivity of the waste water, the type of soil, and the velocity of flow.

4.4.1.2 Closed Conduits

Closed conduits are constructed with various cross-section shapes and from various materials. The selection depends on the quality of the waste water, hydraulic conditions, existing building development conditions, and costs.

Normally, closed conduits are classified according to the material they are made of.

4.4.1.2.1 Concrete and Reinforced Concrete Pipes

Concrete and reinforced concrete pipes are the most commonly used pipes in sewerage systems today, as they are easier to manufacture and cheaper in comparison with other piping materials.

Concrete and reinforced concrete pipes are built with prefabricated parts, or in situ. Fig. 4.4.-1 shows standard cross-sectional shapes with closed profile. Three shapes are normally used: circular, egg shape, and horseshoe, with or without a cunette (LAUTRICH /91/).

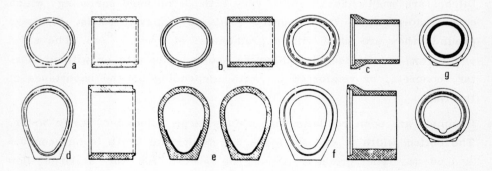

Fig. 4.4.-1: Concrete pipes /91/

a,d - concrete pipes with base and mortise; b - cement pipes without base; c,f - socket pipe without base; e - cement pipe with vitrified clay liner plates; g - concrete ceramic pipe

4.4.2.3 Intake Structures

It often happens, especially in separate systems, that the sewers have to divert rainwater or a watercourse which up to that point runs as an open channel. The connection between the open flow cross section and the closed conduit constitutes the intake structure. The form of this structure is essentially determined by flow conditions in the open channel.

If high velocities occur in the open channel, so that the water is likely to be carrying shingle, or debris, this coarse matter should be retained in a shingle trap. The shingle trap can be dispensed with only if low velocities occur in the open channel. The pit, which should normally retain shingle, then serves as a sludge trap, and it is only a question of time before the sludge begins to putrefy and smell bad.

Fig. 4.4.-9 a shows an example of an intake structure for small flows, and fig. 4.4.-9 b for large flows.

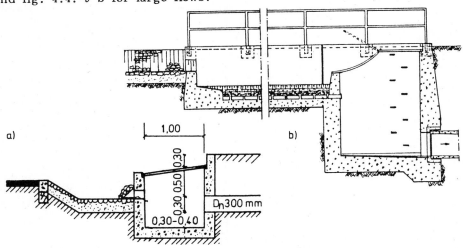

Fig. 4.4.-9: Intake structures
a - for small flows; b - for large flows

If a large watercourse is to be diverted and conveyed within the built-up area in a large closed sewer, a shingle trap and screen can be dispensed with if the sewer is big enough to be entered when the flow is at a low or medium level. Very often, it is then also possible to enter the sewer with a suitable motorized machine and clear it in this way. If a relatively large area has to be drained, and if during a catastrophic storm trees or bushes from this area are likely to be delivered with the first flood wave, the intake should be protected by a coarse screen. At the same time, a large storage area, adequate to hold the oncoming volume of water until the coarse screen can be cleared, should be constructed in front of the intake.

The type of intake structure described here is mainly required for storm sewers. If, in exceptional cases, the surface water is so strongly polluted that it has to be purified in the sewage plant, then the intake structure should not have a shingle trap: in such cases, the sludge accumulation with its unpleasant odor would be noticeable in a very short time. With the screen, the situation is different, in that owing to the pollutant being carried, a mechanically clarifying screen is practically the only one that is required. Under no circumstances should protection of the sewers against coarse matter be neglected.

4.4.2.4 Manholes

Manholes must be placed at regular intervals in straight runs wherever there is a change in direction, pipe size, or gradient, or wherever a junction occurs.

In straight runs, the intervals between manholes on sewers which are too small to be entered, and where there are no high pressure flushing devices, should not exceed 50 - 70 m, and 70 - 100 m for sections which can be entered. Where there is high pressure flushing, distances of up to 160 m are occasionally used.

A manhole (fig. 4.4.-10) consists of base, walls, gully, working chamber, taper section, and cover. They can be made of brick, concrete, or prefabricated parts.

The base can be rectangular or circular. Manholes may have their own base (as in fig. 4.4.10- a, b, c, d) or may have no base of their own, i.e. they open directly into sewers (fig. 4.4-10 e, f). The bottom of the manhole is made of concrete; in situ concrete bases should not be less than 20 cm thick.

The highest part of the bottom should be level with the crown of the sewer on both sides, for pipes of DN 500 internal diameter. If the profile is larger, the height of the floor should be twice Q_{tr} above the water level, but, if possible, 50 cm above the invert.

If possible, the working chamber should be 2.0 m high.

If concrete rings are used in the upper section, the brick or concrete base should be at least 28 cm above the highest crown of the sewers.

Step irons or ladders of noncorrosive material are built into the perpendicular wall (3 st/m).

Bend structures are built like manholes, but the open channel is appropriately shaped. The radius of curvature should be chosen so that the channel runs without interruption. Bends of 90° (fig. 4.4.-11) in a manhole can only be recommended for small sewers. For larger sewers, it is better to use an intermediate manhole (see fig. 4.3.-42 b) or two manholes each with a bend of 45° (see fig. 4.3.-42, c, d). For very big sewers (e.g. radius > 2.0 m), a direct bend is possible.

Changes of diameter should normally take place in manholes. For very big sewers, the diameter change can be done by direct connection (see fig. 4.3.-40).

4.4.2.5 Junctions

The junction of small sewers is made in manholes (fig. 4.4.-12 a); for big sewers, special structures are used (fig. 4.4.-12 b).

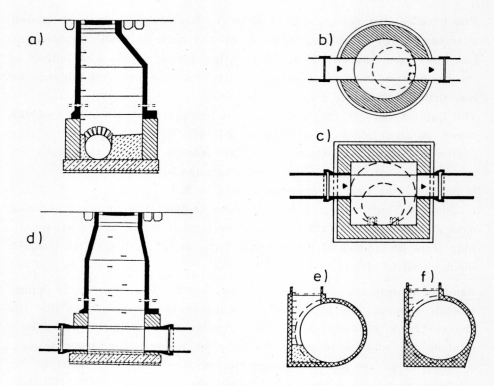

Fig. 4.4.-10: Manholes
a - cross-section; b - circular, top view; c - rectangular, top view; d - connection with sewers; e,f - opening directly into sewer

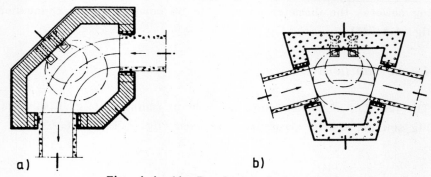

Fig. 4.4.-11: Bend structures
a - sharp bend; b - slight bend

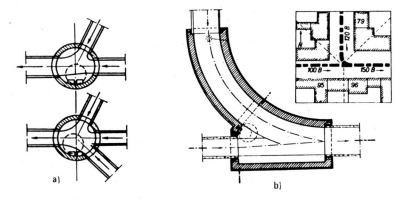

Fig. 4.4.-12: Junctions
a - for small pipes; b - for large sewers

Sometimes these junctions are large constructions (fig. 4.5.-13, SCHOK-LITSCH /151/).

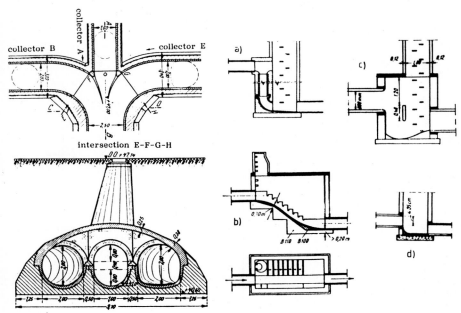

Fig. 4.4.-13: Junctions for very big sewers

Fig. 4.4.-14: Drop manholes
a - with vertical pipe; b - with cascade and channel; c,d - direct fall

4.4.2.6 Drop Manholes

Drop manholes check the velocity of flow at intersections and may also be used to overcome considerable height differences in a short distance (fig. 4.4.-14).

4.4.2.7 Flushing Devices

In sewers where the flow is small, or the gradient low, the prescribed minimum velocity for dry weather cannot be guaranteed and the sewers have to be flushed from time to time. For this purpose, various devices and methods are used.
Fig. 4.4.-15 shows a gate shaft. The sewer is closed with a stop gate, and when the sewage reaches the prescribed height in the shaft, the gate is opened and the rush of sewage flushes the pipe.
Fig. 4.4.-16 shows a portable sewer shutter (LAUTRICH).

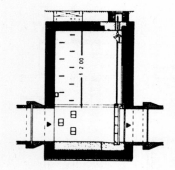
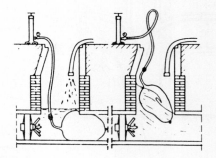

Fig. 4.4.-15: Gate shaft   Fig. 4.4.-16: Portable sewer shutter (LAUTRICH)

Recently, automatic flushing tanks have been developed. The supply of water flushes the sewer at a pressure of $(60 - 80) \cdot 10^5$ Pa.

4.4.2.8 Storm Water Surplusing Works

In combined sewerage systems, not all storm water is carried directly to the treatment plant. Depending on local conditions, part of this sewage is diverted into storm-water tanks, to be treated during dry weather periods, or it may be partly discharged into the receiving water. In order to collect the hydraulic shock load during wet weather, storm-water surplusing works are built. These include
- storm overflows,
- storm-water tanks,
- regulators.

4.4.2.8.1 Storm Overflows

Storm overflows are structures that separate combined sewage into two streams: one that is carried on for treatment in the sewage plant (fig. 4.4.-17), and one that is conveyed to the receiving water or a storm-water tank.

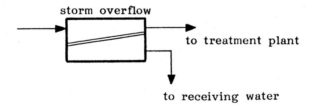

Fig. 4.4.-17: Storm overflow

Storm overflows are constructed and calculated in various ways (see fig. 4.3.-32).
When calculating storm overflows, it used to be assumed that storm water could be discharged into the receiving water as combined sewage with a dilution of 1 : 5. At the present time, such a solution can only be considered for very small discharges. German Standards recommend that as an annual average 90 % of biodegradable and settleable substances in storm water discharges should be taken to the treatment plant, including biological treatment. This requirement implies that storm water in a separate system must also be so treated.

4.4.2.8.2 Storm-water Tanks

Storm-water tanks are structures serving to store water that exceeds the discharge capacity of the sewers that follow it. Water that overflows from the tank is discharged into the receiving water. When the rain stops, the water in the tank is conveyed to the treatment plant. These structures are only built in combined systems.

For short, heavy showers, where the discharge is smaller than the storage capacity of the tank, there is no overflow. During heavier rainfall, at least the initial polluted flow caused by the flushing of the sewers should be retained and, after some delay, carried on to the treatment plant, while only the peak flows are discharged into the receiving water.

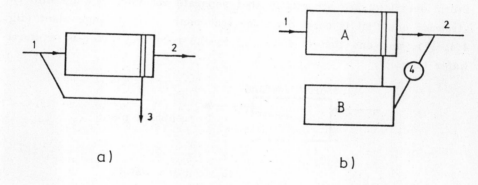

Fig. 4.4.-18: Storm-water tanks
a - collecting basin; b - through-flow basin; 1 - tank overflow; 2 - to treatment plant; 3 - to receiving water; 4 - storm-water tank

Storm-water tanks can be placed in a series connection (fig. 4.4.-18 a) or in parallel (fig. 4.4-18 b). The parallel connection is mostly used when pumping is necessary to empty the tank and convey the sewage to the treatment plant. This enables the greater part of the discharge to flow by gravity.

4.4.2.8.3 Regulators

Regulators are collecting basins used in both combined and separate systems. They store the sewage and discharge it slowly into the sewer, to a pumping station, or to the treatment plant. Through this reduction in flow to pumping stations, the sewers and treatment plants may be relieved, or given smaller dimensions. Regulators normally have an outlet only into the sewer system, but they may have an additional emergency overflow.
Regulators may be constructed as separate tanks or as sewer storage sections (fig. 4.4.-19).

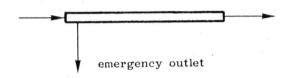

emergency outlet

Fig. 4.4.-19: Regulator constructed as sewer storage section

Open tanks can be used for rainwater. However, for combined sewage, covered tanks or sewer storage sections should be used.

In all cases, regulators should be determined by means of a cost benefit analysis allowing for several possible variants, such as for example large regulator and small treatment plant or pumping station, or large regulator and small pipes downstream of the regulator. Regulators can accordingly influence, directly or indirectly, the costs of the sewerage system. They are economical, and can be recommended for technical reasons in the following cases:
- fulfillment of public requirements for discharging into receiving water,
- reduction of capital expediture and operation costs of pumping stations, in cases where storm water cannot always be discharged by gravity, due to the water level of the receiving water or of the waste water in the sewers,
- junction between new and old sewer sections with inadequate dimensions for discharge of sewage peaks in wet weather,
- junction between open and closed channels.

Fig. 4.4.-20 shows some types of regulations.

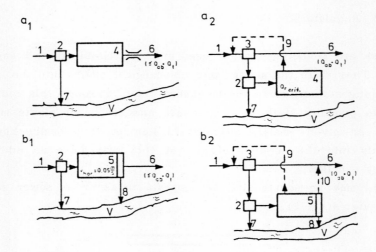

Fig. 4.4.-20: Storm-water tanks

a_1 - regulators in main conduct; a_2 - regulators in by-pass; b_1 - run through basin in main conduct; b_2 - run through basin in by-pass conduct; 1 - inflow; 2 - basin overflow; 3 - separation facility; 4 - regulator; 5 - run-through basin; 6 - discharge; 7 - $Q_{Bü}$ basin overflow discharge; 8 - $Q_{Kü}$ discharge of sewage overflow; 9 - backflow, throtted by pump; 10 - $K_ü$ sewage overflow; V - receiving water; Q_{ab} - flow-rate; Q_t - dry-weather-flow

4.4.2.9 Pumping Stations

Pumping stations must continue to be the exception rather than the rule in sewer systems, since they make operation more difficult and more costly. However, they are required to overcome uphill stretches or to assist the flow on level ground. The need for pumping stations, and their size and location must be determined with regard to general technical and economic conditions. It must be remembered that solid pollutants also have to be pumped together with the sewage.

The most important parts of a pumping station are:
- the wet well for receiving the waste water (sump),
- the pump house, where the pumps are installed,

- the suction and discharge pipes,
- the pumps,
- electrical installations,
- sanitary installations.

The wet well can be built as a separate tank (fig. 4.4.-21 a) or incorporated in the pumping station (fig. 4.4.-21 b). The size of the wet well depends on the pump cycles.

$$V = 0.9 \cdot \frac{Q_m}{Z} \qquad [m^3]$$

where

V = volume of wet well, in m³
Q_m = average discharge, in l/s
Z = cycles per hour, as follows
up to 7.5 kW $Z \leqslant 15$
up to 50 kW $Z \leqslant 12$
over 50 kW $Z \leqslant 10$

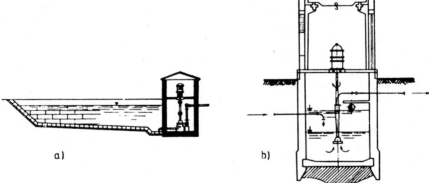

Fig. 4.4.-21: Wet wells
a - as separate tanks; b - incorporated in pumping station

The wet well is provided with a screen, and if required (in large pumping stations) with a comminutor.

In the pump house (dry well) are the water and vacuum pumps, the connections of the suction and discharge pipes, the gate valves, the measuring instruments, and the electric switches.

The <u>suction pipe</u> is normally larger than the discharge pipe. The velocity of flow should if possible not exceed 1.0 m/s, to prevent the water head from separating. The friction losses in suction pipes can be obtained from tables. If there are several pumps, each pump should have a separate suction pipe.

The <u>discharge pipes</u> must also be arranged so that each pump has its own pipe. These pressure pipes can be made of the following materials: cast iron, ductile cast iron, steel, asbestos cement, reinforced concrete, and plastic. To avoid blockages in the pressure pipes, the velocity of flow and the pipe diameter should not be less than the values given in Table 4.4.-1 (LAUTRICH /92/).

Tab. 4.4.-1: Minimum velocities and diameters for pressure pipes /92/

Material to be conveyed	Minimum	
	v (m/s)	$\emptyset$ (mm)
1. Raw sewage, without fine screen or without comminutor	0.7	125
2. Raw sewage with fine screen or with comminutor	0.6	100
3. Coarse screened sewage, minimum settling time 10 min	0.5	60
4. Mechanically clarified sewage	0.4	50
5. Mechanically and biologically clarified sewage	0.3	40
6. Fine sand up to a $\emptyset$ of ca. 1 mm	0.6	50
7. Coarse sand and stones up to a $\emptyset$ of ca. 20 mm	1.0	150

Pumps are installed in dry wells (see fig. 4.4.-21 a, b) or wet wells (fig. 4.4.-21 b), horizontally or vertically.

Various types of <u>pumps</u> are used in sewerage systems. Among the main types are:

- screw-type (Archimedes) pumps (fig. 4.4.-21 d) for low lifts (< 7 m); for higher lifts, two screw-type pumps can be used, one above the other.
- enclosed Archimedes pumps, where the screw is enclosed in a pipe.

- nonclog pumps (fig. 4.4.-22 b), which are especially suitable for untreated sewage, since they permit larger clearances.
- propeller pumps (fig. 4.4.-22 c), which have the advantage that they are more resistant to foreign matter than nonclog pumps and are more efficient. The manometric lift is normally up to 10 m, and in special cases up to 20 m.
- diaphragm lift and force pumps (fig. 4.4.-22 d) are portable and are generally used during construction of the sewer system to remove groundwater with a high sludge content.
- centrifugal pumps are today most commonly used in sewerage systems. A wide range of types is available, e.g. horizontal and vertical dry well pumps, or vertical submersible pumps with integrated or separate motor.

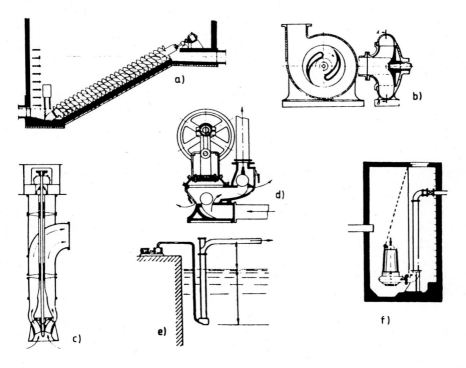

Fig. 4.4.-22: Types of pumps

a - screw-type pump; b - nonclog pump; c - propeller pump; d - diaphragm lift pump; e - mammoth pump; f - submersible centrifugal pump

- mammoth pumps (fig. 4.4.-22 e), also called pneumatic ejectors, convey the sewage by blowing in compressed air, and exploiting the different densities.
- submersible pumps (fig. 4.4.-22 f) can be used for a variety of purposes.

Calculation of the pump size is based on
- the power requirement and type of pump chosen,
- the number of pumps.

The power requirement of the pumps is obtained (fig. 4.4.-23) from the following equation:

$$L_p = \frac{\gamma \cdot Q \cdot H_{man}}{1000 \, \eta_p} \quad [kW]$$

where

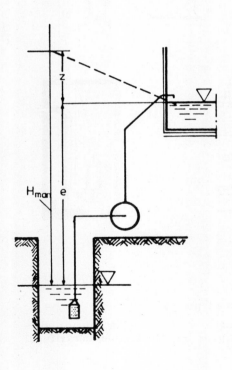

η_p = efficiency of pumps in % according to pump catalogs

L_p = power requirement of the pumps, in kW

Q = discharge, in m³/s

γ = specific gravity of fluid to be conveyed, in kN/m³,

H_{man} = manometric lift in m water level (1 m water level is always equivalent to the unit of pressure $0.1 \cdot 10^5$ Pa) with $H_{man} = e + z$

e = geodetic lift, i.e. the difference in height between pressure and suction water levels, in m

z = flow resistance in the whole suction and pressure pipe line, excluding the pump, in m water level of fluid to be conveyed

In determining the power requirement of the motor, a reserve capacity L_R should be allowed for as follows:

for L_p in kW	L_R
up to 7.5	ca. 50 %, at least 1 kW
7.5 - 20	ca. 25 %
20 - 50	ca. 15 %
over 50	ca. 10 %

Thus, the power requirement of the motor L_M is

$$L_M = L_p + L_R$$

where L_R = reserve capacity, in kW

The motor should be designed for the critical load limit of the coupling between pump and motor, within the set load limits. An appropriate pump can be chosen from pump catalogs on the basis of the computed power requirement.

The number of pumps is obtained using the equation

$$n = \frac{L_{Pr}}{L_p}$$

where

L_{Pr} = total power requirement (including reserve capacity) of the pumps, in kW

L_p = power requirement of the pump chosen from catalog, in kW

The number of pumps should not be too high (about 2 - 3). On the other hand, one should be kept in reserve, e.g.
- for 1 - 3 pumps in operation, 1 reserve pump
- for more than 3 pumps in operation, 2 reserve pumps.

The pump diagram is the ratio of Q to H_{man}. The capacity of a pump can only move on the line $Q - H_{man}$, i.e. a pump can, for example, lift the water higher if the flow is small, and the lift decreases if the flow increases.

On the other hand, the capacity of a pump also depends on the system characteristics (fig. 4.4.-23 a and b, line R).

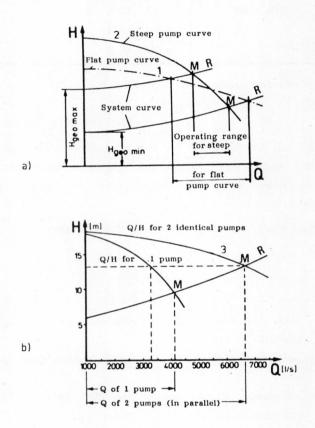

Fig. 4.4.-23: Diagrams for calculation of pumps

a - power requirement for 1 pump; b - power requirement for 2 pumps; R - system curve for pressure pipe; 1 - flat throttle diagram; 2 - characteristic curve for 1 pump; 3 - characteristic curve for 2 pumps

The intersection of these curves, point M, is known as the operating point. It gives the maximum lift - or discharge - of the chosen system. If two pumps are used, discharging into pressure mains, the total capacity is less than the sum of the individual capacities of the two pumps.

<u>Electrical installations</u> must be carried out with special regard for the relevant technical safety regulations.

In large pumping stations, <u>drinking and waste water facilities</u> must be provided for use of the staff and for cleaning the pump house and installations.

4.4.2.10 Sewer Crossing

In sewer systems it is often necessary to cross
- other lines,
- various obstacles (watercourses, railways, or roads).

In principle, it is possible to pass over or under all obstacles. If there are no other factors determining the choice, the most economical construction is chosen.

To pass below an obstacle, a dive culvert (inverted siphon), underpass, tunnel, or (for small sewers) simple wall reinforcements may be used.

4.5 Sewer Construction

Sewer construction must satisfy public regulations with regard to sewerage systems.

For functional reasons, sewer construction is carried out against the planned gradient, i.e. commencing at the lowest point. This method of procedure has several advantages:
- it is easier to check the elevation of the invert,
- completed sections can be put into operation, so that groundwater and waste water can be drained as early as possible,

- connected subsections can be operated, thus ensuring that the development area is kept clean.

Sewer construction involves a considerable amount of surveying work, and this must be carried out with a high degree of accuracy.

The area to be drained must be related to one datum level, normally the national survey datum. Large sewers generally have a small gradient ($\geqslant 0.5$ mm/m), and should therefore be laid with a corresponding degree of accuracy.

During sewer construction, the survey work necessary for verification should be carried out continuously. As far as possible, sewers should be constructed according to scientifically proven methods. Modern machines can make the task easier and more economical, but it is possible to do it just as well using older methods. Difficulties met during pipe laying arise essentially from excavation works and groundwater problems. Both factors have a considerable effect on the cost of pipe laying. The following sections describe the various stages of sewer construction.

4.5.1 First Stage of Construction - Setting Out Sewer Lines

The construction of a sewer system begins with a reconnaissance of the proposed site. The route should be followed on foot, to check the actual conditions on the ground against the plan. It is recommended that the routing of the sewer on the ground should be determined by the authority responsible or the engineer. At the same time, the trigonometric survey points are given to the contractor, and all necessary topographical and geotechnical work can be completed.

When construction work is to include both collector and laterals, the collector is constructed first.

The beginning and continuation of construction work is in many countries subject to strict <u>regulations</u>. For example, the client and the contractor name in writing the supervisors at the beginning of construction and at any change. The supervisors must possess the necessary professional qualifications.

The contractor must make simple daily reports covering the following points:
- number of men employed, divided into groups according to trade, and hours worked by each group,
- type and running times of machines, appliances, pumps and lights used, divided into groups according to type of appliance,
- type of construction work carried out, etc.

<u>Setting out sewer lines</u> begins with driving in locational stakes offset from the center line of the proposed sewer. In addition, a detailed site reconnaissance together with the authority should be carried out, especially when the depth of excavation cannot be measured immediately (e.g. for dive culverts on a river bed, tunneling, in heavy, wet soil, etc.). It is also necessary to determine the datum point and the relation of this to the national survey datum.

Next, the width of the trench is marked on the street surface. The trench width should be determined in the plan, depending on soil conditions, trench depth, type of timbering, and method of excavation (by hand, with machines, or both).

In calculating the trench width, various conditions must be taken into account:
- unsupported trenches with a depth less than 1.25 m need not be calculated;
- the trench width is calculated from the following measurements:
  2 x 15 cm sheeting
  + nominal diameter of pipe, in cm
  + 2 x pipe wall thickness, in cm
  + 70 cm for pipes with outer diameter $\leqslant$ 40 cm or alternatively
  + 40 cm for pipes with outer diameter $>$ 40 cm.

The minimum width b_{min} is graded according to the trench depth

b_{min} = 60 cm, for trench depths $\leqslant$ 1.75 m
b_{min} = 80 cm, for trench depths > 1.75 m.

The Hamburg measurements provide a clear basis for calculation, as shown in Table 4.5.-1 and in fig. 4.5.-1.

Tab. 4.5.-1: Excavation width of trench for vitrified clay and concrete sewers, and sewers built in situ (Hamburg measurements)

Vitrified clay and concrete pipes		Concrete pipes	
DN	Trench width*	DN	Trench width
mm	m	mm	m
150	1.00	500 / 750	1.50
200	1.10	600 / 900	1.60
250	1.10	700 /1050	1.70
300	1.20	800 /1200	1.90
400	1.30	900 /1350	2.00
500	1.40	1000 /1500	2.10
600	1.60	1200 /1800	2.30
700	1.70	1400 /2100	2.60
800	1.90	550 /1000	1.55
900	2.00	700 /1200	1.70
1000	2.10	850 /1400	1.95
1100	2.20	1050 /1550	2.15
1200	2.30	1250 /1800	2.35
1300	2.40	1550 /2000	2.70
1400	2.60	Sewers built in situ	
1500	2.70	550 /1000	1.55
1600	3.10	700 /1200	1.70
1700	3.20	850 /1400	1.85
1800	3.40	1050 /1550	2.00
1900	3.50	1250 /1800	2.40
2000	3.60	1550 /2000	2.70

* width measured at bottom of trench

For braced trenches, 15 cm must be added at the bottom of the trench.

For plastering on sewers built in situ:
addition for DN 550/1000 to 850/1000: 0.30 m
addition for DN 1050/1550 to 1550/2000: 0.50 m.

When the pipe center line and the trench width have been accurately marked on the road surface, the proposed center line and elevation of the pipe for each sewer length is transferred to the trench bottom via three batter boards (sight rails) (fig. 4.5.-2). These batter boards (or "rails") are set at intervals of maximum 30 m, at right angles to the pipe center line, 1.00 to 1.50 m above ground level, and fixed on substantial, free-standing posts.

At least the horizontal board should be planed, and painted in red and white stripes 1 m apart. The upper edge of the board is set by means of accurate leveling at some convenient whole number distance above the sewer invert level.

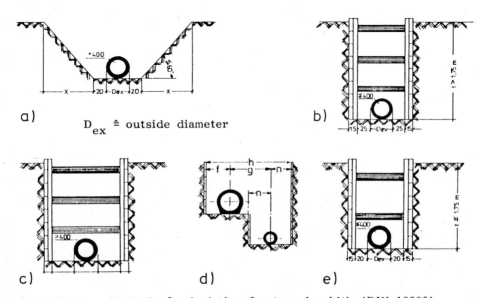

Fig. 4.5.-1: Method of calculation for trench width (DIN 18300)
a - for pipes > DN 400 mm, with slope > 60°, width 30 cm instead of 20 cm; b - for pipes DN ≤ 400 mm, in trenches with depth $t > 1.75$ m; c - for pipes DN > 400 mm; d - for two pipes in same trench; e - for pipes DN ≤ 400 mm, in trenches up to $t = 1.75$ m

If there is a steep gradient, the level of the sight lines is changed by fixing a batter board to both sides of the post.

If excavation is being done by machines, the batter boards are fixed after the initial excavation; the exact invert level is trimmed by hand.

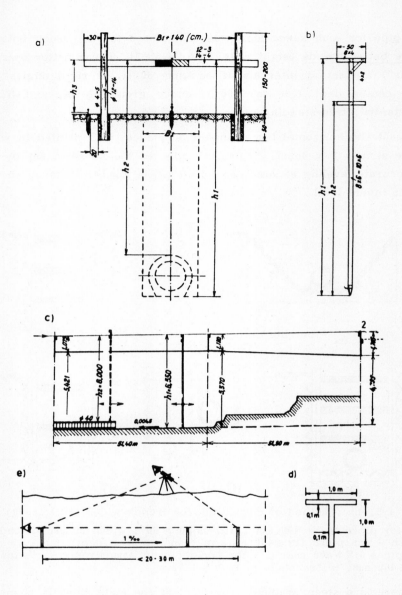

Fig. 4.5.-2: Batter boards for sewer construction

a - general view; b - traveller; c - leveling with cross-staffs; d - cross-staff; e - checking during pipe laying; 1 - nail; 2 - traveller with 2 sight levels

The center line of the sewer is indicated by nails in the middle of the batter boards, their exact position being determined by topographical methods. Then the center line is transferred to the bottom of the trench, using a plumb line from a cord stretched from nail to nail.

4.5.2 Excavation

In built-up areas, trenches are usually cut with vertical walls, so that traffic is not interfered with more than necessary. Outside towns and wherever machine excavation is possible, it is recommended to slope the sides of the trench.

On excavation work, the site should be barricaded and signed in accordance with road traffic regulations. For pedestrians, the trenches must be protected with stout barricades.

Care should be taken that rainwater can drain into existing street gulleys, and access to service pipes must be preserved by seeing that earth does not cover hydrants, cable pits, and other installations. Neighboring buildings may need special measures to prevent collapse or damage.

At the beginning of excavation work, places where service pipes, especially high voltage cables, cross the trench should be accurately located using the plans, or by making a special investigation. It is recommended that representatives of the public utilities be consulted when locating these pipes, since plans cannot always be relied on.

If the sewer is to be laid in a road that is already paved, an agreement must be reached in good time with the highway authority as to how and where the road surfacing materials should be stored, for subsequent replacement.

Excavation begins with removing the road surfacing materials for the planned trench width plus 0.3 m on each side.

Excavation may be done by machine, or by hand, or both. The trench - whichever method is used - is not excavated immediately right down to the final bottom level. The last 10 - 20 cm are usually excavated shortly before the pipe is actually laid, or before casting of the sewer bottom. This is to ensure that the pipes are bedded on firm, undisturbed soil.

The surface at the bottom should be protected against disturbance and kept dry.

In predominantly noncohesive soils, if the bottom has been disturbed or too deeply excavated, it can be made firm again by tamping or vibrating, provided the trench supports allow this.

Today, trenches are mostly excavated by machine. Trench excavators are used, mounted on creeper bands, on tires or even on vehicles, so that there is no need to lay tracks near the trench. To avoid serious accidents, great care should be taken regarding underground service pipes.

Excavation is done by hand in cases where it is uneconomical to use a trench excavator. Where hand excavation is necessary in towns because of underground service pipes, the depth is normally limited to a maximum of 1.50 to 2.0 m. Mixed excavation consists of hand digging and mechanical transport of the excavated material. The earth is dumped with shovels and other instruments directly onto a conveyor belt, which carries it to a waiting transport vehicle inside or outside the trench. If the earth cannot be dumped directly onto a conveyor belt or vehicle, and has to be removed by hand, then one laborer for every 2 m depth is needed.

It is best to deposit the excavated material at one side of the trench, provided there is enough space available in the street. There must still be an area at least 4 m wide left free for motorized traffic, and about 1 m for pedestrians. The space at the other side of the trench can be

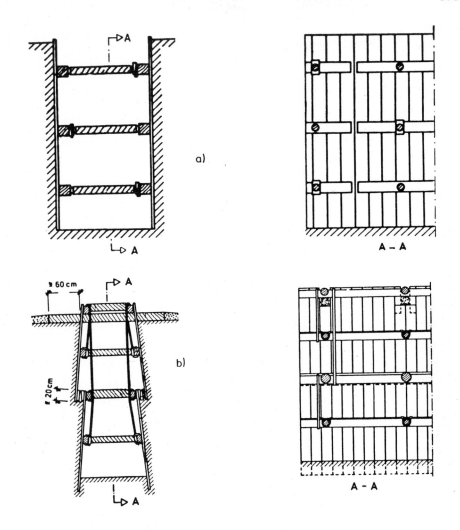

Fig. 4.5.-4: Vertical sheeting
a - with wooden planks; b - with trench sheeting in tiers

<u>Tiered sheeting</u> (fig. 4.5.-4 b). The wooden planks or trench sheeting are placed at a slight incline in several tiers down to the bottom.

<u>Sheet piling</u> is used for unstable soil containing water, when a lowering of the water table is uneconomical and where there is high ground pressure. Wood, steel, and reinforced concrete are used for sheet piling, the degree of resistance depending also on the profile.

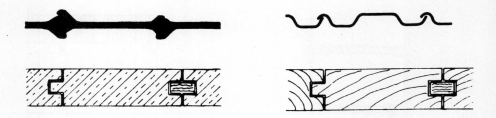

Fig. 4.5.-5: Sheet piling sections

Sheet piling, just like other supports, should generally be fastened together with braces and wales, unless each pile plank is braced individually.

Driving the sheet piling (see also fig. 4.5.- 5) begins between positioned beams. These are placed two along each side of the trench, either at ground level or after excavating to a depth of 1.0 to 1.25 m without supports. They ensure vertical alignment of the sheet piling. The sheet piling is driven with the pile driver to the required depth (h + t) between the two beams. Excavation can always go as far as a depth $\underline{h}$, as the lower end (toe) of the sheet piling should always be at a minimum depth $\underline{t}$ (m) below the trench bottom. The wales should be supported on welded-on brackets or hung on hanger rods or chains.

The anchoring depth is determined for wet ground as follows.
The hydraulic gradient between (1) and (2) is

$$i = \frac{\Delta h}{L} = \frac{\Delta h}{h + 2t + w}$$

and it must be less than the erosion gradient i_{cr}.

A safety margin of n_s should be allowed, giving

$$i \leqslant \frac{i_{cr}}{n_s}$$

where n_s = factor of safety; n_s = 1...2
The critical gradient occurs where

$$i_{cr} = \frac{\gamma'}{\gamma_w}$$

where γ' = specific gravity under buoyancy
       γ_w = specific gravity of the water
with γ' = $(1 - n)$ $(\gamma_{soil} - \gamma_{water})$ [g/cm³]
and   n = the porosity of the wet soil

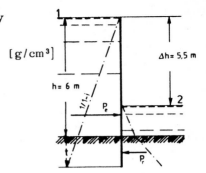

If e.g. γ_{soil} = 2.65 t/m³ and n = 0.4,
then γ' = $(1 - 0.4)$ $(2.65 - 1)$ = 0.999;
with a factor of safety n_s = 1.8, then

$$i = \frac{\Delta h}{h + 2t + w} \leqslant \frac{0.999}{1.8} \ ; \ h + 2t + w \geqslant \Delta h \cdot 1.8$$

$$t \geqslant \frac{1.8 \cdot \Delta h - h - w}{2} \ ; \ t \geqslant \frac{1.8 \cdot 5.5 - 6 - 0.5}{2} = 1.7 \text{ m}$$

<u>Timber sheeting between soldier piles</u>. To line a trench with planks between soldier piles, first the piles are driven in according to static calculations. In unstable soil, they should be driven in to a depth of at least 1.50 m for outer supports, and at least 3.00 m for middle supports.

When the piles have been driven in, excavation can begin. The boards are placed in the profile of the piles and wedged. The piles are braced with vertical timbers and struts.

Fig. 4.5.-6 shows an example of horizontal timber sheeting with H-section soldier piles.

Fig. 4.5.-7 shows an excavated trench with Berlin sheeting using the improved SCHIBLI method (elements) (RÜTHIG).

<u>Other types of sheeting</u>

In sewer construction there is usually no need for sheeting types other than those described above. However, for very large sewers (width over 3 - 4 m) and in suitable soil, other methods can be used.
For example, <u>the sewer walls themselves</u> can be used as <u>supports</u>.

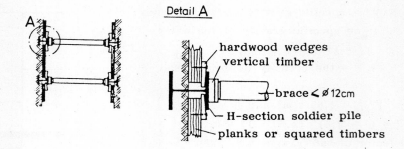

Fig. 4.5.-6: Horizontal timber sheeting with H-section soldier piles

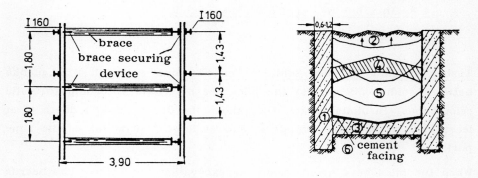

Fig. 4.5.-7: Vertical bracing, Berlin method

Fig. 4.5.-8: Method using sewer walls

This is done by excavating trenches (1) and filling them with concrete (fig. 4.5.-8). When the concrete has set, the soil in between is removed (2), the walls are cleaned and serve then as sewer walls.

There is also the <u>tunnel method</u>, where excavation may be done by hand or by machine.

For small sewers, small earth bridges are normally left standing (B ≤ 1.20 m), in order to keep house entrances free. Sewers are laid below these by the tunnel method.

It is often best to build large sewers with the tunnel method, using the

the pipe. The imported material should be compacted with suitable instruments, so that the pipe is fully supported within the range of the proposed supporting angle.

Concrete bed. If the soil at the trench bottom is not suitable for the construction of a sand or fine gravel bed, if the trench bottom is sloping, or if there is the risk that sand might be washed out by unintentional draining, in this case it may be necessary to bed the pipes throughout on concrete. The thickness of the concrete bed (fig. 4.5.-13 b) should be 5 cm + 1/10 of the nominal diameter of the pipes, but at least 10 cm. The trench must be excavated to the additional depth needed to accommodate the bed. Concrete of at least Bn 150 should be used for bedding. If the soil is unstable, reinforced concrete may be necessary.

The surface of the concrete bed should be shaped to fit the outside of the pipe, so that the pipes are firmly supported within the range of the proposed supporting angle. The bed can be concreted after the pipe has been positioned. If it is made before lowering the pipes, these must be laid on a fresh layer of mortar. To prevent the pipes from "riding" (i.e. moving along the trench bottom) a recess along the bottom line can be provided.

Concrete encasement. To increase the bearing capacity of the pipe, a concrete encasement can be included in the design. When calculating the dimensions, it is important to know whether the concrete will be poured against undisturbed soil or, for example, against sheet piling. In the latter case, the supporting effect of the lateral ground pressure will be reduced once they are pulled.

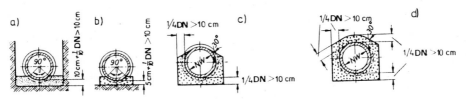

Fig. 4.5.-13: Pipe bedding in non-cohesive soils

Pipes may be completely or partially encased in concrete. Partial encasement is constructed up to springing height (fig. 4.5.-13 c). Complete encasement surrounds the whole pipe (fig. 4.5.-13 d). Concrete of at least B 15 should be used for encasement. Construction joints should be secured with reinforcement bars.

It is often useful if the concrete encasement is divided at suitable intervals coinciding with pipe joints.

Special designs. Bedding on concrete saddles should generally be avoided, since if the sand support is inadequate, the pipe is subjected to bending stresses that have not been allowed for. If it is impossible to avoid using this kind of bed, special static verification will be required for these pipes. The same applies to the bedding of pipes on pile piers. In unstable soil, or if fairly large settlements are anticipated, special measures are necessary, e.g. foundation of the pipes on piles (fig. 4.5.-14), or laying them on a reinforced concrete slab.

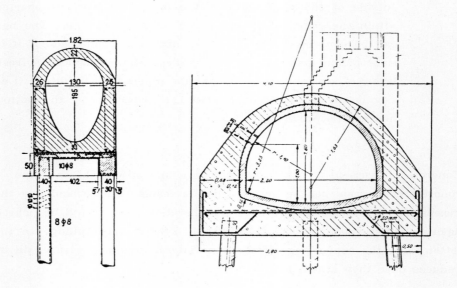

Fig. 4.5.-14: Foundation of pipes on piles

Where there is a transition between soil types of different settling characteristics, protective measures should be taken. For pipes laid beneath sloping ground, the possibility of axial and bending stresses must be considered.

Connections to structures. Connections to structures should be flexible. For manholes, the connection can be built into the wall. If conditions allow, the connection can also be placed at a maximum distance of one meter, measured from the inner surface of the manhole wall in the sewer axis. For other structures, analogous procedures should be followed.

Support and anchorage. Pipes laid in the groundwater should be secured by anchorage or additional loading to prevent buoyancy, if their own weight and the load on them are insufficient. Pressure mains, branches, bends, transition pieces, shutoff devices, etc. that do not make sufficient allowance for longitudinal expansion must be secured in such a way that forces acting on them will be absorbed.

The first consideration in laying finished pipes is that they should follow the proper gradient. In order to place them at the correct elevation, a topographical leveling instrument can be used. To do this, a rod is placed on the trench bottom and then on the pipe invert, and the level checked with topographical instruments. If the elevation is found to be correct at two points 20 - 30 m apart, the intermediate points can be leveled with the aid of 1 m high cross-staffs (see fig. 4.5.-2). Leveling must be done by an experienced person, since correct construction depends on the proper elevation.

Pipe laying is begun at the lowest point of the accurately prepared bed. The pipe is laid with the socket end or groove upstream.

Pipe laying proceeds as follows: the pipes are lowered into the trench, taking all necessary precautions, and first installed "dry" (without making mortar joints) from one manhole to the next, proceeding upgrade. The pipe layer should at this stage match the pipes together (maybe they are not all perfectly round), see exactly how many are needed and check whether the last pipe needs to be shortened (it rarely occurs that only whole pipes are required for one section). When the number, length, and position of the pipes have been established, they can be laid in position. To make sure they are positioned correctly, a cord or thin wire is stretched from batter board to batter board. From this cord, the correct position of the sewer center line is found, using a plumb line, for each pipe as it is laid.

Before the batter boards are used, their elevation must be accurately measured.

With ready-made pipes, it is usual to begin by building the bottom of the manhole. This gives a fixed elevation point and also helps to establish the exact length of pipe needed. However, no part of the pipe (e.g. groove or tongue) should be allowed to project into the clear space inside the manhole.

Proper pipe laying is a matter of particular importance, and should be in the hands of experienced pipe layers.

Before being laid, the pipes have to be transported to the trench. Small pipes can be carried or transported with light mechanical equipment. Heavy pipes have to be transported by stronger equipment. These days, they are usually transported in the excavators used for preparing the trenches. For very heavy pipes, it may be necessary to use special lifting equipment or vehicles (special derrick, portal crane, etc.).

With heavy pipes, it is convenient if transport can be so arranged that the pipes required for laying can be unloaded from the delivery vehicle and lowered into the trench in a single operation.

Light-weight pipes can be lowered into the trench with ropes, and heavy ones with the equipment used to transport them from the storage place to the trench, e.g. a crane truck.

During transport and lowering, it is important to avoid damaging the pipes by careless handling. Therefore, special hooks with a safety catch, ropes, protected belt bands, or other lifting gear should be provided.

During the actual laying of the pipes, especially if the pipes are heavy, it may not always be possible to avoid disturbing the bottom of the trench. Therefore, a compressed concrete slab (at least B 15) over the whole trench width should be provided as a working surface; this should be 5 cm thick for pipes > 400 mm nominal diameter, and 10 cm thick for pipes > 1000 mm nominal diameter. It should be smooth and follow the appropriate gradient.

Before the pipes are laid, they should be checked again visually and by striking to make sure they are sound and undamaged. Any pipe found to be faulty must not be used.

Pipe joints of all kinds should be carefully cleaned and dried before the pipes are laid, and if the sealing method used requires an adhesive coating, this should be applied. Care should be taken that dust, dampness, etc. do not prevent adhesion, and also that the effectiveness of the adhesive coating is not reduced by its being applied too early. If the pipes are to have an external or internal coating, this must be compatible with the sealing material. The manufacturer's directions should be used here as a guide.

If the sealing method leaves the joint with an open space inside and outside, this can be filled with some permanently elastic material. The advantage is that solid matter cannot get into the space, which would prevent movement and make the pipe section inflexible.

<u>Socket pipes</u> must be well supported by the bed along the whole length of the barrel. When laying the pipes, the spigot end should be pushed firmly into the socket. With small pipes, this can be done by hand. Larger pipes may require the use of a fork-type tool, pulley, or hydraulic jacks. In order to allow the pipes a certain freedom of longitudinal expansion (e.g. due to temperature changes), it is recommended that a board or tar board underliner be inserted to separate them from

the concrete base. It is advisable to position the joints in the bed or concrete filling (if used) at pipe joints.

If pipes have to be shortened, they should not be chiseled off, as this might impair the structural cohesion of the material, but should be cut with appropriate tools as far as possible (corundum disks, joint cutters, etc.).

<u>Joint sealing methods</u> for socket pipes depend on the pipe material. A simple method is sealing with grouted or tarred rope and a poured filling. However, this type of joint can only be used for cast iron, steel, and vitrified clay pipes; with asbestos cement and concrete pipes, the sockets may be damaged when the gasket is rammed in (with a caulking tool). Also, with concrete pipes, the jointing material must not be applied hot.

Joints with a gasket ring and poured filling, or with rubber rings ("roll-on" joints) (fig. 4.5.-15), have worked well.

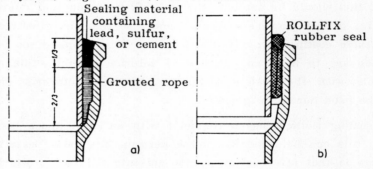

Fig. 4.5.-15: Joints for cast-iron sewer pipes
a) with gasket ring and poured filling; b) with rubber seal ("roll-on" joint)

However, joints should be formed strictly to the maker's instructions.
For large pipes, rubber seals are now almost the only type of joint used, these being provided by the manufacturer.

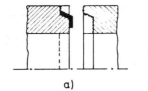

Fig. 4.5.-16: Gaskets for mortise concrete pipes
a) starting position; b) final position

For mortise pipes (fig. 4.5.-16), a joint can be made by applying mortar to the groove and tongue and then placing a strip of mortar around the joint. However, this method does not produce good joints, as the strip cannot be properly made underneath the pipe, and the joint is not absolutely watertight if there is internal water pressure (when the pipe is flowing full).

These days, mortise joints are made with sealing strips. Great care is needed in making these joints, because higher degrees of tolerance are required as the pipe diameter increases.

Before the joint is made, an adhesive coating should be applied. However, this means the sealing strip must be quickly fitted without delay. The rim of the pipe should be neither rough nor damp, and must be free from dust and dirt. The strips must also be clean, and should not be deformed due to the effect of temperature.

Mortise pipes are joined by being pressed together. In addition to strips, KUNZE couplings can be used for joints in concrete pipes (fig. 4.5.-17). A steel ring (4 mm thick) is anchored into the forward edge of the rim of the pipe, and the projecting part covered with a triangular layer of jointing material. The joint is made by pressing the pipes together.

If a pipe is to be fixed into a structure or into in situ concrete, or especially into a manhole, this should be done directly with no intermediate spacer (fig. 4.5.-18).

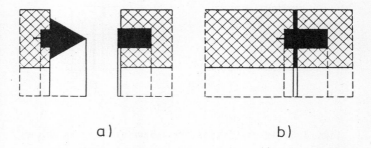

Fig. 4.5-17: Joint with KUNZE coupling
a) before assembly; b) after assembly

Joints at junctions may also not always be very efficient. To avoid this, the trench should be lined with concrete, since it is particularly difficult to make the joints under the pipes watertight.

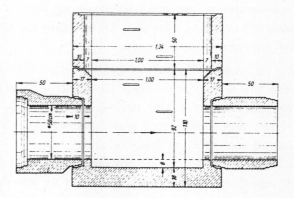

Fig. 4.5.-18: Connection of pipes to manhole /1/

<u>In situ concrete pipes</u> are used for larger diameters, and for sewers where normal joints do not appear efficient enough.
A mobile formwork method is also used (fig. 4.5.-19). After excavation is finished, the in situ concrete base is poured. A layer of asphalt board is laid on the base, so that the concrete pipes do not stick to it, which would prevent them being able to expand under the effect of temperature changes.

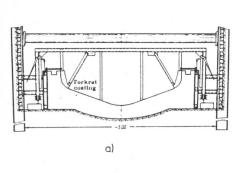 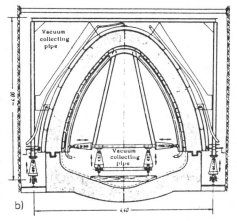

Fig. 4.5.-19: Sewer construction with mobile formwork /1/
a - lowest part of sewer; b - sewer arch, using vacuum concrete method

The formwork and reinforcement are prepared outside the trench. The prewelded iron reinforcement is placed on the board (or sewer bottom). It is lowered into the trench with the aid of a crawler excavator, and held in position with iron blocks. The inside formwork for the pipe is erected. The formwork is made of wood and may be covered with sheet steel on the side that comes into contact with concrete. This enables it to be used several times. When erecting the formwork, its position must be accurately checked by plumbing. After fixing the inner units in place at the right distance from the reinforcement (using spacers and ties), the outer units are erected, but not all together. They are put up in stages as the concrete is poured, so that it can be properly compacted. Fig. 4.5.-20 shows the individual steps of construction /26/.

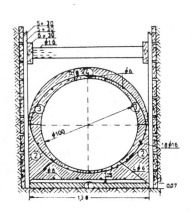

Fig. 4.5-20: Stages in the construction of a reinforced concrete sewer pipe

The concrete should not be poured from too great a height; the best method is to use a trough. A joint should be made every 40 m to allow for interruption of the work and expansion due to changes of temperature. These joints should be closed after about 6 weeks with a concrete strip.

After the concrete is in place it must be covered and wetted with water several times a day to prevent premature drying or setting.

The concrete pipes are plastered inside with cement mortar, and often coated with tar on the outside. A special type of in situ concrete sewer construction is the vacuum concrete method, a licensed construction method from France (RUTHIG). In addition to the unrestricted positioning of the joints, this method has the advantage that the concrete is compact, impermeable, smooth, and strong. Formwork can be made of wood or steel. For awkward section shapes steel is normally used.

The manufacture of brick sewers is an art, unknown unfortunately to many bricklayers. It is a cheap method as far as the material is concerned, and a very economical choice if sufficient cheap labor is available. It could be said that well constructed brick sewers are very durable. The bed (fig. 4.5.-21) for the arch is generally cast of concrete. The sewer is constructed in two parts: first up to springing height, and then to the arch. If the lower part is to be cast from concrete, it should be cast in two or three layers so that it can be well compacted.

If the sewer is built entirely of bricks, then a wood profile should be used for the lower part as well. This must be removed when the mortar has set and before building the arch. The framework for the arch is supported on the lower brick section. The arch is built in two or three layers (rings), depending on the width of the opening. Each ring must be "closed" in itself. After closing the second ring, the framework can be removed, since the first ring can support the others. After the bricks are laid, the joints should be filled with mortar (1:1), or the brickwork should be plastered.

Only Flemish clinker (sewer brick) should be used. These are specially manufactured and fired bricks.

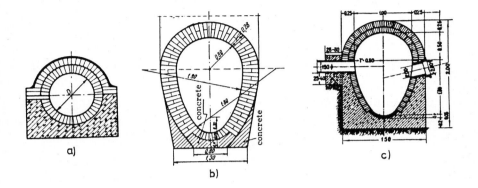

Fig. 4.5.-21: Sections of brick sewers /26/

a) with concrete base; b) entirely of bricks; c) with concrete and basalt shell

4.5.5 Special Structures

In sewer construction, special structures are needed where there are peculiar building conditions or local circumstances.

Of the various types of structures, the following examples are described: sewer outfalls, timbering of large trenches, weirs, and special formwork.

4.5.5.1 Sewer Outfalls

Fig. 4.5.-22 a shows a diagram of a sewer outfall into a water body. The manner of supporting the structure and the bed or bank of the receiving water with sheet piling, walls, etc. should be noted.

Fig. 4.5.-22 b shows a sewer outfall for discharging an effluent below the surface of the receiving water.

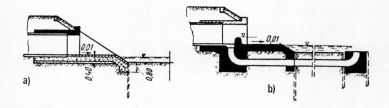

Fig. 4.5.-22: Sewer Outfalls
a) normal construction; b) discharge below minimum water level

4.5.5.2 Sheeting of Large Trenches

It is frequently necessary to support vertical walls in large trenches (e.g. for a pump house) with timber sheeting. Fig. 4.5.-24 shows examples of possible timbering methods and the principle for determining the stresses, which is needed to dimension the struts.

4.5.5.3 Weirs

Weirs are constructed to protect other structures; they serve to keep the site dry. Weirs are designed as earth banks or as sheet pile weirs. Fig. 4.5.-23 shows a sheet pile weir with metal sheet piles.

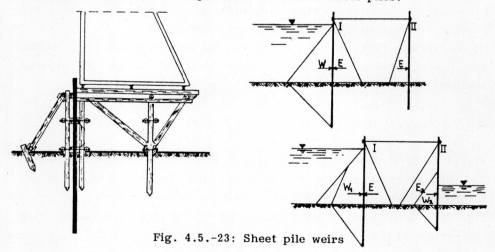

Fig. 4.5.-23: Sheet pile weirs

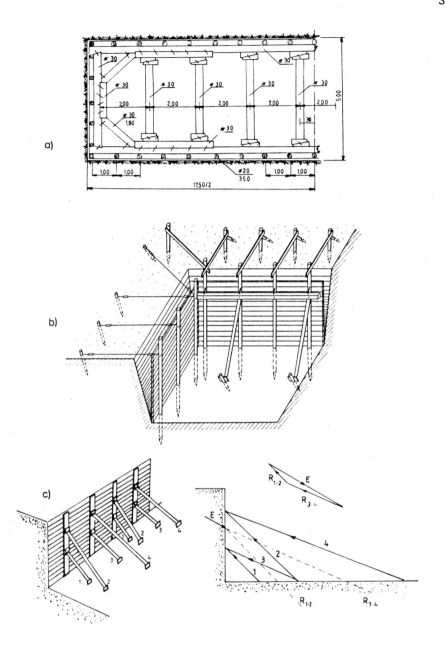

Fig. 4.5.-24: Sheeting of large trenches
a) top view; b) semiplan view; c) wall sheeting and calculation of stresses in raking struts

4.5.5.4 Special Formwork

Special formwork is needed for pouring concrete continuously, e.g. to make the wall of a basin. Fig. 4.5.-25 shows this kind of formwork.

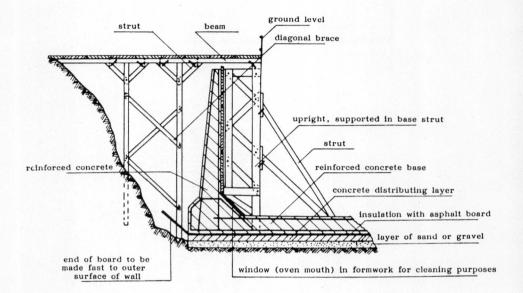

Fig. 4.5.-25: Formwork for basin wall

4.6 Operation and Maintenance of Sewer Networks

Operation of a sewer system begins with the official acceptance of the completed structures.

This is very important and accordingly provided for in all contracts between authorities and building contractors. The official acceptance is documented by a record signed by both parties to the contract. The following points are recorded:
- the precise condition of the structures at the time of the official acceptance,
- the completion of construction work,

Trench bracing with vertical
I-beams and horizontal timber
sheeting

Horizontal timbering in
trench for circular
sewer

Draining pipes in a trench with
horizontal timbering

Outfall structure into Teltow Canal in Berlin

Inside view of upper chamber lock of a dive culvert

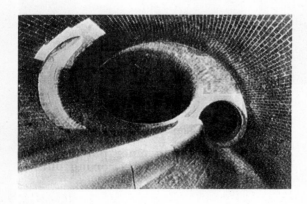

Inside view of upper chamber lock of an emergency outlet dive culvert

- the assuming of responsibility by the authority that is to operate the system.

Those present at the official acceptance include, in addition to the operator (administrating or funding body), the works manager, the building contractor, and - for important projects - the designer also.
There is generally a separate acceptance for each phase of construction, this is necessary because some sections are covered with earth early on and are not visible at the time of the final acceptance. A distinction must therefore be made between partial acceptance and final acceptance.

4.6.1 Official Acceptance of the Sewer System

Partial acceptances consist of one or more acceptances stretching over the whole construction period of the sewer system. Their purpose is to ensure that planning and construction coincide. Inspection is carried out of:
- structures which will not be easily accessible at the time of the final acceptance,
- the execution of construction work.

For mid-progress acceptances during the course of construction, the following should be inspected:
- the depth of excavation, gradient, and bedding,
- dimensions of the foundations (+ base) and all other structures,
- the setting out of the pipes, correct location of street inlets, manholes, storm overflows, culverts, and the entry and exit water level of the different sewers within the structures,
- the correct execution (as regards quality of construction work) of concrete casting, brickwork, and plastering, and the construction of street gutters and storm overflows,
- the watertightness of the pipes.

Leakage tests must be carried out before the trenches are filled. This must, however, be done carefully, and the pipes must be partly covered with earth so that they are not forced out of position by the testing pressure. Leakage tests are carried out with water at a pressure of 5 m water column + 0.1 water column in 10^5 Pa related to the highest wetted point of the pipe. The testing time is 15 minutes. The water added in this time is measured; the individual test values may exceed the values given in Table 4.6.-1 by up to 30 %.

Tab. 4.6.-1: Values for testing leakage in concrete pipes (DIN 4032)

Shape	Nominal diameter	Permissible addition of water in liters per m² wetted inner surface (average values)	
		Types KW-M, KFW-M	Other Types
Circular	100 to 250	–	0.20
	300 to 600	0.08	0.15
	700 to 1000	0.07	0.13
	1100 to 1500	0.05	0.10
Egg-shaped	500/ 700 to 800/1200	–	0.13
	900/1350 to 1200/1800	–	0.10

For all acceptances during the course of construction, records are kept and produced at the final acceptance.

The final acceptance is at the same time the acceptance of operation, and involves quantitative and qualitative inspection.

The quantitative inspection consists of comparing the specified with the actual dimensions:
- longitudinal and cross-sectional dimensions of the sewers,
- number and location of structures.

The qualitative inspection includes inspection of the gradients, plastering, insulation, etc. as regards materials and execution in comparison with official standards and with the plans. Accessible sewers are carefully checked for cracks on the inner wall surfaces and on the invert.

Straight runs and the gradient are checked by accurate leveling on the outside (leveling rods are set up in the manhole), and interior leveling with "cross-staffs" (see Section 4.6.-1).

Cement plaster is checked visually and by striking it lightly with a small hammer; hair cracks up to a maximum width of 0.2 mm are permissible, and the sound should be steady and "full" or "deep".

Before the sewers are put into operation they must be cleaned; rubbish and remains of concrete and plaster must be removed. <u>Inaccessible sewers</u> are inspected using mirrors and a light (fig. 4.6.-1) or television cameras.

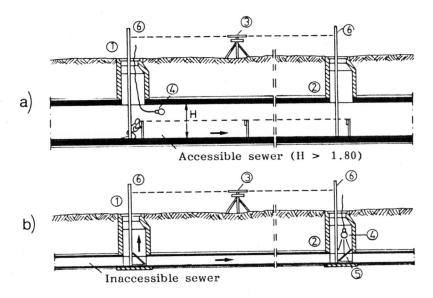

Fig. 4.6.-1: Inspection of sewers
a - accessible; b - inaccessible; 1, 2 - manholes; 3 - leveling instrument; 4 - light; 5 - mirror; 6 - leveling rod

Inspection of <u>pumping stations</u> includes:
- the pump house, as regards correct structural dimensions, quality of workmanship, watertightness of wet well,
- the installations (pumps, pipelines, fittings),
- the mechanical, hydromechanical, and technological transport devices, etc.,
- the automatic controls.

Infiltration should be tested for sewers built in the groundwater.

For sewers laid below public roads, the acceptance also includes approval from the road traffic authorities for replacement of the road surface.

The acceptance commission considers all partial acceptances. It must make decisions concerning possible faults and whether they are to be accounted for by financial deductions.

If the objections constitute a risk for the operation of the sewer system, the affected parts must be replaced or the faults repaired.

The acceptance commission prepares a report and decides on the start-up of operations.

4.6.2 Operation and Maintenance of Sewers and Auxiliary Facilities

The operation and maintenance of the sewers should be organized by the authority or (in the case of industrial sewer systems) by the company concerned. The size and composition of this organization depends on the state of development and the specific problems.

Those departments concerned in the operation of the sewer system should make sure that their staff already have the opportunity to become acquainted with their future duties during the construction period. This ensures a smooth initial start-up.

All maintenance work must be carried out according to fixed yearly and monthly schedules.

4.6.2.1 Operations

The following maintenance work is necessary for the smooth operation of a sewer system:
- sewer inspection,
- sewer cleaning,
- repairs to structures and installations,
- operation of pumping stations.

Sewer inspection should be considered as one of the most important tasks of sewer maintenance. It provides information concerning the condition of the sewers and their auxiliaries and concerning the quality of the sewage. It thus helps to ensure that the sewage flows properly, and to forestall dangers such as interruption of flow, danger to traffic, and damage that might be caused by the quality of the sewage.

Sewer inspection is carried out from the outside and from the inside.

External inspections are carried out every month or every three months (depending on local conditions) by a crew of 3 workers. This crew looks out for all unusual changes that indicate disturbance of sewage flow. So far as possible, each crew should always be made up of the same people, since they are more likely to recognize disturbances in the system due to their experience.

Each sewer is inspected internally at certain fixed intervals, from one to four times a year, depending on its importance and operating conditions; the workers inspect the accessible sewers and note whether there is damage, solid deposits on the invert, and abnormal infiltration or leakage. On the basis of their report, the sewers are then cleaned, flushed, repaired, etc. Inaccessible sewers are inspected using mirrors and a light (see fig. 4.6.-1) or with special television cameras.

The sewage is checked as regards quantity or quality. The quantity of sewage is measured in order to obtain definite information concerning the actual discharge capacity of the sewers. The quality of the sewage is tested by analyses. For normal operation the quality is known; fluctuations could be brought about by the introduction of unusual or toxic wastes, which might have a damaging effect on the sewer and the treatment plant.

Cleaning and flushing of the sewers are determined by the type of construction.

A well planned and properly constructed sewer system does not normally require cleaning or flushing; this applies particularly to combined sewer systems. In such systems, inaccessible sewers should only need cleaning once a year.

Sewers can be cleaned with
- hand equipment in accessible sewers,
- automatic (mech. or hydr.) equipment in inaccessible sewers.

To clean sewers by hand, shovels, spades, etc. are used. The sludge thus collected is carried to the manhole in special small barrows and from there brought out of the sewers.

Inaccessible sewers are cleaned mechanically. High pressure flushing devices ($(60 - 80) \cdot 10^5$ Pa) are commonly used.

Road gullies must be cleaned regularly to prevent blockage and fermentation of the deposits. For this purpose, special eductor-basin cleaners are used.

Sewer repairs can be divided into
- cosmetic or regularly scheduled repairs and
- replacements.

Regularly scheduled repairs are carried out according to the arranged schedule. These are usually of a preventive nature: a real repair is one that is made necessary by unexpected circumstances. The available funds for repair and maintenance must be used economically and rationally. It should be noted that supposed savings on maintenance work may lead in the end to a need for more costly replacements.

The operation of pumping stations must follow the manufacturer's instructions, and they must be used according to need. Yearly, three monthly, or monthly operation schedules should be arranged, so that pumping station managers can work out detailed schedules.

4.6.2.2 Safety Measures in the Sewer System

During construction and operation, strict measures are taken to protect workers from possible accidents, disease, asphyxiation, poisoning, explosions, electric shocks, etc.

For the same purpose, training courses, lectures, conferences, etc. should also be organized, to help workers learn how to protect themselves. This presupposes, however, that there are enough personnel and safety installations to enable the safety requirements to be observed.

Administrative measures and announcements alone are not enough.

Regular health check-ups for the workers are essential, as well as appropriate preventive measures where toxic substances have to be dealt with. Further training of the workers with regard to the safety requirements must take place every year.

5 CONDITIONS FOR THE DISPOSAL OF INDUSTRIAL AND MUNICIPAL WASTE WATER AND SLUDGES

When setting conditions for the discharge of industrial effluents, i.e. limit values corresponding to certain pollution parameters, and making arrangements for checking the quality of the water, a distinction must be made between direct and indirect discharge.

<u>Direct discharge</u> refers to wastes that flow directly into the receiving water via an in-plant sewer system and treatment plant.

<u>Indirect discharge</u> refers to wastes that flow via a public sewer system into a community treatment plant and from there directly into a water body.

The pollution parameters to be checked vary according to the method of sewage removal and treatment (combined or separately in the community treatment plant), and the available receiving water and its quality.

5.1 General Requirements for Drainage of Industrial Effluents in Public Sewerage Systems (Indirect Discharge)

When industrial effluents are treated in community sewage plants, and when they are discharged into the sewer system, the possible effects of special industrial wastes on the treatment plant and on the sewers, i.e. their operation, lay-out, and cleaning capacity, should be considered.

A whole system for the control of direct and indirect discharge of industrial wastes can be seen in fig. 5.1.-1.

General principles for defining the pollution parameters to be checked are:

1. The effluents must not constitute a nuisance or a danger to the environment and especially to sewerage workers (sewer system and treatment plant).
2. The sewerage system must not be damaged. Operation must not be

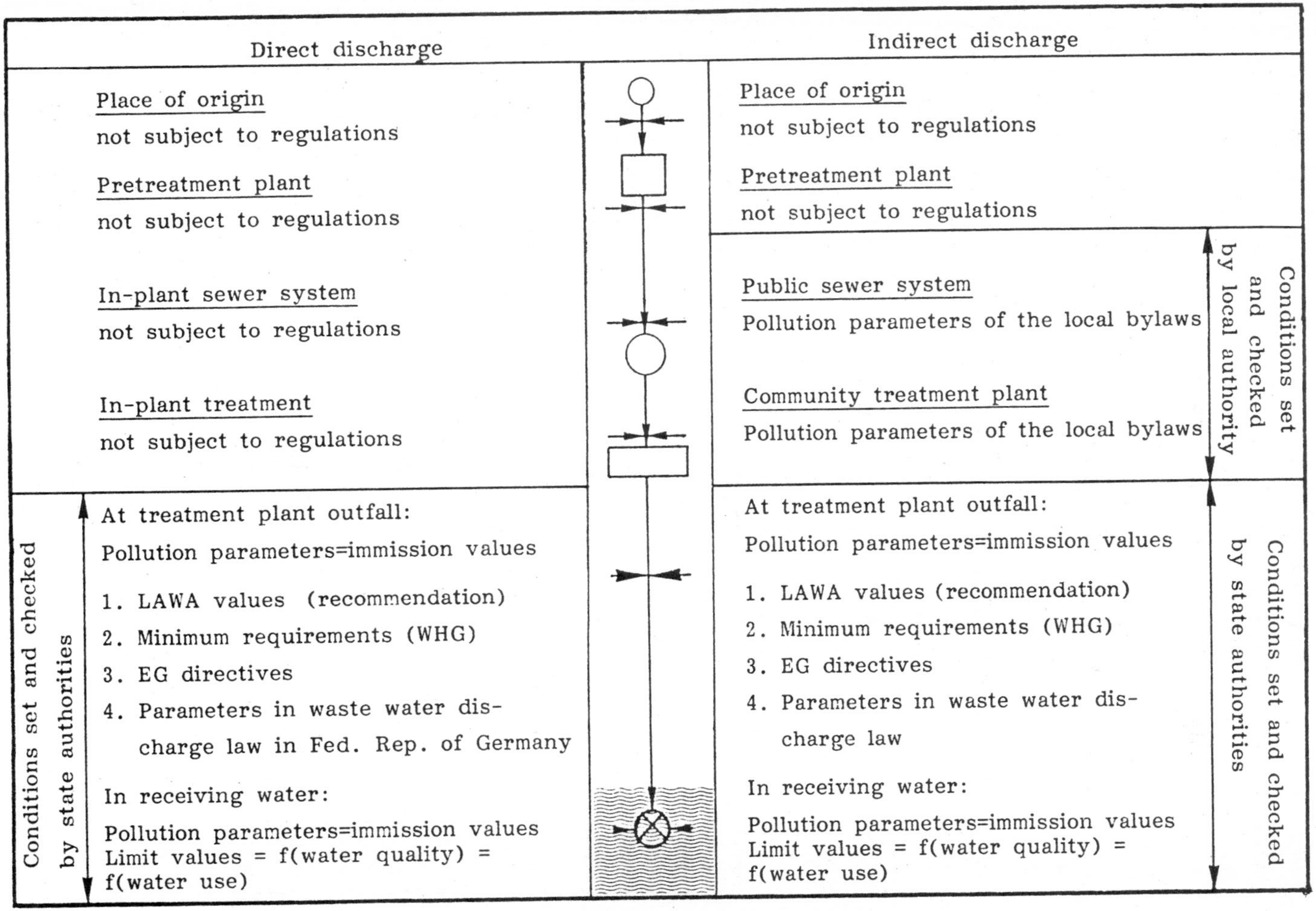

Fig. 5.1.-1 Pollution parameters and their control

disturbed, and efficiency must not be affected.
3. Substances that would pass the central treatment plant but that are not suitable for discharge into the receiving water must be retained.
4. Substances to be discharged at one point should not, when considered together with other neighboring discharges, create a total pollution load exceeding the permitted load for the water body, in order to prevent, for example, excess salination.
5. Total treatment necessary must be minimized.

In general terms it is also possible to define the following areas needing protection (see Table 5.1.-1).

Tab. 5.1.-1: Basic requirements of draft guidelines "Indirect discharge Baden-Württemberg" (July 1977) and ATV Worksheet A 115 (Dec. 1970) /4/

		Areas needing protection
1	Personnel	– against H_2S, HCN, SO_2, CO_2 extreme pH values, excessively high temperatures
2	Structures	– against attack/destruction due to extreme pH values, sulfates, lime-removing carbon dioxide – against deposits of sand, debris, ash, refuse – against crystallization if there is a high slat content
3	Functioning of treatment plant	– against reduction in capacity/stoppage due to excessive concentrations of heavy metals, cyanides, acids, alkalis
4	Quality of receiving water	– against undesirable concentrations of heavy metals, salts, oxygen depletion

5.1.1 General Requirements and Regulations for Sewerage Systems

When industrial wastes are discharged into a municipal sewerage system, they may, due to their special properties, cause great damage in the sewers and in the treatment plant. They may:
- attack or even destroy the normal sewer materials (concrete, reinforced concrete, cast iron), or structures and metal installations, gratings, rakes, etc.;
- leave oil and fat films on the sewer walls, leading to blockage of the sewers and development of anaerobic microorganisms. This may cause the sewage to begin putrefying in the sewers, making treatment in the sewage plant more difficult;
- constitute a physical nuisance and health risk to sewerage workers, due to high temperatures above 35 °C, disturbing substances with intolerable odors and disturbing gases. Settleable solids and toxic substances can interrupt operation, or may even lead to serious accidents.

Sometimes, the anticipated degree of disturbance is so great that normal materials and metal parts cannot be used.

In the Federal Republic of Germany, conditions for discharge into the public sewer system and the public treatment plant are laid down in local bylaws by the municipal authorities.

The ATV Worksheet A 115 /4/ was written to serve as a general guideline for making such bylaws. This Worksheet, "Directions for the Discharge of Waste Water into a Public Treatment Plant", describes general principles for the discharge of waste water into public sewerage systems, and gives in the annex parameters and corresponding limit values for substances contained in the waste water (Table 5.1.-2).

Worksheet A 115 also gives directions for registering and monitoring waste water discharges, and for analyzing samples and evaluating the results. It also gives information showing how this relates to statutory regulations, rules, and standards.

The basic rule is:

It is generally forbidden to discharge into a public sewer system any substances which might block the sewers, which form toxic, foul-smelling or explosive vapors and gases, or which attack construction materials to a serious degree.

This includes the following substances:
- rubble, ashes, glass, sand, refuse, bristles, fibers, leather scraps, textiles, etc. (these solids must not be discharged even after comminution);
- synthetic resin, plastic, cement, calcium hydroxide;
- draff, yeast, latex, bitumen and tar and their emulsions, oil emulsions, lacquers, liquid wastes that will harden;
- petroleum, fuel oil, lubricating oils, vegetable and animal oils, chlorinated hydrocarbons, acids, and alkalis;
- phosgene, hydrogen sulfide, hydrogen cyanide, and hydrazoic acid and their salts, carbides that form acetylene. It is also forbidden to discharge substances which are definitely toxic.

5.1.2 General Requirements and Regulations for Public Sewage Treatment Plants

The consequences of commercial and industrial wastes for community treatment plants depend on the composition of the wastes.

The individual stages of treatment in sewage plants can be affected by the various industrial wastes as follows:

<u>Screens and sand traps</u>: Screens must be made of suitably resistant material, according to the aggressivity of the sewage. All concrete parts must also meet this requirement. Industrial waste water can considerably increase the volume of screenings and grit. In this case, correspondingly larger land fill areas must be provided for, and, possibly, more frequent cleaning.

Efficiency can be adversely affected by coarse floating matter, fibrous material, animal substances, etc. This is particularly so with automatic cleaning and comminuting devices for screenings. It is important that no backwater be formed.

Sedimentation: In primary sedimentation, efficiency is affected by a number of factors:
- High temperatures, e.g. due to discharged cooling waters, reduce the viscosity of the sewage, and this raises the rate of sedimentation.
- The density is chiefly affected by temperature and dissolved substances, e.g. salts; density currents can occur that interfere with sedimentation.
- The rate of sedimentation of lighter floating particles is very low. This seriously affects the sedimentation process. The same effect may be produced when the sewage is diluted, e.g. due to inflow of cooling waters. Among other measures, additional chemical flocculation is required.
- If the volume of sludge increases, the sludge hoppers must be emptied more frequently. This is particularly important with organically polluted sewage, which putrefies rapidly. The hot climate in most developing countries is an important factor here.
- Certain constituents may adversely affect the consistency and flow capacity of the sewage, e.g. fibrous material, but also clay, sand, or coal slime. This makes sludge removal more difficult.

Chemical flocculation and precipitation: This process may be affected by industrial wastes in the following ways:
- The effect of the flocculating agent is bound to a certain pH range.
- The degree and frequency of fluctuations in the amount and quality of the inflowing sewage interferes with the process.
- The process is adversely affected by protective colloids.

Biological treatment: Biological decomposition of the sewage pollutants in a municipal treatment plant may be checked or hampered by industrial wastes due to the following problems:
- The nutrient requirement of the microorganisms is not met. In industrial wastes, there is often a lack of nitrogen and phosphorus, e.g.

Tab. 5.1.-2: Conditions for discharge into a sewerage system /4/

1. General parameters	
a) temperature:	35 °C
b) pH:	6.5 to 10
c) Settleable substances, if separation of sludge is required:	1 ml/l after 0.5 hours settling time
2. Saponifiable oils and fats	250 mg/l
3. Hydrocarbons	
a) Directly separable:	DIN 1999 to be noted separators for light fluids)
b) If removal of light fluids is required beyond using a separator: total hydrocarbons (to DIN 38 409 Part 18)	20 mg/l
4. Organic solvents	
a) Wholly or partially miscible with water and biodegradable: depending on special condition, but maximum standard value always corresponding to degree of solubility.	
b) Halogenated hydrocarbons (calculated as organically bound halogen):	10 mg/l
5. Inorganic substances (dissolved and suspended)		
a) arsenic	(As)	1 mg/l
b) lead 1)	(Pb)	2 mg/l
c) cadmium 1) 2)	(Cd)	0.5 mg/l
d) chromium hexavalent	(Cr)	0.5 mg/l
e) chromium 1)	(Cr)	3 mg/l
f) copper 1)	(Cu)	2 mg/l
g) nickel 1)	(Ni)	4 mg/l
h) mercury 1) 2)	(Hg)	0.05 mg/l
i) selenium	(Se)	1 mg/l
j) zinc 1)	(Zn)	5 mg/l
k) tin	(Sn)	5 mg/l
l) aluminum and iron	(Al) (Fe)	unlimited, provided no technical problems during treatment are anticipated

cont. Tab. 5.1.-2

6. Inorganic substances (dissolved)		
a) ammonium	(NH_4)	
and ammonia	(NH_3)	5 mg/l
b) cyanide, easily set free	(CN)	1 mg/l
c) total cyanide 3)	(CN)	20 mg/l
d) fluoride	(F)	60 mg/l
e) nitrite	(NO_2)	if large loads occur: 20 mg/l
f) sulfate 4)	(SO_4)	600 mg/l
g) sulfide	(S)	2 mg/l

7. Organic substances	
a) phenols volatile in steam (as C_6H_5OH): 5)	100 mg/l
b) dyestuffs:	Only in such low concentrations that after mechanical and biological treatment in a sewage plant the effluent does not visually color the receiving water.

8. Substances that spontaneously consume oxygen	
e.g. sodium sulfite, ferrous sulfate	Only in such low concentrations that no anaerobic conditions are created in the sewer system.

1) If the sludge from the treatment plant is put to agricultural use, assuming this is the best method of sludge disposal, in the sense of recycling, and that the distribution area is conveniently located, then the relevant instruction sheets should be noted, and if necessary a limit must be set to the amount of heavy metals that may be discharged in industrial wastes.

2) Waste water streams containing these substances usually require separate treatment.

3) For very small discharges, no concentration limit is needed.

4) In individual cases, higher values may be permitted, depending on construction materials, dilution, and local conditions.

5) This value may be raised, depending on the type of phenolic substances; for toxic and not easily biodegradable phenols, however, it must be substantially lowered.

in paper and pulp wastes, wastes containing phenol from low-temperature carbonization plants, coking plants, and gas works.

The nutrient balance should if possible have an optimal ratio

$$C : N : P = 30 : 3 : 1 \text{ and } BOD_5 : N : P = 100 : 5 : 1$$

Critical values are degradable $BOD_5 : N = 32 : 1$ and degraded $BOD_5 : P = 150 : 1$

- The pH of the sewage is unfavorably changed. The favorable range is 6.5 to 9. Biological treatment is still possible within the range 5.0 to 9.5. The situation becomes particularly critical when unexpected fluctuations occur.
- The waste water may contain inorganic or organic toxic substances, which may considerably affect the biocenoses and their decomposition efficiency. Highly toxic substances should in all cases be removed before the wastes go to the community treatment plant, i.e. through in-plant treatment.
- Extreme temperatures of < 8 and > 35 °C can have a very negative affect on the life of the microorganisms, and this greatly reduces the efficiency of biological treatment, especially when large changes occur suddenly.
- Oils and detergents may hinder the supply of oxygen, and create anaerobic conditions.
- Various factors can lead to a shortage of oxygen, such as excess growth of microorganisms due to wastes rich in nutrients, for example from dairies, breweries, and distilleries. In this case, additional aeration is necessary in order to supply oxygen.
- High concentrations of iron, very hard waste waters, or the clayey nature of many industrial wastes may hinder the process of decomposition.
- Surge discharges affect the efficiency of biological treatment both by their amount and their concentrations of pollutants.

Disturbances of biological treatment affect the individual processes to varying degrees.

5.2 General Requirements and Regulations for the Discharge of Industrial Wastes into Water Bodies (Direct Discharge)

Cooling waters contain few or no pollutants. They should therefore be conveyed directly to the receiving water, separately from other wastes. However, they introduce heat into the receiving water and cause "thermal pollution", which alters the physical, chemical, and biological processes in the water, which may be speeded up to a certain extent; the oxygen reserves in the water are then used up faster than normal. At the same time, less oxygen is absorbed from the atmosphere. The lower viscosity of the water results in a higher sedimentation rate. Some toxic substances in the water may have a stronger effect due to the higher temperature.

Industrial wastes can contain a large volume of organic pollutants, which constitute a similar load for the water body to that of domestic or municipal sewage. However, they are often hygienically safer, because as a rule they carry no pathogenic germs.

It is a special characteristic of many commercial and industrial wastes that they contain substances that can seriously interfere with the natural biological self-cleaning process in the water body. These substances can be grouped as follows, according to their harmful effects or properties:

- <u>toxic substances</u>, which cause acute or chronic poisoning of the various organisms in the water, or to the consumers of the treated water.
- <u>nuisance substances</u>, which give rise to odor, taste, color, turbidity, and technical problems during treatment, conveyance, and use.
- <u>consuming substances</u>, which upset the oxygen balance in the water.
- <u>nutrients</u>, which cause eutrophication of standing or slow-flowing waters.

Harm caused through the use of the water as a pollutant carrier for commercial and industrial wastes adversely affects other uses, especially potable use, the fishing industry, and agriculture and cattle raising; in addition, it interferes with the self-cleaning capacity of the water and the ecology generally.

Before discharging wastes into the receiving water, harmful pollutants must be reduced by appropriate treatment to a level that will not upset the balance of the water or affect its use.

5.2.1 Classification System for Water Quality

Not all receiving waters are used for the same purposes, and therefore need not all achieve the same quality; rather, different quality characteristics must be attained and guaranteed according to the different uses.

Water quality standards ought to be defined for all kinds of uses. However, at present even in the industrial countries there are still no comprehensive quality standards for all the different uses of water. Such standards are at present being worked out by national and international committees.

Tab. 5.2.-1: Classification of quality of a water body /28/

Class	O_2 - content		O_2 - consumption		BOD_5
	mg/l at 20°C and 760 Torr*	% saturation	mg/l at 20°C	%	mg/l at 20°C
I	8.45 - 8.84	95 - 100	0 - 0.3	0 - 5	0 - 0.5
I - II	7.50 - 8.45	85 - 95	0.3 - 1.1	5 - 10	0.5 - 2.0
II	6.20 - 7.50	70 - 85	1.1 - 2.2	10 - 20	2.0 - 4.0
II - III	4.40 - 6.20	50 - 70	2.2 - 3.8	20 - 40	4.0 - 7.0
III	2.20 - 4.40	25 - 50	3.8 - 7.0	40 - 70	7.0 - 13.0
III - IV	0.90 - 2.20	10 - 25	7.0 - 12.0	70 - 95	13.0 - 22.0
IV	0 - 0.90	0 - 10	12	95	22

*1 Torr = $0.0133 \cdot 10^5$ Pa $1 \cdot 10^5$ Pa = 750 Torr

In order to improve the general condition of surface waters, classes to describe the quality of the water have been introduced in various countries.

Four classes have been set up to classify the quality of flowing waters (LIEBMANN and HAMM); the characteristic parameters of these classes, corresponding to the biological and biochemical properties of the water, are shown in Table 5.2.-1.

The extraction of water for industrial use may possibly still be economical up to class III, but definitely not if classes III-IV or IV predominate.

In certain West European countries, classes II-III are aimed at in critical times, with regard to the general quality of surface waters.

Table 5.2.-2 shows a comparison of the water qualities aimed at in some West European countries.

Tab. 5.2.-2: Requirements for surface waters (Q_{cr})

Country	Water quality	BOD_5 mg/l	O_2 content mg/l	Law
Netherlands	II I - II	5 3	5 8	Minimum quality WVO 1970/JMP
Switzerland	II	4	5	Law for control of water pollution 1971
Poland	II	4	6	Water base law
Sweden	full biological treatment, normally phosphate precipitation			Nature protection 1964 Environmental protection 1969
Fed. Rep. of Germany	II II - III	4 7	6 5	Environmental Control Program 1972 Board of Experts 1976

In some Eastern European countries, e.g. Poland, Rumania, etc., surface waters are divided into 3 classes according to the use of the water. In Poland these are:
- water of quality class I (= 24.1 % of river courses) should meet all requirements necessary for potable use by the population and for the

food processing industry, and should be capable of supporting salmonids (trout, etc.),
- water of quality class II (= 33.7 % of river courses) should be suitable for cattle watering tanks, raising other types of fish, as well as bathing and water transport.
- water of quality class III (= 18.2 % of river courses) can be used in industry (except for the food processing industry) and for irrigation purposes.

A classification system for individual developing countries or groups of countries can only be attempted when detailed information is available concerning the multiple and various conditions and reaction mechanisms, as well as all the purposes the water is used for in the different countries.

Natural water and reclaimed waste water are used in the artificial water cycle for various purposes. These uses and their direct relationship to the problems of the developing countries can be listed as follows:
a) direct or indirect extraction:
   - utilization for human and animal consumption (drinking water, food)
   - utilization in agriculture (irrigation);
   - utilization in industry and commerce (industrialization);
   - utilization as cooling water (power production);
b) utilization for assimilation of waste water (waste water discharge);
c) utilization for fishing industry (food);
d) utilization for recreational purposes (tourism, sport, landscape aesthetics);
e) ecological functions.

For all these uses, quality standards should be defined for each country individually. The selection and permissible concentration limits of the quality parameters depend on the standard of living in each case, and the corresponding required standard of water quality.

The first step should be to set limits for harmful substances, or permissible concentration limits, as a basic essential and not so much related to utilization. The values given are only meant to serve as a guide in this matter.

In general there are two methods for setting limit values:
- immission standards and
- emission standards.

5.2.2 Limit Values for Immission

Limit values for immission are criteria for water quality that describe the conditions in a water body. In the industrial countries, limit values have now been set e.g. for waters intended for potable use and for inland and sea waters used for bathing. These standards have been worked out by the EC states.

The immission value method has the following advantages:
- better overall economy of treatment plants, since sewage treatment is directly related to the quality of the receiving water, with no excessive local requirements as to the degree of treatment.
- regulation of the treatment processes in such a way that the water can be put to the proposed uses.
- consideration of the total pollution of the water body and the self-cleaning dynamics of water bodies.
- possibility of setting different limit values for different treatment plants in order to preserve the ecological balance of the water.

In principle, this method has the following disadvantages:
- there are still no practical directions for the frequency of sampling, for permissible departures from the standards due to flooding or disasters, for extraordinary meteorological or geographical conditions with regard to certain parameters, or for natural enrichment of certain substances, and for shallow lakes ($H < 20$ m) with low water exchange (> 1 year);
- difficulties in determining and checking the discharge into the water of effluents from treatment plants, since the assimilation capacity and quality of the water depend to a large degree on the receiving water flow.

The following immission regulations have been elaborated:
- the draft of the Strasburg Convention in its entirety with the exception of Art. 5,
- the EC directive concerning the quality required of surface water intended for the abstraction of drinking water in the member states, of June 16, 1975, and
- the Council directive concerning the quality of bathing water of December 8, 1975.

5.2.3 Limit Values for Emmission

The principle of emission standards, which is preferred by some industrial states, gives discharge conditions for concentrations at the outfall of a sewer system or treatment plant.

Emission values have the following advantages:
- requirements concerning the quality of treated waste water are the same in all places and independent of receiving water flow;
- the discharge of treated effluents into the receiving water is easy to check;
- the required degree of treatment in sewage plants is easier to determine.

Emission standards have the following disadvantages:
- It is possible that in many treatment plants, even though the emission standards are met, the pollution limits for the receiving water are exceeded;
- operation costs for sewage treatment may be higher since there is no dependent relationship between waste water quality and receiving water flow, i.e. the flow and self-cleaning capacity of the receiving water is not considered, and its reserve purification capacity is not utilized. This is very important in developing countries where there are large rivers; to neglect it is often uneconomical. The treatment capacity of sewage plants can often be kept relatively small.

In principle, waste water need not be treated any more than is required by the self-cleaning capacity of the receiving water.

Many international conventions to control chemical pollution and to protect the seas are based on the emission principle, e.g.:
- the Oslo convention of February, 15, 1972 for the North Atlantic region and the London convention of December, 29, 1972 for all oceans, which serve to control marine pollution due to discharge of wastes from ships and aircraft,
- the Helsinki convention of March 1974 (Baltic Sea) and the Barcelona convention of March, 4, 1976 (Mediterranean), and finally
- the Paris convention of June, 4, 1974 concerning the control of marine pollution due to discharges from land and in this respect also of particular importance for inland waters.

In developing countries where these conventions are recognized and efforts are made to observe them, it is recommended in the interests of unity that the emission principle should also be used for discharge into inland waters.

5.2.4 National and International Standard Values

5.2.4.1 LAWA Emmission Values (Recommended)

The standard values of the Länderarbeitsgemeinschaft Wasser (LAWA) for domestic and industrial waste water give <u>discharge concentrations that can be attained</u> using customary methods. The standard values are related to specific treatment processes and should therefore not be simply equated with requirements for the quality of waste water discharged into a water body /93/.

Some standard values for industries that may be significant in developing countries are shown in the following section. Standard values cannot be set for all industrial wastes, since the composition of wastes from large-scale chemical industries, iron and steel works, coking plants, and some other works frequently varies and fluctuates.

In such cases, expert advice should be sought in setting the values, or values from similar production areas should be taken as a basis.

Domestic sewage

Domestic sewage in dry weather:

a) settleable solids — 0.3 ml/l
b) floating material — none to be visible

Biological treatment:

Full biological treatment

a) settleable solids — 0.3 ml/l
b) floating material — none to be visible
c) putrescibility — negative
d) permanganate value — 100 mg/l

Higher values are permissible if the BOD_5 is below the given value (e.g. for substances which do not degrade easily).

e) BOD_5 — 25 mg/l

Partial biological treatment

If partial biological treatment is sufficient, the requirements may be to some extent reduced; however, at least the following values must not be exceeded:

a) settleable solids — 0.3 mg/l
b) floating material — none to be visible
c) permanganate value — 150 mg/l
d) BOD_5 — 80 mg/l

Domestic sewage combined with storm water:

Preliminary note: It may be assumed that treatment plants are deemed to satisfy the requirements when the mechanical treatment facilities can treat up to $5 \cdot (1 + 4)$ times the dry weather flow, and the biological treatment facilities can treat up to 1.5 to 2 (1 + 0.5 to 1 + 1.0) times the dry weather flow.

With suitable additional measures for storm water treatment it is possible that the given values for settleable solids during wet weather are not exceeded by more than 100 %.

Settleable solids — 0.6 ml/l

	A	B	C	D	E	F
d) permanganate value: mg/l	150	300	400	400	400	–
e) BOD_5: mg/l	50	200	200	200	200	300
f) bleaching agent:			not detectable			

Biological treatment:

	A	B	C	D	E	F
a) settleable solids: ml/l	0.3	0.3	0.3	0.3	0.3	0.3
b) suspended solids: mg/l	20	20	20	30	30	30
c) permanganate value: mg/l	100	100	150	150	200	–
d) BOD_5: mg/l	25	25	25	30	40	30

Higher permanganate values are permissible in certain circumstances, if the BOD_5 values listed are not exceeded.

Wastes from tanneries, fur finishing plants, and leather factories

These wastes should only be treated separately if it is not possible to treat them together with domestic sewage; it is recommended that expert advice be sought.

Mechanical treatment:
a) settleable substances: 0.3 ml/l
b) suspended solids (floating material): none to be visible

Chemical treatment:
a) settleable solids: 0.3 ml/l
b) pH: 6.0 to 9.0
c) permanganate value: 400 mg/l
d) BOD_5: 200 mg/l
e) sulfides (S^{2-}): 2 mg/l
f) total chromium (Cr): 2 mg/l
  as chromate (Cr-VI): 0.5 mg/l
g) substances extractable with petroleum ether: 10 mg/l

Biological treatment:
a) settleable solids: 0.3 mg/l
b) permanganate value: 150 mg/l
c) BOD_5: 80 mg/l

These values may be exceeded if a reduction of 95 % is attained in comparison to the mechanically treated waste water.

Wastes from the petroleum industry

Mechanical treatment:
a) settleable solids: 0.3 ml/l
b) floating material: none
c) substances extractable with petroleum ether   20 mg/l

Chemical treatment:
a) settleable solids: 0.3 ml/l
b) floating material: none
c) pH: 6.0 to 9.0
d) substances extractable with petroleum ether: 10 mg/l
e) sulfides (S^{2-}):   1 mg/l

Biological treatment:
a) settleable solids: 0.3 ml/l
b) floating material: none
c) oil streaks: none
d) BOD_5:   30 mg/l
e) substances extractable with petroleum ether: 5 mg/l
f) sulfides: no longer detectable
g) phenols (volatile in steam):   0.5 mg/l

Coal washing wastes

Mechanical treatment:
a) settleable solids: 0.3 mg/l
b) pH: 6.0 to 9.0

Wastes from the nonmetallic minerals industry and gravel washing plants

The suspended sediment will not be removed by mechanical treatment. If it constitutes a nuisance, it must be precipitated through the addition of appropriate agents.

Special values must be set for industries using chemical additives (e.g. lime, cement) or producing soluble mineral products (gypsum).

Mechanical treatment:
a) settleable solids: 0.3 ml/l

b) pH:	6.0 to 9.0

Chemical treatment:

a) settleable solids:	0.3 ml/l
b) suspended solids:	50 mg/l
c) pH:	6.0 to 9.0

Iron pickling wastes

Chemical treatment:
(neutralization, separation of precipitated metal salts)

a) settleable solids:	0.3 ml/l
b) pH:	6.0 to 9.0
c) total iron (Fe):	2 mg/l

Electroplating wastes

Chemical treatment:
(oxidation of cyanides, reduction of chromates, separation of precipitated metal salts, neutralization)

a) settleable solids:	0.3 ml/l
b) pH:	6.5 to 9.0
c) metals (sum of dissolved and suspended):	
total chromium (Cr):	2 mg/l
as chromate (Cr-VI):	0.5 mg/l
copper (Cu):	1 mg/l
nickel (Ni):	3 mg/l
zinc (Zn):	3 mg/l
cadmium (Cd):	3 mg/l
total iron (Fe):	2 mg/l
d) cyanides (destructible by chlorine):	0.1 mg/l
e) free chlorine:	0.5 mg/l
f) substances extractable with petroleum ether:	10 mg/l

N.B.

ad b): Nickel can only be satisfactorily precipitated with a pH over 9; expert advice should therefore be sought if the waste water has a high nickel content.

ad c): In special cases, lower values for individual metals can be attained by further treatment, e.g. ion exchange.

Ion exchange recirculation plants produce concentrated regenerates. If these are treated using customary methods, the metal content of the discharge should be limited in special cases according to the needs of the receiving water. Expert advice should be sought in these cases.

ad d): Direct determination of the cyanide ion with pyridine/benzidine (Deutsches Einheitsverfahren, System-Nr. D13/3) is for all practical purposes sufficient to detect the cyanide destructible by chlorine, including cyanogen chloride if still present.

If only this direct determination is possible, after distillation with tartaric acid solution, cyanides can to some extent also be detected in complexes that are not destroyed by chlorination and which are also considerably less toxic than free cyanide.

Until a regulation is available in the "Einheitsverfahren", cyanide destructible by chlorine should be determined following an ASTM method:

Two waste water samples are taken; to one of these, with pH 11 - 12, enough NaOCl solution is added to leave an excess active chlorine amount of 5 - 25 mg/l after 1 hour with gentle stirring. The remaining chlorine is then removed with the required amount of sulfite, and the cyanide content determined after distillation with sulfuric acid. The difference between this and a second determination without chlorine treatment gives the amount of cyanide destructible with chloride.

Eloxal wastes

Chemical treatment:
(neutralization; separation of suspended solids)

a) settleable solids:	0.3 ml/l
b) pH:	6.0 to 9.0
c) total chromium (Cr):	2.0 mg/l
as chromate (Cr-VI):	0.5 mg/l
d) fluoride (F^-):	20 mg/l
e) sulfate (SO_4^{2-}):	

ad e):

The sulfate content should be limited according to the special needs of the factory and the receiving water; expert advice should be sought, and regulations must be observed.

Hardening shop wastes

Used hardening salts and solid salt from the floor must not be allowed into the waste water; they must be collected and reused, e.g. by the manufacturer of hardening salts.

Chemical treatment of wastes:
(after detoxication and neutralization)

a) settleable solids:	0.3 ml/l
b) pH:	6.0 to 9.0
c) cyanides (CN^-), (easily set free):	0.1 mg/l
d) free chlorine (Cl_2):	0.5 mg/l
e) nitrite (NO_2^-):	20 mg/l
f) substances extractable with petroleum ether:	10 mg/l

Glass processing wastes

Chemical treatment:
(neutralization with calcium hydroxide! Separation of suspended solids)

a) settleable solids:	0.5 ml/l
b) suspended solids:	50 mg/l
c) pH:	7.0 to 9.0
d) fluoride (F^-)	20 mg/l

Wastes from mineral water, table water, and lemonade production

Mechanical treatment:

a) settleable solids:	0.3 ml/l
b) pH:	6.0 to 9.0
c) temperature:	30° C

Biological treatment:

a) settleable solids:	0.3 ml/l
b) BOD_5	25 mg/l

Wastes from potable and industrial water treatment works

Potable and industrial water treatment works include:
- iron removal plants
- manganese removal plants
- acid removal plants
- bathing water treatment plants
- salt removal plants
- boiler feed treatment plants, including elutriation waters
- plants for backwashing waters from filters and screens

Cleaning waters are not included.

Mechanical treatment:

a) settleable solids:	0.3 ml/l
b) pH:	6.0 to 9.0
c) floating material:	none to be visible
d) temperature:	30° C
e) free chlorine (Cl_2):	0.3 mg/l
f) salts:	

Remarks

ad b): In special cases, a pH value up to 9.5 is permissible, if this is the value of the treated water, as permitted by the statutory regulation for drinking water.

ad f): Ion exchangers produce highly concentrated regenerates. The salt content should be limited in special cases according to the needs of the receiving water; expert advice should be sought.

Wastes from cleaning cloth laundries

These wastes should only be treated separately if it is not possible to treat them together with domestic sewage. Organic solvents used in the plant should not be allowed into the waste water.

Mechanical treatment:

a) settleable solids:	0.3 ml/l

(Higher values are permissible provided the maximum value for suspended solids is not exceeded.)

b) suspended solids:	50 mg/l
c) floating material:	none to be visible
d) pH:	6.0 to 11.0
e) temperature:	30° C

Chemical treatment:
a) settleable solids:	0.3 ml/l
b) suspended solids:	30 mg/l
c) pH:	6.0 to 9.0
d) substances extractable with petroleum ether:	10 mg/l
e) temperature:	30° C

Canning wastes

Preliminary remark:

These wastes should only be treated separately if it is not possible to treat them together with domestic sewage. (Neutralization of the wastes may sometimes be necessary!)

Mechanical treatment:
a) settleable solids:	0.3 ml/l
b) suspended solids:	30 mg/l
c) floating material:	none to be visible
d) pH:	6.0 to 9.0

Biological treatment:

Full biological treatment

a) settleable solids:	0.3 ml/l
b) suspended solids:	30 mg/l
c) BOD_5:	30 mg/l
d) putrescibility:	negative

Partial biological treatment

Storage in ponds, possibly with artificial aeration, counts as partial biological treatment, provided at least the following values are not exceeded:

a) settleable solids:	0.3 ml/l
b) suspended solids:	30 mg/l
c) BOD_5:	150 mg/l

Potato processing wastes

These plants include:

potato steaming plants
production of potato chips and French fries
potato peeling plants
potato canning plants
production of potato puree and flakes

These wastes should only be treated separately if it is not possible to treat them together with domestic sewage.

Mechanical treatment
(possibly with neutralization)

a) pH:	6.0
b) settleable solids:	0.3 ml/l
c) floating material:	none to be visible

Biological treatment

Full biological treatment

a) settleable solids:	0.3 ml/l
b) floating material:	none to be visible
c) BOD_5:	30 mg/l
d) substances extractable with petroleum ether:	5 mg/l
e) putrescibility:	negative

Remarks to d):
This parameter should be determined e.g. for wastes from chips and French fries production.

Partial biological treatment

Storage in ponds, possible with artificial aeration, counts as partial biological treatment, provided at least the following values are not exceeded:

a) settleable solids:	0.3 ml/l
b) suspended solids:	30 mg/l
c) BOD_5:	150 mg/l
d) substances extractable with petroleum ether:	5 mg/l

Starch production wastes

These wastes should only be treated separately if it is not possible to treat them together with domestic sewage.

Mechanical treatment

a) settleable solids:	0.3 ml/l

Tab. 5.2.-4: Minimum requirements for various industrial branches (cont.)

Minimum requirement according to § 7 a WHG / Industry / Product	Type of measurement	Parameter	Settleable solids (single sample)	Filtrable solids (single sample)	COD	BOD_5	Fish toxicity	Cadmium	Total chromium	Trivalent chromium	Bivalent chromium	Lead	Zinc	Iron	Copper
Wood-containing p.	(kg/t)	24h	0.3		5	0.8									
Mainly from used p.	(mg/l)	24h				25									
Real parchment	(kg/t)	24h	0.3		6	1.2									
		24h	0.3		12 − 6										
Clothing	Conc.	2h	0.3	40	280	40	4		2	0.1		3			1
Textile finishing		24h			200	30	3								
Leather finishing	"	2h	0.3		250	25	4		2	0.05					
		24h			200	20	4		1	0.05					
Iron works and steel processing (general)	"	2h	0.5		100							0.5	3	20	
Deviating values for foundries	"	2h	0,8		200										
pipe manufacture	"	2h			200										
lead-coating, patinating	"	2h										2			
tin plate manufacture	"	2h			1.3 (kg/t)										
Nonferrous metallurgical works	"	2h	0.3					1	2		0.5	2	5	20	2
Coking plant	Load+		1												
	Conc.	2h	0.8++		64 (for process water)										
Ore dressing	"	2h	0.3	70											
Nonferrous ore				(mg/l)	20							0.5	3	3	
(submarine, hydrothermal)	"	2h	0.3	70	180x		/4/					0.5	3	3	
Fluorite	"	2h	0.3	70	/80/									3	
Barite	"	2h	0.3	70	80										
Graphite	"	2h	0.3	70	65										
Coal mine	"	2h	0.3	100	100										
coal dressing				(mg/l)											
with briquetting plant	"	2h	0.3	100	200										
Refineries	"	2h	0.3	/180/	/35/	/3/									
		24h		130	25	2									
Glass processing	"	2h	0.5	200 (without silver-plating)											
Discharge 1000 kg/a (raw)		2h		250 (with ")											
		2h	0.5	160 (without ")											
Discharge 1000 kg/a		2h		200 (with ")											
Surface treatment Electroplating and silver-plating	"	2h													

* single sample
** if 50 % of raw material is thermomechanical pulp, then: 5
+ related to 1 t coal charged with 10 % H_2O
++ organic constituents over 10 %
x /140/ because of community treatment plants

5.2.4.3 EC Standards

A complete list of regulations for the protection of water against pollution can be seen in Table 5.2.-5.

Tab. 5.2.-5: EC regulations for the protection of waters against pollution

Water pollution control regulation	Area of application	Date issued	In Force	Emission limits	Immission regulations
0	1	2	3	4	5
Council directive concerning the quality required of surface water intended for the abstraction of drinking water in the member states (75/440/EEC)	Surface water for direct abstraction of drinking water	16.06.1975	7/1977	-	x
Council directive concerning the quality of bathing water (76/160/EEC)	Waters where bathing is permissible	08.12.1975	1/1978	-	x
Council directive on pollution caused by certain dangerous substances discharged into the aquatic environment of the Community (76/464/EEC)	Surface and coastal waters	04.05.1976	5/1976	x	x
Council directive on the quality of fresh waters needing protection or improvement in order to support fish life (78/659/EEC)	Surface water specially designated as fish waters	18.07.1978	8/1978	-	x
Council decision establishing a common procedure for the exchange of information on the quality of surface fresh water in the Community (78/795/EEC)	Surface waters	12.12.1978	6/1979	-	x
Council directive concerning the methods of measurement and frequencies of sampling analysis of surface water intended for the abstraction of drinking water in the member states (79/869/EEC)	Surface water for direct abstraction of drinking water	09.10.1979	11/1981	-	x
Council directive on the quality required of shellfish waters (79/923/EEC)	For costal waters designated for shellfish and snail culture	30.10.1970	11/1981	-	x
Council directive on the protection of groundwater against pollution caused by certain dangerous substances (80/68/EEC)	Groundwater	17.12.1979	1/1981	x	-

It can be seen from this table that directives related to different uses of waters have been issued, perdominantly in terms of immission values.

From both the legal and technical point of view, the immission principle is the best way of safeguarding water quality, but it is expensive and requires a great deal of expert knowledge. In the light of the water use, quality objectives are drawn up, fixing the various requirements that the water must meet. Discharge limits (emission limit values) are set, derived from these quality objectives.

Within the framework of EC supranational law for the prevention of water pollution, binding quality objectives for different types of water use have so far been drawn up and issued in the following directives:
- directive on quality of surface water for drinking (Table 5.2.-6),
- directive on quality of bathing water (Table 5.2.-7),
- directive on the quality of fish waters (Table 5.2.-8),
- directive on the quality required of shellfish waters (Table 5.2.-9).

A distinction is made between mandatory (= imperative) values (I values) and guidelines (G values). For surface water for drinking, the EC further distinguishes values for the various methods of treatment (see Table 5.2.-6) (A values).

A_1 means simple physical treatment,
A_2 means normal physical and chemical treatment and
A_3 means intensive physical and chemical treatment necessary.

Values marked "O" are permissible in exceptional climatic and geographical conditions.

These EC directives cannot be applied in developing countries without further thought. Just as the other guidelines, they can only be used as a general pointer for the development of standards relevant to each specific country.

Tab. 5.2.-6:   EC directive on quality of surface water for drinking

	Parameters		A1 G	A1 I	A2 G	A2 I	A3 G	A3 I
1	pH		6.5 to 8.5		5.5 to 9		5.5 to 9	
2	Coloration (after simple filtration)	mg/l Pt scale	10	20 (0)	50	100 (0)	50	200 (0)
3	Total suspended solids	mg/l SS	25					
4	Temperature	°C	22	25 (0)	22	25 (0)	22	25 (0)
5	Conductivity	µs/cm^{-1} at 20 °C	1 000		1 000		1 000	
6	Odor	(dilution factor at 25 °C)	3		10		20	
7*	Nitrates	mg/l NO$_3$	25	50 (0)		50 (0)		50 (0)
8 1)	Fluorides	mg/l F	0.7 to 1	1.5	0.7 to 1.7		0.7 to 1.7	
9	Total extractable organic chlorine	mg/l Cl						
10*	Dissolved iron	mg/l Fe	0.1	0.3	1	2	1	
11*	Manganese	mg/l Mn	0.05		0.1		1	
12	Copper	mg/l Cu	0.02	0.05 (0)	0.05		1	
13	Zinc	mg/l Zn	0.5	3	1	5	1	5
14	Boron	mg/l B	1		1		1	
15	Beryllium	mg/l Be						
16	Cobalt	mg/l Co						
17	Nickel	mg/l Ni						
18	Vanadium	mg/l V						
19	Arsenic	mg/l As	0.01	0.05		0.05	0.05	0.1
20	Cadmium	mg/l Cd	0.001	0.005	0.001	0.005	0.001	0.005
21	Total chromium	mg/l Cr		0.05		0.05		0.05
22	Lead	mg/l Pb		0.05		0.05		0.05
23	Selenium	mg/l Se		0.01		0.01		0.01
24	Mercury	mg/l Hg	0.0005	0.001	0.0005	0.001	0.0005	0.001
25	Barium	mg/l Ba		0.1		1		1
26	Cyanide	mg/l Cn		0.05		0.05		0.05
27	Sulfates	mg/l SO$_4$	150	250	150	250 (0)	150	250 (0)
28	Chlorides	mg/l Cl	200		200		200	
29	Surfactants (reacting with methyl blue)	mg/l (laurylsulfate)	0.2		0.2		0.5	
30* 2)	Phosphates	mg/l P$_2$O$_5$	0.4		0.7		0.7	
31	Phenols (phenol index) paranitraniline & aminoantipyrine	mg/l C$_6$H$_5$OH		0.001	0.001	0.005	0.01	0.1
32	Dissolved or emulsified hydrocarbons (after extraction by petroleum ether)	mg/l		0.05		0.2	0.5	1
33	Polycyclic aromatic hydrocarbons	mg/l		0.0002		0.0002		0.001
34	Total pesticides (parathion, BHC, dieldrin)	mg/l		0.001		0.0025		0.005
35*	Chemical oxygen demand	mg/l O$_2$					30	
36*	Dissolved oxygen saturation rate	% O$_2$	70		50		30	
37*	Biochemical oxygen demand (BOD$_5$) (at 20 °C without nitrification)	mg/l O$_2$	3		5		7	
38	Nitrogen by Kjeldahl method (except NO$_3$)	mg/l N	1		2		3	
39	Ammonia	mg/l NH$_4$	0.005		1	1.5	2	4 (0)
40	Substances extractable with chloroform	mg/l SEC	0.1		0.2		0.5	
41	Total organic carbon	mg/l C						
42	Residual organic carbon after flocculation and membrane filtration (5 µ) TOC	mg/l C						
43	Total coliforms 37 °C	/100 ml	50		5 000		50 000	
44	Fecal coliforms	/100 ml	20		2 000		20 000	
45	Fecal streptococci	/100 ml	20		1 000		10 000	
46	Salmonella		Not present in 5 000 ml		Not present in 1 000 ml			

I = mandatory.
G = Guide.
0 = exceptional climatic or geographical conditions.
* = see Article 8 (d)

1) The values given are upper limits set in relation to the mean annual temperature (high and low).
2) This parameter has been included to satisfy the ecological requirements of certain types of environment.

Tab. 5.2.-7: EC directive on quality of bathing water

Parameters	G	I	Minimum sampling frequency	Method of analysis and inspection
Microbiological:				
1 Total coliforms /100 ml	500	10 000	Fortnightly (1)	Fermentation in multiple tubes. Subculturing of the positive tubes on a confirmation medium. Count according to MPN most probable number) or membrane filtration and culture on an appropriate medium such as Tergitol lactose agar, endo agar, 0.4 % Teepol broth, subculturing and identification of the suspect colonies. In the case of 1 and 2, the incubation temperature is variable according to whether total or fecal coliforms are being investigated.
2 Fecal coliforms /100 ml	100	2 000	Fortnightly (1)	
3 Fecal streptococci /100 ml	100	-	(2)	Litsky method. Count according to MPN (most probable number) or filtration on membrane. Culture on an appropriate medium.
4 Salmonella /1 liter	-	0	(2)	Concentration by membrane filtration. Inoculation on a standard medium. Enrichment – subculturing on isolating agar – identification.
5 Entero viruses PFU/10 liters	-	0	(2)	Concentration by filtration, flocculation or centrifuging and confirmation.
Physico-chemical:				
6 pH	-	6 to 9 (0)	(2)	Electrometry with calibration at pH 7 and 9.
7 Color	-	No abnormal change in color (0)	Fortnightly (1)	Visual inspection or photometry with standards on the Pt.Co scale.
	-	-	(2)	
8 Mineral oils mg/liter	-	No film visible on the surface of the water and no odor	Fortnightly (1)	Visual and olfactory inspection or extraction using an adequate volume and weighing the dry residue
	≤0.3	-	(2)	
9 Surface-active substances (lauryl-reacting with sulfate) methylene blue mg/liter	-	No lasting foam	Fortnightly (1)	Visual inspection or absorption spectrophotometry with methylene blue
	≤0.3	-	(2)	

cont. Tab. 5.2.-7

Parameters	G	I	Minimum sampling frequency	Method of analysis and inspection
10 Phenols mg/liter (phenol indices) C_6H_5OH	– ≤ 0.005	No specific odor ≤ 0.05	Fortnightly (1) (2)	Verification of the absence of specific odor due to phenol or absorption spectrophotometry 4-aminoantipyrine (4 AAP) method.
11 Transparency m	2	1 (o)	Fortnightly (1)	Secchi's disc.
12 Dissolved oxygen % saturation O_2	80 to 120	–	(2)	Winkler's method or electrometric method (oxygen meter).
13 Tarry residues and floating materials such as wood, plastic articles, bottles, containers of glass, plastic, rubber or any other substance. Waste or splinters	Absence		Fortnightly (1)	Visual inspection.
14 Ammonia mg/liter NH_4			(3)	Absorption spectrophotometry, Nessler's method, or indophenol blue method.
15 Nitrogen Kjeldahl mg/liter N Other substances regarded as indications of pollution			(3)	Kjeldahl method.
16 Pesticides mg/liter (parathion, HCH, dieldrin)			(2)	Extraction with appropriate solvents and chromatographic determination.
17 Heavy metals such as: – arsenic mg/liter As – cadmium Cd – chrome VI Cr VI – lead Pb – mercury Hg			(2)	Atomic absorption possibly preceded by extraction.
18 Cyanides mg/liter Cn			(2)	Absorption spectrophotometry using a specific reagent
19 Nitrates mg/liter NO_3 and phosphates PO_4			(2)	Absorption spectrophotometry using a specific reagent

G = guide.
I = mandatory.

(0) Provision exists for exceeding the limits in the event of exceptional geographical or meteorological conditions.

(1) When a sampling taken in previous years produced results which are appreciably better than those in this Annex and when no new factor likely to lower the quality of the water has appeared, the competent authorities may reduce the sampling frequency by a factor of 2.

(2) Concentration to be checked by the competent authorities when an inspection in the bathing area shows that the substance may be present or that the quality of the water has deteriorated.

(3) These parameters must be checked by the competent authorities when there is a tendency towards the eutrophication of the water.

Tab. 5.2.-8: EC directive on the quality of fish waters

Parameter	Salmonid waters		Cyprinid waters		Methods of analysis or inspection	Minimum sampling and measuring frequency	Observations
	G	I	G	I			
1. Temperature (°C)	1. Temperature measured downstream of a point of thermal discharge (at the edge of the mixing zone) must not exceed the unaffected temperature by more than: 1.5 °C (salmonid) / 3 °C (cyprinid) Derogations limited in geographical scope may be decided by Member States in particular conditions if the competent authority can prove that there are no harmful consequences for the balanced development of the fish population. 2. Thermal discharges must not cause the temperature downstream of the point of thermal discharge (at the edge of the mixing zone) to exceed the following: 21.5 (0) / 10 (0) (salmonid) — 28 (0) / 10 (0) (cyprinid) The 10°C temperature limit applies only to breeding periods of species which need cold water for reproduction and only to waters which may contain such species Temperature limits may, however, be exceeded for 2 % of the time				Thermometry	Weekly, both upstream and downstream of thermal discharge	Over-sudden variations in temperature shall be avoided
2. Dissolved oxygen (mg/l O_2)	50 % ≥ 9 100 % ≥ 7	50 % ≥ 9 When the oxygen concentration falls below 6 mg/l, Member States shall implement the provisions of Article 7 (3). The competent authority must prove that this situation will have no harmful consequences for the balanced development of the fish population	50 % ≥ 8 100 % ≥ 5	50 % 7 When the oxygen concentration falls below 6 mg/l, Member States shall implement the provisions of Article 7 (3). The competent authority must prove that this situation will have no harmful consequences for the balanced development of the fish population	Winkler's method or specific electrodes electro-chemical method)	Monthly, minimum one sample representative of low oxygen conditions of the day of sampling However, where major daily variations are suspected, a minimum of two samples in one day shall be taken	
3. pH		6 to 9 (0) (1)		6 to 9 (0) (1)	Electrometry calibration by means of two solutions with known pH values, preferably on either side of, and close to the pH being measured	Monthly	
4. Suspended solids (mg/l)	≤ 25 (0)		25 (0)		Filtration through a 0.45 μm filtering membrane, or centrifugation (five minutes minimum, average acceleration of 2800 to 3200 g) drying at 105 °C and weighing		The values shown are average concentrations and do not apply to suspended solids with harmful chemical properties Floods are liable to cause particularly high concentrations

cont. Tab. 5.2.-8

Parameter	Salmonid waters		Cyprinid waters		Methods of analysis or inspection	Minimum sampling and measuring frequency	Observations
	G	I	G	I			
5. BOD_5 (mg/l O_2)	≤3		≤6		Determination of O_2 by the Winkler method before and after five days incubation in complete darkness at 20 ± 1 °C (nitrification should not be inhibited)		
6. Total phosphorus (mg/l P)					Molecular absorption spectrophotometry		In the case of lakes of average depth between 18 and 300 m, the following formula could be applied: $$L \leq 10 \frac{\bar{Z}}{T_w}(1 + \sqrt{T_w})$$ where: L = loading expressed as mg P per square meter lake surface in one year $\bar{Z}$ = mean depth of lake in meters T_w = theoretical renewal time of lake water in years In other cases limit values of 0.2 mg/l for salmonid and of 0.4 mg/l for cyprinid waters, expressed as PO_4, may be regarded as indicative in order to reduce eutrophication
7. Nitrites (mg/l NO_2)	≤0.01		≤0.03		Molecular absorption spectrophotometry		
8. Phenolic compounds (mg/l C_6H_5OH)		(2)		(2)	By taste		An examination by taste shall be made only where the presence of phenolic compounds is presumed
9. Petroleum hydrocarbons		(3)		(3)	Visual By taste	Monthly	A visual examination shall be made regularly once a month, with an examination by taste only where the presence of hydrocarbons is presumed

regard to the parameters to be observed. The frequency of sampling must be determined separately for each parameter.

Arrangements for checking waste water discharges should include indications as to what will will happen in the event of objections to the discharged waters: whether samples should be taken more often, and what measures must be adopted in order to ensure that the waste water conforms to the discharge conditions.

Beyond such checking by the local authority, all industrial plants discharging waste water should be advised, or obliged, to be responsible for carrying out their own checks and measurements at the discharge point. The results of these checks should be recorded in logbooks.

The records of results of measures adopted in relation to such checks within the plant should be shown on demand to the operator of the public sewerage system. The logbooks, together with all tracings, etc., if automatic measuring and registering devices be used, should be kept for at least three years, so that past records can be produced when required (e.g. in the event of trouble in the in-plant or the public waste water treatment plant).

In order to ensure comparability of the results of checking by the local authorities and within the factory, both should be done using the same analysis and sampling methods. Particular care should be taken in collecting and keeping the samples.

The discharging body should always be consulted by the local authority when carrying out checks. Sampling must be adapted to suit the specific purpose of the analysis and local conditions. It is important that samples be truly representative. To achieve this, single or random samples are taken to analyse the waste water qualilty at the time of sampling, while collective samples are taken in those cases where, owing to the temporary changes in the waste water quality, mean values have to be established. Composite samples can be collected at 2-hour or 24-hour intervals. It is also important that single samples should not always be collected on the same day of the week or at the same time of day; this ensures fairly representative annual average values.

The required entries in the plant logbook are distinguished according to the type of treatment plant (e.g common municipal treatment plant, pretreatment plants or in-plant waste water treatment).

5.3.2 Checking the Emission Values at the Treatment Plant Outlet

The checking of emission values at the outlet of the community treatment plant, or of the works treatment plant in the case of direct discharge into the receiving water, is in industrial countries the responsibility of the national water authorities, or institutions responsible to them.

As a basic principle, discharge into a public water body should only be allowed if special permission has been obtained from the water authorities.

·This permission can take the form of an official notice under the laws relating to water. This notice sets out the permissible discharge amounts and the relevant emission values to be observed. The notice should have two parts: one concerning the water laws (penal provisions) and one showing the schedule of charges.

5.3.3 Checking the Immission Values

In industrial countries, the checking of immision values, i.e. of the quality parameters to be determined in the receiving water, is also the responsibility of the water authorities. The extent and frequency of checking is determined by the overall water resources plan, especially the proposed uses of the water and corresponding quality requirements.

Measurements and checking of immision values is necessary on the part of the water authorities, both for drawing up a water resources plan and for setting emission values for the direct discharge of industrial wastes and for public treatment plants. The parameters to be checked are determined by water planning, and depend on the quality classes which, in turn, are determined by the proposed uses.

The checking of waste water discharges and the set parameters serves, on the one hand, to ensure that the regulations laid down in the water laws are observed, and, on the other hand, to determine the rates and charges for the use of public sewerage systems and of the public receiving water.

In many industrial countries, exceeding the values laid down in the water laws and in the statutory regulations of the water authorities may have criminal consequences, leading to fines or imprisonment for those responsible.

Municipal authorities may impose fees, to be individually determined, for the use of public sewerage systems.

For the use of the receiving water, the national water authorities may charge fees to cover supraregional water management costs.

5.4 Statutory Regulations for Sludge Usage in Agriculture

In the industrial countries, the problems of sludge disposal have increased due to the forced building of new treatment plants or the extension of existing plants. Constant checking of sludges to be applied to the soil, and of the corresponding soils, is necessary to prevent toxic substances, especially heavy metals, from entering the food chain. The limit values for such harmful substances given in the local bylaws do not apply to the spreading of sludge on soil. This is a special case, for which loads must be calculated on the basis of permanent concentration measurements in the waste water and on its flow rates; from the calculated loads, conclusions can be drawn concerning the heavy metal load of the sludge, taking into account a proportional rate of enrichment in the sludge.

In the Federal Republic of Germany, a statutory regulation concerning the application of sludge to the soil (as of August 1, 1981) has been published on the basis of scientific literature and recent analyses. This regulation lays down the values shown in Table 5.4.-1 as upper limit values not to be exceeded in sludges for the individual heavy metals (the values given in the brackets will become valid when the regulation has been in force for 8 years).

Tab. 5.4.-1: Upper limit values for heavy metal concentrations in sludges intended for application to the soil

Lead:	1,200 mg/kg (600)
Cadmium:	30 mg/kg (10)
Chromium:	1,200 mg/kg (600)
Copper:	1,200 mg/kg (800)
Nickel:	200 mg/kg (100)
Mercury:	25 mg/kg (10)
Zinc:	3,000 mg/kg (200)

The application of sludge to soils used for agricultural or horticultural purposes is forbidden if soil analyses show that the heavy metal concentrations shown in Table 5.4.-2 are exceeded in an average sample.

Tab. 5.4.-2: Limit concentrations in the soil

Lead:	100 mg/kg
Cadmium:	3 mg/kg
Chromium:	100 mg/kg
Copper:	100 mg/kg
Nickel:	50 mg/kg
Mercury:	2 mg/kg
Zinc:	300 mg/kg

If the enrichment factors given in the literature are taken into account for the various metals with biological treatment, it will be noticed that, in spite of observing the upper limit values set in the local bylaws, the heavy metal concentrations in the sludge can be so high as to make agricultural use of the sludge impossible. The following average enrichment factors for heavy metals from raw sewage in the sludge, with biological sewage treatment, have been determined, from which it is possible to calculate the upper limit values for sewage given in Table 5.4.-3 (MÜLLER /116/).

Tab. 5.4.-3: Theoretical limit value in raw sewage /116/

Heavy metal	Average enrichment factor in sludge	Computed limit value raw sewage mg/l
Lead	5000	less than 0.2
Cadmium	4000	less than 0.01
Copper	1000	less than 1
Nickel	2000	less than 0.1
Zinc	3000	less than 1

The significance of sludge use in agriculture, with regard to its nutrient effect, can be seen in Table 5.4.-4. Especially significant is the conservation of commercial fertilizers. The Table gives an overview of the evaluation of the nutrients phosphorus, potassium, magnesium, and calcium in sludge, based on the following assumptions:

1. an annual application of 3000 kg/hectare sludge (dry solid matter),
2. agricultural land with average nutrient demand,
3. a degree of utilization of nutrients in the sludge corresponding in the long run to that of mineral fertilizers,
4. a cereal-root crop ratio of 4:1,
5. prices as in the business year 1978/79.

Tab. 5.4.-4: Economic values of sludge application in agriculture

	Annual demand kg/ha	Supply with 3000 kg/ha dry solid matter	Price DM/kg nutrient	Fertilizing value of sludge DM/ha
Phosphorus as (P_2O_5)	37	100	0.95	35.15
Potassium (K_2O)	48	15	0.50	7.50
Magnesium (MgO)	50	11	0.25	2.75
Calcium (CaO)				
Sandy soils	100	100*	0.10	10.00
Loam and clay	500	100*	0.10	10.00
soils		500*	0.10	50.00

* for wet sludges
** for dewatered sludges pretreated with $Ca(OH)_2$

6 TREATMENT OF WASTE WATER

Proceeding from a general description of environmental problems, and from the basics of water protection and water purification, the general conditions for water purification requirements are to be illustrated as they exist in industrialized countries.

In examining the transferability of these regulations and the prerequisites for their initiation in developing countries, the special conditions in these countries have been investigated. Apart from the available literature from industrialized countries, literature was employed from institutes active in the respective countries.

The necessary degree of purification for a given waste water can be determined by comparison of the allowed waste loads with the pollution arising in municipal waste water and certain types of industries. The required purification in turn determines the selection of the waste water treatment method, which is appropriate for the particular community and industrial structure and practical under the special conditions of the developing countries. Here, not only industrial waste water purification was considered, but also the combined treatment with municipal sewage and the resulting special requirements for pretreatment.

Strong emphasis was placed on demonstrating possibilities as to how useful materials can be recovered and to what extent purified water can be reused, e.g. as industrial water or for irrigation in agriculture. The possibilities for storing sedimented sludge have also been demonstrated. Unfortunately, the negative results in this sector obtained in industrialized countries have not led to the realization of corresponding precautionary measures for new commercial and industrial developments in other countries, intended to avoid harmful consequences. More often the opposite is observed. Industrial concerns evade the stronger environmental protection laws in industrialized countries by establishing production plants in developing countries where no restrictive measures exist.

It must be considered that with the climatic conditions, the effects can be quite different than in the mild regions of western and northern Europe.

Even when the population is less dense in these countries, sufficient water resources exist, and enough space is available that a plant need not be built next to residential areas, these factors do not mean that the dangers to the environment are smaller than in the industrialized countries. A time delay simply occurs until the effects on man are apparent.

A primary consideration must be that the limited water reserves in many countries remain free of contamination. Since sufficient treatment plants for drinking water supply are often not available, a contamination of the surface water and groundwater could have a disastrous effect on public health. Costs saved on the one side due to the lack of water purification by the community and also by industry must be more heavily expended on the other side for health care.

To establish a future-oriented and responsible water supply policy, the aspects of the environmental burden must be made known that arise from the restructuring of employment and industrialization, as well as their resulting dangers.

In each individual case, arguments must be weighed between the need for industrial development of the country and the basic requirement of an environment that is not detrimental to the population or vegetation.

This requires both a responsible planning in the individual countries and also comprehensive advisory and support activities from concerned organizations in developed countries.

Prespective environmental protection measures must be developed and initiated so that,
- the public health,
- the productivity of the natural habitat, and
- the often scarce natural resources

are not negatively influenced as a secondary effect of economic development.

The problems of all concerned should be presented, which could arise from the burden placed on water, but also on the soil and groundwater due to industrial waste water. Furthermore, measures and methods must be indicated showing how these effects could be controlled and which type of purification should be applied. Here, the different industrial strengths of the various regions must be considered. The climatic conditions, the water resources and the housing structure in the individual country are also influencing factors.

Industries are to be obligated to treat their waste waters in such a way that they can be reused in the same plant or elsewhere, or can be diverted to reservoirs when the input requirements are met. In many cases, industrial waste water can be treated together with domestic or municipal sewage water.

This is also preferably recommended by water authorities in different countries. With the combined treatment of industrial waste waters in sewage plants for domestic and municipal waste water, the plant dimensioning and its purification capacity are to be observed. This is in addition to the allowed input values for the sewerage system and possible effects of certain industrial wastes on the sewage plant, i.e. its operation.

Since, in some cases, many organically loaded industrial waste waters can be purified sufficiently with conventional methods, these methods and the limits of their purification capacity are described first.

6.1 Methods of Waste Water Treatment

The treatment of waste water can be accomplished with different methods. All of these methods are based on physical, chemical, and biological phenomena.

The purification mechanisms can be altered in different ways, whereby inevitably, different purifying capabilities result. In connection with the degree of purification chosen, a scaling in the investment costs also exists.

In general, waste water purification for both municipal and industrial waste water includes:
- retaining the pollutant, toxic and reuseable substances, present in waste waters,
- water treatment, and
- separate sludge treatment.

According to the type of treatment used, one distinguishes between treated waters or so-called <u>clarified waste waters</u>, which are either reused or discharged to a receiving water, and <u>the sludge</u>, which is reused, stored, transformed, or incinerated.

The resultant properties of clarified waste water and the sludge depend on the nature of the waste water to be treated and on the treatment method.

The distinguishing criteria of purification methods vary:
- with domestic (municipal) sewage, the methods are classified by the type of water treatment independent of the treatment of separated sludge,
- with industrial waste waters, the main goal is usually the recovery and reuse of raw materials (e.g. phenols, extraction agents, etc.) so that the industrial treatment methods are mostly named according to the reprocessing technique applied.

Fig. 6.1.-7 illustrates some examples of flow diagrams for combined methods of mechanical-biological waste water purification.

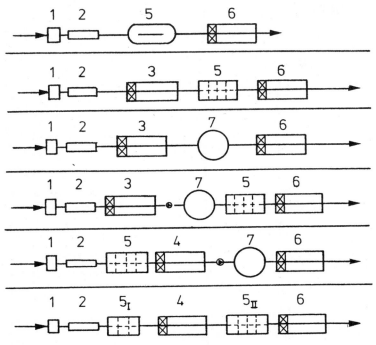

Fig. 6.1.-7: Combined methods of mechanical-biological waste water purification

1 - grating; 2 - sand trap; 3 - preclarification; 4 - intermediate clarification; 5 - activated sludge basin; 6 - secondary clarification basin; 7 - biological trickling filter

Tab. 6.1.-1: Elimination efficiency of treatment methods for municipal waste water /74, 18/

Method	suspended substances	Efficiency in %			
		BOD	COD	P	N
Mech. purification	40 - 70	25 - 40	ca. 15	ca. 15	ca. 7
Biol. purification	85 - 95	85 - 95	ca. 80	ca. 30	ca. 40
Additional reduction of remaining material in %					
Microscreens	20 - 40	5 - 10	5 - 10	-	-
Filtration	50 - 80	10 - 20	5 - 20	-	-
Chemical treatment	70 - 90	50 - 85	40 - 70	50 - 90	0 - 30
Activated coal absorption	50 - 90	ca. 95	ca. 90	ca. 90	ca. 10

The mechanical purification is based on physical properties and includes the separation of sedimentary substances in waste water and their stabilization. These methods are used for preclarification and in exceptions, e.g. for isolated houses, as the final clarification. There are, however, many new purification methods by which the preclarification stage is avoided (grating is always required) or is replaced by special mechanical-biological methods (e.g. adsorption-activation method).

The chemical purification consists of the separation or transformation of settleable, floatable, and dissolved substances through the use of chemicals. These methods are seldom applied to the treatment of municipal sewage, essentially only in special cases for the sterilization of mechanically or biologically preclarified waste water, e.g. from certain hospitals where bath water and waste water are combined, or in general, for a more thorough treatment.

The biological purification employs the activity of certain microorganisms for the oxidation and mineralization of organic substances found in waste water. The waste waters processed with biological purification methods are often preclarified by mechanical purification and almost always preclarified in larger plants.

The final purification has different purposes:
- separation of the nutrients P, N, and K, which can lead to the massive development of microorganisms especially algae in receiving waters (eutrophication),
- the so-called "polishing" of purified waste waters or an additional natural aeration,
- filtration of purified waste waters e.g. to protect the groundwater when the treated water is drained,
- as a security measure for receiving waters with special purity requirements, etc.

The flow diagram (see Fig. 6.1.-2) gives a compilation of the most important elements of waste water purification and their relationship to one another. Furthermore, it also indicates the treatment paths for waste water and sludge and the necessary equipment for sludge treatment and gas usage.

The flow diagram shows only the most important process steps, while the entire installation and the support facilities are not shown.

The purification method is selected with respect to the necessary degree of purification to be achieved. This selection is oriented toward the possible purification capacity (see Tab. 6.1.-1) of each method.

From this table it can also be seen that the capacity of the mechanical method in reducing the organic compounds (BOD_5) is very small, so that for the purification of municipal sewage, biological methods are necessary.

For very small communities, however, a mechanical purification is sufficient under certain conditions, if 95 % of the average monthly flow in the receiving water does not exceed a load corresponding to 7 to 10 persons per 1 l/s of flow-through.

The treatment plant concept is determined through a comparitive technical and economic calculation considering different factors, which are discussed in Sections 6.2 - 6.7.

In the specialist literature, various procedures are given for the selection, however, they can only be seen as general guidelines. The best suited alternative solution must be determined in each case by considering the individual situation.

6.2 Mechanical Purification

In this method, mainly the coarsely dispersed solids are separated out of the liquid phase.

From the large number of possible separation methods, the important ones /136/ are given in Tab. 6.2.-1.

Tab. 6.2.-1: Survey of methods for material separation /136/

I.	II.	III.	IV.
Mechanical (gravity, centrifugal force, pressure, vacuum)	Electrical and magnetical (electrical and magnetic energy)	Thermal (heat energy)	Chemical

1. classing
  a) sifting
    · with screen
  b) wind sifting and flow classing
2. sorting
3. sliming
4. dekanting
5. flotation
6. centrifuging
7. pressing
8. filtering
9. dialyzing
	electrostatical
	a) electric dust filter
	b) electromagnetic
		1. vaporize distillation rectification
		2. drying
		3. extraction
		4. crystalization
		5. adsorption
		6. absorption
		7. diffusion
		8. thermal diffusion
			1. material change and thus change in state of aggregation, followed by separation as in I, II or III
			2. ion exchange
			3. adduct formation (de-paraffinating crude oil with urea)

In treating municipal waste water, mechanical purification is mandatory. It ensures the removal of solid material such as tree limbs and other wood pieces, dead animals, vegetables and other food rests, rags, cloth and cellulose remains as well as other coarse floating or suspended substances, which are usually contained in waste water. This facilitates the following clarification processes, especially the biological stage.

If a pump station is necessary due to the topography at the treatment plant inlet, the separation of the above mentioned material must occur <u>before</u> the pump station.

In the scope of mechanical waste water treatment, essentially two methodologies are applied for municipal sewage. In one case, the flow cross section of the water is divided into numerous small cross sections, whereby the transported solid material is caught on the inlets and must be removed. In the other case, one creates as far as possible, a calm, laminar flow in order to separate with the gravity, since the undissolved contents in waste water have a different density than water. In a special case with turbulent flow, hydrodynamic characteristics are used to separate waste material of different density.

Gratings and screens belong to the first group, while sand traps, oil and fat separators, and sedimentation basins belong to the second. The retained substances are further processed in different ways; material from the grates and screens is sized in communitors, oil and fat can be reused, and the sludge is fermented, stored, or incinerated. Other disposal methods are used depending on the respective regulations.

Coarse separation, sand traps, oil and fat separators, and sedimentation basins are discussed in detail in the following sections.

6.2.1 Coarse Separation

The aim of coarse separation is to remove all contamination that could lead to operational difficulties by clogging pumps. In view of the total treatment, this step is important because the waste water can only then be further treated and processed with relatively little problem. Even when, in percent, it is only a small part of the contaminants, certain precautionary measures must be taken to guarantee the dependable operation of the following steps.

For municipal sewage water, grates and a comminutor for grated material are the main equipment used. For industrial waste waters, various filters and screens are employed in addition.

6.2.1.1 Grating

Grating refers to all kinds of ribbed grates, which are built into the flow cross section. With water passage, coarse material is caught on the grate ribs and must be removed by hand or with suitable mechanical devices. Depending on the spacing between the ribs, one distinguishes between coarse grates, which in waste water technology most often have a spacing of 60-100 mm, and fine grates with spacings in general of 10-25 mm. Since the grating is usually located in the inlet channel, the continual removal of trapped material is a key function for the uninterrupted operation of a plant. For older plants and also new larger plants, one always finds fine grates in addition to the coarse grating.

Fig. 6.2.-1 illustrates a grate with manual cleaning and overflow, which however is only found in small treatment plants. Fig. 6.2.-2 shows a catch grate. The cleaning occurs either with a comb-toothed sheet which meshes with the grate ribs and drags the material out or otherwise with a rake, which is actuated with a cable (Fig. 6.2.-2a), chain (Fig. 6.2.-2b), or a cog-rail. Fig. 6.2.-3 illustrates an arc grate and Fig. 6.2.-4, a comminutor for the trapped material. It is equipped with a cylindrical grate in which the comminutor with a vertical axis is integrated. This kind of comminutor sizes the coarse material to the point that it cannot cause further operational difficulties, but can be removed as sludge for example in the preclarification. The disadvantage of this method is that synthetic materials are also sized. The same is true for wood, cork, and other light materials. At least in plants with sludge fermentation chambers, a thick layer floating at the top can be expected. Therefore, this material is currently removed from the treatment cycle.

As a reference for dimensioning the grating, the remaining flow cross section between the ribs should allow a flow-through velocity of more than 1 m/s. To be certain, one should assume that ca. 30 % of the cross sectional area is blocked. The velocity before the grating should be at least 0.6 m/s to prevent sand from depositing. Theoretically, the passage loss can be calculated as follows from KIRSCHMER /1/,

$$h_v = \beta \left(\frac{s}{b}\right)^{4/3} \cdot \frac{v^2}{2g} \cdot \sin \alpha \qquad [m]$$

where

h_v = water head loss in m
s = rib thickness (usually 8, 10 or 15 mm) in m
b = spacing of ribs in m
v = velocity upstream of the grate in m/s
α = inclination angle of grating against horizontal
β = form factor
    2.42 for rectangular ribs
    0.84 for specially designed forms
    1.79 for round ribs

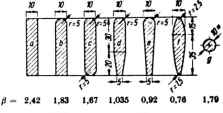

β = 2,42 1,83 1,67 1,035 0,92 0,76 1,79

Rib cross sections

The frictional (head) losses from passage through the grating are usually under 5 cm, while, in contrast, the losses from blockage can lead to a threefold increase in h.

The water content of the trapped material is 0.75 t/m³. The amount (see also Section 6.6.1) varies, whereby hourly fluctuations can mean a doubling in value.

Tab. 6.2.-2 gives a survey of the amount of grate material which arises.

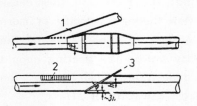 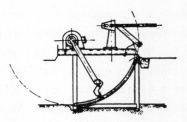

Fig. 6.2.-1: Grate (fixed) with manual cleaning

Fig. 6.2.-3: Arc grate

1 - overflow channel; 2 - overflow; 3 - grate

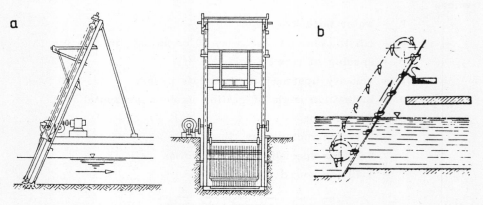

Fig. 6.2.-2: Grate with mechanical cleaning by comb-toothed metal sheet actuated with a) cable; b) chain

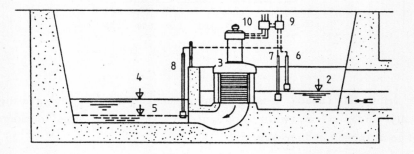

Fig. 6.2.-4: Comminutor for grated material (CONDUX Werke)

1 - inlet; 2 - upper water level; 3 - comminutor; 4 - cylindrical grate; 5 - lower water level; 6 - 10 regulation and control devices, on/off switching

the inside and outside of the drum. When separating chemical flakes, an agent can be added to the spray water (intermittent or continual), which prevents blockage of the microscreen due to deposits. For the same reason, a microbicidal zone is often included after the spray zone when biologically active sludge is being separated. The growth of a potential biological colony can be hindered, for example with chlorine or uv irradiation.

Various formulas are recommended (see Section 6.5.1.2) in dimensioning microscreens. The flow rate Q through the screen cannot be explicitly calculated from this equation. However, since the waste water amount is given at the preengineering stage, the resulting water level difference H can be precalculated when a screen drum is selected with corresponding A and S values from the previously determined filterability index I.

To determine the filterability, a simple "filter meter" was developed with which the index I can be directly found (Fig. 6.2.-7) with the help of a calibration curve, from the passage time T of a given volume of raw waste water V.

Another special case, the closed screen, should be mentioned, which is very often used in processing, especially in the food industry, to remove solids of all kinds. When such screens are clogged, they are usually cleaned by reverse rinsing, and the screened material is then discharged into the sewerage system. This often leads to very high waste water loads.

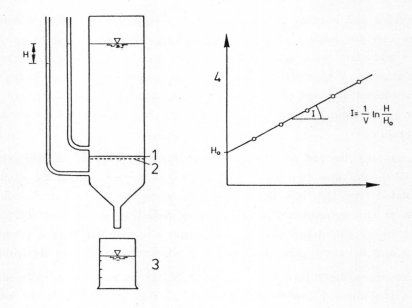

Fig. 6.2.-7: Filter meter
1 - filter cake; 2 - screen mesh; 3 - filtrate volume V; 4 - pressure loss in H

6.2.1.3 Filters

The equipment discussed in the previous section can also be called surface filters when the corresponding mesh size is fine. In contrast, there are the so-called deep filters, i.e. filters with a bed of gravel, sand, hydroanthracite or other materials by which not only the surface, but also the depth of the filter bed participates in the purification.

These filters have only secondary importance in waste water treatment. In general, they are used when a more thorough purification is to be achieved, for example following phosphate precipitation to remove microflakes not settleable with gravity or by the reuse of purified waste water for industrial purposes. Such systems are not applicable to the prepurification of waste water. As with all systems cleaned by reverse rinsing, a reasonable relationship can no longer be reached between forward flow and the reverse rinsing water amounts. The large waste load would also require the repeated change of filter material and further, would lead to the unwanted growth of organisms within the filter bed. Such filters are utilized, however, in rolling mills or other factories in which the industrial water does not contain large quantities of organic contaminents.

Besides these filters for industrial waste water, other types are used that work under pressure or under vacuum. Generally, the following filters are employed in industry:
- filters consisting of coarse sand or fine gravel,
- drum filters, under pressure or without,
- suction filters (vacuum filters),
- disk filters, and
- pressure filters (candle filters).

6.2.1.3.1 Filters with Coarse Sand or Fine Gravel Fillings

Filtration is a physical and physio-chemical process by which suspended or colloidal particles are separated from the liquid phase by penetrating a porous medium and accumulating on the surface or in the pores of the medium. One differentiates between slow filters and fast filters.

For slow filters, the filter bed is the crucial element for the elimination capacity of the filter, and it consists of a sand layer with a certain grain size and depth. All other construction parts are designed so that the processes in the sand bed can run optimally. The raw water to be filtered flows into the filter layer and with increasing penetration, separates the suspended matter into the pores between the sand grains. The pore volume (or the grain size) and the flow velocity must be adjusted to the suspension load so that the particle separation takes place throughout the entire depth of the layer, i.e. the entire filter bed makes up the reaction volume (Fig. 6.2.-8 a). The filtrate is collected for reuse or further treatment.

When the filter bed is spent after a certain operational period, i.e. loaded with filtered particles, the pressure due to backwater is no longer sufficient to force more water through the filter. Then a cleaning process must be carried out which can be termed reverse washing. The flow direction (Fig. 6.2.-8 b) is reversed and filtered water is forced upward at high speed through the filter bed. The sand or gravel layer expands and the relative movement of the individual sand grains to one another causes an abrasive effect that allows the deposited suspensions to be washed out. The washing water is turned off, and the filter process is restarted with the introduction of raw waste water. Until the filter bed and the filter capacity have restabilized, the filtrate is discarded as waste water or it is regenerated. The wash water is sent to the same or a separate treatment plant.

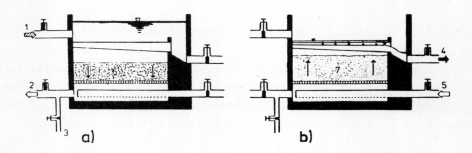

Fig. 6.2.-8 Slow filters

a - filtering operation; b - cleaning operation, 1 - raw water inlet; 2 - filtrate outlet; 3 - start-up filtrate; 4 - wash water outlet; 5 - wash water inlet; 6 - filter bed; 7 - expanded filter bed

The hydraulic calculation in its simple form is based on Darcy's law, which holds for laminar flows as well as groundwater flow:

$$v = k_f \cdot I = k_f \cdot \frac{H}{L} \qquad [m/h]$$

or

$$H = \frac{v \cdot L}{k_f} \qquad [m]$$

where

> H = water head loss in filter in m
> v = entrance flow velocity in m/h
> L = depth of filter bed in m
> k_f = permeability coefficient of the filter layer in m/h

The purification effect of a deep filter is also based on biological processes.

In using filters for industrial waste waters, a reliable precalculation of the filter capacity, pressure differences, operating time, etc. is not possible for flocculent suspensions in raw waste water. Thus a test filter is almost always necessary. The backwater head and filter bed depth should correspond to the original, while the filter cross section should be at least 100 times as large as the maximum grain diameter. As such, the extension of the test results to the full-scale plant is ensured.

In <u>fast filters</u>, the natural limitation of the filter velocity is overcome by a higher head pressure. Pumps are used to feed the raw water and the filter bed is placed in a presssure container (Fig. 6.2.-9). With this arrangement, flow average flow velocities of 20 m/h can be run, with a considerable reduction in the total filter area given according to the equation

$$F_{nec} = \frac{Q}{V} \frac{[m^3/h]}{[m/h]} \triangleq [m^2]$$

However, to maintain a sufficient contact time between filter flow and sand layer, the filter bed depth must be increased to 2 or 3 meters. Pressurized filtration in closed reaction chambers is widely applied following flocculation installations for the secondary purification. In treating circulation water for smelting works, filter velocities of up to 30 m/h have been obtained. The filter capacity can be greatly improved by adding small amounts of a precipitating agent to the waste water.

In practice, filters made with coke, cinders, wood wool, brushwood, and similar materials are often used. They can only be employed for small waste water volumes and are often connected downstream of sedimentation basins. For sludge disposal, they must be washed out or exchanged and as such, they are impractical in operation with only low efficiency. Occasionally certain success has been noted with these filters for oil and fat containing effluents.

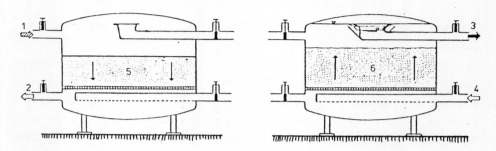

Fig. 6.2.-9: Fast filters

a - filtering operation; b - cleaning operation; 1 - raw water inlet; 2 - filtrate outlet; 3 - wash water outlet; 4 - wash water inlet; 5 - filter bed; 6 - expanded filter bed

<u>Multiple-layer filters</u> (e.g. IMMEDIUM filters from LURGI) consist of defined layers of different grain size and density arranged on top of one another with distinct boundries. An example is the often used two-layer filter with anthracite (d = 1.5 mm, ρ = 1.6 g/cm³) as the upper and quartz sand (d = 0.8 mm, ρ = 2.6 g/cm³) as the lower layer.

6.2.1.3.2 Drum Filters

Drum filters are mainly used for dewatering solids, but also are used for the direct filtration of waste water, especially for fiber residues in the textile, paper, and pulp industry. Thus they not only improve waste water quality, but also fulfill an important economic function, namely the recovery of raw material for reuse.

Fig. 6.2.-10 illustrates a drum filter (e.g. IMPERIAL filter system WACO von H. HOENIG & Co., München).

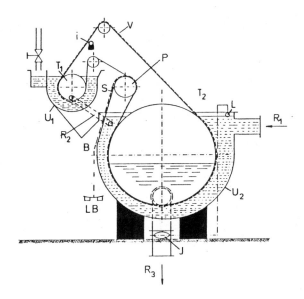

Fig. 6.2.-10: IMPERIAL filter system WACO (von H. HOENIG & Co., München)

T_2 - filter drum; U_2 - filter tank; T_1 - prefilter drum; U_1 - prefilter tank; R_1 - inlet; R_3 - outlet; R_2 - overflow; P - receiving roller; B - slide; i - spray nozzle; J - flow control valve; L - float valve; V - endless screening belt; S - scraper; LB - conveyor belt

The drum with a horizontal axis rotates slowly in a container in which the solids mixture to be filtered is placed. When the waste water passes through the filter cloth, the solid substances separate out and form a filter layer. This is slowly lifted out of the solids mixture by the drum rotation, and is then dewatered with pressrollers, press belts, or suction air and removed. This dewatering is an essential component of filter plants; one obtains a relatively dry residue. To maintain the permeability of the filter cloth, various devices are attached to the individual systems.

Agents can be directly added to the waste water before filtration. The filter apparatus is built as mobile equipment, which can be set up anyplace in the plant. Concerning the working method, one distinguishes between filters operating with and without vacuum.

6.2.1.3.3 Suction Filters

Suction filters have high efficiency. The filter consists of a drum (Fig. 6.2.-11) constructed as a combination of suction cells. As the drum rotates, the cells are sequentially placed under vacuum or overpressure. Thus, the solid material is sucked onto the filter in zone 1, and is dewatered in zone 2. In zone 3 the filter cake is removed with pressure. The cake is then picked up by a scraping device.

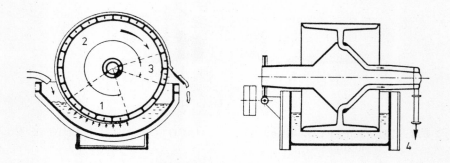

Fig. 6.2.-11: Suction filter /107/
1 - suction zone; 2 - dewatering zone; 3 - removal zone; 4 - filtrate

6.2.1.3.4 Disk Filters

Disk filters are used for both industrial waste water and the usual sludge dewatering. A disk filter consists of a large number of round, disk-shaped filter elements, which are arranged on a hollow horizontal axis placed in a filter tank so that part of the disk is submerged. The disks are divided into sections and have screening plates on both sides, which in turn are covered with filter or sieve cloth. The sections are under vacuum, so that water is continually drawn into the submerged part, while the substance to be filtered remains on the filter surface. The filter cake is lifted and removed by double-sided scrapers before the disk resubmerges.

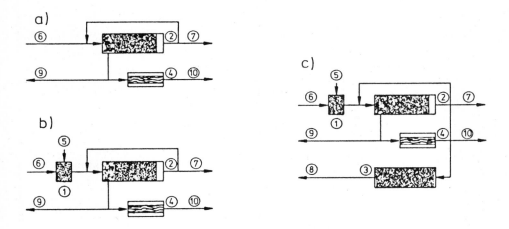

Fig. 6.2.-13: Schematic of the pressure release flotation method
a - total flow method; b - partial flow method; c - recycling method; 1 - flocculation basin; 2 - flotation basin; 3 - storage tank; 4 - scum dewatering; 5 - chemical dosing; 6 - waste water; 7 - clarified water; 8 - return water; 9 - recycling material; 10 - sludge cake

The application of flotation for industrial waste water includes mainly the treatment of oil, fat, and fiber containing waste waters, as well as flocculent waste waters. Included are oil refineries, salad and cooking oil producers, slaughterhouses, meat and fish processing plants, and plants in which semifinished or waste products of these industrial branches are reprocessed, such as soap, gelatin, and glue factories. Mineral oils and fats are primarily contained in the discharges from the iron and steel working industry and from machine-tool and auto producers.

With flotation as industrial waste water treatment, different goals can be attained, e.g. the separation of usable or waste materials, the recirculation of water, or a load reduction by the separation of suspended colloidal substances simultaneously lowering the BOD_5 value.

The purification capacity that can be reached for different types of commercial and industrial discharges can be taken from Tab. 6.2.-3. As far as the attainable purification capacity can be generalized through the characteristic pollution parameters, it can be said that the undissolved substances and those extractable with petroleum ether can be floated out from 80 to almost 100 %. For the bio-chemical oxygen demand, reductions between 50 and 90 % are obtainable, where the higher percentage reduction is reached for waters with predominant oil and fat contents. It generally holds, however, that with the BOD_5 reduction through flotation, at best, a purification can only be expected corresponding to the BOD_5 amount due to suspended substances.

Tab. 6.2.-3: Flotation results for industrial waste waters

Investigation	Raw waste water			Clarified water			Purification effect		
	undiss. subst.	ether sol. fat	BOD_5	undiss. subst.	ether sol. fat	BOD_5	undiss. subst.	ether sol. fat	BOD_5
Type of waste water	mg/l	mg/l	mg/l	mg/l	mg/l	mg/l	%	%	%
Edible oil production	230	460	2900	20	25	94	91.3	94.6	96.8
Margarine production	5000	3900	-	200	40	-	96.0	99.0	-
Oil mill	416	503	-	32	8	-	92.3	98.4	-
Cosmetics production	15000	5405	24500	1800	485	5880	88.0	91.0	76.0
Laundry I	30	138	300	0	30	84	100	78.2	72.0
Laundry II	3469	3014	-	281	475	-	91.9	84.2	-
Wool washing I	4000	2100	970	60	30	90	98.5	96.6	90.7
Wool washing II	8700	4650	2820	81	20	268	99.1	99.6	90.5
Slaughterhouse I	7428	3110	-	712	97	-	90.4	96.9	-
Slaughterhouse II	700	892	1900	10	32	39	98.6	96.4	97.9
Meat processing	970	1706	1540	97	513	277	90.0	70.0	82.0
Poultry processing I	1690	331	1075	275	74	86	83.7	77.6	92.0
Poultry processing II	874	3139	1136	40	18	100	95.4	99.4	91.2
Poultry processing III	357	-	630	91	-	58	74.5	-	90.8
Animal carcass utilization	5353	4614	-	780	775	-	95.4	83.2	-
Tannery I	7792	20590	-	1310	1275	-	83.2	93.8	-
Tannery II	5093	462	2221	384	43	547	92.5	90.7	75.4
Gelatin production I*	-	12300	-	-	10	-	-	99.9	-
Gelatin production II	2580	2825	-	458	315	-	82.9	88.9	-
Soy bean processing	1656	-	3000	42	-	800	97.5	-	73.4
Potato processing	2600	-	2760	60	-	260	97.7	-	90.6
Tomato processing	172	-	276	59	-	168	65.7	-	39.1
Canning factory	1350	-	790	270	-	315	80.0	-	60.1
Fiber board production	1700	-	6170	127	-	3000	92.5	-	51.4

* two-stage flotation system

Flotation systems are based on the following principles:
- natural buoyancy of suspended particles e.g. in oil and fat separators,
- partial diversion of waste water from inlet pipe to foam generator and return to main water e.g. the WOLF Swimmer Foam Trap (Fig. 6.2.-14),

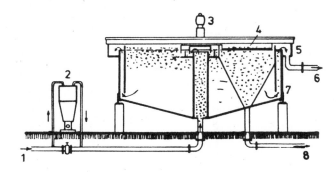

Fig. 6.2.-14: Swimmer foam trap
1 - inlet; 2 - foam generator; 3 - drive; 4 - foam skimmer; 5 - collection channel; 6 - waste water outlet; 7 - fibrous material area; 8 - thickened water outlet

- low-pressure aeration ($0.5-0.7 \cdot 10^5$ Pa), where the fan-generated compressed air is introduced at the bottom of the separator through a dispersing device (permeable plates or perforated tubes). These are used most frequently.
- increased waste water pressure by addition of compressed air at $1.3 \cdot 10^5$ Pa and subsequent expansion at atmospheric pressure with appropriate equipment,
- the waste water saturated with oxygen in the activation basin is exposed to vacuum for 10-15 min in a separate treatment step. The water loses its capacity to dissolve air so that part of the undissolved oxygen vaporizes and carries suspended material with it on its way to the surface.
- addition of precipatating agents, e.g. sulfuric acid, potassium, aluminum sulfate, or foam generators,
- chlorination of the water up to a dosage of 1.5 mg/l.

The separated substances are collected in channels and removed for further treatment.

For the flotation with air bubbles without chemical agents, the chemical and physical surface properties of the suspended particles are of crucial importance. Air bubbles cannot adhere to substances with hydrophilic and tightly knitted surfaces. Mainly fatty, oily, or soapy components or other substances with water repellent surfaces are obtained by waste water flotation.

The "flotation" of heavy particles, which sink due to their density, is used for the extraction of solid mineral substances, e.g. in ore and coal processing.

In the flotation of lighter particles such as floatable, suspendable, or sedimentary substances with a low sinking velocity, a small amount of additional buoyancy is sufficient to force the particles to rise. Fibrous, flocculent, and fatty substances present in many industrial discharges can be separated by "fine bubble flotation" without additional flotation agents because of the better adhering properties of small bubbles. Owing to the accompanying cost savings, the fine bubble flotation is economical for the separation of light substances with hydrophobic surfaces and for thickening activated sludge.

Very fine suspensions down to colloidal size can be flotated after flocculation. Dissolved substances are only accessible to flotation when they are chemically precipitated beforehand. The precipitate is often so fine, however, that before flotation, it must first be coagulated or flocculated.

Suspended solid particles can be floated with adherent air bubbles when the total buoyancy of the air bubbles A_L is larger than the difference between the weight of the rising particle G_T and its buoyancy A_T,

$$A_L > (G_T - A_T)$$

From the numerous different flotation systems, the following sections discuss the oil and fat separators for municipal (domestic) and similar sewage, the mineral oil separators for refineries and other oil industry branches, and gasoline separators.

scraped. When two removal plates are present, they can be manually or automatically controlled depending on the direction of rotation and employed for the particular function. This system, however, can only be recommended when small amounts of separated material is expected.

Further developments, directed at reducing the required elevations or increasing the efficiency of separators, are represented by the parallel plate separator (PPI for parallel plate interceptor) and the corrugated plate separator (CPI), suggested by SIPM (Shell Internationale Petroleum Maatschapij).

The parallel plate separator consists of a welded sheet metal construction (Fig. 6.2.-20), which is completely placed in a rectangular containment structure. A bundle of sheet metal plates, positioned parallel and perpendicularly inclined to the flow direction, are passed horizontally by the flow in such a way that two groups of plates rise at their outer edge, while they sink toward the middle of the container. With the horizontal passage through the plates, oil driven upward from the underside of a plate can only ascend by moving outward, while the heavier sludge particles are directed toward the middle of the system.

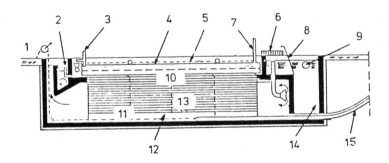

Fig. 6.2.-20: Principle diagram of the parallel plate separator (PPI) from Shell Internationale Petroleum Maatschapij

1 - winch; 2 - water overflow above covers; 3 - overflow valve for collected oil; 4 - covering; 5 - oil-free water surface; 6 - waste water collection tank; 7 - valve; 8 - screening; 9 - inlet; 10 - collected oil; 11 - cable; 12 - sediments; 13 - sheet metal plates; 14 - sand trap; 15 - suction connection for sludge removal

The corrugated plate separator (CPI) is an advanced development of the parallel plate separator and is distinguished by its still shorter length and better efficiency. The parallel and horizontal plate bundle is replaced with a bundle of corrugated plates, which are mounted parallel, but are passed by the flow in an inclined direction from the top downward. The effective sinking or rising path is also greatly reduced in this system, whereby the lighter oil ascends on the underside of the ridges and counter to the flow direction, toward the top. The sludge on the upper surface of the valleys, slides in the flow direction down to the collection area. The most frequently used means of removing the sludge, here in a much smaller volume than e.g. in a preclarification basin at a municipal sewage plant, is the evacuation of the sludge chambers with a tank truck and flexible tubing or piping.

For flotation installations, intermittently operated, it is recommended a pressure saturation to be maintained so that compressed air, pressurized water, or unsaturated water cannot escape when the system is shut down. Otherwise, a temporary instability sets in at the next start-up, characterized by waves and a corresponding fluctuation of the water level and reduction in efficiency. These conditions remain until an equilibrium is reached again in the saturation region, with respect to the circulatory water flow and the air dissolved in it.

6.2.2.1.3 Gasoline Separators

Gasoline separators must be located in front of the sewerage system connection for discharges from garages, auto repair shops, gas stations, and fuel storage facilities to protect the sewer network from explosive gas-air mixtures.

Gasoline separators work on the same principle as fat and oil separators. The German standard DIN 1999 regulates their design, construction, and operation.

Fig. 6.2.-21 illustrates a gasoline separator made of reinforced concrete from DICKERHOFF and WIDMANN /1/.

Fig. 6.2.-22 illustrates a cast iron gasoline separator with the trade name "Servator" from PASSAVANT Werke /1/.

Fig. 6.2.-23 shows a large gasoline separator of the PASSAVANT system series.

Owing to fire danger, the separator requires regular maintenance and careful inspection, particularly the prompt removal of collected fuel.

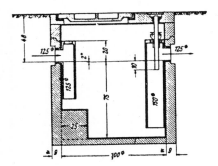

Fig. 6.2.-21 Gasoline separator made of reinforced concrete /1/

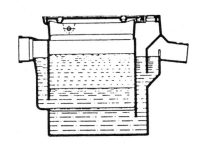

Fig. 6.2.-22 Cast iron gasoline separator /1/

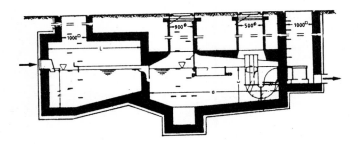

Fig. 6.2.-23 Large gasoline separator, PASSAVANT system

6.2.2.2 Sedimentation Systems

Sedimentation systems are containments with an effluent stream in which the flow velocity is slowed so that settleable substances can sink and floatable substances can collect on the water surface. The separated substances are removed both from the surface and the floor of the basin with suitable devices and, in part, further treated. Finer flocculent material in the waste water, which does not settle due to its specific weight, can be adsorbed by the sinking or rising of settleable or floatable material, respectively. This increases the degree of purification.

The flocculation of the material can be aided by the addition of flocculating agents or by mixing different types of waste water.

The degree of purification of a sedimentation system depends on its design, type of construction, and operation.

The sinking velocity is determined mainly by the diameter and the density of the suspended particles. In general, the velocities are not calculated, but are determined by measurements.

The water temperature also influences the sinking velocity of suspensions.

Table 6.2.-6 gives the sinking velocity at a water temperature of 10° C for the listed solids, from FAIR /48/.

Tab. 6.2.-6: Sinking velocity for the listed solids at a water temperature of 10° C /48/

Substance	Density	Diameter mm						
		1.0	0.5	0.2	0.1	0.05	0.01	0.005
Quartz sand	2.56 kg/l	502	258	82	24	6.1	0.3	0.06 m/h
Coal	1.50 kg/l	152	76	26	7.6	1.5	0.08	0.015 m/h
Domestic sewage suspensions	1.20 kg/l	122	61	18	3	0.8	0.03	0.008 m/h

The sedimentation process, for laminar flow, can be obtained from the diagram, when no mixing occurs between neighboring flow streams. The suspended particles in the waste water stream have horizontal v_h and vertical v_s velocity components. The highest purification efficiency is reached when a given particle settles to point C in a selected settling time period. If it settles before C, for example at C', then the containment is too long, and the extra volume of the basin is useless. If it settles later at C", the containment is too short and an incomplete separation results. The settling process is described by the following equations

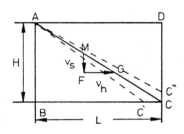

$$F = \frac{Q}{v_s} \quad [m^2]$$

$$v_h = F \cdot H \quad [m^3]$$

$$v_s = \frac{Q}{B \cdot H} \quad [m/sec]$$

$$H = q_f \cdot t_s$$

$$t_s = t_v = \frac{L}{v_h} = \frac{H}{v_s} \quad [s]$$

$$F = L \cdot B \quad [m^2]$$

The triangular relation ABC to MFG gives

$$\frac{H}{v_s} = \frac{L}{v_h} \; ; \; H = \frac{v_s}{v_h} \cdot L \quad [m]$$

where
- F = surface of the basin in m²
- Q = flow rate of waste water in m³/s
- v_s = sinking velocity or surface charge q_f in m/s
- v_h = horizontal flow velocity in m/s
- t_s = settling time in s
- t_v = flow-through time in s
- V = container volume in m³
- B, L = width and length in m

For longer basins, these relations must be taken into account to ensure sufficient sedimentation.

The actual flow in the basin is, however, not laminar, but turbulent; therefore, the settling particles can move opposite the ideal settling direction. To avoid this as far as possible, the flow velocity in sand traps is set at nearly 30 cm/s. To account for the delayed settling in turbulent as opposed to laminar flows, one should, according to HAZEN /1/, multiply the calculated settling surface with a coefficient. This coefficient is the quotient of v_s, the sinking velocity of the sand grain to be settled in calm water, and q_f, the surface charging of the sand trap. Thus the surface of the sedimentation basin is determined by the equation, from HAZEN /1/,

$$F = \frac{v_s}{q_f} \cdot \frac{Q}{v_s} = \frac{Q}{q_f} \quad [m^2]$$

Accordingly,

$$q_f = \frac{Q}{F} < q_{f,allow} \begin{cases} \text{with dry weather} = 18 \ [m/h] \\ \text{with rainy weather} = 36 \ [m/h] \end{cases}$$

or,

$$\frac{Q}{F} = v_s \geq q_f$$

so that the surface charging speed should be less or equal to the sinking velocity.

In practice, the sinking and rising velocities are determined empirically, and the surface charging is given or taken from the literature.

Sedimentation basins are dimensioned through the surface charging and the retention time.

a - The surface area is given by the relation

$$F = \frac{Q}{q_f} \quad [m^2]$$

b - The volume is calculated as

$$V = Q \cdot t_a \quad [m^3]$$

Here

F = surface in m²
V = basin volume in m³
Q = waste water flow rate in m³
t_a = retention time in h
q_f = surface charging in m/h

Currently, two types of sedimentation systems are used in treatment plant engineering:
- sand traps and
- settling tanks.

6.2.2.2.1 Sand Traps

Sand traps are containments used for settling from waste water mineral particles in a size ranging from 0.2 to 2 mm. Thus, primarily those components are separated, which have a negative effect on biological processes and would lead to a useless loading of biological or fermentation containments. In fact, organic substances in waste water are also settled out. With domestic (municipal) sewage, therefore, one obtains not only pure sand.

Concerning the working principles, one distinguishes two types of sand traps: long sand traps and round sand traps.

The <u>long sand trap</u> (Fig. 6.2.-24) consists of sand collection region in its lower part, usually with rectangular cross section, which is bordered by an underlying drainage layer. A trapezoid-shaped channel is located above the drainage section, which facilitates the transport of deposited sludge from the outer edges into the sludge collection area, and aids the flow, such that a velocity of 0.3 m/s is as closely maintained as possible under any flow condition. Holding this value constant is possible only with complicated cross sectional forms of the sand trap or with special spillways at the end of the trap containment.

Fig. 6.2.-25 illustrates a few of these long sand trap profiles, from the ATV, designed to maintain a constant flow velocity ($\leq$ 30 cm/s).

a) Rectangular cross section,

$$h = \frac{v^2 \cdot b^2}{8g} \cdot \frac{1}{x^2} \; ; \quad a = \frac{v^2}{2g} \; ; \quad c = \frac{v^2 \cdot b^2}{8g} \qquad [m]$$

$$x = \frac{v^2 \cdot b}{2 \cdot \sqrt{2g}} \cdot \frac{1}{\sqrt{h}}$$

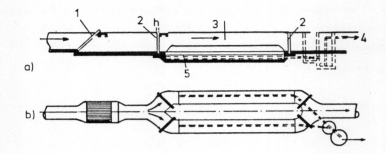

Fig. 6.2.-24: Essener long sand trap (with manual cleaning)
a - longitudinal section; b - plane view; 1 - screening; 2 - retention wall; 3 - sand trap; 4 - to pumps; 5 - drainage area

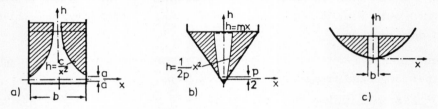

Fig. 6.2.-25: Retention profiles designed for maintaining a constant flow velocity in long sand traps

a - rectangular cross section; b - triangular cross section; c - parabolic cross section

b) Triangular cross section,

$$x = \sqrt{2 \cdot p \cdot h} \quad ; \quad h = \frac{2v}{3pg} x \quad ; \quad m = \frac{2v}{3pg} \qquad [m]$$

c) Parabolic cross section (in practice, trapezoidal form),

$$x = \frac{b}{2} \quad ; \quad h = \frac{6 \cdot v^2}{g \cdot b^2} \cdot x^2 \qquad [m]$$

Here

v = flow velocity in sand trap, as a rule 0.30 m/s
p = twice the distance from the focal point of the parabola to the x-axis

It is practical to build the sand trap in two sections, one in operation, while the other dries out before being manually cleaned.

In larger systems, cleaning the trap is done continuously and mechanically with a clearing bridge. It is possible either to push the sand into a funnel-shaped trench on the inlet side or to pump it continually out to a channel (along the longitudinal axis, see Fig. 6.2.-26).

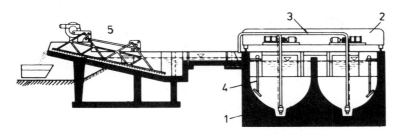

Fig. 6.2.-26: Cross section of the "Duisburg Kleine Emscher" treatment plant of the Emscher water works

1 - long sand trap; 2 - pump conveyor; 3 - suction pump; 4 - aeration piping; 5 - sand classifier

In both cases, a water-sand mixture, which must be separated outside the trap, is removed from the actual sand trap. For this separation, small silolike containers or so-called sand washers are well suited; they work on the settling principle. Flow patterns are generated in these containers, which hold the organic components in suspended motion, while sandy components settle on screens or scraper surfaces. From here, conveyor worms or other removal devices first raise the sandy material out of the water and then transfer it to collection tanks. These additional systems have the great advantage that a pure sand is recovered, free from organic contaminants, that can also be used in construction after intermediate storage.

The aerated sand traps (Fig. 6.2.-27) operate with a much larger volume and very small velocity components in the flow direction. By blowing in air on one side of the basin, a rotational flow in the form of a water whirl is formed whose axis lies in the flow direction. Owing to the high transverse velocity in the water whirl, the organic particles remain suspended, while all heavier components separate in a collection channel. From here, they are withdrawn. Aerated sand traps require a water depth of at least 2 m for forming a good whirl. The basin width and depth should be in the ratio of about 1 to 1.5. The air volume, necessary to maintain a sufficient whirl effect, lies between 10 and 25 cubic meters per running meter and hour. The smaller amount holds for aeration with fine bubbles. The air volume should not be set too conservatively, because decay processes can start in aerated traps, which, with the lack of oxygen, lead immediately to hydrogen sulfide formation and with it, odor emission.

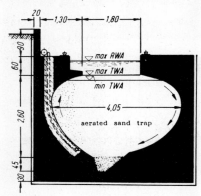

Fig. 6.2.-27: Aerated sand trap at the treatment plant in Heilbronn

The aerated sand trap with fat separator makes other separation processes possible. On the long side of the trap, opposite the air inlets, a second elongated container is arranged parallel to the sand separator. It is connected to the sand trap container often with a slightly sloped floor, and is divided from the sand separation zone in the upper region by so-called submersible walls. When the water whirl stirs up material e.g. fat, fat covered fiber and hair, food rests, etc., it finds its way between the supports into the fat separation container, from which it cannot escape again. The fatty material collects here and is skimmed off the water surface with a special clearing device into a collection trench.

flows of 8 to 24 cm/s, which naturally impede the settling effect.

The most important factors in dimensioning the settling basin are the basin surface area, the surface charge and the flow-through time. A definite relationship exists between these factors, so that they all must be considered when dimensioning the basin.

The construction of the settling basin varies according to the waste water volume, the available building space, the function of the basin with respect to the total treatment system, the flow direction, and the treatment method involved.

Settling basins for municipal waste water are usually classified, according to the flow direction of the treated water, in horizontal (elongated and radial basins) and vertical flow basins. In treating industrial waste waters, horizontal flow basins are built with one, two, or even several levels.

6.2.2.2.2.1 Rectangular Basins

Rectangular basins are made of concrete, reinforced concrete, brick, or earth and are built alone or in groups.

The waste water stream flows horizontally through the basins. The flow and settling process is divided into four zones (Fig. 6.2-30):

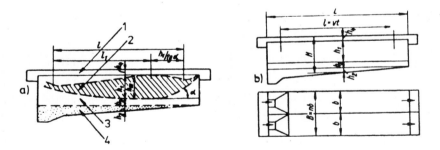

Fig. 6.2.-30: Elongated rectangular basin
a - flow diagram; b - dimensioning scheme; 1 - safety zone; 2 - active zone; 3 - neutral zone; 4 - sludge zone

- the active zone h_1 with an average depth h_m, the most important part of the rectangular basin,
- the sludge zone (depository zone) h_2, where sludge collects,
- the neutral zone h_3, where waste water and sludge mix,
- the safety zone h_4 of the system.

Corresponding to the flow process, the settling process can also be divided into four zones:
- the inlet zone,
- the actual settling or separating zone,
- the outlet zone,
- the sludge zone.

The inlet zone has the function of distributing incoming waste water uniformly over the entire flow cross section (for rectangular basins, the basin width), so that turbulences do not extend into the actual separation zone.

The separation zone itself, within which the settling particles descend at an angle, takes up the largest part of the basin volume. Since, at least in preclarification basins, the processes in this zone follow STOKES law, maintaining laminar flow conditions is a very important requirement. The REYNOLDS number characterizes the flow condition.
Stable conditions exist when a sufficient hydraulic radius is available for a given cross section. For rectangular basins this requirement means that many small, shallow basins should be built, however, this is not possible due to economic reasons.

The outlet zone in turn has the function of removing the clarified water in a uniform way. For rectangular basins this means a uniform approach to the spillway at the end of the basin. If a fat and oil separator is not already foreseen, a retaining wall and a fat removal device is installed before the spillway.

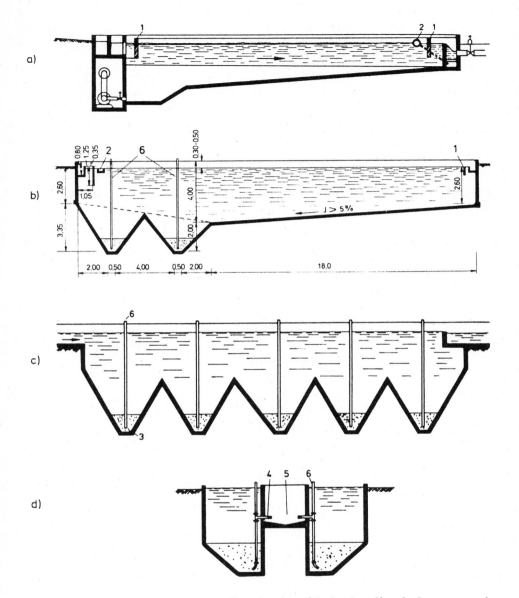

Fig. 6.2.-31: Rectangular settling basin with hydraulic sludge removal
a - with one sludge trench; b - with two sludge trenches; c - with a trenched floor; d - working principle of the hydraulic sludge removal; 1 - submersible wall; 2 - fat and oil separation channel; 3 - deposited sludge; 4 - sludge valve; 5 - sludge removal channel; 6 - piping for the vacuum removal of deposited sludge

The sludge zone should be equipped with hydraulic or mechanical devices for sludge removal.

Fig. 6.2.-31 illustrates basins with trenched floors, which are evacuated with hydraulic pressure differences (or with pressurized water).

Large rectangular basins are equipped with either mobile clearing bridges or chain evacuators. Fig. 6.2.-32 shows a rectangular basin with mechanical sludge removal.

Mobile bridges have the advantage that all driving and control elements are located above water, are easy to reach, and do not suffer from heavy wear. A disadvantage is that their automation requires a relatively complicated control mechanism, which raises or lowers the clearing plate for surface scum or floor sludge and interrupts the lateral motion for these operations.

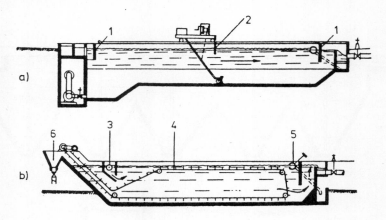

Fig. 6.2.-32: Rectangular settling basin with mechanical sludge removal
a - with mobile sludge removal bridge; b - with chain remover; 1 - submersible wall; 2 - plate for fat and oil collection; 3 - inlet; 4 - chain remover; 5 - device for fat and oil separation; 6 - sludge removal

Fig. 6.2.-33 illustrates a combined basin, the so-called Hamburger basin. A flocculation and activation basin is connected prior to the settling basin (with two collection trenches).

Fig. 6.2.-33: Longitudinal cross section of a Hamburger basin
1 - inlet; 2 - outlet; 3- sludge removal

With the use of chain removers, the chain cog wheels are made of plastic, so that wear on the chain links is hardly observable. The chain mountings are also either plastic or made of a durable alloy. Most recently, even the chain links are manufactured with a very hard plastic, which leads to an extremely low structure weight and excellent rust-resisting properties.

6.2.2.2.2.2 Round Basins

Round basins are found most frequently in waste water engineering. The round basin was developed from long basins, by which the longitudinal cross section is rotated about the perpendicular in the inlet wall. Theoretically, its ecological efficiency must be better than rectangular basins, because, the further the particles recede from the inlet, the larger the flow cross section will be with reduced settling depths, which altogether should lead to a better settling efficiency.

Round basins are made of reinforced concrete with a round base. Waste water, introduced in the center, flows outward to the basin edge and enters a collection channel, which passes it further along. The basins are equipped with a revolving clearing bridge, which requires no controls regarding its movement. Clearing plates for surface scum and floor sludge can be constantly held in working position and, in normal operation, need no adjustment

Domestic sewage contains many substances, especially fat, hair, and fibers, which float to the surface in the preclarification and can be removed by mechanical devices.

The sludge, collected by the scraper plate, falls into a central shaft and under the water pressure in the basin, it is forced into a lateral channel, from which it can be pumped for further treatment. Fig. 6.2.-34 illustrates the flow and design schematics for a round basin. Fig. 6.2.-35 shows a round basin /25/.

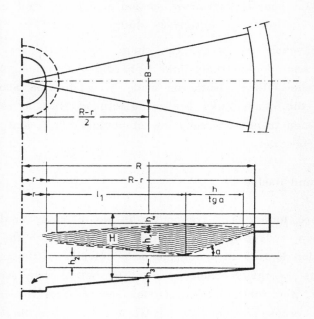

Fig. 6.2.-34   Flow and design schematic for a round basin

In deciding whether a round or rectangular basin should be built, space requirements could be important. Rectangular basins can in general be integrated with other structures to save considerable space, while round basins require a certain amount of open space. In contrast, limiting the number of removal devices is not a valid criterion. Equipping rectangular basins with so-called exchangeable clearers, which is occasionally observed, must be seen as a misunderstanding of savings and a totally false development. Here, several neighboring rectangular basins are equipped with a single remover, which can be transported from one basin to another. Preclarification basins with this installation

531

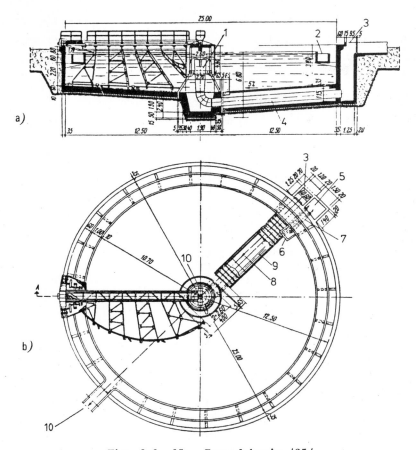

Fig. 6.2.-35: Round basin /25/

a - cross section; b - plane view; 1 - metallic distribution pipe; 2 - channel for clarified water; 3 - inlet shaft for sewage; 4 - sewage pipe; 5 - sludge shaft; 6 - surface sludge shaft; 7 - surface sludge collection pipe; 8 - sewage pipe; 9 - sludge discharge pipe; 10 - outlet for clarified water

cannot be evacuated either continuously or automatically. In addition, the prerequisites for a biological treatment are also worsened.

In comparing the construction costs of round and rectangular basins, it should be noted that the labor share of the costs for concrete forms is so large that there is no appreciable difference between the total construction costs of a round and a rectangular basin. The minimal extra costs for rounded forms are usually already compensated by the lesser amount of concrete needed. Added to this, the static load is smaller for round structures, so that costs for steel reinforcement can also be saved.

Concerning the water depths, the ATV (Abwassertechnische Vereinigung) recommends 1.50 to 3.00 m, the same as for rectangular basins.

6.2.2.2.2.3 Two-level Settling Systems

Various types of two-level sedimentation basins exist:
- combined sedimentation and digestion basins, e.g. Emscher tanks, Üdem tanks,
- sedimentary basins with an intermediate floor.

The Emscher tank, a bilevel settling containment, also called the Imhoff tank, is in use worldwide in small treatment plants as a settling basin and digestion chamber.

The Emscher tank is a compact unit (Fig. 6.2.-36), in which the settling container sits on top of a combined fermentation chamber. The settling material is diverted so that it can slide directly into the fermentation region. Restraints on the slide surfaces hinder the fermentation gas from rising and disturbing the settling process. Emscher tanks are built with round or square profiles.

Properly dimensioned, Emscher tanks are attractive owing to their simple processing methods, low-maintenance costs, and odorless end product. The reliability of these systems, especially in small sizes, has been proven universally. A disadvantage is that the costs for building Emscher tanks are high. Since the fermentation chamber cannot be heated, it must be sized for long retention times. The large fermentation volume, necessary for sites with average yearly temperatures under 10° C, can lead to very large construction dimensions, where, especially, constructing the proper depth is expensive.

Usually, at present, one does not cover the so-called flotation scum region, thereby not being able to collect the fermentation gas. However, it should be taken into account, even in the design stage, that aggressive gases can be expected in the upper regions of the Emscher tank. Even when the concentrations are small due to direct ventilation from the outside, the effects are visible in a short time. Thus, it is necessary to provide protective coatings on concrete and select all metal parts out of suitable material. It cannot be said often enough that galvanized steel is not resistant to the corrosion caused by the aggressive components in the fermentation gas. One must either use stainless steel or protect galvanized structures with additional coatings.

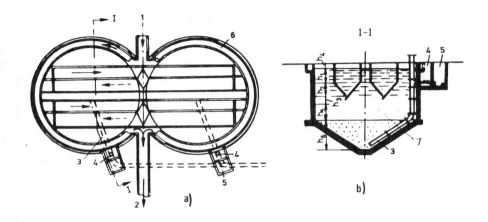

Fig. 6.2.-36: Emscher tank

a - plane view; b - cross section I-I; 1 - inlet; 2 - outlet; 3 - sludge removal; 4 - sludge valve; 5 - sludge shaft; 6 - distribution channel; 7 - fermentation chamber

The settling portion consists of one or two channels in the longitudinal direction, in which the sewage flows horizontally with a low velocity (4 to 10 mm/s). The suspended particles are separated and fall into the lower chamber where they are fermented.

When two wells are combined, as in Fig. 6.2.-36, the distribution channels are equipped with slide valves, which make it possible to vary the flow direction. This is necessary, because the first fermentation chamber (as seen in the flow direction) is always loaded more than the second and, in order to evenly distribute the material to be fermented, directing the flow is necessary.

Larger fermentation chambers are equipped with dividing walls with openings. On the one hand, they strengthen the support of the outer walls and, on the other hand, they discourage the movement of the sludge water or its mixing with settled sludge.

The sludge is withdrawn with pipes in its freshly settled state in small plants and in fermented form in large plants.

To remove the sludge, the sludge slide valve needs to be opened only every one or two months. Thus, Emscher tanks are seen as semiautomatic systems, since they must be manually controlled only periodically.

Tab. 6.2.-9 gives data for the specific fermentation volume per person and the fermentation time for a normally sufficient degree of fermentation.

Tab. 6.2.-9: Fermentation for sludge deposited in an Emscher tank

average yearly temperature °C	specific fermentation volume l/P	fermentation time d
7	75	150
8	65	120
10	50	90

Emscher tanks are dimensioned as follows:
- the settling region as a settling basin with horizontal flow in the longitudinal direction,
- the fermentation chamber according to the degree of fermentation to be achieved.

The Üdem tank (Fig. 6.2.-37) is a round basin in the upper settling region. Sewage in-flow and sludge removal is same as that for a round basin. The sludge slides down to the base and is treated, as in the Emscher tank.

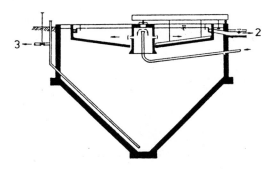

Fig. 6.2.-37: Üdem tank /2/
1 - inlet; 2 - outlet; 3 - sludge removal

Sedimentation basins with intermediate levels, double the effective surface area or reduce the surface charge by one half. With a uniform flow, a better purification efficiency than the common rectangular basin is achieved.

Such settling basins are used exclusively in industry, since for the uniform flow rates, they provide an economic usage of space. Fig. 6.2.-38 illustrates a settling tank with an intermediate floor.

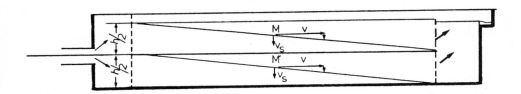

Fig. 6.2.-38: Settling tank with intermediate floor

6.2.2.2.2.4 Lamella Sedimentation Basins

Theoretical considerations regarding the settling process and the resulting requirements for a large hydraulic radius and a shallow basin have led to technological advancements represented here by the so-called lamella sedimentation tank.

To begin with, one can imagine that the efficiency of a rectangular basin could be greatly increased if several, horizontal intermediate levels were inserted. The hydraulic radius would be greatly increased and the vertical path length, traveled by the particles in separating, would be shortened. The path length is inversely proportional to the number of intermediate plates. The practical problem remains, however, as to how to remove the sludge deposited between the intermediate levels. The technical solution thus foresees a slanted arrangement of the levels in the form of a bundle of lamellae positioned at a 60 degree inclination. The flow through the bundle can be either in the same direction or countercurrent to the sludge path. Such basins save considerable space and are also unaffected to secondary influences, such as thermal streams or circular flows caused by the wind. On the other hand, they are only suited to sludge types, which do not tend to collect, bake, or stick, e.g. for the separation of solids from river water in the course of drinking or industrial water treatment. Many cases also exist for industrial and commercial discharges, where a mineral substance is separated. For use with activated slime, satisfactory results are not presently available, while the results are very positive for chemically flocculated sludge (mainly metal hydroxide sludges) from several large plants.

However, the number of systems now in operation is not very large and operating experience is limited. It is obvious that complete removal of coarse particles is necessary for such systems, owing to the manifold structures installed and the numerous dividing edges in the flow stream.

6.2.2.2.2.5 Vertical Flow Sedimentation Basins

Vertical flow sedimentation basins are reinforced concrete or metal basins with a round or square base, which are charged with a vertical flow of waste water, whereby suspended particles settle on the basin floor.

Through gravity, those particles are settled with a sinking tendency larger than their buoyancy. While settling, the particles collide with already present flakes and combine, so that the volume of the particles increases along their path downward.

The waste water is fed into the center of the basin (Fig. 6.2.-39) with a pipe with a deflecting shield mounted on its end. The shield diverts in-coming water upward. The clarified water flows into a channel and is withdrawn.

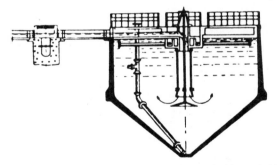

Fig. 6.2.-39: Vertical flow sedimentation basin

Removal of deposited sludge is accomplished with piping under hydraulic pressure (at least 1.5 m water column WS). The sludge is piped to a further treatment station.

To remove hardened sludge from the basin floor, which could impair the withdrawal of fresh sludge, a pressurized water pipe (4 to $6 \cdot 10^5$ Pa) must be installed down to the bottom end of the sludge collection pipe.

The same sizing regulations are used as for the horizontal basins:
- the effective height however is
    $h_1 = v \cdot t$ but ≥ 2.75 m
- the sludge depth h_2 is calculated according to the loading (0.8 l/(P·d) for domestic sewage) and the means of sludge disposal,
- the separation layer $h_2 = 0.4 - 0.5$ m
- the security layer $h_4 = 0.3 - 0.5$ m.

The vertical flow settling basins are seldom (only when space is scarce) used for domestic sewage. Precisely because of their compact form, they are used for low industrial discharge volumes.

6.2.2.2.2.6 Other Types of Sedimentation Basins

For small amounts of domestic sewage, also other basins, not previously discussed, are suitable, e.g. septic tanks, sewage ponds, compact treatment plants, etc.

The treatment of industrial waste water is based on pure chemical or physio-chemical reactions. Here, different types of special containments are used.

The sludge trap can also be considered a small sedimentation tank. This is a settling tank in its simplest form, which does not have to be emptied frequently, depending on the composition of the sludge. If it is a sand-fat mixture, e.g. from an auto washing facility, the trap can be sized larger than that used for a small potato processing plant, because the wash waters also contain a large organic load. To avoid fermentation and its consequences, the sand trap should be cleaned daily. The sludge can be either dumped in a landfill, where the subsequent odor development should be considered, or it can be pumped directly into the sewage plant inlet, by passing the sewer line, to be treated with the normal processes.

6.2.3 Equilibration Basins

Equilibration basins are systems used to mix mechanically waste waters of different volume and concentration.

Such systems are needed in the treatment of waste waters, which occur irregularly or regularly but with different flow amounts and concentrations. The same basins are also often used for mixing waste waters with added chemical reagents.

The installation of equilibration basins is recommended, because with relatively low investment and operational costs, exceptional technical and economic advantages can be realized with respect to the later treatment processes.

The basins are named according to their function in the treatment process, e.g.
- homogenization basins for equilibrating concentration,
- equilibration basins for regulating the flow; the designation is representative for the entire group,
- retainment basins, allowing water storage, when the average flow rate is exceeded,
- protection basins which guard the sewage plant or reservoir for sudden shocks of harmful waste waters, and
- security and control basins installed downstream of the sewage prior to the receiving water to examine the quality of the treated water.

In general, there is little difference in the construction of these systems, but more of a difference in their function and in the waste water composition.

Such systems can be constructed as ponds or as structured basins.

Structured equilibration basins are designed with horizontal, vertical, or tangential inlets. Fig. 6.2.-40 illustrates a horizontal flow homogenization basin and Fig. 6.2.-41 a vertical flow equilibration basin.
Fig. 6.2.-42 shows a rotary system with aeration device manufactrued by H. SCHEVEN in Düsseldorf, which also acts as a flocculation basin. Such mixing tanks with stirring devices are also used to keep additives in suspension, e.g. lime particles.

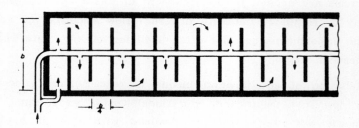

Fig. 6.2.-40: Horizontal flow homogenization basin /2/

Fig. 6.2.-41: Vertical flow equilibration basin /2/

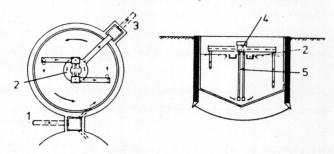

Fig. 6.2.-42: Equilibration basin with rotary stirring and aeration /2/
1 - inlet; 2 - outlet channel; 3 - outlet; 4 - air expulsion and separation chamber; 5 - compressed air

6.3 Chemical Processes of Waste Water Treatment

In waste water treatment, the term chemical processes applies essentially to methods based on purely chemical or physico-chemical reactions. Chemical processes are utilized in waste water treatment to achieve the following:
- neutralization of acidic or alkaline discharges,
- separation of solids that cannot be removed by simple mechanical means,
- separation of colloidal, mostly organic substances and certain dissolved inorganic substances,
- removal of fat and oil residues,
- improvement of the clarifying effect of flotation and filter systems,
- reduction of radioactivity in contaminated waste waters,
- effects achievable by using chlorine or chlorine-oxygen compounds, such as: disinfection, algae growth control, decontamination, deodorization, decolorization, oxidation, and control of fly larvae in biological filters.

Among the chemical processes, the German federal regulations on waste water treatment (ATV) distinguish between reactions with and without conversion.

When the dissolved substances are treated without material conversion, the following physico-chemical reactions occur: adsorption, extraction, membrane separation, distillation.

When the dissolved substances are treated with material conversion, the following chemical reactions are employed: neutralization, precipitation, oxidation, reduction, ion exchange.

When undissolved substances are treated, the following basic processes are common: sedimentation, filtration, flocculation, adsorption, flotation.

6.3.1 Methods of Treatment

The following is a description of the basic processes most common in the treatment of industrial waste waters.

6.3.1.1 Neutralization

Neutralization is a simple chemical reaction. Acidic or alkaline waste waters must be neutralized before they are discharged into a municipal sewerage system. In cases where buffering capability of the receiving reservoir is insufficient, neutralization may also be necessary if the waste water is to be discharged into natural water bodies. Neutralization processes are required mainly in the chemical and metal-processing industries, and also in refineries and other industrial plants.

Basically, a distinction is made among three acid waste water types:
- waste waters containing strong acids whose salts are easily soluble in water (e.g. HCl, HNO_3). Appropriate alkalis or alkaline earths are used to neutralize such waste liquors. Sedimentary sludges do not occur, so that the treatment consists of a simple mixing.
- waste waters containing strong acids whose salts are highly insoluble in water (e.g. H_2SO_4). In these processes, considerable amounts of sludge may occur.
- waste waters containing weak acids (e.g. CO_2, COOH).

Recently, the following groups of neutralizing processes have become common:
- utilization of the buffering capability of the receiving reservoirs,
- mixing the acidic with the alkaline waste waters,
- admixture of chemicals producing alkaline reactions, such as NaOH, $Ca(OH)_2$, etc.
- filtration with neutralizing filter materials such as $CaCO_3$, $MgO-CaCO_3$, etc.

In neutralizing processes, a shift of the pH value occurs, with a resultant flocculation of colloidal constituents in the waste water. In such cases, subsequent treatment in a settling facility is required.

The utilization of the acid-binding capability (buffering capacity) of a receiving reservoir depends on the so-called carbonate hardness of the water it contains.

The following chemical reaction can be given as an example of the process in a receiving reservoir:

$$2\ H^+ + SO_4^{2-} + Ca^{2+} + 2\ HCO_3^- \longrightarrow CaSO_4 + 2\ H_2O + 2\ CO_2$$

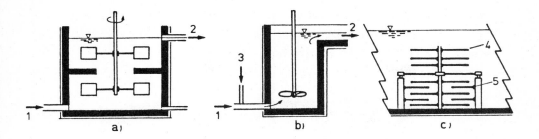

Fig. 6.3.-2: Mixers

a - turbine; b - impeller; c - paddle stirrer; 1 - inflow; 2 - outflow; 3 - precipitating agent; 4 - rotor blade; 5 - stator blade

and initiation of coagulation), the metal salts show a strong adsorption effect on suspended matter. This effect is particularly strong when the particle charge is negative (adsorptive coagulation). The adsorption process occurs on the enlarging surfaces of the growing hydroxide flakes.

In most cases, the coagulated agglomerates remain as small flakes which must be enlarged and made rapidly settleable by the addition of auxiliary flocculation agents.

A coprecipitation of anions always occurs when precipitable anions are present. The precipitation of phosphate by hydrate of lime, iron, or aluminium salts is of great importance in the treatment of waste water because contact between fertilizer phosphate and surface waters can thus be avoided.

Fig. 6.3.-3 shows an aero-accelerator treatment plant produced by the LURGI company.

Specially designed compact units are used when large quantities of industrial waste waters are to be flocculated. These systems principally include various settling chambers provided with equipment for facilitat-

ing the formation and separation and hence the sedimentation of flakes.
There are various types of settling basins, e.g.
- without aeration, with and without recirculation,
- or with compressed-air supply.

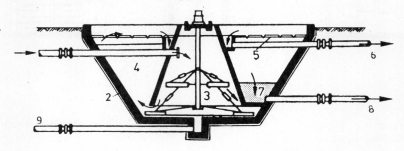

Fig. 6.3.-3: LURGI Aero-Accelerator

1 - inflow; 2 - central basin; 3 - mixing equipment; 4 - settling compartment; 5 - outflow channel; 6 - waste water outflow; 7 - scrapper; 8 - sludge removal; 9 - discharge

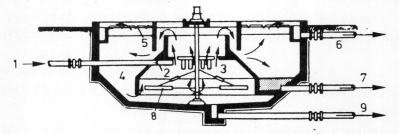

Fig. 6.3.-4: PATERSON Accelerator

1 - inflow; 2 - mixing zone; 3 - contact zone; 4 - settling compartment; 5 - discharge channel; 6 - waste water outflow; 7 - sludge removal; 8 - scrapper; 9 - discharge

The flocculation takes place in a primary (1) and a secondary (2) mixing and reaction zone. This allows the separate addition of additional reagents such as hydrate of lime for Fe(III) precipitation or polyelectrolytes near the impeller to improve the subsequent sedimentation in the suspension filter (3). The waste water comes into contact twice with the sludge. Flocculation occurs in zone 3 where a scrapper (8) prevents a solidification of the settled matter. The sludge is periodically and automatically removed from the settling zone (7). The contact time is in the range from 20 to 40 min and the upward speed is 1.4 to 2.8 mm/s. The determination of reasonable values demands previous testing.

Fig. 6.3.-5 illustrates a "Centrifloc" sludge blanket system.

The dimensioning of the flocculation equipment requires preliminary tests termed JAR tests.

Six beakers with common stirring equipment are used to simulate the flocculation process, including the settlement. The process is monitored by continuous sampling.

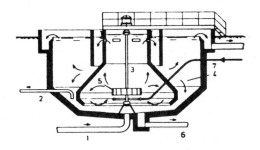

Fig. 6.3.-5: Centrifloc sludge blanket system

1 - inflow; 2 - outflow; 3 - aeration zone; 4 - sludge water; 5 - stirrer; 6 - treated water; 7 - compressed air

Fig. 6.3.-6 illustrates a laboratory equipment used to carry out flocculation tests (JAR test).

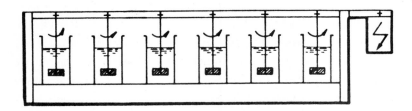

Fig. 6.3.-6: Laboratory equipment for flocculation test (JAR test)

The use of the flocculation process is particularly applicable for waste waters with contaminants largely present in a colloidal form, and for waste waters containing substances which even though truly dissolved, do adsorb to suitable flakes. These include the waste waters from food-processing industries and agriculture (which, however, have the advantage that the organic substances are very well suited for biodegradation), paper and cellulose industry, textile industry, and chemical or petrochemical industries. As waste waters from such industries nearly always occur in major amounts, proven coagulation basins and settling equipment are considered when sedimentation systems are designed.

6.3.1.2.2 Precipitation by Flocculation as a Single Stage Waste Water Treatment, Preliminary Precipitation

The process of precipitation by flocculation can be employed either as a single stage in waste water treatment (direct precipitation) or as a preliminary precipitation step.

The precipitation by flocculation as a <u>single method of waste water treatment</u> is usually sufficient to handle waste waters polluted by substances consisting of coarse and colloidal dispersions. Such waste waters include, for instance, the waste waters from ceramics works and other waters carrying minerals causing turbidity. This method is also sufficient in plants producing high-grade papers if the resultant COD and BOD_5 values are correspondingly low.

With waste waters containing many unsettleable mineral or other inert suspended substances, which, however, also require biodegradation due to their further load with organic and biologically degradable material (e.g. in activation plants), precipitation by flocculation must precede the biological stage when the increase of sludge from the mineral or inert material in the biological stage, without such preliminary precipitation, would exceed the production of biologically active sludge.

Even though the sludges produced in the activation plants have the required sediment volumes and solids contents, their biomass volume is too small owing to the inert solids included. They are therefore not suited to ensure biological purification to the required degree. The preliminary precipitation with aluminium or iron salts and hydrate of lime is successfully used as a separate step, which also protects the biological stages against toxic metals, fats, and oils.

Certain waste water constituents may sensitively interfere with or even inhibit preliminary precipitation. These constituents include substances forming complexes with aluminum or iron, e.g. condensed phosphates, tartaric acid, citric acid, and other natural or synthetic organic compounds. This is why attempts to handle the overload of municipal sewage plants in wine-growing areas in season (harvesting, must production, 1st and 2nd tapping) by preliminary flocculation have proved a failure.

6.3.1.2.3 Secondary Precipitation

In preliminary precipitation compounds or substances are often removed that must be present in subsequent activation steps or that would add other benefits to the process.

Examples are an excessive removal of phosphorus compounds vital for biodegradation, the reduction of waste water buffering capacity by conversion of hydrogen carbonate into permanent hardness, or the precipitation of so-called carbonate hardness by using excessive quantities of hydrate of lime in preliminary precipitation (decarbonization) according to the equation

$$Ca(HCO_3)_2 + Ca(OH)_2 \rightarrow 2\ CaCO_3 + 2\ H_2O$$

An excessive separation of solids that improve the properties of activated sludge may also occur as a result of the preliminary precipitation. The opinion is that nearly any waste water subjected to chemical treatment prior to biological clarification would tend to create bulking sludge caused by sphaerotilus, also, in rare cases, by actual sewage fungi.

Secondary precipitation is the technically better and more efficient solution in cases where preliminary precipitation is not definitely required. Compared to preliminary precipitation, which may accompany the usual preliminary clarification process, the secondary treatment requires investments for mixing, reaction, and sedimentation basins.

A variety of highly loaded waste waters can be taken as a practical example to demonstrate, in simple laboratory tests, that prolonged biological treatment facilitates or is even essential for flocculation, and that the precipitation is safe and reliable after a thorough biological purification.

The secondary precipitation process has the distinct advantage that it produces decisive improvements in the quality of the discharge from the biological stages. With highly purified biological discharges, this process offers improvements which are often impossible by preliminary waste water precipitation before the biological treatment stages. These include a nearly complete phosphate removal, the removal of suspended and turbidity producing matter initially formed in the stage of biological treatment, and the flocculation or precipitation of organic metabolic products from the biological stage. In addition, secondary precipitation considerably reduces the germ content in waste water.

Secondary precipitation can be realized without additional structures for sludge separation using the biological filter method. In this process, the discharge from the biological or trickling filters is recirculated to dilute the waste water and in order to maintain the minimum surface input on the upper surface of the biological filter. In contrast to the activation process, however, the sludge supplied to the secondary sedimentation basin is not involved in the aerobic step of process in the normal biological filter operation. In addition, the sludge volumes arising from the biological filter process preceding the secondary sedimentation basins are negligibly small when compared to those arising at the same location in activation systems. Therefore, the discharges from biological filters may be treated upstream of the secondary settlement basins with greater amounts of flocculating agents than are used for the discharges from activation basins, where in nearly all cases unfavorable effects on the purification efficiency could be observed as a result of the addition of major quantities of flocculating agents.

In the secondary precipitation of biologically treated waste waters the added quantities are within the range from 10 to 80 mg Fe^{3+} or Al^{3+}/l. The added iron or aluminum quantities result in a weight increase of ca. 2.3 to 2.5 g dried matter at 105° C for 1 g of Al and ca. 1.6 to 1.8 g dried matter at 105° C for 1 g Fe.

The weights of the flocculated and precipitated substances are in addition.

The freshly precipitated hydroxides are highly aqueous so that solids contents of only 6 to 20 g dreid matter/l are obtained in the sedimentation basins, depending on the type and composition of the coprecipitated substances.

The addition of chlorides and sulfates of aluminum and iron causes an increase of the chloride and sulfate concentrations in the treated waste water. The consequences of such increases must be considered. Since both anions are easily soluble in water and even though dissolved sulfate may be precipiated in the form of calcium sulfate, this possibility does not apply for the metal salt quantities used due to the very high solubility (1438 mg SO_4/l of water at 20° C).

An increase of the sulfate concentration may, for example, entail negative effects in terms of odor in the treatment plant (desulfurization under anaerobic conditions) and the anaerobic alkaline fermentation in sludge digestion. Chlorides are a very important factor in the metal corrosion caused by the substances contained in the water.

If auxiliary flocculation agents must be added in the simultaneous, preliminary, or secondary precipitation processes, the quantities are mostly within the range from 0.1 to 0.5 g/m³.

Iron salts, hydrate of lime, and polymers are principally added to improve the dehydration properties of sewage sludges, except in the waste water treatment.

6.3.1.2.4 Simultaneous Precipitation

The simultaneous precipitation process is intended to effect a reduction of the phosphate contents simultaneously with the biological treatment in activation processes. The realization of this process in practice is extremely simple, since only iron or aluminum salt storage and dosing equipment are required. The salts may be dosed and added in a dissolved form to the inflow to the activation basin, in the activation basin, or to the discharge supplied to the secondary sedimentation basin. Additional dosing of hydrates of lime may be necessary.
In practical operation, it is expedient to determine the most efficient point of addition, using simple flexible piping and only then should the permanent stationary installations be made. The dosing quantities are in

the range from 8 to 30 mg of iron or aluminum per liter. The phosphate contents are reduced by 80 to 90 % of the total quantities discharged into the activation stage. The population correction factors in relation to the phosphorus contents (mg P/l) are:

inflow into the treatment plant	15
in the mechanically pretreated waste water	13.5
in the mechanically treated waste water	
without simultaneous precipitation	10.5
with simultaneous precipitation	1.05 to 2.1

for 200 liters of waste water per person and day.

Often no or only negligible increases of the sludge volumes result from the simultaneous precipitation process whereas the dry weight of the solids increases with an increase in the amount of flocculating agents. One explanation may be that the hydroxide flakes produced in activation systems remain there much longer than they would stay in preliminary and secondary precipitation systems, and that, in the final analysis, they are shriveled, nonhydrous products of precipitation.

The addition of iron salts into activation stages offers a considerable advantage when waste waters containing hydrogen sulfide or its water-soluble salts are treated. The problems with these waste waters are that in the treatment process the biodegradation is disturbed and that more or less offensive odors occur during aeration. When there are no or insufficient quantities of S^{2-}-utilizing organisms in the activation systems, an S^{2-}-proportion corresponding to the concentrations in the inflow occurs in the activation stage, which is comparatively slowly reduced simply by chemical oxidation with elementary oxygen. The result is an intoxication of the other fermenting organisms, along with a reduction of the clarification level. And, moreover, the escaping air contains hydrogen sulfide so that it is toxic and corrosive, and has an offensive odor.

A frequent or permanent S^{2-} content in the waste water may, on the other hand, initiate a mass development of filamentary sulfur bacteria which reorganize the activated sludge into the unwanted bulking sludge. These phenomena may be encountered by the continuous dosing of small quantities of iron salts (5 to 10 g Fe/m³) into the activation stage. The activated sludge is thus provided with iron hydroxide, which converts sulfides to iron sulfides that are insoluble in water so that they are eliminated before the further biological utilization begins. Iron sulfide reacts with oxygen slowly, forming iron (III) hydroxide and elementary sulfur so that, in the overall reaction, iron practically assumes the role of an oxidation catalyst in the following reaction:

$$2\ S^{2-} + O_2 \longrightarrow 2\ O^{2-} + 2\ S\ .$$

When small iron quantities are continuously added, the activated sludge becomes ferruginous, so that large iron quantities are constantly available for the described sulfide conversion in the entire system.

6.3.1.3 Adsorption

Adsorption is a term defining the accumulation or agglomeration of highly dissolved substances (adsorbed matter or adsorbate) to the surface of a solid (adsorbing substance or adsorbent). In waste water treatment, this term mainly refers to the accumulation of substances dissolved in water on the inside surface of activated-charcoal particles and thus to their removal from the water flow to be clarified. The adsorption of suspended particles to the grain surface in a sand filter is also an essential factor. When the accumulated substances are retained by intermolecular forces (van der WAAL forces), the term "physical sorption" is applied. A desorption of the accumulated substances can be achieved by the application of appropriate counteracting forces, i.e. by introducing energy into the system.

The chemical sorption effects the agglomeration of the substances to the surface of the adsorbing matter as a result of chemical reactions with the surface molecules, so that this process is most often irreversible.

Activated charcoal is the most important industrial adsorbent. The structure of its pores can be influenced by suitable control of the manufacturing process so that an optimum adsorption of certain substances can be achieved.

In this production process, the organic raw material is first converted

to coal and is then "activated", which means that the cavities filled with tar in the carbonization process are hollowed out so that coal with a very large inside surface (600 to 1200 m²/g!) is produced. These surfaces formed by pores of different size (micropores with d = (4 to 15) $\cdot$ 10^{-4} µm) represent the sorptive potential of the activated charcoal. For practical application in waste water treatment, this active carbon is produced in two forms:
- in the form of a powder (d = 50 - 70 µm) which allows dosing into the waterflow directly,
- in the form of a granulate substance (d = 1- 4 mm) which is suited to fill filter columns.

When the active carbon has been coated with adsorbed matter to an extent that the equilibrium concentration c becomes too high, it can be regenerated, which means that the adsorptive capacity of the surface can be restored by desorption. Various methods are available for desorption:
- thermal regeneration where high temperatures (750 - 1000 °C) are applied to burn the accumulated substances, with simultaneous addition of oxidation gases to reactivate the pores of the sorbent matter. The regeneration cycle is repeated several times, and the pores expand while the load capacity sinks.
- chemical regeneration where solvents and oxidants (e.g. concentrated sulfuric acid) are used in the attempt to effect desorption. The resulting selectivity and the high cost, however, have so far prevented its application on an industrial scale.
- biological regeneration where activated charcoal is used and biological slime forms on the granules of the adsorbing substances during waste water purification. This phenomenon, which was an undesired effect at first, is now being used in testing microbial decomposition of the organic adsorbates. This method is still in the trial phase and similar experience gained in the treatment of freshwater is being applied.

Owing to the possibility of reactivation, granulated active carbon is preferred. The grain diameter is usually in the range from 1 to 4 mm,

and allows an application that roughly corresponds to quick sand filtration. When activated charcoal is used to charge a filter bed, three mechanisms of removal take place, depending on the mode of operation:
- adsorption of dissolved substances from the waste water,
- filtration of suspended and colloidal particles,
- biological decomposition of accumulated organic substances.

Fig. 6.3.-7 illustrates open filter beds with different layered structures.

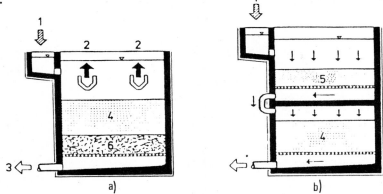

Fig. 6.3.-7: Open filter beds with different layered structures
a - single-layer filter filled with activated carbon; b - double-layer filter with separate floors for quartz sand and activated carbon
1 - inflow; 2 - flushing water discharge; 3 - screenings discharge; 4 - activated carbon; 5 - quartz sand; 6 - supporting gravel layer

Fig. 6.3.-8 illustrates a closed active-carbon pressure filter.

The adsorption method is used in water purification when the water contains only small quantities of substances that are difficult to eliminate, e.g. coloring, odorous, and flavoring substances. The same principles also apply to the treatment of industrial waste water. Excessive waste water pollution would result in rapid clogging of the adsorption surfaces so that large quantities of the adsorbing agent or frequent and expensive regeneration of the adsorbing agent would be required.

Adsorption methods, mostly in the form of active carbon adsorption, are applied in fields such as roller cooling-water treatment in rolling mills

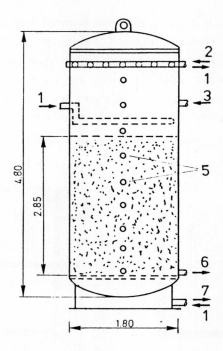

Fig. 6.3.-8: Section of a closed active-carbon pressure filter
1 - flushing water; 2 - inflow; 3 - charcoal addition; 4 - flushing nozzles; 5 - sampling connectors; 6 - charcoal removal; 7 - outflow

for removal of fat and oil to allow cooling-water recirculation; in ice production in breweries to treat the discharges from machinery rooms to allow recirculation; in dyeing plants to purify the dye-loaded waste liquors using active carbon in powder form with a density of 0.8 kg/m³ and with subsequent filtration to remove the charcoal; in phenol production to purify the gas-scrubbing liquors at 60 to 70° C; in gasoline synthesis to purify the condensation water.

6.3.1.4 Extraction by Solvents

The extraction process makes use of the capability of some substances (solvents) to dissolve in certain liquids but not in the waste water in question. Thus the liquids carrying the dissolved substances to be removed can be separated and recovered from the waste water.

Benzol, butyl acetate, isopropyl ether, and other products are used as solvents.

The method is used primarily in cases where the removed substances are to be reused, e.g. in phenol extraction from the waters discharged by coke-oven plants.

The methods of extraction may be different, depending on how the solvents are contacted with the waste water, e.g. in simple or multiple contact, in counterflow, etc.

The method may be applied both in towers lined with Raschig rings, and in installations comprising several stages, with countercurrent flow, provided with mixing basins and separators. In spite of the slightly higher cost of operation and maintenance, extraction centrifuges are coming into use because of their substantially reduced space requirements.

6.3.1.5 Membrane Separation (Ultrafiltration, Reverse Osmosis)

The separation by means of membranes is a method that allows the removal of dissolved substances and even electrolytes from aqueous solutions. Unlike conventional filtration, where the solids or constituents in colloidal solution remain on the filter medium (except for losses bound in the filter cake by adsorption) while the total liquid volume is available in a reusable form, the membrane separation processes of hyper- and ultrafiltration are so designed that the volumetric flow is split into a partial flow enriched with dissolved substances and a partial flow with a low content of dissolved substances.

Hyperfiltration and ultrafiltration are distinguished by the operating pressure and thus also by the membranes to be used and the systems to be treated. In hyperfiltration, which is also termed reverse osmosis, real solutions are used, which are cycled through suitably dense membranes in pressurized equipment with pressures as high as $100 \cdot 10^5$ Pa. The membranes repel the salts but allow the solvent to pass through. The ultrafiltration method is principally distinguished from this process only by the fact that the maximum pressures are as high as $10 \cdot 10^5$ Pa and that the systems to be treated contain macromolecules or even solids. Appropriate membranes are used accordingly.

It is not advisable to apply membrane separation techniques in general for the treatment of waste water discharged from an industrial plant manufacturing a wide variety of products. In such a case conventional techniques are preferred in order to obtain a product that can be discharged into the receiving reservoir or a concentrate extract in the form of a conveyable or compacted sludge that can be processed for deposition in a sanitary landfill. In contrast, the membrane separation produces two liquid streams of different pollutant concentration. Concrete concepts on where the permeate and the concentrate are to be deposited or used should therefore be developed. It has proven advantageous to install the membrane system as the last step in a process producing a defined amount of waste water.

The ultrafiltration process is employed, to an ever-increasing extent, to treat oil emulsions which are used in the metal-processing industry as drilling, cutting, and drawing oils. Used oil emulsions are concentrated up to 50 %, and the concentrate is then incinerated, having a heat value of roughly 12.5 MJ/kg; the permeate containing the emulsifier can be reused.

The ultrafiltration process is also applied in the food industries for concentrating dairy and starch-containing discharges. Proteins for feeding can be produced from the concentrate; the present feeding protein oversupply, however, sets economic limits to this technique. The reduction of the biologic oxygen demand of the waste water so treated is an advantage in any case. Table 6.3.-1 shows some results achieved in the food industry.

Tab. 6.3.-1: BOD values for waste water discharged from food processing industry, and retention values

Waste water	BOD_5 (mg O_2/l)	Retention values %
carbohydrates	18250	99.7
potato starch	12700	85.1
vegetable proteins	21500	99.6
cheese whey	45000	99.3

Whereas the ultrafiltration technique is applied to treat waste water discharged by certain branches on an industrial scale, hyperfiltration systems are predominantly operated with pilot plant dimensions. Many results have only been obtained with laboratory equipment. Results are available from tests for the recovery of silver salts from photographic developer solutions. Particular attention has been paid to toxic metal salt solutions discharged by the electroplating industry. The volumetric flow for detoxication plants could be reduced by concentrating the discharges; results are available for solutions of the cyanide complexes of the metals Ni, Cu, and Zn, and also EDTA complexes and the behavior of chromium compounds have been studied.

Waste waters from fermentation processes involve a very high oxygen demand. In tests on fermentation water, hyperfiltration techniques were used to obtain a permeate containing only low-molecular organic substances. A practical solution was found for concentrate elimination.

The waste water from a yarn-manufacturing plant has been treated in the same manner. Waste waters containing dyestuffs were treated on a laboratory scale; both a good zinc salt retention (97 %) and dyestuff retention (95 %) were observed.

6.3.1.6 Distillation

The distillation method is based on a change of the waste water characteristics by evaporation of liquids and by their condensation. As most of the mineral and organic substances are difficult to evaporate, a high-quality clarification can be achieved.

Owing to the energy costs and the extensive installations, distillation is expensive.

6.3.1.7 Oxidation

Oxidation processes are used to remove unwanted constituents having a strong reduction capability. These techniques are applied, for instance, to treat waste waters containing hydrogen sulfide, mercaptans, and other sulfur compounds.

Atmospheric oxygen and chlorine are the most commonly used oxidizing agents. The treatment of the waste waters discharged from electroplating plants may be considered as a typical example of the technical application of oxidation and reduction processes.

Atmospheric oxygen is used to oxidize organic substances with a low reduction power. A satisfactory rate can be achieved only at a certain temperature and pressure. To obtain it, the pressurized oxidation with oxygen is applied at temperatures of 110° C and at $3.5 \cdot 10^5$ Pa while the wet incineration takes place at temperatures between 225° C and the critical temperature of water (347° C) and at pressures of about $(50$ to $150) \cdot 10^5$ Pa. Fig. 6.3.-9 shows a schematic of the ZIMMERMANN method of wet incineration.

Ozone is the strongest oxidizing agent. In waste water technology ozone has only been applied to a limited extent since the required ozone quantities are very large, demanding large investment of capital. Between 9 and 15 kWh of electricity are required per kg O_3 produced.

Partial oxidation, however, has advantages for the conversion of substance mixtures difficult to decompose biologically, to substances which are accessible to biodegradation. Cyanide oxidation is also possible by means of ozone.

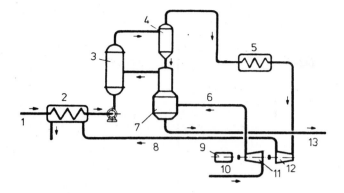

Fig. 6.3.-9: Schematic of the ZIMMERMANN wet incineration method
1 - waste water inflow; 2 - preheater; 3 - reactor (270 °C, 100 10^5 Pa); 4 - separator; 5 - superheater; 6 - compressed air; 7 - saturator; 8 - waste gases, condensation products; 9 - generator; 10 - air; 11 - compressor; 12 - impeller; 13 - outflow

Chlorine is used as a gas to oxidize alkaline waste waters. When minor waste water quantities are to be treated, sodium hypochlorite or an aqueous solution of chlorine can also be used as the oxidating agent. This method is applied for cyanide oxidation (6.8 kg chlorine per 1 kg cyanide ions) under alkaline conditions.

6.3.1.8 Reduction

According to MEINCK /107/, the reduction technique is suitable for waste waters containing easily reducible noxious constituents. The toxic chromic acid in waste waters from chrome-plating works can, for instance, be reduced by sulfurous acid or the salts thereof and by divalent iron salts and can be converted into a form that can be precipitated with lime. An excess of reducing agents, in turn, results in a corresponding oxygen consumption in the waste water. Chlorine-containing waste waters from the chemical industry can be reduced by means of activated charcoal.

6.3.1.9 Ion Exchange Process

In the ion-exchange process certain ions are absorbed from an electrolyte solution while an equivalent quantity of other ions is given off to the solution.

The ions may be either cations or anions, and, accordingly, a distinction is made between cation and anion exchangers. Most exchangers are polymer substances whose polymer structure carries certain fixed substances. Apart from the synthetic-resin types, exchangers on an inorganic basis (zeolites) are also available.

The ion exchange process produces the reversible binding of the waste water substances to an auxiliary material, in a manner similar to that involved in adsorption. Regeneration of this material produces a concentrated phase of waste water substances, while it is low in foreign matter.

The following exchanger types are common for waste water purification:

<u>Cation exchangers</u>: resins with sulfonic acid and carboxylic acid groups are capable of exchanging cations. They bind copper ions, for instance, releasing sodium or hydrogen ions in exchange.

<u>Anion exchangers</u>: resins with amino groups are examples of anion exchangers. They bind sulfates or cyanides, for instance, in exchange for chloride or hydroxyl ions. Ion exchangers with a macroporous resin structure are predominantly used for waste water treatment. These resins are provided with micropores produced by polymer cross linkage [diameter $(5 \text{ to } 15) \cdot 10^{-4}$ µm], and additional macropores resulting from a special production method, with a diameter of $(150 \text{ to } 500) \cdot 10^{-4}$ µm. The strength of this resin structure is thus substantially increased, particularly its resistance against the osmosis load occurring when concentrated solutions are treated. The selection of the particular exchanger type depends on the pH value of the solution. Anion exchanggers can absorb in both the strongly and the weakly acid range, and in neutral and slightly alkaline solutions for strongly basic resins up to about pH 10 and slightly basic resins up to ca. pH 8.

The range of application of cation exchangers is

        pH 1   - 13   for strongly acid resins
        pH 4   - 9   for slightly acid resins
        pH 1.5 - 10   for chelate-forming resins.

The group mentioned last possesses a marked selectivity in terms of heavy-metal ions, distinctly dependent on the pH value. Even small pH modifications are enough to improve the exchanger's selectivity substantially.

The most important application is the removal of metal ions from waste waters discharged from galvanizing works, the chemical industry, mines and ore extraction, as well as by the removal of radioactive constituents from waste water. The advantage lies in the fact that the exchange-active groups of the ion exchanger are chemically fixed to the resin structure so that they do not release any organic substances to the waste water. Experience has shown that the ion exchange process can be used to enrich heavy metals when the respective metal occurs in small concentrations. In general, the concentration should be less than 3 g/l, preferably below 500 mg/l. Hence, the ion-exchange technique can be expedient when unwanted metal ions must be removed from solutions rich in salts. It is also useful for the enrichment of pit waters, groundwaters, or production waters, particularly also when organic chelating agents are present, and, moreover, also for removal of residual concentrations after a preliminary purification.

Ion exchangers have been successfully applied in the electro-plating and anodizing industry for preparing galvanic baths and rinsings. They permit the recovery of valuable metals such as Cu, Cd, Ni, Cr, Au. The water so treated can often be reused, which results in a reduction of the water discharge. The iron contained in pickling baths may be removed after a suitable pre-treatment by means of ion exchangers.

Special requirements are made on the recirculated water in the production of electronic components. Here it is important that the water be free of surfactants and other organic substances. The application of activated charcoal has failed to provide satisfactory results. The use of ion exchangers (e.g. Amberlite XAD-2), however, was suited to meet the given requirements.

6.3.1.10 Other Methods of Physico-Chemical Treatment

6.3.1.10.1 Stripping

This technique is applied particularly in mineral oil refineries and in the chemical industry for expelling easily volatile constituents from waste water, e.g. hydrogen sulfide, mercaptans, ammonia, and organic solvents. The stripping technique is also the basis for a dephenolization method used in gasworks and for the treatment of coke-oven discharges.

Stripping may be accomplished with steam, flue gases, or inert gases. Flue gases are mainly used for stripping strongly alkaline waste waters since they not only expel the volatile substances but also simultaneously neutralize waste water. The sulfurous acid and the carbon dioxide contained in the flue gas contribute to this effect.

The process, however, may also constitute a nuisance to the surrounding environment. In such cases the incineration of the materials in a boiler or their admixture to the air supply for an incinerator is advisable.

6.3.1.10.2 Aeration

Aeration is necessary in numerous waste-water treatment methods. Aeration should be applied in the pre-treatment of commercial waste waters, by atomizing or spraying, by mechanical means (paddle wheels, brush aerators, impellers, or the like), by blowing compressed air in a fine, medium, or coarse bubble form, or through aeration columns, in activation processes. Aeration is also applied in many applications to expel solvents (e.g. in the pharmaceutical industry) and other volatile organic compounds (e.g. in the petrochemical industry).

Iron (III) chloride, solid, $FeCl_3 \cdot 6 H_2O$
The salt is available in well sealed barrels or plastic bags. When air containing water is contacted with the material, the crystals deliquesce as a result of hygroscopicity. (It should be kept dry.)

The commercially available material contains:
- iron   20.5 % by weight Fe
- chloride   38.5 % by weight Cl
- crystallization water   41.0 % by weight H_2O

Iron (III) chloride solution, roughly 40 % by weight $FeCl_3$ in an aqueous solution.
The commercially available material contains:

	in 1 liter	in 1 kg
iron, g	196.9	137.7
chloride, g	375.1	262.3
density, kg/l	1.43	

Iron (III) sulfate solution, roughly 20 % by weight Fe_2O_3 as $Fe_2(SO_4)$ in an aqueous solution.
The commercially available material contains:

	in 1 liter	in 1 kg
iron, g	209.9	139.9
sulfate, g	541.4	360.9
density, kg/l	1.5	

Iron chloride sulfate solution, roughly 41 % by weight $FeClSO_4$ in an aqueous solution.
The commercially available material contains:

	in 1 liter	in 1 kg
iron, g	184	122.7
sulfate, g	316	210.7
chloride, g	117	78.0
density, kg/l	1.5	

Iron (II) sulfate hepta-hydrate, solid, $FeSO_4 \cdot 7 H_2O$
This salt is inexpensive when iron and steel pickling plants with cor-

responding pickling-bath regeneration equipment are located nearby, since this material is usually considered a waste product from the pickling process.

Compared to the trivalent iron [which precipitates quantitatively as iron (III) hydroxide, $Fe(OH)_3$, at pH values near the neutral point in waste waters containing no chelating agents or flocculation preventing substances], the divalent iron is still highly dissolved in this pH range. Oxidation to form trivalent iron is therefore required; this process is realized with chlorine in many applications.

The iron sulfate hepta hydrate is dissolved in tanks with acid-resistant cladding, by adding water while stirring. In another tank the concentration desired for dosing is set by the controlled addition of water, and the solution is then pumped from the second tank through control pumps to the point of use. The stock solution should be set to concentrations of roughly 400 g $FeSO_4$ 7 H_2O/l, while the ready-to-use solution should be set to concentrations of ca. 200 g $FeSO_4$ 7 H_2O/l. A check of the concentration by density measurements using a densimeter is normally sufficient.

The chlorine is withdrawn in the form of a gas from flasks or barrels, and is converted to a chlorine-water mixture by means of refining water. This mixture is then introduced into the dosing line for iron-2-sulfate heptahydrate. After combining the two solutions, the mixture passes through a small reaction tank for a few minutes to ensure the quantitative iron oxidation.

Following the equation

$$2\ Fe^{2+} + Cl_2 \longrightarrow 2\ Fe^{3+} + 2\ Cl^-$$

127.5 g of chlorine are necessary per 1 kg $FeSO_4$.
1 kg of $FeSO_4$ 7 H_2O of a good product of industrial quality contains 0.2004 kg iron.

Another method of oxidizing divalent iron consists in the oxidation of the iron by aeration in the waste water to be treated, followed by admixture dosing. To ensure that the reaction runs quantitatively within a few minutes, it is necessary to keep the pH value in the reaction compartment above 7.5.

6.3.2.2 Other Iron and Aluminium Salts

Industrial salts are commercially available in large quantities, which are salt mixtures of aluminum and iron, containing slight admixtures of titanium. They are mostly water-soluble sulfates with low contents of constituents insoluble in water.

Various commercial products are known, e.g. with 8.20 % by weight Al and 1.75 % by weight Fe, or with 6.88 to 7.41 % by weight Al and 2.80 to 3.50 % by weight Fe.

The effective metal quantities of the individual salts can be correlated with reference to the indicated iron or aluminium percentages, e.g. for determining a reasonable purchase price. Per 1 kg of iron one obtains:

    4.878 kg $FeCl_3 \cdot 6\ H_2O$, solid
    7.262 kg $FeCl_3$, 40 %, liquid or 5.978 liter, 40 %, liquid.

When industrial-grade material is used, the type and quantity of the admixtures are very important because they could contribute to the discharge of pollutants into the waste water under unfavorable conditions. With iron (II) sulfate hepta-hydrate of industrial quality (produced from steel pickling) the admixtures should remain within the following concentration ranges when the material is used for waste water treatment (all items in % by weight):

copper	up to 0.003	bismuth	up to 0.001
manganese	up to 0.1	antimony	up to 0.001
nickel	up to 0.005	arsenic	up to 0.001
chromium	up to 0.005	free sulfuric acid as H_2SO_4	up to 0.4
lead	up to 0.05	water (without crystall.	
tin	up to 0.01	water)	up to 1.4
cadmium	up to 0.01	residues insoluble in water	up to 0.15

6.3.2.3 Commercial Polymer Flocculation Agents

Polymer flocculating agents is the generic term defining organic compounds with long chains. When they carry groups accessible to cationic or anionic dissociation, they are termed polyelectrolytes; polymers carrying no charge are called nonionic.

Examples of such flocculating agents:

<u>Anionic polyelectrolytes</u>: polyacrylic acid, partially hydrolyzed polyacrylamide, pectin, alginates.

<u>Cationic polyelectrolytes</u>: polyethylene imine, poly(diallyldimethyl ammonium) chloride.

<u>Nonionic products</u>: polyacrylamide, polyethylene oxide.

The products are marketed by the manufacturers in different polymerization levels and physical states. The manufacturers deliver, along with each product, exhaustive literature containing information on applications, the production of the ready-to-use solutions, the dose levels, the points of admixture, the introduction into waste water or sludge, the physical data concerning shelf life and handling, storage conditions, toxicology, and ecology.

6.3.2.4 Hydrate of Lime

Hydrate of lime or slaked lime is commercially available in different quality grades. The main constituent is calcium hydroxide, $Ca(OH)_2$. The industrial quality grades common in waste water treatment applications contain variable amounts of magnesium, aluminum, iron, and silicic acid. Hydrate of lime is available in a finely ground form. As the solubility of $Ca(OH)_2$ in water is very low [it is 1.65 g $Ca(OH)_2$/l at 20° C], a suspension, generally called lime wash or lime water, is used when the material is dosed into the waste water.

The lime wash is normally produced in isolated batches of 100 to 200 kg of slaked lime per m lime wash. It must be stirred constantly in the dosing containers in order to prevent the suspension from setting. When small quantities are required, it is common to prepare individual batches of bagged material in vessels containing the quantity required for 1 to 2 days. For larger quantities it is common to deliver the material in trucks, to stock it in silos, and to distribute the current demand through dry dosage equipment. These devices discharge the powder into a small container with a float-controlled water inflow and a continuously operating stirrer. Centrifugal pumps are provided to

the easier it is to achieve a high metabolic rate. Related to the respective diameter of the organism, the quotient

$$\frac{\text{surface}}{\text{volume}} = \frac{O}{V} \frac{cm^2}{cm^3}$$

or the ratios $O_a : O_p$ (see Table 6.4.-2) between different organisms and a volume of prokaryotes of equal size can be given in relation to the mean prokaryote size of 1 μm.

Tab. 6.4.-2: Relationship between the outside surface of the body of different animals and a prokaryote size of equal volume /33/

Organism	Volumes	outside surface of the body O_a	outside surface of the prokaryotes O_p	$O_a : O_p$
dimensions		cm^2	cm^2	
prokaryotes	1 μm³	$6 \cdot 10^{-8}$	$6 \cdot 10^{-8}$	1 : 1
eukaryotes	1000 μm³	$6 \cdot 10^{-6}$	$6 \cdot 10^{-5}$	1 : 10
mouse	0.1 dm³	130	$6 \cdot 10^{5}$	1 : 46000
man	0.1 m³	$13 \cdot 10^{3}$	$6 \cdot 10^{9}$	1 : 460000
cow	0.5 m³	$38 \cdot 10^{3}$	$30 \cdot 10^{9}$	1 : 790000
elephant	3.0 m³	$13 \cdot 10^{5}$	$18 \cdot 10^{10}$	1 : 1400000

WILKINSON and SCHLEGEL use the oxygen consumption rate per unit mass as a measure of the energy production in a body living under aerobic conditions. They quote the productiveness of a cow and of a quantity of yeast of the same weight as an example, measuring the productiveness by the protein production per day. A cow of 500 kg live weight, for instance, produces some 500 g of protein per day. Within the same period, an equal quantity of yeast is able to produce up to 50,000 kg of protein, which is one hundred thousand times as much. In other words, 5 g of yeast can furnish the same protein production as a cow does. Sufficient nutrient quantities or sufficiently degradable waste water quantities must, of course, be available to supply the microorganisms with their appropriate nutrients. This example leads to the statement that, with the same biomass, the surface ratio is in the range of

1 : 800,000, whereas the productiveness ratio is ca. 1 : 100,000.

The normal activated sludge in aerobic activation systems is a mixture of prokaryotes and eukaryotes. Here, too, the bacteria are at the beginning of the feeding chain. The aerobic biological treatment of waste waters is supported by this group of organisms. Organisms can fundamentally absorb only substances from a solution, which have the nature of a nutrient, whether they are suited to build up the body's own tissue or capable of producing energy.

A distinction is made between the two main groups of heterotrophic and autotrophic microorganisms:
- heterotrophic organisms (animals, bacteria, fungi) and
- autotrophic organisms (plants).

Heterotrophs utilize organic material to produce energy through oxidation or respiration (energy metabolism) with a conversion of the organic material, e.g. sugar to CO_2 and H_2O. A substantial percentage of the absorbed organic material is used to synthesize the body's substance (growth metabolism).

Autotrophs synthesize organic substances from CO_2 and H_2O through photosynthesis, using the energy of sunlight. Nature has developed still further processes for the synthesis of organic substances, e.g. through chemosynthesis, which, however, will not be discussed here in greater detail since they are of only minor relevance in waste water treatment.

The natural uncontrolled environment comprises a closed cycle between synthesis and breakdown, which results from the interaction of the bioactivities of heterotrophic and autotrophic organisms. Autotrophs build up organic substances from inorganic matter, and release oxygen. The oxygen is used in the oxidation of organic substances into inorganic matter.

Table 6.4.-3 shows some aspects of the heterotrophic and autotrophic microorganisms according to SCHLEGEL /149/.

Tab. 6.4.-3: Synthesis and breakdown of organic substances /149/

HETEROTROPHIC MICROORGANISMS decompose fats, proteins, and carbohydrates, e.g. $$C_6H_{12}O_6 + 6\ O_2 + 38\ ADP + 38\ (P)$$ $$6\ CO_2 + 6\ H_2O + 38\ ATP$$
AUTOTROPHIC MICROORGANISMS synthesize organic matter through photosynthesis: $$6\ CO_2 + 6\ H_2O + LIGHT\ QUANTA$$ $$C_6H_{12}O_6 + 6\ O_2$$

The first formula in Table 6.4.-3 shows that 38 adenosine triphosphate (ATP) units of energy are produced from the breakdown of 1 mole of sugar in respiration. ATP is the universal carrier of chemical energy between energy producing and energy consuming reactions in the cell. ATP may be regarded as a sort of energy currency within the cell.

Since the nutritional components in commercial and industrial waste waters include predominantly organic matter, low heterotrophic organisms are mainly used. The basic principle of biological waste water treatment consists in the conversion of a majority of dissolved substances to the biomass of organisms so that these substances can be removed from the waste water as a sediment. This process requires energy. One part of the nutrients is oxidized, i.e. "mineralized" and dissolved again in an inorganic form, e.g. as nitrates, sulfates, CO_2, and the like. These substances, too, then undergo conversion by bacteria of a different type (e.g. Nitrosomonas and Nitrobacter) to synthesize the mass of the bacteria which, settling in sedimentation basins, can then be separated from the waste water.

The bacterial cell is divided into two cells under favorable environmental conditions, which separate from each other more or less quickly. This process of separation or fission lasts only some 20 minutes with heterotrophic bacteria. The growth rate, however, does not remain at this level or else the mass of the bacteria would be 4,000 times the weight of the Earth within 24 hours.

The biomass, formed by single-cell or more complex microorganisms, must be supplied with nutrients sufficient to enhance the development of microorganisms, i.e. to provide the microorganisms with cell components and energy. The activity of the microorganisms is the more intensive, the greater the quantity of organic substance is. Normally, the mineral substances vital for cell development, e.g. potassium, magnesium, phosphorus, and sulfur, are available in domestic waste water in quantities sufficient to cover the microorganisms' demand.

Nitrogen and phosphorus, however, are often available in small quantities, and in some industrial waste waters the quantities of these substances are often insufficient. Consequently, these substances must be replenished. The values recommended in the German federal regulations on waste water treatment (ATV) quote the general quotient BOD : N : P = 150 : 5 : 1 as the minimum ratio between the organic substances (BOD) percentage, nitrogen and phosphorus, with BOD : N : P = 100 : 5 : 1 being defined for waste water and 90 : 5 : 1 for the sludge. According to SIERP, the mean nutrient contents per 1 m³ of municipal sewage may be assumed to be as follows:

        80 g of nitrogen (N),
        20 g of phosphorus in the form of P_2O_5,
        60 g of potassium in the form of K_2O.

6.4.1.2 Processes in Metabolism

The relationship between the nutrients available and the mass of organisms determines and controls the metabolic processes and along with them the growth stages and the oxygen demand.

In a batch test, microorganisms are studied in a known nutritive solution, which remains constant since there is no supply or discharge. According to SCHLEGEL /149/, the following stages in the growth of organisms can be noted (Fig. 6.4.-2):

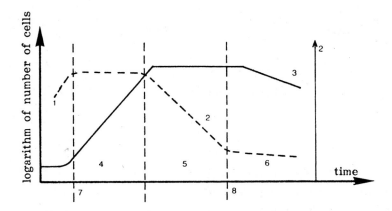

Fig. 6.4.-2: Stages of growth of the microorganisms and oxygen consumption /149/

1 - start-up period; 2 - O_2 consumption; 3 - number of cells 4 - exponential stage; 5 - stationary stage; 6 - dying stage; 7 - substrate respiration; 8 - endogenous respiration

Initial Stage. The enzymes that are necessary to break down macromolecular compounds are formed. An exact determination of the oxygen consumption has so far not been made.

Exponential Stage. This phase is characterized by an excessive nutrient quantity available so that the growth rate is determined by the maximum rate of division. Substrate respiration prevails (breakdown of the organic pollutants by respiration).

Stationary Stage. The number of cells does not increase since the quantity of nutrients available is limited now. The internal respiration activity (respiratory process of breakdown of the cell substance) increases, whereas the substrate respiration is continually decreasing.

Dying Stage. The nutrients available have now been totally consumed. The microorganisms oxidize their own cell substance, which they consume as a last nutrient reserve. The oxygen consumption is mainly determined by internal respiration.

More complex organisms, however, should not be precluded from biodegradation. The soil cultures irrigated with waste water and the biological filters contain, for instance, different flies, worms, spiders, etc. which, in their turn, feed on the bacterial substrate. Occurrence of such organisms is almost precluded, however, in activated sludge basins where a uniform flow is maintained.

6.4.1.2.1 Factors Influencing the Biological Processes in the Treatment of Commercial Waste Waters

The biological decomposition normally takes place under aerobic conditions. But anaerobic processes are also employed to treat commercial waste waters.

When dissolved oxygen is available in sufficient quantities, the decomposition process is principally aerobic. The biodegradation processes involves two stages: the rapid adsorption of organic substances to the cell mass and the relatively slower oxidation or stabilization step involving the actual breakdown of the organic substances to CO_2 and H_2O, the two final products of the aerobic decomposition. One portion of the organic contents is converted to the bacterial body substance, too, i.e. organic substance is resynthesized.

The biodegradation rate in domestic waste water depends on many factors, specifically on the organic load, the sludge volume, the number of organisms, the useful oxygen volume, the oxygen supply, and the temperature. In commercial waste waters, the nature of the contents, their composition, the nutrient ratio, and the pH value are also decisive.

Oxygen is vitally important. The systems are supplied with oxygen by the introduction of air. As little as 5 to 15 % of the oxygen supplied by the introduction of air can be used for biological activities so that sufficient continuous aeration must be ensured. Any interruption in the air supply may either inhibit or even stop the decomposition process, depending on its duration.

The waste water temperature should not vary, otherwise these variations may influence the biological processes. The biological activity is assumed to roughly double when the temperature rises by 10° C. Results obtained in practical operation, however, have shown that the temperature influence is reduced when the sludge load decreases. Fig. 6.4.-3 illustrates the influence of temperature on the decomposition by activated sludge on internal respiration and maximum substrate respiration according to SAWYER.

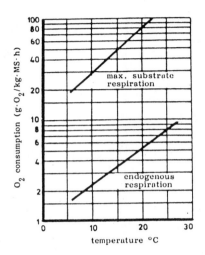

Fig. 6.4.-3: Oxygen consumption (related to 1 kg of activated sludge, per hour) for maximum substrate respiration and internal respiration, as a function of temperature

Nutrient quotient. Commercial waste waters often contain too little or too much nitrogen or phosphorus in relation to the BOD_5 factor. In these cases, a special treatment or the addition of nutritive salts becomes necessary. Moreover, the waste water must contain sufficient quantities of the other organic substances, nutritive salts, and trace elements such as sodium, potassium, calcium, magnesium, iron, manganese, copper, and sulfur, which are vital for the development and reproduction of cells.

The pH value should be within the range from 6.0 to 8.5. Under certain conditions, a biological treatment can also be effected in a pH range between 5.0 and 9.5. Still higher pH values are also possible in exceptional cases. A biological treatment was still achieved, for instance, with a pH value as high as 12 in tests made with textile discharges.

The composition of waste waters must correspond to the capabilities for degradation by the microorganisms. Some substances have a very high rate, others a low rate, of biological decomposition. The rate depends on the adaptability of the processing microorganisms. The adaptation may result from the synthesis of the necessary enzymes or groups of enzymes, or from the selection of the microorganisms so that biocenoses are created that are best suited for the nutrients available or the environmental conditions. Adaptation may occur in both ways simultaneously.

6.4.2 Processes of Decomposition

6.4.2.1 Aerobic Decompositon of Organic Carbon Compounds

The aerobic carbon consumers are extremely important and vital in waste water treatment. They use elementary oxygen, for instance, to decompose d-glucose according to the equation

$$C_6H_{12}O_6 + 6\ O_2 \longrightarrow 6\ CO_2 + 6\ H_2O + 2822\ kJ$$

Useful organic carbon compounds such as carbohydrates, fats, and proteins are rapidly consumed for growth and metabolic processes and thus removed from the waste water. A particular index of this process is the decrease of COD and BOD_5.

Microscopic fungi are occasionally involved in this chain of events, too.

With an appropriate oxygen supply, the aerobic decomposition of biologically utilizable organic carbon compounds usually results in carbon dioxide, water, and cell substance. Initially organically bound nitrogen is then available mainly in the form of ammonia (NH_4^+).

With the following approximate loads, the decomposition down to BOD_5 values ranging at 20 mg O_2/l (complete treatment) is achieved:

activation system $\qquad\qquad B_R$ roughly 1.0 kg $BOD_5/(m^3 \cdot d)$
biological filter $\qquad\qquad\quad B_R$ roughly 0.6 kg $BOD_5/(m^3 \cdot d)$
biological disk filter, total surface load $\quad$ max. 10 g $BOD_5/(m^2 \cdot d)$

6.4.2.2 Nitrification

The conversion of organically bound nitrogen to ammonia-nitrogen as a result of anaerobic and aerobic processes of decomposition in sewerage systems, reservoirs, and preliminary clarification basins as well as in biological purification stages creates the conditions for the existence of ammonia and nitrite oxidants, which are also termed nitrificants or nitrifying bacteria.

Ammonia is converted to nitrate in two stages: in the first step, bacteria of the Nitrosomonas species oxidize ammonia to nitrite according to the equation:

$$NH_4^+ + 1.5\ O_2 \xrightarrow{Nitrosomonas} NO_2^- + H_2O + 2\ H^+ + 230\ kJ$$

with the subsequent step of bacteria of the Nitrobacter species oxidizing nitrite to nitrate according to the equation:

$$NO_2^- + 0.5\ O_2 \xrightarrow{Nitrobacter} NO_3^- + 75\ kJ.$$

Many aerobic treatment stages with an incomplete nitrification process contain nitrite and nitrate in varying proportions, in addition to ammonia, in the final discharge. The finally discharged flow from thoroughly nitrifying systems no longer contains detectable nitrite quantities, since both processes are mass-equivalent and simultaneous, which means that the nitrite produced by Nitrosomonas is instantly oxidized by Nitrobacter to nitrate and thus cannot be determined. The two equations may be combined as follows:

$$NH_4^+ + 2\ O_2 \xrightarrow[Nitrobacter]{Nitrosomonas} NO_3^- + H_2O + 2\ H^+ + 305\ kJ$$

This example demonstrates that ammonia and nitrite oxidants always cooperate in nature. They use the chemical energy released in the oxidation processes to assimilate carbon dioxide (CO_2) as their exclusive carbon source. Their growth thus depends also on an accessible CO_2 content.

The equation reveals that the nitrification is a process producing mineral acids, which must be balanced by the addition of alkaline agents if the alkalinity of the waste water is insufficient for automatic neutralization.

Nitrifying bacteria have a reproduction time between 10 and 30 hours, which means a rate considerably lower than that of the carbon-decomposing microorganisms. They do not become active before the carbonaceous compounds have largely been decomposed and only after the biological oxidation of sludge proceeds as a result of insufficient quantities of dissolved carbonaceous nutrients, with appropriate oxygen supply. These conditions prevail in the following systems:

activation systems	up to	$B_R = 0.5$ kg $BOD_5/(m^3 \cdot d)$
biological filter systems	up to	$B_R = 0.2$ kg $BOD_5/(m^3 \cdot d)$
biological disk filter systems	up to	6 g $BOD_5/(m^2 \cdot d)$

An effective nitrification seldom occurs in systems with normal and high loads. Compared to the decomposition of organic carbon compounds, the nitrification process depends on temperature to a much greater extent. The nitrification rate of a particular system is increased in a linear function with a water temperature of up to 30° C, approximately. The ammonia arriving in the receiving water reservoir may display many noxious effects there.

When the biological-chemical requirements are met, the nitrification process in the reservoir is more or less complete. The necessary oxygen consumed in this process is withdrawn from the water so that the oxygen content may drop below the minimum levels demanded by fish. As a result of nitrification, the nitrogen keeps its fertilizing effects on algae and water plants. Nitrate impairs the use of the surface waters (receiving reservoirs) for drinking water production. The recommended level of the "International Standards" published by the World Health Organization is 45 mg NO_3^-/l in drinking water. Contents as low as 10 mg NO_3^-/l are deemed unacceptable for baby food because of the risk of methemoglobinemia.

When the nitrification process comes to a total or partial stop at the nitrite stage, the fish are endangered, since nitrite quantities as low as ca. 10 to 20 mg NO_2^-/l are highly toxic for fish. Such nitrite levels may also render the conditioning of raw water for drinking water supplies more difficult, in addition to being dangerous to humans.

A. SCHIFFERS and collaborators report a particular problem caused by the biological nitrification. The Ensdorf power plant on the Saar uses river water to supplement a cooling water cycle, which gives off heat in a large cooling tower with natural air draft cooling. This layout gave rise to enormous difficulties mainly due to the fact that the very high NH_4^+ contents in the Saar river water, from industrial waste waters, reached peak values up to 100 mg NH_4^+/l and were nitrified under the conditions of the circulation system (warm water, cooling tower similar to a biological filter), which produces a favorable environment for nitrifying bacteria. The circulation system turned acidic and therefore strongly corrosive to the building materials, since the carbonate hardness was insufficient to bind the acid formed in the nitrification process. Therefore expensive measures had to be taken.

The biological conversion of ammonia to nitrate, which is completed in water treatment plants, is but a partial solution to the problem, since the fertilizing effect of the nitrogen is retained and the nitrate contents must be deemed unacceptable for drinking water supply from the viewpoint of water usage.

In exceptional cases, nitrate is helpful in natural waters, i.e. in cases where other waste water discharges into the water reservoir overburdening the oxygen system. In these cases, facultatively anaerobic bacteria utilize the oxygen bound in the nitrate to oxidize the substrate (nitrate respiration or denitrification) and prevent the subsequent fermentation processes and the sulfate respiration, which form toxic and mephitic substances, at least until the total nitrate supply has been consumed. 4.57 g of oxygen are consumed for the oxidation of 1 g of ammonia-nitrogen (NH_4^+-N) equal to 1.2879 g of NH_4^+ according to the equation:

$$NH_4^+ + 2\ O_2 \rightarrow NO_3^- + H_2O + 2\ H^+.$$

With a mean quantity of 8.5 g NH_4^+-N/(p·d) in waste water discharged from a mechanical preliminary clarification system, the oxygen demand will then be 38.8 g which is nearly as high as that of BOD_5 with 40 g/(p·d).

The presence of organic, biologically degradable substances inhibits the development and activity of nitrifying bacteria. Apart from the low reproduction rate of these bacteria, this fact might be one reason why the nitrification processes take place only in the range with a very low sludge load as here a further-going breakdown of the organic compounds is combined with a high sludge age.

High ammonia concentrations have a toxic effect on Nitrobacter in an alkaline environment. Nitrosomonas improves the environmental conditions for Nitrobacter in two ways: ammonia contents and pH value are reduced. An excessive pH reduction, however, e.g. to values below 6.5, has an inhibiting effect, too. Many substances can interfere with nitrification at very low concentrations where the decomposition of organic compounds is not yet disturbed. These substances include heavy metals such as copper, nickel, zinc, and chromium, and also hydrogen sulfide, mercaptans, aliphatic amines, and others. With nitrite contents of ca. 150 mg/l and higher, inhibiting effects on the conversion of nitrite to nitrate by Nitrobacter can be observed.

All these facts contribute to the frequent application of the nitrification process in waste water treatment with NH_4^+ contents up to ca. 70 mg NH_4^+/l, i.e. mainly in the treatment of municipal waste waters. This process is simple and relatively easy to control.

This is not yet true for waste waters with very high ammonia contents such as waste waters discharged from rendering plants, coking plants, refuse pyrolyzation plants, and similar installations. The processes in water treatment facilities of such plants often show a frequent variation of NH_4^+, NO_2^-, and NO_3^- in different concentrations. Occasionally, nitrite contents > 100 mg of NO_2^-/l with only little or no nitrate occur, sometimes NH_4^+ occurs alone, while in other cases only small quantities of NH_4^+, no NO_2^- and large amounts of NO_3^- may be detected. These phenomena can also be observed when the other parameters, e.g. oxygen supply, sludge load, temperature, and pH value, are considered optimum for complete nitrification. Certain aids such as reduction of the pH value by mineral acids (shift of the pH dependent NH_4^+/NH_3 equilibrium) or the pH increase by alkalis to levels in the range from 6.5 to 7.0 may be expedient.

6.4.2.3 Denitrification

Various other bacteria which are widespread in nature and which are also involved in the biological waste water treatment can utilize both free oxygen dissolved in the waste water and oxygen bound in nitrites and nitrates to break down organic carbon compounds. The bacteria use this property when free oxygen is not available. Then they use the oxygen bound in nitrites or nitrates just like free oxygen in the complete oxidation of organic carbon compounds to carbon dioxide and water directly, without detectable intermediate products. One example is the various bacteria of the Pseudomonas species that oxidize carbohydrates, organic acids, and alcohols either aerobically with free oxygen or anaerobically with nitrite and nitrate oxygen. The oxidation of d-glucose in the anaerobic conversion with nitrate oxygen, for instance, takes place according to the following equation:

$$5\ C_6H_{12}O_6 + 24\ NO_3^- \rightarrow 12\ N_2 + 24\ HCO_3^- + 6\ CO_2 + 18\ H_2O + 12{,}856\ kJ$$

In the nitrite/nitrate respiration, the oxygen is released in the form of a gas and escapes from the waste water into the atmosphere which contains ca. 79 % by volume of nitrogen. This is a suitable method to remove the fertilizer nitrogen from the waste water. Moreover, energy is saved in activation systems with aeration equipment provided that the process is diligently controlled, e.g. with intermittent aeration or creation of low-oxygen or oxygen-free zones (simultaneous denitrification). The acid-forming nitrification processes and the alkali-forming denitrification processes are kept in equilibrium so that the environment, on the whole, remains in the neutral pH range.

This applies also to nitrification-type biological filter systems in which pumps return the nitrate-carrying discharge into zones with a low oxygen supply (preliminary clarification, low-oxygen parts of the trickling filter) for denitrification. When nitrogen is contained in the biomass of the biological filter, slime and denitrification occur, and the total nitrogen contents in the outflow from biological filter systems is reduced by

ca. 12 to 41 % for $B_R = 0.6$ kg $BOD_5/(m^3 \cdot d)$, and
ca. 18 to 57 % for $B_R = 0.2$ kg $BOD_5/(m^3 \cdot d)$, both relative to the inflow.

Presently, efforts are being made to control and master the nitrification/denitrification processes, which are more or less occasional and uncontrolled. The activation process is particularly suited to achieve this aim, since the nitrate, which inevitably occurs as a by-product in low-load activation systems, is simultaneously removed by denitrification.

Nitrites and nitrates of an origin other than biological can also be removed from the waste water by biological denitrification.

The following prerequisites must be met for denitrification to take place:
- The presence of nitrites and nitrates;
- an intimate mixture with activated sludge and utilizable organic substrate without oxygen supply, i.e. by slowly moving stirrer or paddle equipment. When intermittent aeration is provided, the intervals between the aeration phases should be 30 minutes (operation of a small number of the revolving agitators, however, must continue to prevent the sludge from settling). A separation of nitrate- and substrate-carrying water from activated sludge should not occur during this period.

Denitrification may also occur in the secondary clarification basin. This is achieved especially when, for any reason (e.g. interim overburdening), the extent by which decomposition of the organic carbon compounds normally occurs is not achieved for short periods in nitrifying systems. In these cases, the released nitrogen gas bubbles adhere to sludge particles so that they give rise to sludge lift and to the formation of floating sludge in the final clarification basin. This floating sludge is called denitrification sludge. Owing to its nature, this phenomenon occurs particularly in those systems whose volumetric load is in the weak to normal range, which means that they operate under unstable conditions.

The provision of a compartment equipped with a slowly moving stirrer or paddle system may be a remedy when this compartment is arranged between the activation and the secondary clarification stages as an anaerobic denitrification basin. Dwelling times should be 30 to 40 minutes.

Table 6.4.-4 is a survey of the microbiological processes for nitrogen removal from municipal waste waters according to BRINGMANN.

Tab. 6.4.-4: Sequence of the microbiological processes for nitrogen removal from municipal waste waters

type of nitrogen compound	microorganisms converting the nitrogen compounds	products of the microbiological conversion of the nitrogen compounds	type of conversion	reaction changes due to conversion	redox range of the conversion process
organically bound carbon	ammonifying bacteria	ammonia nitrogen (NH_4^+)	ammonification		aerobic
ammonia nitrogen (NH_4^+)	nitrifying bacteria (Nitrosomonas)	nitrite nitrogen (NO_2^-)	nitrification	acidification	aerobic
nitrite nitrogen (NO_2^-)	nitrifying bacteria (Nitrobacter)	nitrate nitrogen (NO_3^-)	nitrification		aerobic
nitrate nitrogen (NO_3^-)	nitrate-reducing bacteria (Escherichia coli and others)	nitrite nitrogen (NO_2^-)	denitrification		anaerobic
nitrite nitrogen (NO_2^-)	nitrite-reducing bacteria (Pseudomonas fluorescens and others)	nitrogen (N_2) dinitrogen monoxide (N_2O)	denitrification	alcalization	anaerobic

6.4.2.4 Desulfurization

Apart from the anaerobic decomposition of sulfurous amino acids (desulfurization), the formation of hydrogen sulfide or its salts by exclusively anaerobic bacteria utilizing the oxygen bound to sulfate in the anaerobic respiration (desulfurization) is an important source of hydrogen sulfide formation in an oxygen-free environment. In this process, the minerally bound oxygen is used by bacteria to oxidize organic carbon compounds in a manner comparable to denitrification. The oxygen bound in the sulfate is consumed in respiration, with formation of H_2S. Its relationship to hydrogen sulfide is a function of the pH value.

With pH values below 7, H_2S is mainly produced,
with pH values of about 7, H_2S and HS^- are formed in roughly equal proportions,
with pH values higher than 7, HS^- or S^{2-} mainly occur.
The equation for the d-glucose conversion is as follows:

$$C_6H_{12}O_6 + 3\ SO_4^{2-} \longrightarrow 6\ HCO_3^- + 3\ H_2S + 295\ kJ.$$

Hydrogen sulfide is a mephitic substance and highly toxic. It is not stable under aerobic conditions. It is either chemically oxidized to elementary sulfur, or sulfur-oxidizing bacteria oxidize it through the sulfur stage to sulfate. The desulfurization process occurs in an oxygen-free environment only when there is no oxygen available from nitrites and nitrates. This step can therefore be prevented by the controlled addition of nitrate. Aeration or the addition of iron (III) chloride can also be a remedy wherever this is possible.

The anaerobic sludge stabilization by methanogenic fermentation is inevitably linked to the desulfurization process. High sulfate contents in the sludge water, however, may inhibit or prevent the methanogenic fermentation unless countermeasures are taken.

6.4.2.5 Fermentation Processes

Not only desulfurication but also the anaerobic decomposition of organic matter may result in fermentation and in the release of mephitic substances such as some low fatty acids, amines, and hydrogen sulfide. Other final fermentation products include alcohols, ketones, CO_2, and H_2.

The fermentation processes in oxygen-free zones in sewer systems and treatment plants after completion of the denitrification process are mostly uncontrolled. They are enhanced by long dwelling times in nonaerated zones and by the composition of the substrates. The formation of organic acids is accompanied by a reduction of the pH value, which may drop to levels as low as 5 and below. The preceding fermentation processes often impede the aerobic clarification since some metabolic products of the anaerobic microorganisms may have inhibiting effects. The formation of bulking sludge can be frequently observed in

When pH changes have to be expected in a range that does not ensure safe and reliable operation of the biological treatment stages appropriate facilities must be installed or at least provided for the measured addition of neutralizing agents.

6.4.3.3 Temperature

Aerobic biological processes do not generally create any problems within the temperature range from + 10° C to + 25° C, which is common in waste water treatment. The multitude of organisms balances the activities, increasing or decreasing as a function of temperature within this range, when the treatment works will have been correctly dimensioned. Temperature decreases will thus entail only a slight reduction of the plant clarification efficiency.

It has become apparent, however, that various secondary phenomena may occur, e.g. improved clarity of the outflowing treated waste water and reduced excess sludge production in the activation step during warm weather. With temperatures below + 10° C or above + 25° C, its influence should be considered in the selection of the method of treatment and of the engineering equipment and facilities.

It is known that heat is released from the metabolic process of aerobic microorganisms. This heat, however, is irrelevant for the operation of the normally used methods of treatment and with the low substrate concentrations of the waste waters to be treated. With high substrate concentrations and when particular aerators with low heat dissipation and isolated reservoirs are used - possibly in combination with a supply of pure oxygen instead of air for further reduction of the dissipated - heat - the biologically generated heat may be used to increase the temperature of the substrate to be decomposed. This may contribute to the accelerated decomposition of the organic compounds and to a hygienization effect.

When the conversion rate of aerobic microorganisms generally rises in the temperature ranges common for waste water, and when good decomposition results are obtained with aerobic-thermophilic microorganisms even at temperatures between 60 and 65° C, other factors, too, should be included into the considerations when the suitability of such a method is assessed.

The substrate to be processed must be continuously available in order to maintain the temperature levels. Since the clarification efficiency of the biological methods depends on the thorough separation of the sludge-type solids produced, a check should be made to determine whether the flocculating capacity of these solids, and thus their separation from the treated waste water, at higher temperatures will be retained.

The increased temperature of 30 to 35° C, which has been recommended for the treatment of phenol-carrying waste water to increase the purifying effects, has not been confirmed in all cases, e.g. it is not applicable in the biological treatment of phenol-carrying waste water discharged from coking plants. Many years of operation of three large activation works for the treatment of such waste waters in the Saarland has shown that maximum operating reliability and the best clarification effects are ensured within the temperature range from 20 to 28° C, while short intervals of temperature reduction by 4 to 5° C may result in unsatisfactory operation (strong foaming, reduced purification efficiency). "Short interval" is to be understood to apply to changes within 24 hours.

With temperatures in excess of 28° C in such activation systems, considerable difficulties have always been encountered because the activated sludge has lost its flocculating capacity and was discharged along with the water as a sediment rich in solids. On the other hand, substrate and sludge respiration increases to an extent that the oxygen demand can no longer be reliably covered by the atmospheric air, which means that the process changes from aerobic to anaerobic conditions.

When aerobic processes are provided to decompose the carbon compounds and, additionally, to convert the ammonium to nitrate in a biological method (nitrification), one should be aware that the temperature has a substantially stronger effect on this conversion. This fact must be definitely taken into account in the layout and dimensioning of waste water treatment works.

The temperature increase influences on the reduction of the dwelling periods in systems for aerobic and anaerobic stabilization of sludges containing organic solids that are accessible to biodegradation.

Short-term variations of the waste water temperature may give rise to density currents in clarifiers, mainly at points where such variations may occur due to irregular and nonuniform admixtures of particularly warm or cold waste waters.

6.4.3.4 Necessary Minerals

Microorganisms require a number of minerals or elements, e.g. nitrogen, phosphorus, sulfur, potassium, sodium, calcium, magnesium, and iron, supplemented by so-called trace elements such as zinc, copper, manganese, boron, and vanadium, for maintenance and reproduction.

Nitrogen and phosphorus particularly are required in considerable quantities. In relation to the BOD_5 value, a demand of 3 to 5 kg N and 0.5 to 1 kg P per 100 kg BOD_5 is calculated in practical plant operation. The relationship in pretreated domestic waste water is roughly 100 kg BOD_5 : 21.3 kg N : 6.8 kg P.

The volume of degradable carbon compounds available in domestic waste water is too small to link nitrogen and phosphorus completely into the organisms, so that major quantities of these substances remain in the waste water even after the usual totally biological treatment, with the result that the outflow from the treatment plant has a fertilizing effect on surface waters (eutrophication).

Commercial and industrial waste waters often have insufficient quantities of nitrogen and phosphorus in relation to their contents of organic degradable substances; therefore, a combined treatment of these different types of waste water may be economically expedient and necessary to protect unpolluted water.

When industrial plants are required to treat their own waste waters, compounds of nitrogen and/or phosphorus are usually added. This applies, for instance, to wood-fiber board production, paper mills and cardboard production, coking plants, the beverage industry, the chemical and pharmaceutical industries, and the like.

Ammonium salts, ammonium hydroxide, and urea supply nitrogen, whereas phosphoric acid and the water-soluble salts of orthophosphoric acid supply phosphorus. The applicable additives are chosen based on their easy handling (storage, solubility, possibility of dosing) and reasonable price.

1 kg P is contained in:

2.36 liter H_3PO_4, industrial quality, green, density some 1.7 kg/l; 79 % H_3PO_4

3.01 liter H_3PO_4, industrial quality, green, density some 1.53 kg/l; 69 % H_3PO_4

11.54 kg Na_2HPO_4 12 H_2O

12.35 kg Na_3PO_4 12 H_2O

4.31 kg $(NH_4)_2HPO_4$

1 kg N is contained in:

3.87 kg NH_4Cl, industrial quality

4.75 kg $(NH_4)_2SO_4$, industrial quality

5.35 liter ammonia hydroxide, density some 0.907 kg/l; 25 % NH_3

2.15 kg urea, industrial quality.

Anaerobic treatment methods involve a certain N and P demand, which might be similar to that of the aerobic methods.

Whereas adding N- and P-compounds is necessary to achieve a successful biological treatment in numerous practical applications, only very rare cases have become known where the other aforementioned elements are missing so that they would be relevant.

In order to assess the the N- and P-balance, an analysis should be made to determine the ammonium and total nitrogen contents, and the orthophosphate and total phosphorus percentages. When the level of ammonium nitrogen and the orthophosphate phosphorus appear to be too low for BOD_5 degradation, while the total nitrogen and phosphorus contents have been calculated and found to be sufficient, degradation tests can be made to determine whether or not the nitrogen present as ammonium nitrogen or the phosphorus bound as orthophosphate phosphorus may be utilized in the biological decomposition.

Plants adding N- and P-salts should monitor these salts in their final discharged outflow, until sufficient knowledge of the quantities that should be added has been gained. Following such a procedure, it has been possible to reduce the P-admixture for the biological treatment of phenol-carrying waste waters discharged from coking plants to a mean value of 0.24 kg relative to 100 kg BOD_5.

6.4.3.5 Inhibiting Agents and Toxicants

Domestic waste waters are easy to clarify in biological processes. The sludges from these processes are not difficult to treat in subsequent supplementary stages. However, many commercial and industrial waste waters have pollutants that lower the clarifying efficiency of biological waste water treatment plants. Examples of such waste waters are the discharges from metal surface finishing processes in galvano-chemical methods, and waste waters from iron- and steel-pickling plants, accumulator charging stations, printing shops, photographic establishments and laboratories, copy shops, chemical cleaners, pesticide- and herbicide-producing plants, disinfectant production, as well as coke-oven plants, tanneries, and leather-processing industries.

The addition of acids, lyes, liquors, cyanides, and other volatile mephitic or toxic substances must be prohibited not only to protect the plant maintenance workers and the neighboring residential areas, but also to protect the sewerage system. Such a prohibition is generally part of the public rulings and regulations issued by municipalities and communities.

In addition, substances that, in excess of defined threshold concentrations, may have deleterious effects on downstream treatment stages must not be released into the waste waters. This applies mainly to silver, copper, zinc, chromium (especially chromate), nickel, cadmium, and mercury compounds in dissolved and undissolved forms.

The operating temperature should not exceed 30° C in the preliminary chemical treatment of the waters discharged from the respective plants before they are discharged into the public sewerage network. The pH value should be in the range of 6.5 to 9. The settleable substances should not exceed the maximum of 0.3 ml/l after sedimention for 2 hours. 0.1 mg CN should be the critical value for the easily releaseable cyanides. Chromate reduction to trivalent chromium with mandatory precipitation of this element together with the other metals should be required

6.4.3.5.1 Determination of Threshold Values for Substances

The definition and determination of threshold values for substances that may inhibit or interfere with the biological treatment processes is very difficult. The inhibiting and toxic effects on microorganisms do not only depend on the absolute contents of the respective substances, but are influenced by many concomitant phenomena and conditions.
- Metals, if toxic, have specifically toxic effects when they are dissolved. Their solubility is affected by the pH value, by the anions present, and by the buffering capacity of the waste water to counteract pH changes. Sulfides or compounds carrying S-H groups are particularly difficult to dissolve. On the other hand, the reaction of the enzyme S-H groups with heavy metals damages the enzymes and the basic bacterial metabolism.
- When several substances are present simultaneously, where each substance has toxic effects on bacteria, the total toxicity does not always equal the sum of the isolated toxic effects but may correspond to either a multiple or a fraction of the total.
- Some toxic substances forming complexes with organic compounds are released when these compounds undergo biodegradation. Thus they develop toxic effects only in the decomposition stage in biological treatment systems. Examples of such phenomena are the deactivation of copper by peptone or of cyanides by formaldehyde, and their release when peptone or formaldehyde are biologically decomposed.
- The bacterial growth inhibition caused by many toxic substances may be reduced or eliminated by the supply of a sufficient nutrient quantity.
- In the course of their gradual adaptation the microorganisms can tolerate higher concentrations of many toxic substances. Poison such as cyanides, phenols, and numerous other organic compounds are then able to be biodegradated.

Strict compliance with the aforementioned local rules and regulations must be observed to avoid damage to the biological activities in waste water treatment plants. Appropriate pollution control facilities must be available to industrial plants and shops, e.g. in the form of central detoxification and neutralization systems for the electroplating and galvanizing industry.

Experience has shown that, when many of those plants are located in a town or community, noxious discharges into the main sewers must be expected in spite of local regulations and monitoring. In such cases it is advisable to protect the installed sewer tubings and reserved zones in the treatment plants against corrosion by acids. The effects of such incidents can then be remedied by special processes in the treatment facility.

The normally common precipitation of the settleable substances by sedimentation is effected in combined neutralizing, flocculation, and sedimentation basins. The sewage sludge produced is concentrated in filter presses without any further biological treatment.

RUDOLFS and his collaborators have published an extensive survey of toxic agents with their effects on aerobic and anaerobic biological processes and on the self-purification capacity of running waters, together with the results of the BOD analysis.

The following approximate values outline the magnitudes and ranges where damage to the biological processes in waste water treatment plants can be expected:

copper	1 - 3 mg Cu/l
chromium $^{3+}$	10 - 20 mg Cr/l
chromium $^{6+}$	3 - 10 mg Cr/l
cadmium	3 - 10 mg Cd/l
zinc	3 - 20 mg Zn/l
nickel	2 - 10 mg Ni/l
cobalt	2 - 15 mg Co/l
cyanide	0.3 - 2 mg CN^-/l
hydrogen sulfide	5 - 30 mg S^{2-}/l

These values are supplemented by the following descriptions of some examples from actual operations:

6.4.3.5.2 Disturbance of the Methanogenic Fermentation of Sewage Sludge in Municipal Sewage Treatment Plants

Such disturbance could be recognized by the decreasing sludge digesting gas formation, by the dropping pH value in the contents in the septic tank, and by the reduced methane gas proportion in the sludge digesting gas. Analyses of the contents of the sludge digester are given in Table 6.4.-5:

Tab. 6.4.-5: pH reduction in the septic tank

	S town	T town	K town
pH	6.85	6.7	6.5
total solids g TS/l	47.7	51.8	53.8
containing (% by weight)			
Cu	0.10	traces	0.10
Zn	1.38	0.81	1.11
Pb	n.d.	n.d.	0.33

(n.d. = not determined)

In all three cases zinc alone or zinc and lead were the cause of the disturbance. When the source of the discharges had been identified and when such illegal discharges had been stopped, the normal digestion process was resumed. It should be noted that for town S the digestion compartments had to be emptied completely and a new fermentation process started. In the two other cases a charging pause of several days was sufficient to resume the digestion process, with external energy being supplied to maintain the temperature range of 30 to 33 °C in the septic basins and with frequent recirculation. Only sludges free of zinc or zinc combined with lead could be charged into the tanks, or else the disturbances would have reoccurred.

The methanogenic fermentation is inhibited by nonferrous metal contents as low as 100 mg/l of crude sludge; when their proportion rises to approximately 300 mg/l or even higher, the fermentation process is completely stopped. The concentrations are 0.3 % and more for inhibition and some 1 % and higher for complete disappearance of the methanogenic fermentation, both related to percentage by volume in the dry sewage sludge substance.

6.4.3.5.3 Disturbance of the Aerobic Decomposition in the Sludge Activation System in Municipal Sewage Treatment Plants caused by Chromate

Chromate contents in the inflow to the treatment plants as a result of the illegal discharge of a chromic acid carrying galvanizing bath were measured in the sewage treatment works in the town of SG. The discharge into the highly loaded municipal activation system lasted for approximately one hour. The solids contents in the activation basin were 5.5 g of dry substance per liter. 100 kg CrO_4^{2-}, dissolved in 500 m³ waste water, were discharged into the activation process for one hour. The maximum inflow concentration level was 270 mg CrO_4^{2-}/l. The reaction in the activation system was very strong foaming, with a reduced decomposition efficiency. The elimination of organic matter, measured by the potassium permanganate consumption between the inflow and outflow points of the activation system, followed the scheme given below:

before the chromate discharge	65 % approx.
several hours later	25 % approx.
one day later	40 % approx.
two days later	45 % approx.
three days later	55 % approx.
four days later	60 % approx.

This pattern demonstrated that the clarification efficiency of the activation system had suffered a long-lasting negative influence.

6.4.3.5.4 Disturbances in the Biological Purification Processes

Disturbance of the biological purification processes may also be caused by substances that do not involve any direct inhibiting or toxic effect. Even though aluminum or iron compounds are not toxic by nature, if large quantities are introduced after their conversion to voluminous precipitation products (hydroxides), they may cause an adverse effect in the purification process, reducing the purification efficiency. Consequently, the sludge volumes produced in activation systems may be so greatly increased that the resulting frequent removal of excessive sludge leads to the exhaustion of the biologically active material, eventually reducing the purification efficiency to its minimum.

Moreover, an unwanted pH reduction may occur in waste waters with a low acid-forming capability.

Waste waters with an imbalanced load may promote the development of organisms that disturb the clarification processes. This holds mainly in cases where a large proportion of discharged waste waters stem from sugar refineries, dairies and milk-processing plants, fermentation works, fruit juice and lemonade production. In most cases the biocenoses in aerobic processes are dominated by filamentous organisms. They produce light and voluminous sludges with poor settling and concentration properties (bulking sludges). The clarification efficiency of activation systems deteriorates as a result of the considerable reduction of the solids contents in the activation basins and thus of the mass of the active substance.

Likewise, filamentous sulfur bacteria, which oxidize sulfide under aerobic conditions, may give rise to the formation of bulking sludges in activation systems.

Finally, it should be noted that sulfate-reducing bacteria very easily absorb certain substrates (desulfurication). For example, a direct correlation between the discharge of whey or must and wine carrying substrates, on the one hand, and the formation of hydrogen sulfide could be detected in some sewage treatment plants. With an overload of such substrates, the hydrogen sulfide formation was considerable even in the oxygen-supplied activation basins. The result was strongly annoying emissions and complaints from the plant's neighbors.

6.4.3.5.5 Disturbance of the Purification Processes by Heavy Metals

Heavy metals may also cause disturbance of the clarification process. Tables 6.4.-6 and 6.4.-7 are used to assess of the influence of some heavy metals in different concentrations on the biochemical oxygen demand and the oxygen consumption in mixtures of activated sludge and waste water. These tables were originally published in a study by H. HEUKELEKIAN and I. GELLMANN. They used WARBURG equipment for their analyses.

BOD_5. Provided that the system is appropriately dimensioned, stabilization may take place in the activation stage in combination with the waste water purification. Otherwise, separate basins can be provided for this purpose.

Stabilization under anaerobic conditions with the help of methanogenic bacteria is carried out in heated or unheated digestion compartments. The final products are essentially methane and carbon dioxide.

Sludge treatment, both aerobic and anaerobic, improves the dehydration properties of the sludges and reduces the sludge volume. The methanogenic fermentation process may also be used to purify even highly loaded waste waters.

6.4.4.1 Systems for Biological Waste Water Treatment

There are three main methods for the biological treatment of commercial and industrial waste waters (see Fig. 6.1.-1):
- natural methods, i.e. large volume, natural and biological methods such as the utilization of agricultural waste waters and waste water ponds. These methods are characterized by the use of a natural environment or an environment copied from nature, combined with the economic utilization of the products. These methods have two objectives, (1) the purification of the waste water and (2) the conversion of the directly utilizable pollutants. The disadvantage of these methods are that they require a large space and discharge areas;
- artificial methods, i.e. biological methods in compact systems such as biological filters and activation systems;
- combinations of the preceding methods. Introduction of waste water directly into the receiving water is possible only when minimum requirements are met. The permissible emission thresholds for discharges into natural waters are so low (see Section 5) that the introduction of untreated commercial waste waters is practically impossible.

It was previously common practice with domestic waste waters, after initial mechanical purification, to accept a discharge standard as high as 30 liters/sec per inhabitant (MNQ). Such a high load on natural waters is no longer permissible.

6.4.4.1.1 Utilization of Agricultural Waste Water

6.4.4.1.1.1 Fundamental Principles

The land treatment of waste water means supplying agricultural and forest areas with waste water in order to improve the yield from agriculture, to replenish groundwater supplies, and also to protect the natural water and the environment.

The application of waste water in agriculture has been known since ancient times. Up to the end of the 1950s, this waste water use had also been common in the Federal Republic of Germany. With continuously increasing waste water quantities, however, ever larger land areas were required, and after the introduction of new standards for the discharge of effluents into natural waters, which involved strict statutory demands on the discharge quality, this type of waste water utilization in agriculture has declined. Now, this method is almost exclusively used by small towns and villages.

Under the influence of light, air, and the sun's energy, the pollutants in waste water are mostly decomposed, while, at the same time, the desired moistening of the soil is effected.

This method of purification is common all over the world, particularly in warm climates, such as those that occur in the developing countries. This method is based on the requirement that a widespread distribution of the waste water is possible for agricultural utilization on cultivated and forested lands. The waste water is allowed to seep into the soil, or it passes through a soil layer roughly 1 meter thick, whereupon it is collected in a drainage or other collector system for introduction into a natural water body.

Special seepage areas allowing higher seeping rates are often provided for the land treatment of waste waters. These areas are charged after tillage of the fields or when excessive moisture or frost or other factors prevent the waste water distribution on the fields.

6.4.4.1.1.2 General Conditions

Industrialization has also had certain impacts on agricultural methods. Apart from the basic demand that water be available in sufficient quantities, an increasing number of chemical agents such as pesticides and fertilizers are used in agriculture today. The waste waters can often be expediently utilized particularly within the framework of these advanced agricultural techniques.

Water for agricultural purposes is normally supplied from surface or subsurface waters whose salt concentrations must not exceed 1 g/l. Tests have shown that 2 g/l are possible under certain conditions. Waters exceeding this threshold are noxious to plants and to the soil, because, after a couple of years, a salt crust covers the soil, which then becomes dry and can no longer be cultivated. Then recultivation is necessary, which means that the soil is "washed" or "rinsed" with low-salt water by irrigation. Such measures usually demand an extended drainage network in the field being irrigated.

The water used for large fields is normally pumped from the large rivers whose volume of water is so large that they are not too strongly contaminated with bacteria. When, however, the water quality involves a health risk, the water must be passed through sedimentation and sterilization systems. When organic contamination is considerable, such systems should also be installed in order to prevent further deterioration of health conditions. It should be noted that agricultural areas are always located near settled areas, so that not only the farmers but also the other inhabitants in a particular region may be infected with various diseases.

At a drinking water consumption of 250 liter per person per day (l/P·d), the nitrogen concentration in domestic waste water is about 20 mg/l. Half of this nitrogen is utilized by plants when agricultural waste water treatment methods are applied. The purified discharges from biological filters show a nitrate concentration of ca. 10 mg/l and a nitrite concentration of 0.5 mg/l, whereas the discharges from activation basins contain 7 - 8 mg/l of nitrate and 0.5 mg/l of nitrite. The domestic waste waters, moreover, contain 6 - 7 mg/l of phosphate and 12 - 15 mg/l of potassium.

According to SIERP /158/, domestic waste waters contain the substances given in Table 6.4.-8.

Tab. 6.4.-8: Valuable substances contained in domestic waste waters /158/

	N mg/l	P as P_2O_5 mg/l	K as K_2O mg/l	organic matter g/(P·d)
raw waste waters	12.8	5.3	7.0	55.0
biologically treated waste waters	10.9	2.8	6.7	19.0
digested sludge	1.3	0.7	0.2	20.0

The drainage or seeping fields have a considerable purifying efficiency. They not only mineralize the organic substances, but also contribute to the killing of bacteria. These processes take place in the topmost soil layers or in the sublayer some centimeters below the top layer. This population reduction of bacteria is increased by solar radiation. While the water is seeping into the soil, up to 90 % of the bacteria are destroyed along a seepage distance between 1 and 7 m. Research studies, however, have demonstrated that the number of coliform bacteria at the surface of the fields sprayed with waste water is only slightly reduced after 10 to 15 days. This means that the watering must be interrupted before harvesting and that the food intended for human consumption must be processed to destroy pathogenic microorganisms.

The arriving waste water is diverted to alternate fields with other suitable crops during the tillage and harvesting periods. This leads to a seasonal or annual alternation in irrigation. In the event that there are no alternatives for diverting the waste waters to other crops or seeping fields, the sewage must be stored on suitable lands, e.g. when the farmers do not immediately utilize the fields and are willing to make them available for this purpose. The areas used to store waste water or for soil filtration purposes may be different from one year to the next. When cultivated soil is actually used as a storage area, the result will be a reduced yield. It is therefore economically reasonable to make use of any possibility of discharging the waste water on soil not suited for agricultural purposes, in order to store the sewage.

Between the annual growth periods in soil cultivation, however, major quantities of waste water can be passed to the fields. A nonrecurring waste water supply of 3 - 5 cm (maximum 6 cm) in height is then possible. There should be a pause of 4 to 6 weeks between two consecutive watering periods.

Tab. 6.4.-9: Suitability of industrial waste waters for agricultural utilization /76/

I	II	III
Agricultural utilization to be recommended provided that suitable land is available nearby	Agricultural utilization to be recommended under certain conditions	Not suited for agricultural utilization
breweries, distilleries, malting plants, yeast production plants, retting pits, potato chip production plants, vegetable canneries, jam and marmalade as well as fruit syrup production plants, dairies, wood grinding plants, potato starch production plants	sugar refineries, rice and corn starch production plants, leather glue production plants, bone glue production plants, knacker's yards, slaughterhouses and meat canneries, tanneries, margarine production plants, sauerkraut pickling plants, paper mills, cardboard and strawboard mills, sulfate pulp production plants, textile industry (bleaching plants, mercerizing plants, finishing works, fulling mills, dyeing shops and plants, printing plants etc.), wool washing shops, copper rayon industry, fish meal industry and fish canneries, wood saccharification industry, mining	varnish and dyestuff production plants, soap production plants, industries producing inorganic heavy chemicals, pharmaceutical industry, metal industry, sulfite cellulose production plants, viscose rayon mills, low-temperature carbonization plants, gas works, by-product production from coking plants, generator gas facilities, wood carbonization industry, laundries, buna production plants, explosives production plants, coal washing plants, mineral oil industry

Not all crude waste water, however, is suited for a direct utilization by soil irrigation or seepage. Table 6.4.-9, from JACOBITZ /76/, illustrates the suitability of industrial waste waters for utilization in agriculture.

Tab. 6.4.-10: The U.S. and West German standards applicable to the agricultural utilization of waste waters /179/

	United States	Federal Republic of Germany
fruit and wine production	clarified waste water, no spraying, no use of fruit falling down	no spraying of neighboring lands
textile fiber and seeds production	clarified waste water, spraying or irrigation	pretreatment in screening and settling basins, a biological treatment and chlorination are required for sprayed waters
harvested products for human consumption, which are treated to destroy pathogenic germs	clarified waste water for spraying. Biological treatment and disinfection required for irrigation (maximum: 23 coliform-type microorganisms per 100 ml)	spraying must be stopped 4 weeks before harvesting
harvested products for consumption in raw state	2.2 coliform-type microorganisms per 100 ml. Disinfected waste water, filtered for 100 units maximum turbidity for spraying, unless treated by coagulation	potatoes and cereals; spraying restricted to the blooming season

Fig. 6.4.-9 shows a survey of the oxygen absorption through the pond surface at 20° C as a function of the oxygen deficit (from the German federal regulations on waste water treatment). This graph illustrates that, depending on the wind strength and the depth of the pond, an oxygen absorption between 1 and 15 g/(m²·d) is found at a saturation deficit of 50 %. Since calm conditions are important for oxygen absorption, a value of more than 4 g/(m²·d) should not be assumed.

The passage time is in the range of several days to weeks. Consequently, the ponds are superior to the artificial methods in confined spaces in terms of capacity of impact loads and concentrations.

Moreover, the operation of a minor nonaerated pond with water depths between 0.6 and 0.8 m for final clarification has become common. Under favorable conditions, fish can be bred in these ponds.

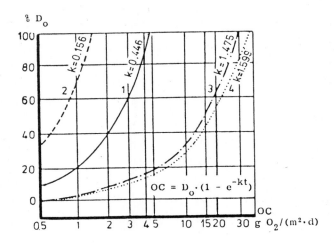

Fig. 6.4.-9: Oxygen absorption through the pond surface at 20° C as a function of the oxygen deficit /2/

1 - large lake (acc. to IMHOFF); 2 - small pond without wind influence (acc. to IMHOFF); 3 - small pond with wind influence; 4 - sedimentation pond (acc. to O'CONNOR/ECKENFELDER); D_o - oxygen deficit; OC - oxygen absorption

6.4.4.2.3 Aerated Waste Water Ponds

Aerated waste water ponds are used to reduce the organic substances in raw or mechanically treated waste water. Artificial sedimentation basins or sedimentation ponds are provided for the mechanical waste water treatment.

Compared with nonaerated ponds, the artificial aeration offers
- a higher absorption capacity, better distribution, and better utilization of the oxygen,
- more uniform distribution of the contaminants and microorganisms throughout the pond volume,
- control of the oxygen supply and the clarification efficiency or a certain independence from natural uncontrollable factors.

A distinction is made between the completely aerobic and facultatively aerobic ponds. The first type includes a circulation so intensive that all the sludge is maintained in floating condition and is carried off with the water runoff. The minimum aeration efficiency is rated at 20 W/m³ of pond volume. The outflow needs final clarification by sedimentation. When the sludge is returned from the final clarification stage, the system is designed as an activation system with long aeration times (long-term aeration).

The facultatively aerobic pond system, however, is much more widespread. The aeration rate is only 3 to 4 W/m³ of the pond volume. This is sufficient to keep the water in circulation and supply it with oxygen, whereas this rate is too low to keep the sludge continuously floating. Instead, the sludge settles at the bottom in uniform layers. It thus acts like a stationary, biologically aerobic active sludge on the nutrients passed through it, comparable to the slime in biological filter systems. The water depths are between 2.5 and 3 m, not including a free bottom zone of some 0.5 m. The pond design is usually rectangular.

A volumetric load of 20 to 30 g BOD_5/(m³ d) is recommended as a basic value in dimensioning. The duration time, however, should not be less than 5 days with high inflows from external sources.

The total volume should be divided, where possible, into two ponds in tandem arrangement. The accumulation of a few tenths of a meter will enhance the simultaneous treatment of rainwater.

The aerator is intended to serve two purposes: introduction of oxygen and water circulation. Oxygen quantities between 1 and 1.5 g O_2/g BOD_5 are sufficient. Depending on the arrangement and design of the aerator, 1 to 3 W/m³ are sufficient to furnish the necessary circulation effect.

Aeration at intervals - e.g. 15 min air supply on, 15 min off - has also proven to be expedient for operation. However, during the night, in the morning, and at times where the oxygen concentration drops below 4 mg/l, the aerator operation should continue to operate. This ensures a proper adaptation to the load and the decomposition efficiency of the pond.

In the event that the waste water pond is not provided with a separate final clarification stage, a particular zone should be provided in the pond for this purpose. The final clarification zone or a separate final clarification system should be designed for duration times of 1 to 2 days.

The deposited sludge must be removed every 2 to 5 years.

6.4.4.2.4 Polishing Ponds

So-called polishing ponds (also termed polishing lagoons) are installed for further clarification downstream of biological filter and activation systems.

Polishing ponds are frequent in the Federal Republic of Germany. They are single-stage systems even though it may also be better to divide the total volume and distribute it to several ponds in tandem arrangement.

For municipal waste water with a BOD_5 of 25 mg/l, the oxygen demand of the polishing ponds is met by the oxygen absorbed through the pond surface and by biogenic aeration. With a greater inflow pollution and duration times less than 2 days, artificial aeration may be necessary.

6.4.4.2.5 Safety Ponds

Safety lagoons may be thought of as a kind of polishing pond that becomes necessary when industrial waste waters are treated. When highly noxious waste waters are fed into the receiving reservoir after treatment, the provision of a safety pond is recommended. These waste waters are then allowed to stay in these additional ponds for 2 to 5 days (depending on their contamination). This period can be used to examine the pollution level of the treated waste waters, and possibly to initiate environmental protection measures if necessary.

6.4.4.2.6 Waste Water Fish Ponds

Waste water fish ponds can be considered to be a particular type of waste water ponds. According to the definition in the German federal regulations on waste water treatment (ATV), waste water fish ponds are ponds installed or modified for the purpose of waste water treatment. They are charged with diluted waste water after preliminary mechanical treatment, so that the resulting oxygen and nutrient conditions will allow economical breeding of edible fish. Waste water ponds may also be provided for preliminary water treatment. Carp and tench are the species best suited for this method of fish production.

An area requirement of 5 $m^2/(P \cdot d)$ or a load of 2000 P/ha per day is recommended for the dimensioning and design of waste water fish ponds. The resulting BOD_5 load is 8 g/m^2 or 80 kg/ha and day. The water to be clarified should remain in the pond for ca. 2 days. The inflow water depth should be about 0.5 m, whereas the outflow water depth should be 1.5 to 2 m.

6.4.4.3 The Biofiltration Process

The biofiltration process is used to clarify the waste water with the help of aerobic microorganisms by decomposing the pollutants. These microorganisms live on a solid material (packing) in biological or trickling filters. The sewage is continuously passed over the microorganisms to feed them, while the sewage contents are decomposed.

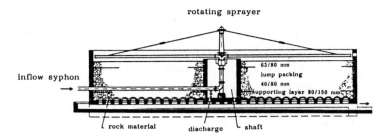

Fig. 6.4.-11: Low biological filter

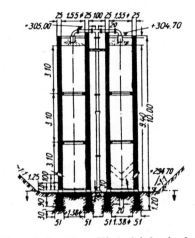

Fig. 6.4.-12: High biological filter

The biological filters must be carefully packed before operation begins. It is also important that the packing material be washed immediately before charging, e.g. on vibrator screens, when natural rock material (lava, lumps of rock, coarse gravel) is used. This step is necessary to remove constituents such as sand, which can be washed out, in order to avoid clogging, settling and malfunction of the systems installed downstream. Experience has shown that the discharged outflow from lump packings occasionally contains fine granular material. Thus it is advisable to provide a small gravel screen in the outflow channel of the biological filter.

When the filter operation is started with normal charging, the biological slime will eventually grow depending on the composition and temperature of the waste water. In the first few weeks after start-up, biochemical decomposition of organic carbon occurs only with domestic waste water. Nitrification begins only after a longer period of operation. Nondegradable substances are treated even later. Spring is the best time to start the operation of a biological filter system.

6.4.4.3.2 Components of Biological Filters

Biological filters are composed of the following main components:
- the structure,
- the packing material, and
- the waste water distribution system.

The structure is composed of, in most cases, a vertical, cylindrical concrete wall. A separate central cylinder accomodates parts of the distribution system. The structure is normally open at the top and bottom, which causes chimney draft effect that provides sufficient air, and thus oxygen, for the bacterial flora. In some cases, e.g. with strong odors, covers for the biological filters are installed. These types of filters may require artificial aeration by ventilators or similar equipment. The discharge air requires deodorization, e.g. with a scrubber system that washes the air with oxidizing solutions.

Biological filters may have a round or rectangular (square) floor design. Circular biological filters can be irrigated with rotating sprayers, which distribute the waste water uniformly. In the rectangular (square) filters, however, there is no passage of waste water through the corner areas so that these filters require appropriate distribution equipment (jets, channels, mobile steel piping on motor-driven vehicles).

The walls of the biological filter are made of brick, concrete, or reinforced concrete. In small communities and, generally, in any place where cheap packing material is available, the biological filter materials may be deposited without supporting walls, in heaps corresponding to the natural sloping angle of the packing material.

The bottom of the biological filter is made of concrete or reinforced concrete. It contains the channels and gutters that collect the treated waste waters. The spacing between these collectors ranges from 1.50 to 4.00 m. The outflow channel should have a grade of 0.5 to 1 % or more. The central channel should be accessible from both ends.

The draining floor serves various purposes in operation:
- supporting the packing material,
- collecting and discharging the treated waste water,
- supplying sufficient air into the filter.

Prefabricated structural elements, brickwork, or concrete elements are used to build the hollow floor. These elements should be provided with apertures corresponding to at least 1 % of the floor space of the biological filter, and a cavity 0.15 m high, to ensure an appropriate air supply.

The packing material must serve two purposes. It is used as a support for the biological slime, and it must form pores for aeration. Mechanically and chemically resistant rock or slag materials of different grain sizes are commonly used in filter packing. The material is arranged in three layers:
- the supporting layer on top of the draining floor, 80 to 150 mm high, the grain size being between 60 and 100 mm,
- the effective layer having a height of 1.60 m in low-load biological filters, with a grain size of 30 to 50 mm. The effective layer height is 3.50 to 4.00 m in trickle-type filters, while the grain sizes range from 40 to 80 mm, or 40/63 or 63/80;
- the distribution layer - at the filter top - 0.20 m high, with grain sizes between 20 and 30 mm.

Only the height of the effective layer varies with the different filter heights, whereas the heights of the protective and distribution layers are as stated above.

Because of the risk of clogging the packed biological filters, there are limits to inflow concentration and volumetric load. When the lump size is increased or when the flow-through rate is made considerably higher to avoid this difficulty, the treatment efficiency is reduced. These interrelations have been the subject of more than 30 years of searching for better packing material with respect to the cavity volume or the specific surface area or both, possibly accompanied by a reduction in weight. Studies have been made, for instance, with calcined clay rings, "Raschig" rings, hollow blocks, recently also hollow cylinders of foamed material and used tires, cans, pieces of pine wood, duckboards and wooden grating, wire meshing, peat packed in layers in so-called "biological fibrous filters" and corrugated asbestos. Experiments with increasing the contact time resulted in the installation of nearly horizontal corrugated plates with "meandering grooves".

In the early 1950s, U.S. experts developed packing elements in biological filters that consist of profiled plastic plates (including polystyrene, PVC, polyurethane, polyethylene). The plates are compact and easy to transport. They are assembled on site into packages of some 60 x 60 x 120 cm (depending on the brand) and arranged in the filter in layers staggered by 90°.

The waste water distribution equipment must uniformly distribute the waste water to be treated over the entire surface of the packing material. This utilizes the total treatment efficiency of the filter, while the air required by the aerobic microorganisms is available to the packing material between the irrigation intervals.

The waste water may be distributed by two different methods:
- through stationary jets with high pressure heads (in excess of 1.5 m) or channels requiring a dosing basin (feeder compartment) to control the interruption of the waste water supply (Fig. 6.4.-13). These jet systems are used in small treatment systems.
- through mobile equipment such as rotating sprayers (Fig. 6.4.-11), with troughlike side guides or tilting troughs (Fig. 6.4.-14).

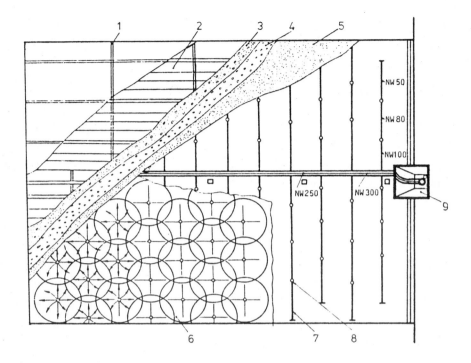

Fig. 6.4.-13: Trickle filter with nozzles

1 - bottom beams; 2 - hollow blocks; 3 - drainage layer; 4 - effective layer (packing material); 5 - distribution layer; 6 - moistened area; 7 - distribution piping; 8 - nozzles; 9 - dosing basin

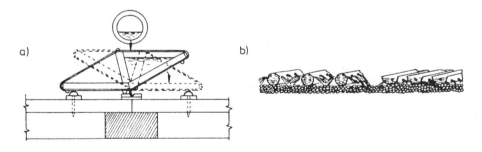

Fig. 6.4.-14: Water distribution equipment
a - tilting troughs; b - stationary side guides

6.4.4.3.3 Biological Filter Design

When a biological filter system is planned, primary attention should be given to the appropriate design of the flow pattern, the inclusion of the filter into the elevation chart of the treatment plant, and the determination of the type of the packing material (crushed rock, scorified lava, plastic elements).

The biological effect of the filters is largely governed by the following factors: waste water quality, temperature, surface charge, BOD_5 volumetric load, height of the packing, grain size, filter aeration.

The flow pattern determines the clarification process and thus the treatment level that can be achieved. It is thus determined as a function of the required extent of treatment; Fig. 6.4.-10 illustrates the various options. Option (a) is suited for small treatment plants, options (b) and (c) are especially suited for low-oxygen needs or already digested inflow. In case (d) the inflow is not regenerated for preliminary treatment. Example (e) is applied for large waste water quantities. Version (f) entails an additional load to the preliminary and final clarification stages, while in arrangement (g) the recirculation flow has a biological "activation" effect. Case (h) involves a structural reduction if final clarification may be omitted or if the waste water may be introduced directly into the receiving reservoir.

Including the biological filter into the elevation chart of the treatment plant may lead to different solutions regarding the installation site, the delivery head, and the size and the necessity of the pump station.

The selection of the packing material depends on the plant's objectives. Another criterion in the decision-making process is the investment, where not only engineering parameters but also economic aspects should be considered.

The waste water quality decisively influences the clarification efficiency of the biological filters or the outflow concentration, the $KMnO_4$ consumption, and the percentage efficiency of total nitrogen decomposition.

The waste water temperature affects the thermal balance in the reactor compartment, while the outside air temperature is of less importance. The effects on the clarification process are different. There is a varying decrease of the biological activity with temperature in the range from 4° C to 30° C. The biological activity of more complex organisms stops below 10° C, where the risk of clogging arises.

The surface charge $(m^3/(m^2 \cdot h) = m/h)$ is a measure of the irrigation intensity under normal charging conditions; this factor affects the contact time

$$\frac{t}{H} = \frac{c_n}{q_F^\delta}$$

where:
t = contact time (h)
H = packing height in the filter (m)
c_n = constant
q_F = surface charge (m/h)
δ = coefficient

Studies with different c_n's have furnished values between 0.408 and 0.82. The relationship

$$\frac{t}{H} = c_n \cdot q_F^{-2/3}$$

is given here as an example of an empirical mean value, with δ depending on the grain size, the shape, and the surface of the packing, on the thickness and type of the biological slime, or directly on the inflow concentration and the volumetric charge.

RINCKE /140/ has developed a nomogram (Fig. 6.4.-15) for the practical dimensioning of biological filters that considers various factors. In this example, the influence of the inflow concentration, the surface charge, the height of the packing, the volumetric load and the temperature on the percentage clarification rate and the outflow concentration are combined for the biological trickle filter.

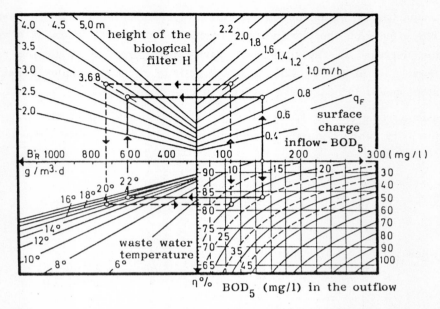

Fig. 6.4.-15: Efficiency diagram for trickle-type filters /140/

Biological filter systems for municipal and domestic waste waters are dimensioned for normal conditions. If, however, the sewerage system also collects heavily contaminated industrial and commercial waste waters or collects these discharges alone, the suitable treatment scheme or the required clarification efficiency must be determined empirically.

The essential parameters applied in the dimensioning of biological filters are as follows:
- the volume of the biological filter or the number of biological filters,
- the dimensions of the biological filters (height, diameter), the surface charge, the charging sequence, and recirculation volumes where applicable.

The waste waters to be handled in the biological filters must have been treated in preclarification basins to an extent that the sedimentary solids have been eliminated. This is done to prevent clogging of the nozzles, the openings in the rotary sprayers, or the filters, and to reduce the oxidation capacity of the biological slime. At the same time, the concentration peaks can be balanced within certain limits as early as this stage.

The biological filter volume is dimensioned on the basis of a BOD_5 volumetric loading B_R expressed in kg BOD_5/m^3 of filter volume per day ($kg/m^3 \cdot d$), which is chosen in accordance with the clarification level to be attained. Thus, the following packing volume will result:

$$V = \frac{BOD_5 \text{ inflow} \quad [kg/d]}{BOD_5 \text{ volumetric load} \quad [kg/(m^3 \cdot d)]} \quad [m^3]$$

The BOD_5 inflow per day is calculated either from the number of inhabitants and the population correction factors, or from reliably measured data, for a given inflow concentration S_o or an inflow concentration influenced by recirculation:

$$V = \frac{S_{om} \cdot Q}{B_R} \quad \text{or} \quad V = H \cdot F \quad [m^3]$$

where:

- V = the biological filter volume, in m^3
- H = filter height, in m
- Q = the mean flow rate, in m^3/d
- B_R = the BOD_5 volumetric load, in $kg/(m^3 \cdot d)$
- S_{om} = the mean inflow concentration per hour, in $kg/(m^3 \cdot h)$
- S_{ot} = the mean inflow concentration during the hours of the day, in $kg/(m^3 \cdot h)$
- q_A = the mean surface loading per hour, in $m^3/(m^2 \cdot h)$
- r = coefficient corresponding to the recirculation rate in relation to the recirculated volume
- F = the surface area of the filter, in m^2
- A = surface of the biological filter mass, in m^2

For example, assuming a 14 hour peak, the following filter height is obtained:

$$H = \frac{14 \cdot q_A \cdot S_{ot}}{B_R} \quad [m]$$

and for additional recirculation operation:

$$H = \frac{14 \cdot q_{A(1+r)} \cdot S_{ot}}{(1+r)B_R} \quad [m]$$

where:

$q_{A(1+r)} = \frac{Q}{A}$ is the surface loading, including a constant recirculation volume, in $m^3/(m^2 \cdot h)$

Previously, the volumetric load of the biological filter was controlled mainly by the waste water quantity. For domestic waste waters, the following values could be assumed as mean values:
- for low-load biological filters 1.0 m^3 of waste water per 1 m^3 of packing material (1.0 m^3/1.0 m^3)
- for biological trickle filters 3.0 m^3 of waste water per 1 m^3 of packing material (3.0 m^3/1 m^3)

This calculation was correct because only domestic waste water was involved, and there was no continuous passage through the contact filter. The so-called "contact" technique was used only with a single or a double loading supplied to the biological filters, irrespective of the contamination level in the waste water.

Later, when the biological filters were continuously loaded, the <u>oxygen demand</u> (or BOD_5) of the daily waste water amount per unit filter volume was also considered as a dimensioning parameter.

The following volumetric loads apply to mean conditions:
- for low-load filters   200 g $BOD_5/(m^3 \cdot d)$
- for high-load filters   600 g $BOD_5/(m^3 \cdot d)$

Since the BOD_5 values of the waste waters are usually not available in the design stage, the number of inhabitants to be connected to the network is used to define the anticipated loading. With a drinking water consumption of 200 l/person per day and for 60 g of BOD_5 per person per day in the untreated waste water, the waste water will still contain 40 g $BOD_5/(P\ d)$ after sedimentation in the pretreatment (sedimentation) basins. This results in an admissible volumetric load of

$\frac{200\ g/(m^3\ d)}{40\ g/(P\ d)}$ = 5 Pcorr for low-load filters, and

$\frac{600\ g/(m^3\ d)}{40\ g/(P\ d)}$ = 15 Pcorr for high-load filters.

Such screening is expedient mainly when the individual structures of the system should be kept small and when the objective is the possible accommodation of the secondary treatment units in covered compartments. Measured results have shown that the outflow values are substantially better than the values obtained when secondary sedimentation basins are used, and that the results almost approach the maximum values achievable from our present engineering and economic viewpoint (secondary sedimentation basin with downstream sand filters).

The high sludge density leads to a very high sludge age. On the one hand, this means that systems with immersed biological filters may be so dimensioned that nitrification occurs when desired. On the other hand, the high sludge age involves a very thorough decomposition of all those compounds that belong to definitely unstable groups so that the excess sludge production is usually slightly less than the sludge production in an activation system. In any case, however, the volume of the produced sludge is smaller because the high density of the flakes results in a better dehydration level in the static thickening stage.

Immersed biological filters are widely used because of the advantages of this technique (including a strongly adhering and thickly growing biological slime, low depth, and economical operation). Apart from the simple and inexpensive design, this method involves a possible variation of the clarification process. When highly concentrated waste waters are to be treated, for instance, it is possible to include intermediate treatment basins at any point. The biological treatment can be continued, after an intermediate phase, with any other clarification process. Immersed biological filters can also be used successfully for the final treatment of rotting waste water. It even permits an economical decomposition of mineral oil products contained in the waste water. To give an example: 99 % rates of petroleum product elimination have been achieved.

One factor in the dimensioning of the immersed biological filters is the ratio between the basin volume and the overgrown surface of the disks, which has a decisive influence on the treatment efficiency.

The dimensioning values usually indicated are based on the assumption that the basin must contain ca. 4 to 6 l of water per 1 m² of rotating overgrown surface, or that 10 m³ of air and 0.3 m³ of biomass are required per 1 m³ of domestic waste water after a preliminary mechanical treatment, in order to achieve a 90 % clarification efficiency. The BOD_5 surface load is a decisive value that influences the efficiency rate and the dimensions of biological disk filters.

The values recommended in Table 6.4.-15 can be used to estimate the dimensions, with A_S corresponding to the surface of a disk.

Tab. 6.4.-15: Recommended values for disk filter dimensioning (according to /2/)

BOD_5 decrease (%)	80	90	95
number of rolls	2	> 3	> 4
A_S (m²/P)	1.0	2.0	3.0

The commercially available disks have the following surface areas (considering both sides):

with 2.0 m diameter A_S = 5.9 m²
with 3.0 m diameter A_S = 13.0 m²

30 disks with 3 m diameter or 34 disks with 2 m diameter are attached per meter of shaft length. After thorough start-up, the system's power consumption is about 75 W/m shaft length for disks with 3.0 m diameter, and 50 W/m shaft length with disks of 2.0 m diameter.

Another common estimate is as follows:

The required total disk surface is estimated from the daily BOD_5 supply and the specific surface load:

$$A = \frac{S_{om} \cdot Q}{B_A} \quad \left[\frac{kg/d}{kg/(m^2 \cdot d)}\right] = [m^2]$$

where:

A = the required disk surface, in m²
S_{om} = mean daily BOD_5 in the inflow, in kg/d
B_A = BOD_5 surface load, in kg/(m²·d), with the BOD_5 load of

the first roll not exceeding the value of 60 g/(m²·d) for domestic waste water, and 40 g/(m²·d) for rotting waste water.

For BOD_5 treatment rates of 85 % at least three, and for 90 % rates at least four rolls should be provided in tandem arrangement.

The following B_A values are suggested as dimensioning parameters for the treatment of domestic waste water. The result is a thorough nitrification while the minimum requirements are met:

- minimum requirements $\quad B_A = 8$ g/(m²·d)
  the same surface load of $B_A = 8$ g/(m²·d) is also envisaged for systems including less than 500 persons.

- thorough nitrification $\quad B_A = 4$ g/(m²·d).

The same quantities of excess sludge production as in the biological filter methods with comparable treatment efficiency rates must be expected, i.e.
0.8 kg/kg for minimum requirements, and
0.8 kg/kg for thorough nitrification.

The calculated water amount can be corrected when strong waste water quantity variations (m³/h) or BOD_5 supply variations (kg BOD_5/h) must be expected. In such cases, correcting factors are applied for more precise calculation, to consider the specific flow rate q (m³/min) as a function of the number of persons connected to a specific network:

Persons	Correction Factor for q (m³/min)
> 10,000	1.0
10,000 - 5,000	1.1 up to 1.2
5,000 - 1,500	1.2 up to 1.3
1,500 - 400	1.3 up to 1.5
< 400	1.5

This means that the corrected inflow is calculated as follows:
$$q_{corrected} = \text{correction factor} \cdot q \quad [m^3/min]$$

When sewage plants are provided for less than 400 inhabitants, ca. 50 %

more disk surface must thus be provided than would be necessary in large-size plants, so that a balanced outflow can be achieved even with strongly intermittent waste water discharges. Industrial plants which discharge pulsating loads of concentrated waste waters should also be considered by the inclusion of a suitable correction factor, provided that such pulsating load periods will regularly last for more than 5 hours with more than three times the mean load.

The basin volume (net) is calculated according to the equation:

$$V_B = 0.32 \, D^2 \cdot (L - m \cdot e) \qquad [m^3]$$

where:

V_B = the basin volume (net), in m^3
D = the disk diameter, in m
L = the roll length between the external walls, in m
m = the number of disks on a shaft
e = the thickness of the disks, in m

The number of disk revolutions is calculated as follows:

$$n = \frac{6.37}{D} \cdot (0.9 - \frac{V_R}{Q_h}) \qquad [rev./min]$$

where:

n = the number of disk revolutions per minute. Sufficient bacterial growth on the biological filter is only achieved when the following minimum rotational speeds are observed as a function of the disk diameter:
  for 2.0 m diameter n higher than 2.05 rpm
  for 3.0 m diameter n higher than or equal to 1.368 rpm
Q_h = the dimensioning flow rate of the inflow, in m^3/h.

The passage time is calculated according to the equation:

$$T = \frac{Q_h \cdot 24}{V_B} \qquad [h]$$

6.4.4.5 Sludge Activation Systems

Sludge activation systems are used for the biological treatment of waste waters with activated sludge.

6.4.4.5.1 General Fundamentals

The sludge activation systems comprise aeration basins, secondary clarification basins, and aerating equipment. Microorganisms clarify the waste water. They eliminate and partly respire the organic matter contained in the waste water, or convert the organic matter into sedimentary biomass.

The activated sludge is mainly composed of flocculating microorganisms (specifically bacteria and protozoa), and is mixed with dissolved oxygen and waste water. This ensures that the microorganisms in the activated sludge are kept in suspension and are maintained in constant contact with both the organic pollutants in the waste water and the oxygen.

Biological processes cause the organisms to absorb the organic matter from the waste water for the conversion or respiration to biomass.

This process converts part of the organic matter present in the waste water in a dissolved and a colloidal form to a solid form so that the phases can be separated later by settling in the secondary sedimentation tank.

The carbonaceous organic pollutants as well as the nitrogenous compounds can be oxidized (the latter to the nitrate level).

The waste water is aerated in concrete or reinforced-concrete basins, whose shape depends on the method of aeration, the type of waste water inflow, and the method used to introduce the activated sludge.

These basins are called aeration or activation basins. Pumps are provided to recirculate the required sludge from the secondary sedimentation tanks to maintain a sufficient quantity of biomass.

Fig. 6.4.-17 is a schematic of the sludge activation technique from BÖHNKE /31/.

The sludge activation method is presently the highest-grade technique for the biological treatment of waste water. Its prominent advantages are:
- safety in terms of the clarification level to be achieved in the treated waste waters, since the most relevant influencing factors such as supply of waste water, oxygen, and bacteria mass (activated sludge) can be controlled;
- higher efficiency compared to the biological filters, owing to the increased independence of temperature;
- shorter start-up periods of the sludge activation systems (less than 2 weeks), compared to the biological filters (4 to 6 weeks);
- definitely no odor and no flies.

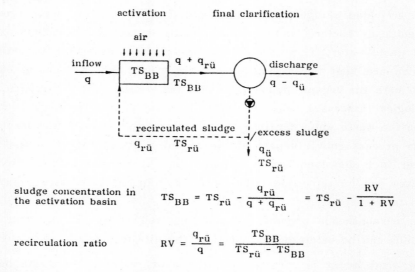

sludge concentration in the activation basin
$$TS_{BB} = TS_{rü} - \frac{q_{rü}}{q + q_{rü}} = TS_{rü} - \frac{RV}{1 + RV}$$

recirculation ratio
$$RV = \frac{q_{rü}}{q} = \frac{TS_{BB}}{TS_{rü} - TS_{BB}}$$

Fig. 6.4.-17: Schematic of the basic sludge activation process /31/
TS_{BB} = total solids of the sludge in the activation basins; $TS_{rü}$ = total solids in the recirculated sludge

6.4.4.5.2 The Activated Sludge

Activated sludge is produced by the growth of various bacteria and other microorganisms when waste water contains sufficient quantities of nutrient-rich matter and sufficient oxygen is available (i.e. under aerobic conditions). Since the microorganisms rapidly consume the dissolved oxygen, additional oxygen must be supplied from the atmosphere or the waste water/sludge mixture must be constantly aerated. The sludge consists of flakes at and in which the biological processes take place. These processes are similar to those in the surface film of biological filters. In spite of extensive research work, a difference between the mineralizing organisms in the biological filter slime and the organisms developing on the flakes in the activated sludge could not be determined. This confirms that the biological processes in the filters and in the sludge activation basins are fundamentally the same with the distinction, however, that the microorganisms in the biological filters develop in the form of a surface film that "adheres" to the packing material (as in land irrigation methods), whereas, in the sludge activation basins, they develop on the flakes "floating" in the waste water/air mixture (just like the self-purification of natural waters).

The activated sludge comprises a biocenosis*), a rich flora of aerobic bacteria, some fungi and bacteria-eating protozoa, ameba (cilia, flagellata). The presence of many protozoa means that a high-quality sludge is being produced. A well-activated sludge is characterized by the presence of bacteria and mobile cilia.

The following can be stated about the various biocenoses of the activated sludge:

- High-load systems and systems in the start-up stage or overburdened systems show strongly developed bacteria flakes with zoogloea. They are only associated with flagellata and ameba.
- Normal-load systems with a sufficient O_2 supply include many different protozoa species in addition to the bacteria flakes.
- Low-load and insufficiently loaded systems include bacteria flakes of a reduced size, mostly with a darker coloring. Many protozoa species are present, but the populations are rather reduced.

---

*) biocenosis is a balanced association of plants and animals in the same environment (biotope)

In general, activated sludge has a slightly chestnut-brown color, the smell of soil, and a high percentage of water, it is easily digestible and is influenced by inhibiting and toxic substances when certain limits are exceeded.

The water content of the activated sludge ranges from 98.5 to 99.3 %. When the moisture exceeds 99.5 % (it may become as high as 99.75 %), the sludge tends to bulk, which is an indication of spore formation, so that flakes are rarely formed. Moreover, the age of the sludge and the presence of chromium and arsenic have a certain influence on the bulking-sludge phenomenon.

The sludge index (I_{SV}), after MOHLMANN, indicates the size of the volume (VS) of 1 g of dry substance (TS) of the activated sludge after 30 minutes settling. In other words: the sludge volume of the waste water after 30 minutes settling, which contains 1 g of dry substance, is as follows:

$$VS_R = TS \cdot I_{SV}$$

In practical operation, this index is determined as follows. A cylindric jar 60 mm in diameter is filled with 1 l of the waste water/sludge mixture from the sludge activation basin. After 30 minutes, the volume of the sediment is determined, e.g. 300 ml/l. The dry substance of this sludge was determined, for instance, to be 6 g/l. The sludge index is calculated as follows:

$$I_{SV} = \frac{300}{6} = 50 \text{ ml/g}$$

or expressed in another way: the sedimented sludge with a water content of 98 %, for instance, has a sludge index of

$$I_{SV} = \frac{100}{100-98} = 50 \text{ ml/g}$$

Other sludge indices are also common in the technical literature.

The sludge index and the sedimentation rate are highly relevant parameters in the sludge activation process.

The sludge index of a well-flocculated activated sludge ranges between

6.4.4.5.3.1 Aeration with Atmospheric Air

Fig. 6.4.-18 /26/ shows three examples of various types and techniques of aeration: a - pneumatic aeration for the introduction of compressed air, b - mechanical aeration with circulation of the water and contacting same with more air, and c - combined aeration.

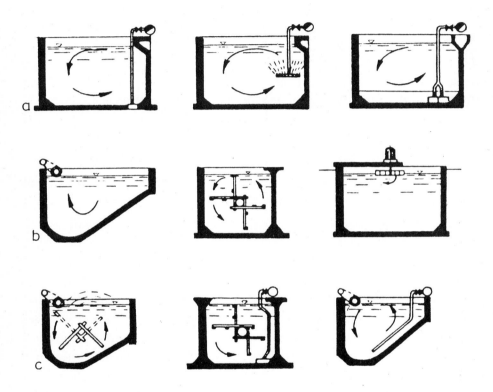

Fig. 6.4.-18: Examples of various types and techniques of aeration
a - pneumatic; b - mechanical; c - combined

The introduction of air into the water is not only a way of introducing oxygen, but also contributes to keeping the sludge particles in suspension so that the sludge does not settle in the basin and hence remains aerobic. This requires a water circulation rate of 0.15 m/s.

6.4.4.5.3.2 Procedural Methods

The sludge activation method demands thorough preliminary treatment. Provided that a safe process stability is ensured, a preliminary treatment may be generally omitted, however, in small-size plants and in compact treatment systems (smaller than ca. 10.000 m³ of waste water per day).

The Milwaukee sewage plant and some treatment plants in New York are exceptions, they do without any preliminary treatment. SETTER and HEUKELEKIAN even hold the opinion that this pretreatment would be less important in the activation method than in any other biological treatment.

However, in actual practical operation, preliminary treatment is usual in sewage plants with volumes in excess of 10,000 m³/d. This is the only way for them to remove fat and oil upstream of the activation basins.

The waste water is aerated by several methods, which depend on the treatment level to be achieved in the waste water, on the sludge stabilization, the inflow of waste water and sludge into the activation basin, and on the specific modes of operation. Fig. 6.4.-19 (a) illustrates the schematic of the usual aeration method applied in the biological waste water treatment, the so-called "total" (in the U.S. also called "conventional") or complete clarification: The waste water undergoes a preliminary treatment in 1 and is introduced into the aeration basin 2 where it is aerated for approximately 4 hours until the minimum requirements for the introduction into the receiving reservoir are met. The waste water is passed through the secondary sedimentation basins 3 where the sludge settles. One part of the settled sludge is recycled by pumps as recirculation sludge into the activation basin. To a limited extent, this admixture makes it possible to adapt the sludge volume in the activation basin to the requirements of the activation process. The excess sludge produced is either pumped from the secondary sedimentation basins directly into the sludge treatment stage or it is passed through the sedimentation basin 1. Here, in the pretreatment basin (sedimentation basin), it is also used as a flocculating agent, which occasionally improves the sedimentation process by as much as 25 %.

Fig. 6.4.-19 (b) is a schematic of a partial treatment process (50 to 75 % BOD_5 elimination) with shorter aeration time (or separate regeneration of the activated sludge). This system was operated for the first time in New York by GOULD /59/ and his collaborators (modified aeration). The aeration period is reduced from 4 to 2 hours in this system. The sludge content of 1.5 to 3 kg TS_R/m^3 of the activation basin volume decreases to 0.5 kg TS_R/m^3 of the basin volume. The activated sludge is circulated for a short time, it has a low sludge age, consists almost exclusively of bacteria while groups of more complex organisms such as protozoa do not occur. This technique provides for continued sludge aeration in so-called regenerators (in the U.S. also called "activated aeration"). This process occurs in basins designed as aeration basins or as separate tanks (see Fig. 6.4.-19 c).

Fig. 6.4.-19 (d) is a schematic of the so-called step aeration (GOULD system) with distributed waste water supply either at the basin inflow (a) or along the entire length of the basin (b). This arrangement provides for distribution of the high oxygen supply, which is necessary at the basin inflow in the conventional methods, all over the activation basin. The oxygen supply can be utilized more economically. Moreover, comparatively uniform conditions are created for the microorganisms.

Fig. 6.4.-19 (e) is the schematic of an arrangement including the distributed waste water inflow with activated sludge distributed in the basin, which ensures a comparatively expedient exploitation of the microorganisms' activities.

Fig. 6.4.-19 (f) is the schematic of a process with complete mixture for small waste water quantities; this system is employed in many cases for commercial waste waters.

New aeration schemes and types are being developed such as high-load aeration basins. These can be compared to the high-load biological filters. The waste water is thoroughly aerated for a short period of 0.5 to 2 hours, a preliminary treatment may be omitted or the duration time

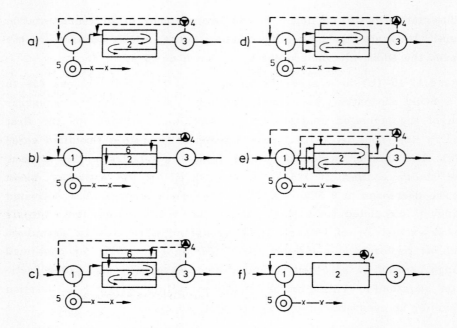

Fig. 6.4.-19: Methods applied in activation basins
1 - preliminary clarification; 2 - aeration basin; 3 - secondary clarification basin; 4 - pump station; 5 - digestion compartment; 6 - regeneration basin; a - conventional (common) regeneration basin; b - partial treatment with integral regeneration basin; c - distributed waste water inflow; d - distributed waste water inflow; e - distributed waste water and sludge inflow; f - complete mixture

can be reduced to 0.5 hour. The duration time in the secondary sedimentation basins is also reduced to 2 hours. The outflowing discharges have a low clarification level, while the sludge is comparable to that produced in the pretreatment basins.

Another system (e.g. KRAUS) provides for mixing the recirculated sludge with digested sludge and digested sludge liquor, with the subsequent introduction of the mixture into an aeration basin.

In various municipal sewage plants in the United States one portion of the completely clarified waste water is mixed with waste water discharged from a pretreatment stage. The result is a partly treated waste water (see Fig. 6.4.-19 b) but with increased sludge production and a higher chlorine consumption for disinfection.

Fig. 6.4.-20 illustrates frequently used versions of the oxidation ditch process (termed orbital aeration in the United States). The circular ditch provides for an appropriate energy utilization with a comparatively low cross-section of passage (depth of water 1 m, 3 to 5 m of water level width), while both nitrification and denitrification processes are integrated. The excess sludge is stabilized to an extent that it may be dehydrated on a drying bed without offensive odors.

Fig. 6.4.-20 (a) shows an accumulating ditch for intermittent operation. Fig. 6.4.-20 (b) shows a double ditch for alternating operation, where ditch I may operate, for example, as an aeration volume and ditch II as a secondary sedimentation basin, and vice versa. Operation of this system is also intermittent. Fig. 6.4.-20 (c) illustrates a ditch with a double channel for two-way passage, which is continuously operated. Fig. 6.4.-20 (d) is the schematic of a ditch with a secondary sedimentation basin, which is common for concentrated waste waters.

With some highly concentrated commercial waste waters one single aeration step is insufficient. So several biological steps must be installed.

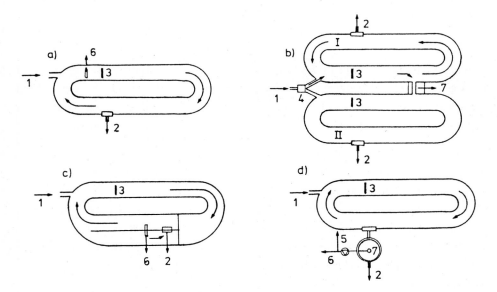

Fig. 6.4.-20: Oxidation ditches

a - accumulating ditch; b - double ditch for alternating operation; c - ditch with double channel for alternate passage; d - ditch with secondary clarification basin; 1 - inflow; 2 - starter, designed as slide weir for the ditch types a, b and c; 3 - roll aerator; 4 - inflow distributor; 5 - sludge recirculation; 6 - excess sludge removal; 7 - final clarification tank

6.4.4.5.3.3 Aeration with Pure Oxygen (Gassing with Oxygen)

The technique of gassing with oxygen is a practical alternative to the conventional activation method:
- The method of oxygen gassing offers advantages for the easily degradable organic substances and for highly concentrated waste waters with a high oxygen demand, since a higher oxygen supply is possible than with the introduction of air.
- With highly odorous waste waters, the oxygen supply technique should have priority, since a possibly required waste gas treatment can be very easily achieved owing to the covered basins and the low waste gas quantities.
- In very large plants and with confined space available, the application of pure oxygen has advantages since the permitted load may be higher. This entails reduced basin volumes and thus savings of space and costs.

In many cases, the application of pure oxygen and the higher oxygen concentrations in the activation basin have a positive affect on the sedimentation characteristics of the activated sludge.

An oxygen concentration between 0.5 and 2.0 mg O_2/l in the activation basin is generally deemed sufficient for complete treatment. In practical operation, however, it has been noted that the oxygen concentration and the treatment efficiency depend on the oxygen consumption of the activated sludge (sludge activity). In low-load systems, the oxygen consumption is very small. Oxygen concentrations as low as 0.5 mg O_2/l are obviously sufficient to ensure an oxygen supply into the inner pores of the flakes with an optimum treatment efficiency.

This is not always the case with high-load systems. For instance, desulfurization can be noted even with oxygen concentrations between 1 and 2 mg O_2/l in activation systems receiving easily degradable substrates (waste water from wine production, dairies, and the like) at a high load (i.e. B_R above 1.5). This phenomenon is due to the fact that under such conditions bound oxygen (and also oxygen from sulfates) is used for decomposition. In one case, the desulfurization process could be totally stopped by increasing the oxygen concentration from 1 to 2 up to 4 to 5 mg O_2/l, which, however, can be achieved only when the through-flow rate (load) is reduced while the defined aeration rate is maintained.

When oxygen is supplied instead of atmospheric air, a good practice which has become common in many activation systems, the oxygen supply can be increased to almost five times the original quantity, at the same efficiency rate. Even in highly loaded systems, the high oxygen percentages, which are necessary for an optimum clarification efficiency, can be maintained at values higher than 4 mg O_2/l.

The equipment types used produce oxygen from atmospheric air at the sewage plant site. The oxygen can be introduced into the covered activation basins (which are loaded with a very slight positive pressure of ca. 5 hPa, and which are mostly designed to comprise several compartments in tandem arrangement) by means of a compressed-air rotor aerator or a compressed-air centrifugal aerator. In these systems, the oxygen is supplied from the production unit or a liquid-oxygen storage tank through the pressure line. The oxygen is delivered in liquid form to small plants and is stored on site. System operation with 6 to 10 mg O_2/l has become common. According to the information obtained from a company distributing oxygen-operated activation systems, the economic situation (investments and operating costs) becomes comparable to the conditions in common aeration systems with a plant size in the range of 8000 kg BOD_5 and more per day; with higher loads, the economy is even better.

The closed basin methods are more widespread. The basin cover creates a separate gas volume above the water level. The activation basin is also often subdivided by partitions into two to four interconnected compartments (so-called cascade arrangement). Fig. 6.4.-21 shows the schematic of such a layout. The waste water, the recirculated sludge, and the oxygen are supplied into the first compartment. Oxygen-rich gas is continuously recirculated from the gas compartment into the waste water/sludge mixture. Surface aerators or compressors with filter cartridges or gassing stirrers are used for this purpose. Each of the adjoining compartments is supplied with the gas/waste water/sludge mixture from the preceding compartment. The waste gas is discharged from the final compartment. Less than 10 % of the introduced oxygen quantity is discharged along with by-products (CO_2 and N_2), while the clarified waste water is removed through a baffle (e.g. UNOX, LINDOX, OXYAZUR IV, OASES, and SIMPLOX methods).

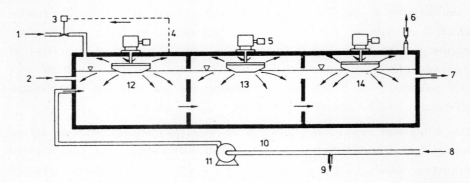

Fig. 6.4.-21: Schematic of a system for pure-oxygen gassing, "UNOX" system, with an activation basin comprising three compartments and surface aerators

1 - oxygen supply; 2 - waste water; 3 - pressure control; 4 - pressure signaling; 5 - surface acrator; 6 - waste gas removal; 7 - to secondary clarification basin; 8 - from secondary clarification stage; 9 - excess sludge; 10 - recirculated sludge; 11 - sludge pump; 12, 13, 14 - aeration compartments

Compared to the system of aeration with atmospheric air, the method of aeration with pure oxygen offers some advantages:

- The activity of the biomass is higher, the biomass settles more rapidly, and thickening is better than in sludges from systems operated with atmospheric air. Activation and secondary sedimention basins may therefore be designed smaller than the tanks usually used in air-operated systems.
- The efficiency invariably remains at a high level even with varying load conditions, owing to the oxygen supply rate.
- Spray water, aerosols and foam cannot affect the environment.
- The waste gas quantity is only ca. 1 % of the waste gas amount produced in conventional air-aerated systems. The removal and treatment of such a low waste gas quantity can also be controlled so that specific odor problems can be managed.
- Owing to the increased sludge volume (mean TS_{BB} - 5 g TSS/l) in the activation basin, high volumetric loads are possible. The overall dimensions of the activation basins can therefore be kept smaller. A

triple volumetric load is possible with the same efficiency in treatment.
- Owing to the low sludge volume index, better sedimentation properties can be expected. Therefore, the extent of secondary treatment can be reduced (I_{SV} = 50 - 80 mg/l).
- Smaller excess sludge volumes are produced. The quantitative reduction may be given in the range of 10 to 20 %. This means corresponding savings for all the downstream sludge treatment equipment such as thickeners, digestion towers, and sludge dehydration systems, including the associated labor, energy, and equipment costs.
- Apart from special cases where oxygen can be supplied to the site through a pipe line or in a tank truck, it must be assumed that the required oxygen supply is produced directly in the sewage plant.
- When pure oxygen is used for gassing in both the single-stage and the double-stage process, nitrification is possible in open systems.
- Considering the present state of research and development, savings can be expected for high waste loads, i.e. municipal waste waters in sewage plants with corrected population values in the range of 100,000 and 200,000 and more, in terms of the total energy used for oxygen production and introduction.

The method of gassing with pure oxygen, however, also involves some disadvantages:
- A considerable share of the operating expenses in water treatment works is always allotted to the cost of oxygen introduction. When pure oxygen is used, the oxygen production cost must be added.
- Higher costs must be expected in consideration of the additional equipment needed for measuring, alarm, and protection, which are involved in the pure-oxygen method. This amount, however, accounts only for a comparatively small share in the total cost of water treatment operation.
- The majority of the methods demands that the activation basins be covered, which incurs additional expenses. The price per m³ of useful volume in covered compartments is roughly 2.5 to 3.5 times as high as that of open basins.
- The fact that the basins are covered, in combination with the low waste gas quantity, must be deemed an advantage particularly when the treatment works are located near residential areas. Odor emissions and the spreading of aerosols are thus avoided. An economic

advantage, however, cannot be derived directly since suitable provisions have so far not been realized in conventional plants so that comparisons could not be made.

6.4.4.5.4 Aeration Equipment

A wide variety of aerator systems is available for aeration in the sludge activation methods. Their selection depends on the composition of the waste water and on the engineering and economic conditions of construction and operation of the sludge activation plants.

In the following, some details of the three predominant methods of aeration are explained: pneumatic (pressurization) aeration, mechanical (surface only) aeration, and combined aeration systems.

6.4.4.5.4.1 Pressurization

Aeration by pressurization is the oldest method that is presently in use in most plants. This method involves the introduction of compressed air through distribution networks into the aeration basin.

The shape of the basin and the type of the aerator equipment are harmonized in consideration of the waste water and air distribution systems.

The basins are designed only as normal aeration basins (Fig. 6.4.-22) or as basins for intermittent aeration and sedimentation (Fig. 6.4.-23).

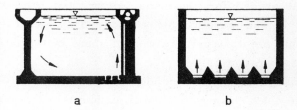

    a   b

Fig. 6.4.-22: Usual aeration basins with compressed-air supply
a - lengthwise along the walls; b - in a furrowed basin

The cross section of the usual basins is mostly rectangular, their depth is normally 2.5 to 5 m. The necessary contact time is achieved when the basins are designed as rectangular elongated basins taking into consideration the waste water flow rate. The overall length is restricted to 150 m approximately. At least two units are installed so that safe operation is ensured. The number of channels per unit may be even or odd. The bottom of the basin is designed in accordance with the aeration equipment: introduction of the compressed air along the walls (Fig. 6.4.-22 a) or with furrows; the latter design, however, is used only very rarely (Fig. 6.4.-22 b).

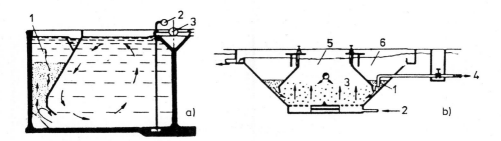

Fig. 6.4.-23: Aeration sedimentation basins

a - with conventional activation basin; b - "Oxycontact" type (DEGREMONT, France); 1 - sludge; 2 - compressed air; 3 - waste water; 4 - excess sludge; 5 - aeration zone; 6 - sedimentation zone

The aeration and the downstream secondary sedimentation basins are also installed in the form of a single channel.

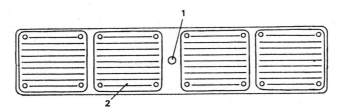

Fig. 6.4.-24: Frame for removable (screw-fastened) porous plate aerators

1 - hole for compressed air line connection; 2 - screw-in porous plates

Both elevated and deep-seated aerators are provided for aeration.

The deep-seated aerators are installed in the bottom (plate aerator, see Fig. 6.4.-24) or up to 0.3 m above the bottom, if possible.

The aerators above the bottom may be so arranged that they can be removed or tilted outward. This allows inspection and cleaning.

Elevated aerators such as the INKA System (Industrikemiska Aktiebolaged, Fig. 6.4.-25) consist of stainless steel or plastic tubes immersed by ca. 0.8 m.

The domes, which are in use specifically in Great Britain, are suited to introduce waste water and air or air alone.

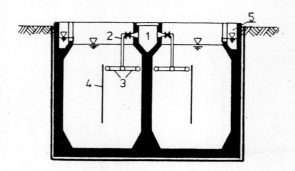

Fig. 6.4.-25: INKA aeration system

1 - air channel; 2 - air supply; 3 - perforated pipes; 4 - wooden partitioning (Ekran); 5 - waste waster inflow channel

The larger the total surface (or the smaller and the more uniform the bubbles), the more effective is the aeration process in theory. Depending on the bubble diameter and shape the rising rate ranges between 0.2 and 0.3 m/s. The size of the bubbles in the different aeration basins is not uniform since, on the one hand, tiny bubbles combine into bigger ones, or, on the other hand, larger bubbles are split into tinier ones as a result of turbulence and circulation. Therefore, the bubble size is classified by the diameter of the air outlets as follows:

- small, spherical bubbles with a diameter less than 1.5 mm; these bubbles rise in straight lines;
- medium lentil-shaped bubbles with a diameter between 1.5 and 18 mm; these bubbles rise along helical paths, in oscillating or swinging movements;
- big mushroom-shaped bubbles with a diameter up to 120 mm.

Tables 6.4.-16 (fine-bubble aeration), 6.4.-17 (deep-seated medium-bubble aeration), 6.4.-18 (elevated medium-bubble aeration, INKA system) and 6.4.-19 (gross-bubble aeration) list recommended values of the oxygen supply in various types of aeration; from EMDE.

Tab. 6.4.-16: Recommended oxygen supply values for fine-bubble aeration

	Favorable conditions O_2 supply	Favorable conditions O_2 yield	mediocre conditions O_2 supply	mediocre conditions O_2 yield
	$\dfrac{g\ O_2}{Nm^3 \cdot m}$	$\dfrac{kg\ O_2}{kWh}$	$\dfrac{g\ O_2}{Nm^3\ m}$	$\dfrac{kg\ O_2}{kWh}$
pure water	12	2.2	10	1.7
operational conditions	10	1.8	8	1.3

The gross energy expenditure is 6 Wh/Nm³ of air and per meter of introduction depth for mediocre conditions. The ratio is 5.5 Wh/Nmm for optimum conditions.

Tab. 6.4.-17: Recommended oxygen supply values for deep-seated medium-bubble aeration

	favorable	favorable	mediocre	mediocre
	O_2 supply	O_2 yield	O_2 supply	O_2 yield
	$\dfrac{g\ O_2}{Nm^3 \cdot m}$	$\dfrac{kg\ O_2}{kWh}$	$\dfrac{g\ O_2}{Nm^3}$	$\dfrac{kg\ O_2}{kWh}$
pure water	7	1.4	6	1.1
operational conditions	5.5	1.1	4.5	0.8

The gross energy expenditure is 5.5 or 5.0 Wh/Nm³ of air and per meter of introduction depth for mediocre or optimum conditions due to the lower outlet resistance (compared to the fine-bubble aeration system).

Tab. 6.4.-18: Recommended oxygen supply values for elevated medium-bubble aeration

	favorable		mediocre	
	O_2 supply	O_2 yield	O_2 supply	O_2 yield
	$\dfrac{g\ O_2}{Nm^3}$	$\dfrac{kg\ O_2}{kWh}$	$\dfrac{g\ O_2}{Nm^3}$	$\dfrac{kg\ O_2}{kWh}$
pure water	9	1.8	8	1.5
operational conditions	7.5	1.5	6.5	1.2

The energy expenditure is ca. 5.5 or 5.0 Wh/(Nm³·m) at an immersion depth of 0.8 m.

Tab. 6.4.-19: Recommended oxygen supply values for gross-bubble aeration

	favorable		mediocre	
	O_2 supply	O_2 yield	O_2 supply	O_2 yield
	$\dfrac{g\ O_2}{Nm^3}$	$\dfrac{kg\ O_2}{kWh}$	$\dfrac{g\ O_2}{Nm^3}$	$\dfrac{kg\ O_2}{kWh}$
pure water	6	1.2	5	0.9
operational conditions	4.5	0.9	4.0	0.7

The energy expenditure is ca. 5.5 or 5.0 Wh/(Nm³·h·m).

6.4.4.5.4.2 Surface Aeration

The basis of the surface aeration technique is the mechanical introduction of oxygen at the water surface. This method is also used when ponds, lakes, and natural waters are aerated. The surface aerators produce simultaneous circulation flows that mix the activated sludge with the water and prevent sludge sedimentation. The cylindrical and centrifugal aerators are the most frequently used types among the numerous aeration units.

<u>Cylindrical aerators</u> are available as brush aerators and barred rolls (e.g. air-lift rotors). The brushes or steel bars are attached to a horizontal axis. The assembly is rotated so that the brush or bar elements are repeatedly immersed into the water surface to spray the water, thus contacting it with more air. At the same time, the water is circulated.

There are various ways of providing cylinders with agitating elements. Instead of the piassava brush originally used by KESSENER (therefore the term "brush aeration" was formerly used), now comb-type steel sections (flat bars, angle sections, etc.) are generally used in radial arrangements ("barred cylinders"), e.g. the air-lift rotors, or in concentric or paraxial arrangements ("cage cylinders") (Fig. 6.4.-26).

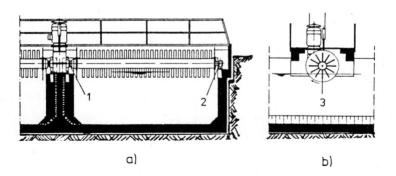

Fig. 6.4.-26: Air-lift rotor
a - longitudinal section; b - cross-section;
1 - flexible connector; 2 - thrustbearing; 3 - air-lift rotor

The cylinders are installed in longitudinal basins or circulation tanks. The longitudinal basins must have a suitable design (Fig. 6.4.-27).

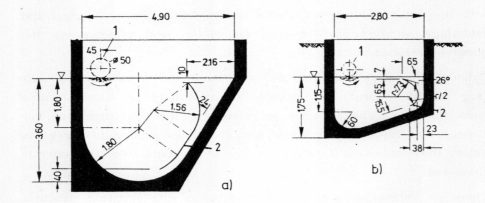

Fig. 6.4.-27: Longitudinal basin with PASSAVANT roll aerator
a - deep-end basin cross-section; b - flat-end basin cross-section; 1 - plate roll; 2 - baffle

A circulation tank comprises a closed elongated channel. The width of the channel depends on the usual rotor lengths between 3 and 9 m. The basins are designed to have a depth up to 4 m. When the water is deeper than 3 m, the negative influence of surface-active substances (e.g. surfactants) in the waste water on the velocity of flow in the area of the basin bottom must be considered.

Rotary aerators are composed of a rotary element, which rotates about a vertical axis at the water surface. These aerators are usually arranged in the center of the associated basin floor area. They can be suspended from a crossbar or platform, or fixed to the water surface by means of floating bodies (pontoons).

Many companies offer rotary aerators. They comprise the aerator rotor, the gear unit and transmission system or the shaft bearing, and the drive unit.

The aerator rotors are made of steel plates provided with angle steel elements (e.g. Vortair, Simcar system - Fig. 6.4.-28), rotors (e.g. Gyrox System, GEIGER - Fig. 6.4.-29), impellers (e.g. BSK, AMC - Fig. 6.4.-30) without or with a feed pipe (Fig. 6.4.-31).

The diameters of the aerator rotors range between 0.5 and 4 m. Table 6.4.-20 gives recommended values for the oxygen supply by rotary aerators.

Tab. 6.4.-20: Recommended oxygen supply values for rotary aerators

	favorable O_2 supply $\dfrac{kg\ O_2}{kWh}$	mediocre O_2 yield $\dfrac{kg\ O_2}{kWh}$
pure water and under operational conditions	1.8-2.2	1.3-1.8

In view of the variations of the water level and possible load variations, it is necessary to provide for a 1.5 to 2 safety factor when the equipment elements are designed.

The oxygen supply and especially the oxygen yield of a specific rotary aerator is also determined by the shape and size of the activation basin or the components and installations in the basin. In small basins with only 2 to 4 m spacing between the rotary aerators and the basin wall, better oxygen yields are achieved than would be obtained in large basins with 5 to 10 m spacing. When the spacing is smaller, the air bubbles introduced at the water surface are carried into deeper water layers with a higher intensity due to the radial surface flow toward the basin wall. The number of revolutions of the rotors may vary. Depending on the specific method of aeration, it may be roughly 20 to 70, in some cases it may even be higher than 100 rpm.

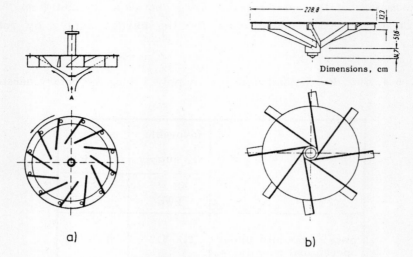

Fig. 6.4.-28: Rotary aerators with plates
a - Vortair; b - Simcar

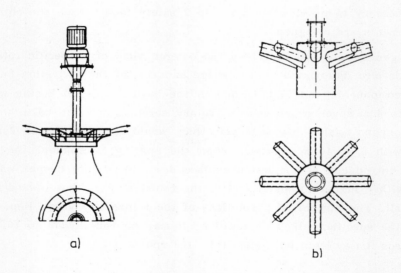

Fig. 6.4.-29: Rotary aerators with rotors
a - Gyrox; b - Geiger

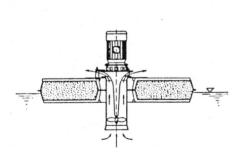 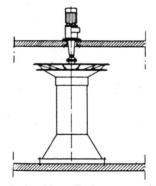

Fig. 6.4.-30: Rotary aerator with turbines (INFILCO)

Fig. 6.4.-31: Rotary aerator with line pipe

6.4.4.5.4.3 Combined and Other Aeration Equipment

Highly loaded industrial waste waters often demand a higher oxygen introduction efficiency.

Combined methods may provide for a sufficient supply of the required oxygen volumes by simultaneous aeration by means of fine-bubble aeration techniques and a paddle or by means of rotary aerators combined with compressed air (Fig. 6.4.-32).

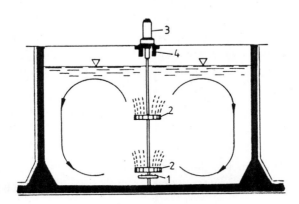

Fig. 6.4.-32: Combined DORRCO method
1 - compressed-air distributor; 2 - rotors; 3 - driving motor; 4 - bridge

Examples of other aerators are known as ejector aerator, jet aerator, counterflow aerator (Fig. 6.4.-33).

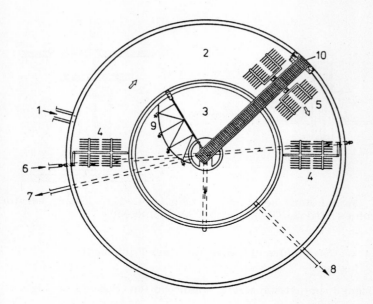

Fig. 6.4.-33: Circular counterflow basin (SCHREIBER)
1 - inflow; 2 - activation basin; 3 - secondary clarification basin; 4 - stationary aerator equipment; 5 - rotary aerator equipment; 6 - air; 7 - sludge recirculation; 8 - discharge; 9 - rake; 10 - mobile bridge

6.4.4.5.5 Design of Sludge Activation Systems

The following data are basic in the dimensioning of activation systems:
- BOD_5 load of the biological stage:

$$B_B = B (1 - \eta_{VK}) \text{ in kg } BOD_5/d,$$

where:

B = BOD_5 load in the inflow to the water treatment plant, in kg BOD_5/d

B_B = BOD_5 load of the biological stage, expressed in kg BOD_5/d

η_{VK} = clarification efficiency in pretreatment (%)

- waste water quantity in dry (TW) and rainy weather (RW):

$$Q_{TW} \text{ and } Q_{RW}; \text{ in } m^3/h$$

- clarification efficiency (%) to be achieved, and hence selection of the sludge load: B_{TS}, in kg BOD_5 inflow/(kg·TS·d)

(TS ≙ dried matter)

Fig. 6.4.-37 shows the process system occurring in such a plant design. When a recirculation system is provided, the resulting process advantages and the ensured sufficiently high surface charge can contribute to a further decisive stabilization of the entire purification process. Such a combination requires a highly effective preliminary clarification, otherwise the biological filter will become clogged.

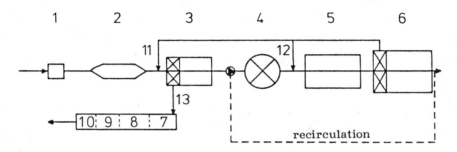

Fig. 6.4.-37: Biological filter combined with acti-vation processes
1 - rake; 2 - sand screen; 3 - preliminary clarification basin; 4 - biological filter; 5 - activation basin; 6 - secondary clarification; 7 - preliminary thickener; 8 - digestion compartment; 9 - secondary thickener; 10 - sludge dehydration; 11 - excess sludge; 12 - recirculated sludge; 13 - digested sludge

RINCKE and DAHLEM point out to the fact that the sludge with a predominantly absorptive effect should be removed from the highly loaded biological filter stage in an intermediate clarification step before a prolonged biological reaction would produce redissolution. Moreover, in the event of contamination, the buffering effect of the high-load biological filter stage may remove the contaminants eliminated by adsorption, together with the excess sludge from such intermediate clarification. Besides, the provision of an intermediate clarification stage increases the sludge age and thus the possibility of accelerated nitrification in combination with a higher rate of elimination of the slightly degradable substances.

In practice, however, the intermediate clarification stage is often not provided, since the rinsed-off biological filter sludge may promote the biological purification efficiency of the activated sludge, on the one hand, and improve the sedimentation characteristics of the sludge in the final clarification stage, on the other hand. When the process is so arranged and when the sludge concentration is possibly raised to 6 - 10 g/l, the associated oxygen demand must be considered in the activation stage. The biological filter sludge has an oxygen demand slightly higher than that involved in the recirculation sludge in the activation stage. According to ROSEN /33/, an increase by ca. 5 % of the oxygen demand without an intermediate clarification stage is normally expected.

The biological filter will have to be designed for a volumetric load of ca. 0.8 to 1.0 kg $BOD_5/(m^3 \cdot d)$. In consideration of the statutory discharge concentrations, however, a sludge load of 0.3 kg $BOD_5/(kg\ TS \cdot d)$ is deemed acceptable when the second biological stage is designed. When, however, the dimensioned B_{TS} load is about 0.15 kg $BOD_5/(kg\ TS \cdot d)$, there is not only a noticeable reduction of the ammonia compounds but also a particularly increased elimination of the slightly degradable substances, which must be stressed. The weaker loading of the second biological stage is the more favorable solution in relation to the reliable compliance with the minimum requirements and under the specific aspect of further improvement of pollution control.

6.4.4.5.6.5 Biological Filter Method with Preceding Activation

The method that includes an activation stage preceding the biological filter stage (Fig. 6.4.-38) is presently applied much more often than the process combination in the reversed arrangement. The reason for this more widespread application most probably is the fact that the first high-load activation stage has a sufficient capacity of buffering surge loads. So far, in practice, the high-load activation stage has been preceded by a preliminary clarification stage. LINDNER /33/, however, emphasizes that this process combination may go without an efficient preliminary clarification. With the first biological stage affording the main purification effect in this arrangement, the subsequent biological filter has the function of final residual purification, polishing, and nitrification.

A further oxygen enrichment and removal of suspended matter must be deemed an advantage.

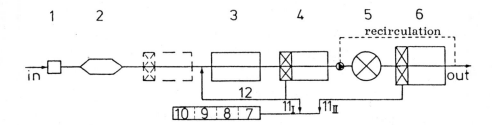

Fig. 6.4.-38: Trickle filter method with preceding activation
1 - rake; 2 - sand trap; 3 - activation basin; 4 - intermediate clarification basin; 5 - biological filter; 6 - final clarification; 7 - preliminary thickening; 8 - digestion tank; 9 - subsequent thickening; 10 - sludge dewatering; 11 - excess sludge; 12 - recirculated sludge; 13 - recirculation

HARTMANN, WILDERER and NAHRGANG /94/ suggest a sludge load of 0.7 to 1.0 kg BOD_5/(kg TS·d) in the first stage with such a process arrangement. According to their concept, a reduction of the load below this value does not enhance reliability. Also in view of an ensured sufficient substrate supply for the adjoining biological filter, it appears appropriate to provide the load values of the first activation stage in the range acceptable here.

Various studies have shown that there is no distinct functional relationship between the BOD_5 elimination efficiency of the biological filters and the volumetric load B_R. Apart from the further significant parameters such as the biological filter height, the packing material, or the surface charge, however, the amount of the BOD_5 inflow concentration is an important influence on the efficiency of the biological filter. With BOD_5 inflow concentrations as high as 40 mg/l and volumetric loads of ca. 100 - 200 g BOD_5/(m³·d), the BOD_5 efficiency ranges between 15 and 50 % in the systems studied. Volumetric loads higher than 200 g BOD_5/(m³·d) might hardly be normally achieved with such low inflow concentrations. The reason of such small elimination efficiencies is most probably due to an incomplete development of the individual metabolic cycles in these cases, which results from the low supply of organic matter. The efficiency is better with higher inflow concentrations. An increase in the volumetric load, however, results in a reduction of the BOD_5 elimination efficiency while the process characteristics are similar to those in comparable single-stage biological filter systems. Moreover, the BOD_5 inflow concentrations are within the range which is usually deemed expedient in single-stage biological filter systems.

When the two methods (i.e., biological filter preceding activation and biological filter with preceding activation) are compared to each other, it becomes very clear that the biological filter as a second biological stage is much more responsive to low inflow concentrations whereas the activation as the second stage displays a positive compensation effect. For this reason, it appears to be advisable to set the discharge from the first biological stage to 60 - 120 mg BOD_5/l. The attached biological filter should then be dimensioned for a volumetric load of 200 - 300 g $BOD_5/(m^3 \cdot d)$. The hydraulic load of the biological filter should be ca. 0.70 m/h for biological filter heights of 3.50 m, with this process arrangement. With such load conditions not only noticeable nitrification but also compliance with the statutory discharge concentrations can be expected.

6.4.4.5.6.6 Downstream Polishing Ponds

In contrast to the waste water ponds, the polishing ponds are supplied with waste water discharged from a completely biological purification. The purpose of these ponds is a continued natural purification. The polishing ponds thus have an important supplementary function in a multistage chain of processes.

Polishing ponds have a triple effect.
- They serve as mixing and equalizing basins in consideration of both hydraulic loading and variations of concentration. When used as sedimentation basins, they normally retain the total of the sedimentary matter still present.
- With a well functioning secondary clarification, the polishing pond is mainly a volume having biological effects. The decomposition activities of the bacteria supplied to the waste water in the activation basin or in the biological filter are further continued.
- With a slightly overburdened biological stage or sedimentation stage, the provision of subsequent polishing ponds may be an expedient solution for the sanitation and efficiency enhancement, in terms of both process engineering and economic aspects, of biological waste

water treatment systems, provided that sufficient space is available. With a correspondingly long duration time of the waste water in the polishing ponds, another noticeable reduction of the BOD_5 or COD discharge values can be expected.

It should be considered that considerable quantities of algae may be present in the discharges from polishing ponds during the vegetation periods. These algae result in a comparatively high BOD_5 and COD discharges. The COD concentration, for instance, increases by ca. 1 mg/l due to the biomass of an algae amount containing 10 µg chlorophyll.

Excessive algae growth can be avoided when the waste water duration in the pond is less than 2 days while the hydraulic efficiency (quotient of the actual duration and the theoretical duration) is not less than 0.5. Elongated, narrow, or meandering ponds have a higher efficiency than wide ponds.

Regarding the purification efficiency of finishing ponds, operational experience is available that can be used to assess their efficiency. The BOD_5 and COD efficiencies and discharge concentrations depend on the theoretical duration time and the inflow concentration in each case. A mean efficiency of 61 % and a mean COD of 42 % are achieved with BOD_5. Thus, discharge concentrations between 10 and 15 mg BOD_5/l and 40 - 60 mg COD/l can be achieved in the range of normal inflow concentrations.

The sedimentary substances, which are still contained in the discharge from secondary clarification, are normally completely retained. The concentration of undissolved substances is mostly reduced to 5 mg/l. The total germ count in low-load polishing ponds is reduced by one to four decimal powers, while more than a negligible phosphate elimination cannot be expected.

In the dimensioning of polishing ponds, the duration time, the pond depth, and the surface load are the decisive parameters.

In view of BOD_5 elimination, a duration of 2 days is deemed favorable since a remarkable improvement of BOD_5 degradation cannot be determined with longer duration times. When the objective is nitrogen oxidation, a duration of $t_R = 5$ d and a BOD_5 surface charge of 4 g BOD_5/(m²·d) must be rated. Nitrate reduction and denitrification involve surface loads of 5 g BOD_5/(m²·d), so that oxygen-free sludge/water contact surfaces can be achieved for denitrification at least during the night.

The most expedient water depth ranges between 1 and 2 m. Anaerobic conditions can thus be avoided. The area requirement is determined by the required surface, plus 50 % for the grade areas and the edge strips. It is roughly 0.3 m²/P. The grade inclination should range between 1 : 2 and 1 : 3. Only strongly permeable soils require sealing.

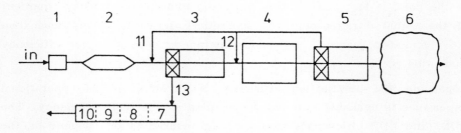

Fig. 6.4.-39: Polishing ponds arranged downstream
1 - rake; 2 - sand trap; 3 - preliminary clarification; 4 - activation basin; 5 - secondary clarification; 6 - polishing pond; 7 - preliminary thickening; 8 - digestion tank; 9 - secondary thickener; 10 - sludge dewatering; 11 - excess sludge; 12 - recirculated sludge; 13 - digested sludge

When the efficiency of existing biological filter or activation systems is insufficient or when receiving reservoirs have a very high responsiveness, the downstream arrangement of polishing ponds is a simple and reliable method of improving and stabilizing the discharge from a waste water treatment plant. When an activation stage is provided in larger-scale systems, it is normally expedient to extend the activation stage to a sludge load of $B_{TS} = 0.3$ kg BOD_5/(kg TS·d) and to improve the secondary clarification stage in compliance with the new dimensioning regulations, in order to minimize the ground area required.

The downstream arrangement of polishing ponds is fundamentally to be considered only when the necessary areas can be made available. One approach to arranging the necessary extensions in the form of polishing ponds may, for instance, be when polishing ponds are installed on sites of former sludge drying facilities, in cases where the sludge removal of the conventional sewage plant is converted to wet sludge discharge or sludge dehydration by machines.

6.4.4.6 Anaerobic Waste Water Treatment

The anaerobic waste water treatment is restricted to those organic substances that are accessible to the attack by facultatively or inherently anaerobic organisms. Waste waters rich in carbohydrates and fat, which occur predominantly in the production of semi-luxuries, are particularly suited for anaerobic purification.

The matter is decomposed in several stages. Even slight disturbances (pH variations, toxicity, heavy metals, etc.) may result in a collapse of the alkaline digestion process, while the anaerobic decomposition of substances remains in the acidic phase.

The anaerobic process is substantially slower than the aerobic process, since the methanogenic bacteria have a very low growth rate. Because of the resulting long duration time (up to 30 days), a single-stage anaerobic waste water digestion process is economically advisable only for small waste water quantities. Otherwise, the area requirement and the digestion tank costs would become excessive.

According to DOWNING /33/, anaerobic waste water treatment is more economical for waste waters with strong organic contamination and a COD in excess of 4000 mg/l, since this process does not incur high aeration costs. The production of the energy-rich methane gas, the low cell production (sludge volume produced), and the stabilization of the sludge are additional advantages.

A satisfactory aerobic purification is obviously possible in some cases only by way of anaerobic decomposition. Such modified decomposition characteristics must presumably be reduced to the fact that anaerobic microorganisms supply or form other enzymes under anaerobic conditions, that are able to initiate biochemical processes of decomposition for substrates which are not accessible to aerobic utilization.

The contacting process, which is an improved anaerobic activation process, must be mentioned as one of today's possibilities in process engineering. Activation is the main aspect in this process. This means that an attempt is made to set the conditions in the reactor in a way that a maximum of anaerobic activated substance will be produced which can reach the maximum activity possible of metabolic processes under the respectively set conditions. The sludge recirculation method is used to achieve this biological activity. Circulation is also required so that the substrate will be contacted with the microorganisms /159/.

Favorable conditions in terms of both process engineering and microbiology may also contribute to a considerable reduction of the reaction times.

In recent years, the anaerobic purification in the so-called up-flow reactor (see Fig. 6.4.-40) has been accomplished in the Netherlands. The waste water flows through the reactor from the bottom upward to a supplementary sedimentation basin in the upper section of the reactor. Experience has shown that the sludge is already settling when the water rises, so that the sedimentation section is provided for no purpose other than safety. The waste water inflow at the bottom and the internal gas production result in circulation.

Owing to the residual pollution in the discharge, a single-stage anaerobic waste water treatment is normally not possible. Industrial waste waters discharged from a preliminary digestion stage, however, can be supplied to a sewage plant without any restrictions.

The waste water digestion process is applied even today in individual clarification cesspools.

Where plenty of land is available and where the climate is favorable in the United States, waste waters from food, chemical, or petrochemical industries undergo anaerobic purification in waste water ponds.

Waste waters from sugar refineries are still partly decomposed under anaerobic conditions in dumping ponds in Germany.

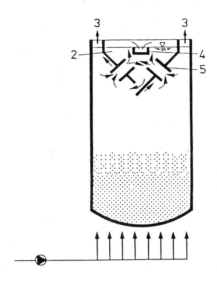

Fig. 6.4.-40: Schematic of an up-flow reactor
1 - distributed inflow; 2 - sedimentation section; 3 - sludge digestion gas; 4 - discharge channel; 5 - inflow pipe

The foregoing statements show that none of the single-stage systems is sufficiently efficient except an activation system with B_{TS} below 0.15 or a low-load biological filter. Since the domestic and industrial waste water load is increasing and since the operating costs of the aforementioned systems are high, there is a trend toward two- or multistage systems.

6.4.4.7 Final Clarification

The efficiency of biological and biological-chemical waste water treatment systems is influenced by the efficiency of the final clarification stage. This final clarification efficiency depends on
- the specific properties of the substances to be sedimented, e.g. the sedimentation and thickening characteristics of the activated sludge, expressed by the sludge index I_{SV} (l/kg) and by the reference sludge volume VS_R (l/m³),
- by the hydraulic conditions (flow processes), and
- by the coagulation and flocculation processes.

Since the sludge properties are determined specifically by the preceding stage, it is not possible to give general dimensioning data for the dimensioning of the final clarification stage. It is rather necessary to differentiate between
- final clarification basins downstream of biological filter systems, and
- final clarification basins joining activation systems.

6.4.4.7.1 Final Clarification Basins Downstream of Biological Filter Systems

In contrast to the dimensioning recommendations for final clarification basins in activation systems, which have been drafted in detail from the process engineering combination of activation and final clarification processes, considering the specific sedimentation and thickening properties, only relatively wide ranges of dimensioning recommendations have so far been made available for the final clarification basins in biological filter systems. One of the reasons is that the sludge from biological filters has good sedimentation and thickening properties due to its low porosity and void water content, so that the final clarification stage can be normally dimensioned by considering hydraulic criteria.

The solids concentration in the inflow into the final clarification stage of biological filter systems normally ranges between 95 and 130 g/m³, which corresponds to ca. 1.5 to 4 % of the solids contents in the discharge from activation basins. As a result of strong load variations and external influences such as temperature and mixed water loading, the solids contents fluctuate considerably.

The solids flow balance in the biological filter system is considerably affected by mixed water loading. When the dimensioning inflow is admissibly increased by 100 %, the dry solids contents in the inflow into the final clarification basin increases by up to 100 % owing to the doubled surface load of the biological filter, the halved preliminary clarification time of the dry weather inflow, and the solids load increase from the sewerage network. It is not seldom that concentrations of filterable substances in the inflow into the final clarification basin in amounts of 200 g/m³ and more are achieved /24/. In spite of the relatively small solids concentrations of the rinsed-off biological filter sludge and the intermittently occurring strong variations in concentration, the biological filter sludge still has good sedimentation and thickening properties, so that the thickened sludge from the final clarification stage contains solids in amounts between 40 and 80 kg dried matter/m³.

The sedimentation rate and the time elapsing until the sedimenting substances arive on the bottom sludge zone, are of great importance, so that the surface charge and the theoretical flow-through time must be considered as the decisive dimensioning parameters.

The separation processes in final clarification basins with horizontal flow or with vertical flow without a filter are essentially influenced by the sludge flocculation process. Owing to the low solids concentration of the biological filter sludges, the formation of sedimentary particle sizes is determined particularly by the molecular coagulation in the inflow section. The molecular coagulation is promoted by the increased creation of turbulence when mixed water is introduced, and when the recirculation ratio is increased.

Since the sedimentation rate and the settling time are criteria important for the quality of the sedimentation process, the final clarification basins of biological filter systems are dimensioned as a function of the surface charge q_F and the theoretical flow-through time t_R, which, for dry weather inflow (Q_{TW}), correspond to the following equations:

$$q_F = \frac{Q_{TW}(1+RV)}{F_{NK}} \qquad [m/h]$$

$$t_R = \frac{V_{NK}}{Q_{TW}(1+RV)} \qquad [h]$$

The recirculation ratio (RV) should be RV = 1 + 1 at maximum for packings with both rocky and synthetic packings. The more the mixed water inflow $Q_{TW}(1+m)$ increases, the more the quantity of recirculated water can be gradually decreased. This volume reaches the zero level when the mixed water inflow corresponds to the double dry-weather inflow. Thus, all charging conditions extend only up to the maximum charge for dry weather, which means that an almost constant solids concentration must be expected in the outflow from the biological filter. Accordingly, the following equations furnish the decisive dimensioning parameters for mixed-water inflow:

$$q_F = \frac{Q_{TW}(1+m)}{F_{NK}} \qquad [m/h]$$

$$t_R = \frac{V_{NK}}{Q_{TW}(1+m)} \qquad [h]$$

In accordance with the above equations for q_F and t_R, the dry-weather inflow (Q_{TW}) at $(1+m) < (1+RV)$ determines the surface (F_{NK}) and the useful volume (V_{NK}) of the final clarification system whereas, at $(1+m) > (1+RV)$, the mixed water inflow is the important factor.

In view of compliance with the minimum requirements, the solids discharge from the final clarification basin in biological filter systems must be restricted to 20 g/m³. As becomes evident from Table 6.4.-24, the surface charge in horizontal-flow final clarification basins should remain below 0.8 to 1.0 m/h for $Q_{TW}(1+RV)$ or $Q_{TW}(1+m)$, depending on the flocculation ability and the solids concentration in the inflow. To achieve this condition theoretical flow-through times of 3.0 h or even less should be rated. According to the relation between the mean useful depth and these parameters

$$T_{NK} = t_R \cdot q_F \qquad [m]$$

the water depth is calculated to range between 2.4 and 3.0 m. If the marginal depth is not less than 2.0 m, the overflow charge may be 10 m³/h per meter of outflow length according to Table 6.4.-25.

Owing to the improved solids retention in the filter of vertical-flow final clarification basins with suspended filter, the admissible surface charge can be increased by 30 %. The thus applicable range from 1.0 to 1.3 m/h relates to the mean filter cross section. As a result of the time-independent particle accumulation in the filter, the establishment of the duration time is not required. On the other hand, basins without a filter are subject to the same criteria as those applying to horizontal-flow units. With surface cross sections of 9.0 m and more, however, the establishment of the duration time is not required as a result of the basin geometry.

Tab. 6.4.-24: Surface charge and theoretical flow-through time for the dimensioning of the final clarification basins of biological filter systems /33/

	surface charge q_F m/h	theoretical flow-through time t_{NK} h
horizontal flow	0.8 - 1.0	3.0
vertical flow without suspended filter	0.8 - 1.0	3.0
with suspended filter	1.0 - 1.3	-

For biological filter systems dimensioned for more than 500 persons + population correction factor, the surface charge of the final clarification basins under dry-weather inflow conditions can be restricted to q_F = 1.0 m/h. The flow-through times should be longer than 2.5 h under dry weather inflow conditions, and 1.5 h when mixed water is supplied.

Tab. 6.4.-25: Surface charge and final clarification time in the final clarification basins of small biological filter systems /33/

Sewage plant size	surface charge q_F m/h	final clarification time t_R h
< 50 P	0.4	3.5
50-500 P	0.4-0.6	3.0
500 P + p corr.factor	1.0	2.5-1.5

6.4.4.7.2 Final Clarification Basins in Activation Systems

Up to the beginning of the 1970s, final clarification basins had been dimensioned considering almost exclusively the calculated flow-through time and hydraulic values. The density currents and particularly the sedimentation characteristics of the activated sludge were not considered.

Only the fundamental studies by STOBBE, PFLANZ, MERKEL, KALBSKOPF, and others /31/ resulted in a characterization of the sedimentation features of the activated sludge by way of the sludge index. This index does not define the sedimentation of an individual flake but the settling of the sludge level and thus the impeded sedimentation process.

Fig. 6.4.-41 is a schematic of the sedimentation process in a final clarification system. Four zones are created:
- freshwater zone,
- separation zone,
- storage zone,
- thickening zone.

Below the topmost freshwater zone, a zone prevails where the inflowing waste water/activated sludge mixture is deposited in layers corresponding to its density. At the same time, the impeded sedimentation process begins. In the zone below, the settling rate is retarded by a layer of denser sludge, which builds from the bottom upward. The sedimentation zone passes into the compression zone, where the mutually supporting sludge flakes exert a pressure on the water in the pores, pressing the water out. This compression stage is slow. The bottommost zone immediately above the bottom of the basin is determined by the radius of operation of the sludge scraper. When there are strong variations in the inflow, e.g. as a result of precipitated water, the final clarification basin must assume the additional task of storing the activated sludge displaced from the activation basin. To this end, a storage zone is provided.

The separation of the mixed inflow into its two main constituents, "activated sludge" and "biologically purified waste water", is the basis for the sequence of events defining the secondary clarification downstream of activation systems. After the required reaction time, the waste water or mixed water flow, which carries residual substances, must be removed from the final clarification basin. The concentrated sludge is supplied to the aeration tank /156/ where it is again utilized for the biological purification. Fig. 6.4.-42 shows a schematic of these processes.

Moreover, depending on the intensity of the recirculated flow, the sludge accumulation, the excess removal, and the hydraulic charge, different patterns of storing active dry substance are required in the final clarification stage.

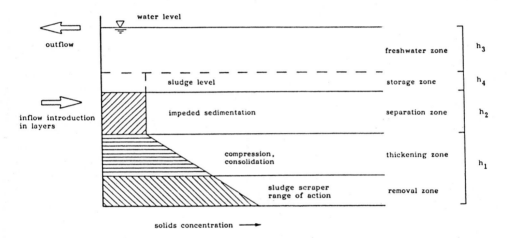

Fig. 6.4.-41: Sedimentation processes in a final clarification basin

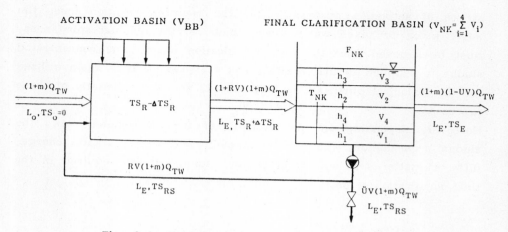

Fig. 6.4.-42: Flow of the recirculated sludge

L_O = substrate concentration in the inflow, in g/m³; L_E = substrate concentration in the outflow, in g/m³; TS_O = solids content in the inflow, in g/m³; TS_R = total solids content in the activation basin, in kg/m³; TS_{RS} = total solids content in the recirculated sludge, in kg/m³; TS_E = solids content in the outflow, in g/m³; TS_R = shifted solids contents for mixed-water inflow, in kg/m³; ÜV = excess sludge ratio; F_{NK} = surface charge of the final clarification, in m³; V_1 = thickening zone, in m³; V_2 = separation zone, in m³; V_3 = freshwater zone, in m³; V_4 = storage zone, in m³; T_{NK} = mean depth of the final clarification, in m; Q_{TW} = dry weather inflow, in m³/h; m = mixing factor for mixed water inflow, related to Q_{TB}; RV = recirculation ratio, in %

6.4.4.7.2.1 Final Clarification Basins with Horizontal Flow

In <u>final clarification basins with horizontal flow</u> the clarification efficiency depends mainly on the flocculation ability of the activated sludge. The flocculation process due to molecular coagulation starts as early as the inflow stage into the secondary clarification stage, and reaches its maximum efficiency in the inflow zone with a highly turbulent flow. Depending on the inflowing volume of activated sludge, zones of turbulent mixture near the inflow arise, which influence the formation of sedimentary macro flakes. These zones expand, the greater the individual factors become in the relationship

$$Q_s = (1 + RV) \cdot (1 + m) \cdot Q_{TW} \cdot (TS_R - \Delta TS_R) \cdot I_{SV} \quad [m^3/h]$$

where:

- Q_s = inflow through-flow rate, in m^3/h
- RV = recirculation ratio, in %
- m = mixing factor for mixed-water inflow
- Q_{TW} = dry-weather inflow, in m^3/h
- TS_R = total solids content in the activation basin, in kg/m^3
- ΔTS_R = shifted solids contents in the mixed-water inflow, or the difference in the solids contents in the activation basin in dry weather and rainy weather, in kg/m^3
- I_{SV} = sludge index, in kg/m^3

As can be seen from this equation, the flocculation volume and thus flocculation itself is not only determined by the inflow- and recirculation flow-dependent hydraulic parameters, but also depends, to an important extent, on the solids contents and the sludge index of the supplied waste water/sludge mixture, since both parameters characterize the concentration, the size, and the weight of the flakes that prevail immediately before the final clarification stage is charged. Even though zones in the inflow into the final clarification stage, with a turbulent flow-through, as shown, are an important prerequisite for the efficient separation of solids, excessive turbulence must be avoided; otherwise, macro flakes already formed in the inflow can break apart again. This aspect must be considered particularly when the inflow zone and the recirculation sludge control system are designed.

Following the formation of macro flakes under turbulent conditions in the inflow zone, the subsequent solids separation occurs under laminar conditions in the clarification volume, which comprises the separation and freshwater zones. Owing to the low flow rate, the macro flakes settle toward the surface of the bottom sludge layer. As a result of the flow coagulation, additional suspended matter is adsorbed to the macro flakes in the course of sedimentation. This process, however, requires a sufficiently high macro flake concentration and sufficiently dimensioned clarification volumes required by the laminar flow conditions.

The influence of flocculation (molecular coagulation) and secondary flocculation (flow coagulation) on the separation of solids in final clarification basins with a horizontal flow arrangement in activation systems is based on the dimensioning references given in Working Sheet A 131 of the German federal regulations on waste water treatment (ATV) /4/. When the resulting consequences for the surface, the depth, and the volume of the final clarification facilities are correspondingly considered, a noticeable improvement of the discharges can be expected. A distinction must be made among the following conditions of operation.

a) Dry weather inflow with a restriction of the discharged solids to TS_E = 30 mg/l at maximum. In such a case, a solids shift must not occur from the activation stage into the final clarification stage, i.e. $\Delta TS_R = 0$.

b) Mixed water inflow (sewage and rainwater) at a low load capacity of the receiving water, in correspondence to a maximum of discharged solids TS_E = 30 mg/l, while the solids content in the activation basin remains constant ($\Delta TS_R = 0$).

c) Mixed water inflow at a low load capacity of the receiving water, corresponding to a discharged solids maximum of TS_E = 30 mg/l and a solids content in the activation stage, which is reduced by ΔTS_R ($\Delta TS_R > 0$).

d) Mixed water inflow at a higher load capacity of the receiving water, corresponding to a discharged solids maximum of TS_E = 60 mg/l while the solids contents in the activation stage remain constant ($\Delta TS_R = 0$).

e) Mixed water inflow at a higher load capacity of the receiving water, corresponding to a discharged solids maximum of TS_E = 60 mg/l while the solids contents in the activation stage are reduced by TS_R ($\Delta TS_R > 0$).

The final clarification stage is dimensioned in two steps. First the required basin surface is determined and then the useful depth is established. The reference sludge volume VS_R of the activated sludge

$$VS_R = (TS_R - \Delta TS_R) \cdot I_{SV} \qquad [l/m^3]$$

is used as a controlling parameter forming the basis of calculation. Depending on the sludge load and the organic industrial waste water percentage, the sludge index I_{SV} (l/kg) must be rated according to Table 6.4.-26. For activation systems without preliminary clarification, for preliminary clarification times shorter than 0.3 h, or for oxygen gassing systems, the lower limits may be used.

The assessment of the solids shift ΔTS_R from the activation stage into the final clarification stage is determined by the above conditions of operation. In the event of dry weather inflow, a shift of the solids must not occur, i.e. ΔTS_R must equal zero for this inflow condition. For long rainwater inflow (which will presumably occur more frequently as a result of the discharge of rainwater basins) ΔTS_R is determined to be the difference between the dry substance contents in the activation basin with dry weather and rain weather inflow. This difference is calculated as follows:

$$\Delta TS_R = TS_{R,TW} - TS_{R,RW} \qquad [kg/m^3]$$

In order to avoid an excessive reduction of the decomposition efficiency the maximum displacement from the activation stage should be as low as 1.3 kg/m³ for $TS_R = 4.3$ kg/m³, or 30 % of TS_R with $TS_R = 4.3$ kg/m³ /24/. Moreover, the recirculation ratios for the dry weather and wet weather inflow must be appropriately controlled in accordance with the known equilibrium conditions

$$RV_{TW} = \frac{TS_{R,TW}}{TS_{RS,TW} - TS_{R,TW}} \qquad [\%]$$

or

$$RV_{RW} = \frac{TS_{R,RW}}{TS_{RS,RW} - TS_{R,RW}} \qquad [\%]$$

(Fig. 6.4.-34). The dry substance contents in the recirculated sludge may be rated as

$$TS_{RS,TW} = \frac{1200}{I_{SV}} \qquad [kg/m^3]$$

for dry weather inflow, and

$$TS_{RS,RW} = \frac{1200}{I_{SV}} + 2 \qquad [kg/m^3]$$

for wet weather inflow, or it can be taken from Fig. 6.4.-43 as a function of the sludge index.

When the admissible solids shift is utilized, particularly smaller sludge volume charges, higher bottom concentrations as a result of increased sludge deposition, and reduced turbulence in the inflow zone are achieved by recirculation quotas for the mixed water inflow that are as small as possible. The available supplementary sedimentation zone, moreover, ensures a considerably improved sedimentation activity under dry weather inflow conditions. The latest studies have shown, on the other hand, that considerable additional loads of the biological treatment stage occur mainly after periods of filling and discharging of the rain overflow basins in the sewerage network. These additional loads are in the range of 180 to 390 % of the mean dry-weather pollution. For this reason the sludge load required for purification should be maintained by largely avoiding a shift of solids within the sludge cycle upstream of low-load receiving waters. On the other hand, the provision of a shift-dependent storage zone is recommended because of the improved sedimentation capability, when the depth of the final clarification stage is dimensioned /24/.

When an impairment of the purification efficiency due to sludge transport is feared, the recirculation sludge transport must be so controlled that $\Delta TS_R = 0$, with consideration of a dry substance share in the recirculation sludge under rainy weather conditions, which is increased by 2 kg/m³ (Fig. 6.4.-43). Accordingly, a constant dry substance share may be retained in the activation stage. A maximum recirculation ratio of 1.5, related to the dry-weather outflow, should be considered as the threshold for the transport of recirculated sludge (this applies also to the foregoing case where $\Delta TS_R > 0$, with recirculation smaller than that for $\Delta TS_R = 0$). Assuming that the charge into the sewage plant in rainy weather is normally $Q_{RW} = 2\ Q_{TW}$, the maximum recirculation ratio $RV_{RW} = 0.75$ is obtained for wet weather. The recirculation ratio for rainy weather, which is required for the constant solids contents, is calculated to be

$$RV_{RW} = \frac{TS_{RS,TW} - TS_{R,TW}}{(TS_{RS,TW} + 2) - TS_{R,TW}} \cdot RV_{TW} \qquad [\%]$$

when the recirculation ratio of the dry weather inflow is known.
As a function of the reference sludge volume decisive for dry weather conditions, i.e.

$$VS_R = (TS_R - 0) \cdot I_{SV} \qquad [1/m^3]$$

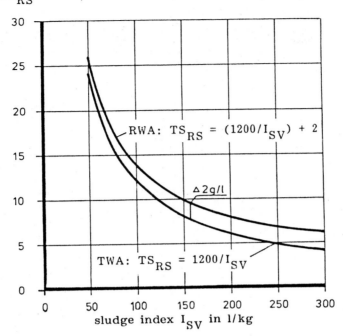

Fig. 6.4.-43: Determination of the solids contents of the recirculated sludge, as a function of the sludge index (RWA = wet weather outflow, TWA = dry weather outflow)

the specific admissible surface charge $q_{F,adm}$ (m/h) is obtained, which must be observed when the final clarification surface is sized. The q_F value is given by the lower graph in Fig. 6.4,-44. For $VS_R = 240$ l/m³, a surface charge of $q_F = 1.6$ m/h must not be exceeded. Moreover, the prevention of contents of filterable substances in the outflow discharged from the final clarification stage higher than 30 mg/l, demands that a sludge volume charge SV_F

$$SV_{F,adm} = 300 \text{ l/(m}^2 \cdot \text{h)}$$

be observed. The equation

$$SV_F = q_F \cdot VS_R$$

is used to arrive at the second condition governing the admissible surface charge:

$$q_{F,adm} = \frac{300}{VS_R} \quad [\text{m/h}]$$

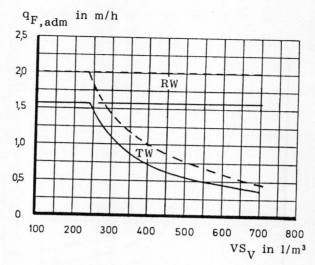

Fig. 6.4.-44: Calculation of the surface charge
——— horizontal through-flow; - - - - vertical through-flow
RW = wet weather; TW = dry weather

The smaller value of $q_{F,adm}$ is used to calculate the required surface of the final clarification stage as follows:

$$F_{NK} = \frac{Q_{TW}}{q_{F,adm}(VS_R, SV_{F,adm})} \quad [m^2]$$

When for the <u>rainwater outflow</u> the dry substance contents in the activation stage remains constant due to a corresponding adaptation of the recirculation ratio (see above) ($\Delta TS_R = 0$), the admissible surface charge must be determined for an admissible solids discharge of $TS_E \leq 30$ mg/l, in the same manner as for the dry-weather outflow. In other words, the surface charge is determined for a reference sludge volume $VS_R \geq 240$ l/m³ to be $q_{F,adm} = 1.6$ m/h or $q_{F,adm} = 300/VS_R$, or for a reference sludge volume = 240 l/m³ it must be determined to be $q_{F,adm} = 300/VS_R$ m/h from the lower graph in Fig. 6.4.-44 or the admissible sludge volume charge $SV_F = 300$ l/(m²·h).

Since the supplied water quantity is greater than the volume supplied in dry weather, the required surface of the final clarification stage is determined according to the following equation:

$$F_{NK} = \frac{Q_{RW}}{q_{F,adm} \, (VS_R, \, SV_{F,adm})} \quad [m^2]$$

When in wet weather dry substance is displaced from the activation stage within the admissible limits ($\Delta TS_R < 0$), the reference sludge volume decisive for the admissible surface charge is obtained according to the following equation:

$$VS_R = (TS_R - \Delta TS_R) \cdot I_{SV} \quad [1/m^3]$$

which means that it is reduced by the share $\Delta TS_R \cdot I_{SV}$. When an admissible solids discharge of $TS_E \leq 30$ mg/l is taken, the admissible surface charge rises accordingly, as is illustrated by the lower graph in Fig. 6.4.-44, or it is calculated from the admissible sludge volume charge according to the equation

$$q_{F,adm} = \frac{300}{VS_R} \quad [m/h]$$

In such a case, the required basin surface is determined according to the relationship

$$F_{NK} = \frac{Q_{RW}}{q_{F,adm} \, (VS_R, \, \Delta TS_R, \, SV_{F,adm})} \quad [m^2]$$

When sizing the final clarification stage for <u>wet weather through-flow</u> it is assumed that a solids outflow up to $TS_E = 60$ mg/l is permitted, a surface charge limit of $q_F = 1.6$ m/h must also be observed for reference sludge volumes in excess of 240 l/m³, in order to ensure sufficiently long through-flow times.

The required mean useful depth of horizontal-flow final clarification basins of activation systems comprises the following depth zones (see Fig. 6.4.-41):

- h_1: thickening zone, including removal zone,
- h_2: separation zone,
- h_3: freshwater zone,
- h_4: storage zone.
- For h_1, the thickening zone,

$$h_1 = \frac{(TS_R - \Delta TS_R) \cdot I_{SV}}{1000} = \frac{VS_R}{1000} \quad [m]$$

- For h_2, the separation zone where the sludge/water mixture is distributed and separated, a depth $h_2 = 1.0$ m should be considered for dry-weather outflow.
- For h_3, the freshwater zone, a minimum depth of $h_3 = 0.50$ m should be provided.
- For h_4, the storage zone, a depth h_4 should be considered for rainwater outflow because of possible sludge shift from the activation stage into the final clarification stage. The depth of the storage zone results from the relationship:

$$h_4 = \frac{TS_R \; V_{BB} \; I_{SV}}{500 \; F_{NK}} \quad [m]$$

(V_{BB} - Volume of the activation stage)

The overall depth $h_{tot} = \sum_{i=1}^{4} h_i$ must be provided over two thirds of the flow path of horizontal-flow final clarification basins with an inclined basin bottom. The minimum depth in this area should be 2.50 m. A marginal depth of 2.0 m should be the minimum for circular final clarification basins.

In order to avoid sludge rising, e.g. as a result of denitrification processes, the sludge storage time should be restricted to 0.5 to 1.0 h. The sludge storage time applying to dry-weather conditions can be established in a first approximation by the relationship

$$t_s = \frac{h_1 \cdot F_{NK}}{RV_{TW} \cdot Q_{TW}} \cdot (h) \leqslant 0.5 - 1.0 \quad [h]$$

The individual dimensioning parameters required to size horizontal-flow final clarification basins are represented in the survey in Table 6.4.-26, together with the required equations.

6.4.4.7.2.2 Final Clarification Basins with Vertical Flow

As long as the ratio between the flow path length and the useful depth is 10 : 1 in rectangular basins, and 5 : 1 in circular basins with a central inflow, a horizontal through-flow prevails. With depth-to-radius ratios = 1 : 3, a vertical through-flow prevails. Values between should not be considered because of the instable flow charcteristics resulting from such conditions.

Tab. 6.4.-26: Dimensioning parameters for horizontal-flow final clarification basins /33/

		DRY-WEATHER OUTFLOW Q_{TW} $TS_E \leq 30$ mg/l $\Delta TS_R = 3$ mg/l	WET-WEATHER OUTFLOW $Q_{RW} = Q_{TW} \cdot (1 + m)$				
			$TS_E \leq 30$ mg/l		$TS_E \leq 60$ mg/l		
			$\Delta TS_R = 0$ mg/l	$\Delta TS_R > 0$ mg/l	$\Delta TS_R = 0$ mg/l	$\Delta TS_R > 0$ mg/l	
Sludge index ISV	l/kg	waste water with low industrial share / waste water with high industrial share			$B_{TS} > 0.05$: 100 – 150 ; $B_{TS} < 0.05$: 75 – 100 / 150 – 200 / 100 – 150		
Reference sludge volume VSV	l/m³	$VSV = TS_R \cdot ISV$	$VSV = TS_R \cdot ISV$	$VSV = (TS_R - \Delta TS_R) \cdot ISV$	$VSV = TS_R \cdot ISV$	$VSV = (TS_R - \Delta TS_R) \cdot ISV$	
Dry substance content in the activation stage	kg/m³	depends on dimensioning of the activation stage	$TS_{R,RW} = TS_{R,TW}$	$TS_{R,RW} - TS_{R,TW} = \Delta TS_R$	$TS_{R,TW} = TS_{R,TW}$	$TS_{R,RW} - TS_{R,TW} = \Delta TS_R$	
Admissible solids transport	kg/m³	0	0	$\rightarrow TS_{R,TW} > 4.3$; $TS_R < 1.3$; $\rightarrow TS_{R,TW} < 4.3$; $TS_R < 0.3\, TS_{R,TW}$	0	$TS_{R,TW} > 4.3$; $TS_R < 1.3$; $TS_{R,TW} < 4.3$; $TS_R < 0.3\, TS_{R,TW}$	
Dry substance content of the recirculated sludge	kg/m³	$TS_{RS,TW} = \dfrac{1200}{ISV}$			$TS_{RS,RW} = \dfrac{1200}{ISV} + 20$		
Required recirculation ratio	–	$RV_{TW} = \dfrac{TS_{R,TW}}{TS_{RS,TW} - TS_{R,TW}} \leq 1.5$	$RV_{TW} = \dfrac{TS_{RS,TW} - TS_{R,TW}}{(TS_{RS,TW} + 2) - TS_{R,TW}} \leq 0.75$	$RV_{TW} = \dfrac{TS_{R,RW}}{TS_{RS,RW} - TSR,RW} \leq 0.75$	$RV_{RW} = \dfrac{TS_{RS,TW} - TS_{R,TW}}{(TS_{RS,TW} + 2) - TS_{R,TW}} \leq 0.75$	$RV_{RW} = \dfrac{TS_{R,RW}}{TS_{RS,RW} - TS_{R,RW}} \leq 0.75$	
Admissible volumetric sludge charge	l/m²·h		300		600		
adm. surface at VSV ≤ 240 charge ≥ 240 $q_{F\,adm.}$	m/h m/h		1.6 or $\dfrac{300}{VSV}$		1.6 or $\dfrac{600}{VSV}$		
Surface of the final clarification basin F_{NK}	m²	$F_{NK} = \dfrac{Q_{TW}}{q_{F,adm}(VSV;SV_F)}$	$F_{NK} = \dfrac{Q_{RW}}{q_{F,adm.}(VSV \cdot SV_F)}$	$F_{NK} = \dfrac{Q_{RW}}{q_{F,adm.}(VSV \cdot \Delta TS_R \cdot SV_F)}$	$F_{NK} = \dfrac{Q_{RW}}{q_{F,adm.}(1.6 \cdot SV_F)}$	$F_{NK} = \dfrac{Q_{RW}}{q_{F,adm.}(1.6 \cdot \Delta TS_R \cdot SV_F)}$	
Depth of the final clarification basin h	m	$h_1 = \dfrac{VSV}{1000}$; $h_2 = 1.0$	$h_3 \geq 0.5$	$h_4 = \dfrac{\Delta TS_R \cdot V_{BB} \cdot ISV}{500\, F_{NK}}$; $h_{tot.min} = 2.50$			
Sludge storage time l_S	h		$l_S = \dfrac{h_1 \cdot F_{NK}}{RV_{TW} \cdot Q_{TW}} \leq 0.5 - 1.0$				
Volume V_{NK}	m³		$V_{NK} = F_{NK} \cdot h_{tot}$				
Theoretical flow-through time	h	$t_{TW} = \dfrac{V_{NK}}{Q_{TW}}$			$t_{RW} = \dfrac{V_{NK}}{Q_{RW}}$		

The clarification efficiency in <u>final clarification basins with vertical flow</u> is not only the result of flocculation processes and sedimentation under gravity but is also frequently influenced positively by providing a suspension filter. This arrangement provides for a separation of the solids and a residual clarification through the inflow plane. One part of the supplied sludge volume first migrates upward in order to enrich the suspension filter with solids, while the position and thickness of the filter varies as a function of the load. Owing to the continuously arriving flow, the supplied particles are converted into sufficiently heavy flake structures in the filter, which is terminated at the top by the

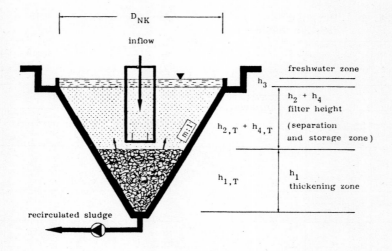

Fig. 6.4.-45: Final clarification basin with vertical flow

separation zone. The sedimentation rate of the flakes is sufficient to settle into the funnel end, represented as a thickening zone in Fig. 6.4.-45, against the flow.

As in horizontal-flow units, the sizing of final clarification basins with vertical through-flow in activation systems requires two steps. A distinction of the aforementioned five modes of final clarification must be made to determine the admissible surface charge, since the reference sludge volume varies according to the relationship

$$VS_R = (TS_R - \Delta TS_R) \cdot I_{SV} \qquad [1/m^3]$$

as a function of the mode of operation, or since there are different limits applicable to the admissible sludge volume charge SV_F $(1/(m^2 \cdot h))$. In this calculation the fact must be considered that the admissible surface charge values may be rated higher by 30 % for uniform hydraulic charging, as far as the suspension filter above the inflow plane is concerned. As a function of the determined reference sludge volume, $q_{F,adm}$ can be determined from the upper graph in Fig. 6.4.-44.

For verification, the admissible surface charge is to be determined through the sludge volume charge $SV_{F,adm} = 400 \; 1/(m^2 \cdot h)$ to be observed in vertical-flow final clarification tanks:

$$q_{F,adm} = \frac{400}{VS_R} \qquad [m/h]$$

The smaller value of the admissible surface charge is used to calculate the required surface of the final clarification basin for dry-weather conditions or wet weather outflow with a solids outflow of $TS_E < 30$. When an increase of the solids outflow under rainy weather conditions up to 60 mg/l is deemed acceptable with efficient receiving waters, the maximum surface charge allowed is 2.0 m/h, or $q_{F,adm} = 800/VS_R$ for reference sludge volumes $VS_R = 240$ l/m³ from the admissible sludge volume charge $SV_{F,adm} = 800$ l/(m²·h). The required surface F_{NK} of the final clarification stage is calculated using the allowed surface charge and the inflow. This surface value must be rated as the surface of the separation zone in vertical-flow final clarification basins.

The required useful depths (theoretical values determined by the effect, not yet related to the specific funnel arrangement) and the volumes of the individual zones are determined as in the case of the zones in horizontal-flow basins.

The theoretically required useful depth h_1 of the thickening zone is obtained from the relationship

$$h_1 = \frac{(TS_R - \Delta TS_R) \cdot I_{SV}}{1000} = \frac{VS_R}{1000} \quad [m]$$

while the required volume of the thickening zone is calculated to be

$$V_1 = h_1 \cdot F_{NK} \quad [m^3]$$

The theoretically necessary depth of the storage zone is determined from the relationship

$$h_4 = \frac{\Delta TS_R \cdot V_{BB} \cdot I_{SV}}{500 \cdot F_{NK}} \quad [m]$$

The necessary storage volume is obtained as the product from the useful depth and the final clarification basin surface:

$$V_4 = h_4 \cdot F_{NK} \quad [m^3]$$

A useful depth of

$$h_2 = 1.0 \quad [m]$$

should be considered for the separation zone under dry-weather outflow conditions in compliance with the dimensioning regulations of Working Sheet A 131 of the German federal regulations on waste water treatment (ATV). For rainy weather, the useful depth may be reduced to 0.5 m provided that the depth of the storage zone h_4 is ca. 1.0 m. The volume of the separation zone is

$$V_2 = h_2 \cdot F_{NK} \qquad [m^3]$$

Useful depths of

$$h_3 \geq 0.5 \qquad [m]$$

are required for the freshwater zone, which ensures the reliable retention effect of the final clarification basin under external influences (wind, solar radiation) and under hydraulic surge loads.

The volume of the freshwater zone is

$$V_3 = h_3 \cdot F_{NK} \qquad [m]$$

Since the vertical-flow final clarification basins are normally designed as funnel-shaped circular basins (Dortmund wells), the basin dimensions must be so determined that the volumes required as a function of the desired effect will actually be achieved. This means that the actual depths of the individual zones, related to the funnel geometry, must be determined through the stated required volumes (Fig. 6.4.-45). The required diameter of the circular basin is calculated to be

$$D_{NK} = \sqrt{\frac{F_{NK} \cdot 4}{\pi}} \qquad [m]$$

In order to ensure that the activated sludge will slide down along the walls of the funnel, the funnel is normally provided with an inclination, relative to the horizontal plane, of 60 or m : 1 $\geq$ 1.7 : 1.
The actual thickening depth must be

$$h_{1,T} = \sqrt[3]{\frac{V_1 \cdot 3 \cdot m^2}{\pi}} \qquad [m]$$

in order to ensure the volume V_1 of the thickening zone in the tip of the funnel, which is required as a function of the desired effect.

exchange, break point chlorination, and as
phosphorus elimination (nutrient for algae) by way of
microbial P-check, algae-P-elimination, simultaneous precipitation, separate chemical precipitation, selective ion exchange.
- <u>elimination of dissolved inorganic substances</u> (salts):
ion exchange, electrodialysis, hyperfiltration, distillation (flash evaporation).

6.5.1 Removal of Suspended Matter

The sedimentary solids are usually removed from the waste water by sedimentation. Simultaneously, nonsedimentary matter is adsorbed to medium and large size biological or chemical flakes and thus separated from the waste water.

The sedimentation normally takes place in sedimentation basins. These conventional clarification basins can be improved by various installations (e.g. tube contact filters or settling tubes, plates, also termed laminar separators, algae filter layer) and by microstrainers, quick sand filters, flotation, ultrafiltration, or hyperfiltration.

6.5.1.1 Installations in the Sedimentation Containment

Equipment installed in the sedimentation containment allows for an improved separation or an increased charging at the same efficiency. Examples of such installations are laminar separators, tube contact filters, algae filter layers, etc.

<u>Laminar separators</u> are mounted at a spacing of 25 to 50 mm (see Fig. 6.5.-2). The waste water passes through the oblique plates from the top downward. Sludge and waste water are removed separately at the bottom. The laminar separators are recommended for municipal waste waters only when flocculating agents (such as lime for phosphate precipitation) are employed. Fig. 6.5.-2 is a functional section through a tube contact filter or settling tube for installation in a sedimentation basin /67/.

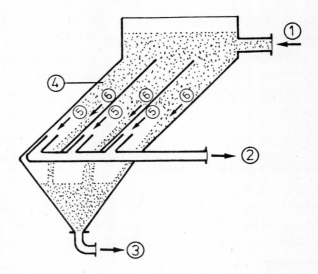

Fig. 6.5.-2: Functional section through a laminar separator /67/
1 - inflow; 2 - water; 3 - sludge; 4 - laminar zone; 5 - freshwater discharge; 6 - sludge collector funnel

Tube contact filters (settling tubes). Settling tubes are mounted in the outflow zone of rectangular horizontal-flow final clarification basins. The square plastic tubes (ca. 5x5 cm and 60 cm long) are mounted in the form of a screen (Fig. 6.5.-3) with a slope of 5° to 60° relative to the horizontal.

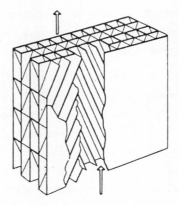

Fig. 6.5.-3: Functional section through a plastic settling tube

The thus separated sludge falls down to the bottom where it is collected by a chain remover and transported toward the sludge funnel through which it is carried off. Fig. 6.5.-4 illustrates a final clarification basin provided with a settling tube arrangement in the outflow zone /67/.

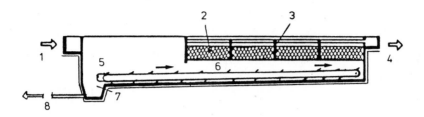

Fig. 6.5.-4: Final clarification basin with settling tube arrangement in the outflow zone /67/

1 - inflow; 2 - settling tube arrangement; 3 - baffles; 4 - outflow; 5 - sludge separation zone; 6 - chain remover; 7 - sludge funnel; 8 - sludge discharge

SLECHTA and CONLEY recommend the following values as operating parameters:

surface charge	1 - 5 $m^3/(m^2 \cdot h)$
surface load	150 - 400 $kg/(m^2 \cdot d)$
sludge concentration	1 - 5 g/l
sludge volume index	35 - 135 ml/g

6.5.1.2 Microstraining

The microstrainer or microsieve is a slowly rotating drum. It is a cage-type supporting structure covered with a very fine sieve cloth (Fig. 6.5.-5). 23 μm is the mesh size presently used. At a drum diameter of 3 m, this mesh corresponds to a number of revolutions of 4 - 5 rpm or a peripheral speed up to ca. 30 m/min.

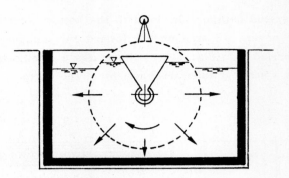

Fig. 6.5.-5: Section through a microstrainer

The following equation may be taken as a basis in microstrainer dimensioning:

$$H = \frac{m \cdot Q \cdot C_1}{A} \cdot e^{n \cdot Q \cdot I/S} \qquad [m]$$

where

- H = water level difference
- Q = filtered water quantity per unit time, in m^3/min
- C_1 = initial cloth resistance, related to the through-flow rate, in $m/(m \cdot h)$
- A = effective submersed strainer surface, in m^2
- I = strainability index, in m^{-1}
- S = strainer surface submersing per unit time, in m^2/min
- m = 0.544, in min/m^2
- n = 0.856
- I = strainability index

The explicit through-flow quantity Q cannot be calculated from this equation. However, since the waste water volume arising is, as a rule, a predetermined design parameter, the resulting water level difference H can be predicted from the previously determined strainability index I, after the selection of a standard strainer drum with suitable A and S values.

A simple "filtrameter" has been developed to determine strainability. This meter allows for a direct conclusion of the index I from the through-flow time T of a predetermined raw water quantity V through a calibration curve (see Fig. 6.2.-7).

6.5.1.3 Filtration

The filtration technique may be employed in various elimination processes. It precedes the remaining treatment stages.

Filters are predominantly important when discharges from conventional sewage plants are introduced into waters from which drinking water is derived. The removal of suspended matter not only improves the visible properties of the discharge, e.g. appearance, turbidity, etc., but also reduces the oxygen demand or oxygen consumption. Sewage plants for which BOD_5 discharge values below 10 - 15 mg/l are demanded cannot be operated without a further "mechanical treatment" stage.

Filtration techniques can be distinguished by whether the filtering effect occurs at the surface, in one plane, or with three-dimensional distribution through the depth of the filter bed. Whereas microstrainers belong into the first group, simple sand filters display a slight three-dimensional effect. Genuine three-dimensional filters are the so-called multilayer filters where the larger part of the filter layer is composed of coarser-grain material sometimes in graded layers (e.g. Hydroanthrazit 1.6 - 2.5 mm), while a thinner layer of finer material (e.g. quartz sand 0.7 - 1.2 mm) is provided at the filter end. Since the three-dimensional or deep filtration is distinctly superior to the other methods owing to its considerably longer service life and often better effects, this technique will be discussed mainly.

The removal of a suspended particle may be subdivided into
1. the transportation, which, in its turn, depends on hydraulic and physical factors, and
2. the accumulation, which is a purely chemicophysical process.

A survey of the mechanisms and process parameters in waste water filtration is given below.

The term transportation of matter may apply to all those processes that take place when solid particles are conveyed to the filter material. When flow processes are involved to deliver the solids, a laminar flow is normally assumed to exist in the filter medium.

In these processes, particles may be contacted with solid surfaces at the points where the flow lines are reflected at a filter grain.

Another possibility is the transportation by the solids' inherent mobility. In such a process, the particles carried in the water, undergo a downward acceleration (sedimentation), which depends on their density relative to the water. This effect depends on the density of the flakes and becomes relevant only with particle sizes above 2 µm. Other effects are also possible, but reference is made here to the original literature.

In contrast to accumulation, the pure straining effect of a three-dimensional filter is low. A straining effect can be achieved only in the following manner:

1. The particles are larger than the pore diameter, which is hardly the case with modern three-dimensional filters.
2. The particle concentration is so high that the particles mutually support each other when passing through a pore (silo effect). In such a case, the filter should be complemented by a preceding sedimentation stage, since, with a high particle concentration, filtration is usually not economical.
3. Small particles are caught in the edges between the filter grains without accumulation to the grains. They are clamped, so to speak. This effect is hardly relevant.

The first theoretical model for the filter flow goes back to DARCY. In his tests with a vertical-flow filter, he found the following relationship

$$Q = F \cdot k_f \cdot J \qquad [m^3/h]$$

where

Q = quantity of waste water charged per unit time, in m^3/h
F = cross-sectional area of the filter, in m^2
k_f = permeability coefficient, in m/h
J = pressure gradient in the filter, in parts per thousand

Since the permeability coefficient k_f depends on both the respective liquid and the filter bed used, the specific permeability k_o is introduced, which is influenced merely by the filter bed.

$$k_o = \frac{K \cdot g}{\nu} \qquad [m^3/s]$$

where

K = the specific permeability, in m^2

ν = the kinematic toughness, in m^2/s

The permeability coefficient k_o in m/s indicates the particular volumetric flow (in m^3/s) that flows through an area F of 1 m^2 of the porous medium when the vertical tube level slope J normal to this area A equals 1.

The specific permeability K (m^2) divided by ν (m^2/s) of the flowing medium, indicates the particular quantitative volume Q_p (kg/s) that flows through an area of 1 m^2 of the porous medium when the pressure differential equals 1.

Strictly speaking, the validity of DARCY's law is restricted only to the purely laminar flow zone. The permeability coefficient in this zone is constant, so that a linear interdependence between the filter rate and the pressure loss arises.

6.5.1.3.1 Sand Filters

Section 6.2.1.3 contains a survey of the techniques used in waste water treatment engineering. A fundamental distinction is made between accumulative and dry filtration with their different modifications.

The actual advantage of the dry filters, i.e. to allow processes of biological degradation apart from the filtration effects, can be controlled by an additional filter flow with air or gaseous oxygen. The introduced oxygen allows for aerobic biological processes throughout the filter.

A waste water filter may be charged either in an open system or in a closed pressurized system. The illustration indicates the possible applications of the open or closed technique for upward and downward flow operation. With the open downward filtration system, the throughflow quantity is practically always limited by the filter resistance. In the course of filter operation, this resistance increases and causes an accumulation of untreated water above the surface of the filter bed. The possible accumulation height is usually 1 m at maximum.

The pressure filtration permits higher charging rates and a longer filter operation. Pressurized filters are normally operated with a maximum charging pressure of some meters water column.

Upward filtration is also possible in a pressurized form. This effect is achieved by the arrangement of a grid a little below the surface of the filter bed. This grid or screen is permeable for the filter medium, while the filter material is retained. In the open technique, this material is pressed from the bottom upward with a corresponding filter resistance.

Filter beds with granular material are purged by an upward flow at a rate increased to a multiple of the filtration rate. During the purging operation, three processes are overlapping. First, the filter bed is loosened, then the covered individual grains can be cleaned by rubbing them against each other, and finally the contaminating particles removed from the filter material are carried off.

Filter purging, also termed back-flow purging of the filter, is carried out either as a process using water or a combined operation using air and water.

A satisfactory filter bed purification demands a minimum purging rate. This rate depends on the density of the filter material, on the grain size, and on the grain distribution throughout the filter bed. However, it does not depend on the filling density of the filter material and on the height of the filter bed. Extensive test series have furnished the following efficient purging speeds for quartz sand of various grain diameters:

$$0.5 - 1.0 \text{ mm}; \quad v_{RS} = 45 \text{ m/h}$$
$$1.0 - 2.0 \text{ mm}; \quad v_{RS} = 70 \text{ m/h}$$
$$1.5 - 2.0 \text{ mm}; \quad v_{RS} = 94 \text{ m/h}$$
$$2.0 - 3.0 \text{ mm}; \quad v_{RS} = 100 \text{ m/h}$$

The combination of air scavenging and water purging provides for a more intensive rubbing and cleaning of the filter grains and thus a purging water quantity lower than that used for water purging alone.

Filter layers (Fig. 6.5.-6) consisting of filamentous algae are used to remove biological flake residues in the outflow zone of the final clarification basins. To this end, two nettings are mounted 100 or 200 mm below the water surface. Between these nettings, a dense coherent layer of algae and deposited sludge is formed, which acts like a degradation filter. The mesh of the nettings is ca. 10 mm (upper) or 1 mm (lower). The hydraulic load should be ca. 0.5 m^3/m^2 up to 2 m^3/m^2 (peak load 5 m^3/m^2). The filter is normally purged when the pressure loss is 200 mm (HELMER/SEKOULOV /67/).

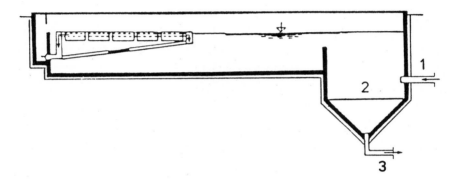

Fig. 6.5.-6: Schematic of an algae filter layer technique /67/
1 - inflow; 2 - sludge zone; 3 - returned sludge (ultra-filtration)

6.5.1.3.2 Land Filtration

The land filtration of waste water into the subsoil must also be considered to be a step of further thorough purification. Projects dealing with these problems are presently under development in many countries. A significant effect of this technique is the replenishment of the groundwater.

Good preliminary purification of the waste water is an important aspect in this technique, since it is a means to prevent difficulties during the infiltration phase (clogging, etc.).

The starting point for all further measures is a reservoir into which the waste water is pumped from a sewage plant. This reservoir may be considered functionally as a polishing pond. The oxygen demand of such ponds is met by diffusion through the pond surface and, to an even greater extent, by the biogenous oxygen production when biologically purified waste water with a certain residual BOD is used. One portion of the nutrients P and N may be included into the biomass and thus possibly be removed from the waste water. When the water is kept in these ponds for 2 days or less, excessive algae growth with its specific problems does not occur. Experience gathered in the practical operation of polishing ponds has shown that less than 5 mg/l of suspended matter are achieved, so that this treatment stage must be considered very important in the final clarification of waste water. The water should not be removed from the reservoir directly at the surface, since the algae concentration may be higher there.

Additional treatment should provide for possibilities of continued biological degradation apart from the elimination of residual suspended matter. This may be achieved by means of a biological disk or rotary biological filter with subsequent quick filtration.

6.5.1.3.3 Ultrafiltration and Hyperfiltration

The ultrafiltration technique is based on the separating effect of a semipermeable membrane through which the waste water is pressed under pressure. Suspended and colloidal matter cannot pass through the membrane and is thus retained, whereas dissolved substances are not eliminated by this filter. The required pressure is in the range of ca. 10 to 20 m water column.

The hyperfiltration technique (reverse osmosis) is based on the same principles. In this method, the membrane is practically impermeable even to molecules, except for water molecules. The required pressure is higher than 60 m water column. This method is presently not used in the treatment of municipal waste waters in view of the high costs (see Section 6.5.4.3).

The permeation of substances through the membranes is suited for water of roughly 10^{-7} g/(cm²·s). Therefore, large areas are required to achieve economical flow rates. Ultra-filtration membranes of aliphatic polyamides may be used up to 60° C.

A general treatment of industrial waste waters by means of membrane separation is not possible. The reasons lie in the operating conditions of the waste waters and in the effects of the individual substances on the membranes or their concentrations. The permeated matter may be discharged into a receiving water or used for industrial purposes, whereas the pollutant-carrying concentrate undergoes a conventional treatment, for instance, aiming at the achievement of depositable material, or the concentrate is incinerated. In other cases, the concentrate may be treated to recover valuable substances. The membrane separation may also be used to recycle the permeated matter or the concentrate. Following an intermediate separation step, however, the recovery of both volumetric flows is also possible.

Ultrafiltration is a technique used to retain organic substances when waste waters from the food industry are treated. Table 6.5.-1 shows the BOD_5 retention in waste waters from the food industry according to STAUDE /67/.

Tab. 6.5.-1: BOD_5 values in waste waters from the food industry, and retention (in %) after the membrane separation

Waste water	BOD_5 (mg O_2/l)	R(%)
carbohydrates	18 250	99.7
potato starch	12 700	85.1
vegetable proteins	21 500	99.6
cheese whey	45 000	99.3

The recovery of valuable substances is much more sensible. For example, the skimmed milk is concentrated by ultrafiltration, while cheese formation is so controlled that the total protein contents are precipitated /STAUDE in 67/. The ultrafiltration technique is also used industrially to crack oil emulsions.

The ultrafiltration technique allows for a concentration of drilling oil emulsions up to an oil concentration of 50 %. A further concentration is not advisable, since, owing to the permeation of the emulsifiers through the membrane, the emulsions are cracked while an oil film blocks the membranes. The thermal value of a thus produced oil concentrate is ca. 715 Joule/kg.

New installations for electrosubmersion painting are almost always equipped with ultrafiltration units. The ultrafiltration provisions separate a paint-free rinsing solution from the submersion painting bath solution. This rinsing solution is then used to rinse the painted pieces, while the rinsing water loaded with paint not required is recirculated into the dip-painting bath.

6.5.2 Elimination of Dissolved Organic Substances

Two types of dissolved organic substances can be distinguished by their biochemical properties:
- the substances accessible to biological degradation, and
- the refractive substances or substances only slightly or not at all accessible to microbial attack.

The substances accessible to biological degradation can be mineralized by means of biological processes such as trickle filter or activation techniques. The usual measure to describe their elimination is the BOD_5. The refractive organic substances can be characterized only by means of the COD and the contents of organic carbon TOC.

These summarizing parameters are predominantly suitable to characterize domestic (municipal) waste waters while other parameters are additionally required to describe the industrial waste waters.

Two alternatives must be distinguished for the application of the physicochemical processes of elimination of organic substances:
- the application joins a mechanicobiological treatment stage. This

waste water (BOD_5 from 5 to 20-25 mg/l) contains nearly exclusively refractive substances (COD from ca. 40 to 60 mg/l).
- The application of adsorption, e.g. as a second stage in a chemical flocculation process. In this step, the organic substances are removed only to an extent that they can be flocculated and separated in the form of suspended and partly also colloidal particles. The degradable substances are reduced by ca. 60 % while the refractive substances are reduced by ca. 30 %, i.e. much less than in the first case.

The only method presently used is the activated charcoal adsorption. Other processes such as the chemical oxidation by means of ozone or by way of reverse osmosis are applied too, but are, so far, only rarely used on a large scale.

6.5.2.1 Adsorption by Activated Charcoal

Activated charcoal is produced from organic raw materials such as wood or bones, by dehydration, carbonization, and activation at a high temperature under vacuum conditions. During the activation, fine pore cavities and channels are produced so that the activated charcoal is given a very large internal surface. The specific surface of commercial charcoals for waste water treatment ranges from ca. 600 to 1200 m^2/g.

When the waste water is contacted with the activated charcoal, the substances are retained at the adsorption surfaces in different concentrations. Biological substances are adsorbed as well, so that a microbial degradation can develop in the pore cavities of the charcoal grains. This effect, on the one hand, reduces the sorption areas available, while, on the other hand, organic matter contained in the waste water can be additionally eliminated. As a result of these two phenomena, a saturation charge is therefore determined in activated-charcoal columns that is by far higher than the merely physical adsorptive capacity of the carbon.

Moreover, fine suspended particles and colloids are deposited on the charcoal grains.

When activated charcoal is used in the final clarification of waste waters after a mechanical-biological treatment, most of the organic substances are eliminated.

When the adsorption of activated charcoal is used as the second stage in a physicochemical purification process, comparatively large quantities of degradable and refractive organic substances must be eliminated. This application creates anaerobic conditions combined with the formation of hydrogen sulfide. Only the addition of nitrate into the inflowing waste waters is a suitable means to prevent this effectively.

In the course of operation, the adsorption capacity of the activated charcoal decreases, so that its sorptive capacity needs regeneration beyond a limit defined by the economics. In practical use, the regeneration of the sorptive capacity can be accomplished only by thermal regeneration processes in which the charcoal is treated in multifloor, (mostly in) drum-type or fluidized-bed furnaces at temperatures between 750 and 950° C.

The activated charcoal is used in two different ways:
- dosed addition of the charcoal in the form of a powder, and
- contact columns with activated charcoal in the form of a granular substance.

The <u>powdered</u> activated charcoal has a grain size between ca. 10 and 50 µm. It permits quick diffusion and adsorption of substances dissolved in the waste water. It is added, in a mixing step, into the flocculation stage, the sedimentation stage, and possibly filtering stage (in rare cases). The purification of waste waters discharged from a mechanical-biological treatment demands dosed quantities of 100 to 200 mg/l of activated charcoal, which results in very high cost. Another 15 to 20 % loss arises in the regeneration step, so that this method is seldom used.

The <u>granular</u> activated charcoal has a mean grain diameter of 0.8 to 2.0 mm, so that it can be used as a packing material in contact columns resembling sand filtration. The activated charcoal has a specific surface area of 600 to 1000 m^2/g. It can be regenerated in situ with continuous reuse in a cycle. Fig. 6.5.-7 illustrates a schematic of the reactivation of loaded activated charcoal in a regeneration cycle according to the SWINDELL-DRESSLER COMPANY /67/.

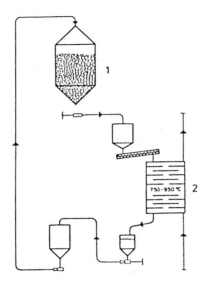

Fig. 6.5.-7: Schematic of the reactivation of loaded activated charcoal in a regeneration cycle /67/
1 - activated carbon column; 2 - regeneration furnace

Various methods are used to operate contact columns. Examples are stationary bed adsorbers, sliding bed adsorbers and suspension bed adsorbers.

In the stationary bed adsorbers, the activated charcoal is packed into the contact column. The waste water flows through this packing. The system structure and operation are simple. Since the system operation must be stopped for the regeneration and back-purging times, two or more columns are necessary.

In sliding bed adsorbers, the activated charcoal continuously moves down inside the column, it is removed at the bottom and supplemented with regenerated carbon at the top. The waste water flows upward through the packing in countercurrent at a rate of 14.5 m/h, and is discharged at the top. The contact time is ca. 30 min. The advantage of this technique lies in the constantly high purification level of the discharge, in the full utilization of the adsorption capacity, and in the possibility of system automation. These advantages are offset by the considerable investment necessary.

Fig. 6.5.-8 is a section through the sliding-bed adsorber in the experimental Lake Tahoe installation /67/.

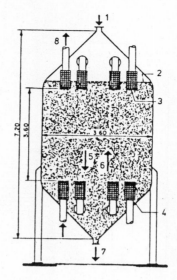

Fig. 6.5.-8: Section through the sliding bed adsorber of the Lake Tahoe sewage plant /67/

1 - charcoal inlet; 2 - charcoal packing; 3 - discharge filter; 4 - inflow filter; 5 - charcoal; 6 - water; 7 - charcoal outlet; 8 - discharge

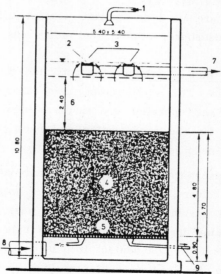

Fig. 6.5.-9: Fluidized-bed adsorber /67/

1 - charcoal addition; 2 - discharge screen; 3 - discharge collector; 4 - water; 5 - perforated filter bottom; 6 - maximum expansion; 7 - discharge; 8 - inflow; 9 - removal of activated charcoal

The <u>suspension bed adsorber</u> (fluidized-bed adsorber) is charged from the bottom with an inflow rate so high that the granular filling is constantly kept in suspension.

Fig. 6.5.-9 illustrates the section through a suspension bed adsorber which was designed for the Vallejo sewage plant /67/.

The techniques using activated charcoal provide for elimination rates of ca. 70 to 80 %, a BOD_5 reduction by more than 60 %, the achievement of 10 to 15 mg COD/l in the discharged outflow and 4 to 6 mg TOC/l in waste waters after a preliminary mechanical-biological treatment.

The experience gained in the operation of American systems /67/ can be used to indicate the following specific quantities of activated charcoal required:
- for the outflow from a mechanical-biological sewage plant with lime packing and multilayer filtration
    25 - 50 g/m^3 of waste water
- for the outflow from a mechanical-biological sewage plant after sand filtration
    50 - 70 g/m^3 of waste water
- for the outflow from the chemical flocculation stage
    180 - 220 g/m^3 of waste water

6.5.2.2 Oxidation by Ozone

Ozone has many different effects on the waste water according to MAJUMDAR and SPROUL /67/:
- decolorization and improvement of odor and flavor,
- reduction of turbidity, of the contents of suspended matter, BOD_5, and COD,
- break-up of detergents and other substances that are not at all or hardly accessible to biological decomposition,
- disinfecting effects on bacteria and viruses.

Since only part of the organic matter contained in the waste water is oxidized, its elimination requires a further step.

Relatively expensive equipment is necessary to produce ozone. It is therefore common to treat domestic (municipal) waste waters without ozone.

Fig. 6.5.-10 is a schematic of an ozonization system according to MEINCK /107/.

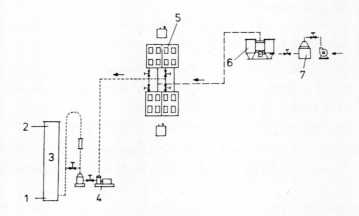

Fig. 6.5.-10: Schematic of an ozonization system /107/
1 - waste water inflow; 2 - waste water outflow; 3 - reactor tower; 4 - air compressor; 5 - ozonization system; 6 - air dryer; 7 - cooler

6.5.2.3 Desorption of Volatile Substances

The term desorption applies to the expelling of a gas from a liquid below the boiling point.

Organic substances dissolved in the waste water can be desorbed (stripped) from the aqueous phase provided that they have an appropriate volatility. This problem is generally not involved in municipal waste waters.

Fig. 6.5.-11 illustrates a stripping tower for 57,000 m³ of waste water per day for tower desorption of ammonia (see also Section 6.5.3.1.2).

6.5.2.4 Reverse Osmosis (Hyperfiltration)

The reverse osmosis (see also Section 6.5.1.3.3) is presently under trial as a new method of separating organic substances. Owing to the low throughput of the systems, however, its applicability to domestic (municipal) waste waters is doubtful. It is already in practical operation, however, to treat filtration products discharged from the thermal sludge conditioning stage. This technique provides for a possible re-

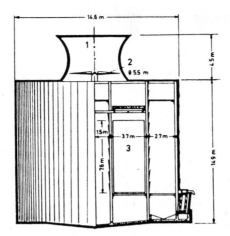

Fig. 6.5.-11: Stripping tower for ammonia desorption from the waste water, in Orange County (USA)
1 - exhaust air downdraft; 2 - fan; 3 - contact filter packing

duction of the $KMnO_4$ consumption from 3,000 to 3,500 mg/l down to 600 to 1200 mg/l.

In hyperfiltration, the integral and the asymmetrical cellulose acetate membranes have turned out to be best (STAUDE /67/). Depending on the shape of the equipment where they are applied, the membranes are produced as flat or tubular structures of different diameters, up to hollow fibers. Their different forms must be suited for application under pressures as high as $100 \cdot 10^5$ Pa. A certain reduction of their permeability (compaction) due to the reduction of water in the membrane with an increasing pressure is unavoidable. The more the intermolecular cross linkage increases due to the rising temperature, the more the permeability of cellulose acetate membranes is reduced as a result of the formation of hydrogen links. Withstanding the pressure, however, requires pressure-resistant elements. This complex unit including the membrane is called a module /67/.

6.5.3 Elimination of Nutrients

An increase of uncontrolled discharge of waste waters into the natural waters results in an extraordinary growth of water plants. The excessive fertilization is termed eutrophication (formerly also called "the blooming" of waters). The word "eutrophication" derives from the two Greek words "eu" (good) and "trophos" (nourishment), which means that the excessive enrichment of a eutrophicating water with nutrients results in a correspondingly excessive production of biomass. The constant growth and loss of algae and other water plants has a negative effect on the water quality (e.g. as a result of excessive oxygen consumption), on the treatability of the water, e.g. for drinking water supply, and on the environment due to nauseating odor.

Numerous research projects have lead to the result that unhindered growth of vegetation is possible only when an optimum balance is achieved between all the cell components (C, O, H, N, P, S, trace elements), the climatic conditions (intensity of the light, temperature, wind), as well as other natural environmental factors (catchment area, erosion, depth, currents, riparian zone, inflows into and outflows from the lake).

All these elements must be considered in the struggle against the eutrophication factors. Since, however, measures against the majority of the elements in nature are not possible, the efforts are generally concentrated on the restriction of the substances supplied by human activity. It has been found that a limitation of growth is mainly determined by the available concentrations of the elements C, N, and P, and sometimes of iron as well, and that the restriction of P and N is a good solution. According to WUHRMANN /67/, 0.01 to 0.05 mg/l of phosphorus and 0.1 to 2.0 mg/l of nitrogen are considered to be concentration limits beyond which excessive algae growth commences.

Nitrogen is introduced into the natural waters through agricultural discharges (agriculture and livestock breeding), domestic (municipal) and industrial waste waters. According to HELMER, SEKOULOV /67/,

6.5.3.2.2 Phosphorus Elimination by Physicochemical Processes

The physicochemical processes of phosphorus elimination have been used as an obligatory third purification stage in Sweden, Switzerland, the United States, and other countries for a long time.

Phosphorus contents as high as 40 % or even more occur in domestic waste waters in the form of inorganic polymolecular phosphates such as pyrophosphates, tripolyphosphates, trimetaphosphates which mainly result from detergents. Small quantities, ca. 10 %, are present in the form of organic phosphates and are adsorbed to and co-eliminated with the flakes formed by chemical precipitation. The adsorbed material occasionally results in a detrimental effect on the sludge index and thus in sedimentation or separation problems.

Iron (II), iron (III), aluminum, and calcium ions are used as precipitating agents.

6.5.3.2.2.1 Iron Phosphate Precipitation

Divalent iron Fe^{2+} forms flakes that cause problems in sedimentation. The precipitation properties are improved by treatment with gaseous chlorine (Cl_2), or by aeration, with oxidation to trivalent iron Fe^{3+}. The trivalent iron Fe^{3+}, used in the form of iron chloride ($FeCl_3$) or iron (III) sulfate $Fe(SO_4)_3$, forms slightly soluble iron phosphate $FePO_4$, which improves flocculation and the sedimentation properties, while polyphosphates and organic phosphorus are eliminated simultaneously.

The waste water alkalinity is important in the phosphate precipitation with iron and aluminum salts. An insufficient alkalinity during the precipitation reaction results in P or Al elimination far below the optimum.

6.5.3.2.2.2 Aluminum Phosphate Precipitation

The aluminum ion Al^{3+} forms well sedimenting flakes. It is therefore used mostly in the form of aluminum sulfate $Al_2(SO_4)_3 \cdot 18\ H_2O$ for the phosphate precipitation.

The mole relationship between Al and P is 2 : 1. In practice, ratios between 3 : 1 and 1.5 : 1 are necessary to eliminate the phosphate completely.

6.5.3.2.2.3 Calcium Phosphate Precipitation

Owing to its low price, calcium or lime is the most widespread precipitating agent in waste water purification among all the agents used for phosphorus elimination.

The calcium phosphate precipitation begins with a process of softening the waste water, forming a calcium carbonate precipitate. Only after the formation of 60 to 80 % of the calcium carbonate does the calcium phosphate precipitation commence. It is thus necessary to determine first of all the relationship between lime and alkalinity (e.g. according to WUHRMANN Ca : HCO_3 = 1 : 1.5).

The calcium precipitation is carried out in the United States at pH levels of 11, with coprecipitation of the magnesium also present in the waste water.

After the calcium precipitation, the turbidity of the discharge is only very low. At the same time, the waste water is also softened, which is excellent for reuse.

The phosphate precipitation is recommended for the mechanical or the combined mechanical-biological sewage plants.

6.5.3.2.2.4 Methods of Treatment

Phosphate precipitation is used as the third purification stage in the conventional mechanical-biological sewage plants. Three techniques can be distinguished by the type of dosing location: preliminary precipitation, simultaneous precipitation, and subsequent precipitation.

In <u>preliminary precipitation</u>, the precipitating agent is added upstream of the sedimentation basin. The preliminary precipitation influences the

composition of the waste water, so that the discharge contains phosphorus concentrations as low as 1 to 3 mg P/l. In such an arrangement, it is recommended that the chemical dosing is automatically controlled and the pH value is monitored.
In simultaneous precipitation, the precipitating agents are added directly into or immediately upstream or downstream of the activation basin. This treatment process also serves to precipitate the phosphate, which is then removed together with the excess sludge.
Iron chloride ($FeCl_3$) is used nearly exclusively as the precipitating agent in practice. Common doses are in the range of 10 to 20 mg Fe/l. Two dosing quantities (one for the day and one for the night) are determined empirically, so that excessive doses are avoided that could lead to the inhibition of the biological processes. From an engineering viewpoint, this process is the easiest to realize and can be economically used in the existing sewage plants.

The subsequent precipitation stage is an independent treatment unit installed downstream of the final clarification basins to reduce the phosphate concentration to the required limit (e.g. 1.0 mg P/l). This technique makes use of chemical agents to remove also other substances occurring in the waste water. Aluminum salts are preferred in subsequent precipitation.
The technique of subsequent precipitation requires reactor basins in addition to dosing and mixing units. Compact systems are frequently installed for this purpose, which include integral flocculation and sedimentation basins (such as the Centrifloc Clarifier, the Accentrifloc Clarifier of the PCI company).

6.5.4 Elimination of Dissolved Inorganic Substances

Even though, as a result of the common water usage, dissolved inorganic substances are introduced into the waste water beyond their prevailing natural extent, such substances are discharged into the sewage plants mainly from industrial enterprises.
Owing to the eutrophicating effects of some of these substances (e.g. P, N), on the one hand, and as a consequence of technical and economic reasons, on the other hand, these inorganic substances must be retained.

Simple methods of demineralizing water have been known for a long time, such as distillation, calcium soda processes, etc. In waste water treatment, however, only methods on an industrial level are considered. The ion exchange methods, the electrodialysis, and the hyperfiltration should be discussed here as examples.

6.5.4.1 Ion Exchange Methods

The salts occurring in waste water in a dissociated form may be retained and thus removed from the waste water by the exchange for hydrogen (cations) and hydrdoxyl (anion) ions. The ion exchangers remove the corresponding minerals from the waste water to be treated. The exchange capacity of the ion exchanger decreases with an increasing service life, until regeneration becomes necessary.
There are two basic groups of ion exchangers, natural (zeolites) and synthetic (resin) exchangers.

The appropriate resins must be selected for an efficient waste water treatment, i.e. resins that are suited to retain the substances from the waste water.
Industry uses the ion exchangers very frequently for both waste water treatment and particularly for the retention of various valuable materials from industrial processes. Ion exchangers are normally not used in the treatment of municipal waste waters. The reasons for this are the high costs incurred by the treatment and the regeneration, and the storage of highly concentrated regenerated material, which may occur in quantities as high as 10 to 15 % of the discharge.
The containers for ion exchangers are cylinders filled with the appropriate granular exchanger material instead of sand, in a manner comparable to a closed pressure filter for drinking or industrial water conditioning.

6.5.4.2 Electrodialysis

This method is discussed in Section 6.3.1.10.8.
At present, it is used to demineralize brackish water for the industrial water supply. This method, however, has so far not been considered appropriate in practical municipal waste water treatment.

6.5.4.3 Hyperfiltration

In reverse osmosis, a pressure P higher than the osmotic pressure is exerted on the solution having the higher concentration. The result is a counterflow or negative current (from the high-concentration solution toward the pure water).

The current that can be achieved with such an osmosis process is as follows:

$$F = K (P - \Delta\pi)$$

where

F = flow rate, in $1/(m^2 \cdot h)$
K = the membrane-specific efficiency, in $1/(m^2 \cdot h \cdot 10^5 \text{ Pa})$
P = the corresponding pressures, in 10^5 Pa

Fig. 6.5.-14 illustrates the schematics of the osmotic currents.

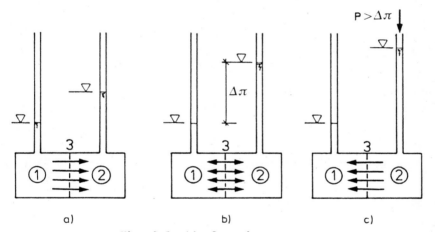

Fig. 6.5.-14: Osmotic currents

a - osmosis; b - osmotic equilibrium; c - reverse osmosis (hyperfiltration); 1 - pure water; 2 - salt solution; 3 - semipermeable membrane

BAYLEY and WAGGOTT /67/ used the hyperfiltration techniques to achieve the values given in Table 6.5.-3 in waste water purification.

Tab. 6.5.-3: Hyperfiltration test results with municipal waste water /67/

concentration mg/l	inflow $Q_z = 100\%$	concentrate $Q_k = 25\%$	product $Q_p = 75\%$
evaporation residue	751	1927	151
suspended matter	3	2.5	9
COD	22	80	4
organic carbon	11.5	26	3
BOD_5	2	8	2

Since the concentrate needs further processing this technique must be deemed uneconomic for the treatment of municipal waste waters.

Industry uses hyperfiltration techniques for various treatments including:
- the retention of mercury salts (up to the 50 % concentration, the mercury retention was 99 %),
- for the concentration of heavy-metal ions from rinsing waters discharged from galvanizing operations,
- for the retention of the silver complexes in waste waters from photographic operations,
- for the selective separation of inorganic compounds such as nitrates
- for the separation of organic macromolecules.

6.6 Treatment of Commercial and Industrial Waste Waters

Various methods and techniques are used to remove commercial and industrial waste waters:
- introduction into the public natural waters,
- connection to municipal drainage networks,
- seepage,
- agricultural utilization,
- elimination of the waste water.

All the aforementioned techniques, except elimination, demand that the waste waters be treated to an extent that they comply with the different requirements applicable in each case (cf. Chapter 5 and Section 6.4.3.2).

The selection of the respective method of treating commercial and industrial waste waters in suitable large-scale plants depends on the composition of the waste waters, on the magnitude of the discharge quantities, on the purification level to be achieved, on the possible reuse of the waste water, and on the residual substances in the waste waters.

A considerable part of the various types of waste waters, however, can be sufficiently purified when the usual methods are applied, which are common to treat domestic (municipal) waste waters. There are various other types of waste waters, e.g. those discharged from the chemical industry, which require a specific chemical or electrolytic treatment to retain the pollutants.

Chapter 2 discusses the formation and the quantities of commercial and industrial waste waters.

6.6.1 Inorganic Industrial Waste Water

6.6.1.1 Waste Waters from Mortar, Sandy Limestone, Cement and Porcelain Production Plants

When the lime is slaked in mortar-producing plants, alkaline waste waters are produced. They contain mainly unbound and bound lime with a pH of ca. 10. The required waste water treatment is restricted to neutralization, to the introduction of flue gases when necessary, and to the final sedimentation in sufficiently sized sedimentation basins.

The waste waters from sandy limestone producing plants are composed of oil-carrying cooling waters and condensation products from the hardening vessels. These waste waters are produced in uniform quantities and are characterized by high temperatures and strongly alkaline properties (pH higher than 10). They contain undissolved and dissolved constituents in quantities higher than 1 g/l, apart from organic substances, and occasionally oil in quantities exceeding 10 mg/l. The purification includes an extensive cooling with a subsequent thorough deoiling step, and final sedimentation in sedimentation basins.

Cement-manufacturing plants produce waste waters predominantly when the dust formed in the individual steps of the operation is removed in a wet condition.

It is common practice to discharge the waste waters to deads and seep them there. The disadvantage of this method is that the residual substances in these waste waters are washed out in rather strong rain so that they are introduced into the receiving reservoir. Therefore, large waste water collector basins must be provided.

Electric filters are also used to retain the dust so that the waste water quantities produced are decreased.

The porcelain and china-producing plants as well as the ceramics industry (e.g. crockery, earthenware, tile, and ceramic tube production plants) discharge both liquid and solid wastes.

- The waste waters often carry coarse-grain minerals and broken pieces so that a preliminary purification by screen systems and coarse-grain removers is required.
- Discarded faulty charges contain large quantities of easily sedimented substances, which can neither be pumped nor removed after a short settling or thickening time. The occurring hydroxide sludges can be kept in a condition suitable for pumping or removal only by the addition of considerable amounts of flocculating agents (iron and aluminum salts).
- Owing to the auxiliary agents used in operation, the charge condition of the turbidity-causing matter changes. This always requires an advance dose and preliminary reaction of flocculating agents (iron or aluminum salts) with ca. 5 to 10 minutes duration time in order to allow for a continuous operation with the same strongly anionic auxiliary flocculating agent.

These waste waters result in water turbidity in the receiving water and cause deposits on the bottom of the natural water. Mechanical and combined mechanical-chemical methods are used to clarify the waste water.

The methods of mechanical clarification include the sedimentation basins or also the waste water ponds suggested by LIEBMANN /94/, where the sedimentation rate ranges between 0.004 and 0.015 m/h (Fig. 6.6.-1).

Flocculating agents (such as lime, limewash, aluminum sulfate, or synthetic flocculating agents) are used in the mechanical-chemical processes. A preliminary clarification of the waste water is economical since

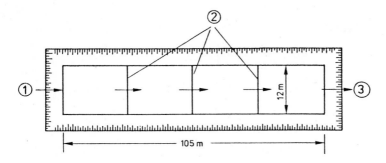

Fig. 6.6.-1: Waste water pond
1 - inflow; 2 - partitionings; 3 - outflow

the best operation of the flocculation process is achieved at suspended matter concentrations below 1,000 to 3,000 mg/l.

Fig. 6.6.-2 illustrates the schematic of a mechanical-chemical sewage plant for a ceramics manufacturing plant /94/.

Electric filters are used to remove the dust.

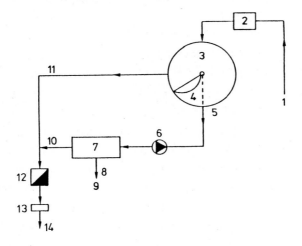

Fig. 6.6.-2: Schematic of a mechanical-chemical sewage plant for a ceramics producing plant /94/

1 - inflow; 2 - flocculating agent conditioning unit; 3 - sedimentation basin; 4 - scrapper; 5 - sludge containing 28 - 30 % suspended matter; 6 - pumps; 7 - filter press; 8 - wet residues; 9 - transport to the deads; 10 - filtration product; 11 - clear water; 12 - flowmeter; 13 - pH meter; 14 - to the receiving water

6.6.1.2 Waste Waters from Metal-Processing Industries

6.6.1.2.1 Waste Waters from Ironworks

It is now common to use waste waters from ironworks through recirculation, which has turned out to be an efficient operation. Two main cycles can be distinguished in ironworks:
- the blast furnace-cooling cycle including the cooling system for the fan equipment, and
- the gas cleaning cycle.

The waste waters produced by the ironworks and by the flue gas purging operation contain - apart from small quantities of cyanogen, sulfur, and phenols - ashes, iron oxide and silicic acid slag and ore particles, aluminum, carbon, as well as calcium and magnesium compounds. These waste waters contain high concentrations of chlorides and contaminants purged out of the flue gas as well, such as zinc compounds, cyanides, and ammonium in varying percentages. The recirculated water is treated by means of the following types of equipment:

- blowing out the carbonic acid by aeration upstream of the sedimentation basins. Calcium and zinc hydrogen carbonates are converted to carbonates in this step, and are then precipitated as solids. They are converted to well sedimenting sludges together with the other undissolved substances by the addition of flocculation agents (iron or aluminum salts) and auxiliary flocculating agents (organic polymers). Then these sludges are removed from the sedimentation basins in a continuous operation. The sludges are dehydrated in specific equipment whereupon they are deposited in landfills.
- cooling of the waste water after its chemical-mechanical treatment, in cooling towers, and reused in flue gas, preliminary, and secondary coolers.

The flue gas purging waters used in such recirculation cycles are purified just like the recirculated water (Fig. 6.6.-3, according to ZUR /185/).

Alkali cyanide, one of the contaminants, is formed in certain zones within the blast furnace from alkalis charged together with the coke, the ores and additives, the coke-bound carbon and the nitrogen in the furnace blast. Another possibility of alkali cyanide formation is the reduction of alkali carbonates, which are introduced into the lower furnace section by the charges sliding downward.

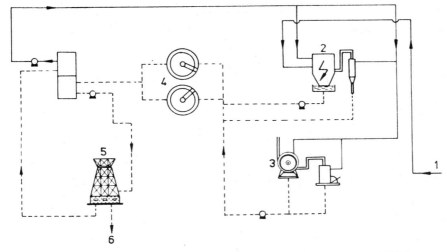

Fig. 6.6.-3: Purification in a recirculation system /185/
1 - inflow; 2 - electrogas purification; 3 - Theisen equipment; 4 - circular clarification basin; 5 - cooling towers; 6 - receiving water

Apart from the alkali cyanide, the flue gas contains hydrogen cyanide and dicyan. These cyanogen compounds are discharged together with the flue gas and are partly introduced into the flue gas purging water. The cyanogen concentration in the blast furnace increases, the closer the high-temperature zones are relative to the furnace throat. The production of foundry pig iron, hematite pig iron, ferrosilicon and ferromanganese is associated with an increasing formation of cyanides.

The cyanide concentrations are as high as several hundreds of milligrams CN^-/l of recirculated water. A conversion of the cyanide to harmless polymers by the addition of formaldehyde to the recirculated water in the loop under certain conditions (pH value set to 8 - 10, long duration times) is possible. The other treatment methods such as oxidation of the cyanide with chlorine or ozone, or the precipitation in the form of iron cyanide, are difficult and are not yet fully satisfying.

The furnace slags are granulated, which means that a vesicular sharp-edged slag sand is produced, which is used to produce Portland blast-furnace cement and metallurgical brick.

Slag dumps in ironworks occasionally furnish seepage waters with a strongly alkaline reduction and with a high concentration of soluble sulfides of calcium, potassium, or sodium. These waste waters should not be introduced into surface waters or into the groundwater. They must undergo an appropriate waste water treatment process.

6.6.1.2.2 Waste Waters from Steelworks and Rolling Mills

Waste waters from steelworks and rolling mills (converters), which contain oil and fat in addition to sinter, must not be introduced into the receiving water without being treated.

The coarse sinter is collected directly in sinter channels below the mill train. Fig. 6.6.-4 according to SIERP /158/ shows a BAMAG sinter collector. This equipment introduces and circulates the waste water at the outside wall. The sinter falls down, is collected on the bottom, and is removed. The treated waste water can be reused.

The fine roll sinter and the fine dust are collected in simple sedimentation basins. Submersed partitionings retain the oil and fat in the same sedimentation basins, and discharge the retained matter in collector channels or shafts. If necessary, chemicals such as aluminum sulfate, iron salts, lime, caustic soda solution, etc. are added or compressed air is possibly blown in. In special cases, further steps of purification are provided. For instance, small quantities of flocculating agents (e.g. 12 g iron chloride and 5 g 30 % sodium silicate per m^3) are added and diatomaceous earth filters, wood fiber, slag, coke or, for higher demands, even sand filters or suspension cartridge filters are used.

The settled viscous, greasy sludge, which is pasty due to oil residues, is removed by the bottom scrapper and by pumps, and then dried on beds.

In some cases, very strong foaming was noted in the water loops for exhaust gas treatment in cupola furnaces. The reason of such foaming could be clearly determined to be the use of scrap packages from domestic garbage. The scrap packages contain many cans containing fat and oil residues, which pass into the recirculated water where they cause excessive foaming beyond control, obviously with microbiological processes of decomposition and degradation contributing to this effect.

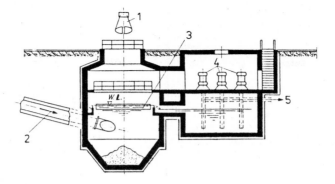

Fig. 6.6.-4: BAMAG sinter collector /158/
1 - grabber; 2 - sinter water inflow; 3 - submersed circular wall; 4 - pumps; 5 - outflow

6.6.1.2.3 Waste Waters from Machine Tool Production and Mechanical Workshops

The oil-carrying waste emulsions are introduced into the public sewerage networks or into the receiving water. Prior to their discharge they must be thoroughly deoiled. This means that they must first be collected on the site of their production. Then the emulsion is dissociated and then the oil must be separated.

The dissociation may take place in a system operating either intermittently or continuously. The treatment includes the addition of certain acids, salts, or dismulsifiers. It may be supported by the supply of heat.
The dissociated oil is separated from the aqueous solution.

Fig. 6.6.-5 shows the schematic of a continuous equipment for the dissociation of waste emulsions of drilling oil, as manufactured by DIDIER-Werke AG.

6.6.1.2.4 Waste Waters from Metal-Pickling Plants

Two groups of the numerous methods of purifying waste waters from pickling plants can be distinguished. The first group includes those methods that provide for an elimination of the pickling baths by way of neutralization, and the second group includes all the techniques aimed at a recovery of the acids and metal salts from the baths.

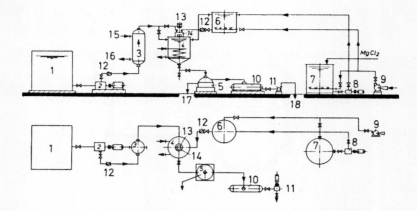

Fig. 6.6.-5: Schematic of a continuously operating dissociation equipment for waste emulsions of drilling oil, manufactured by DIDIER-Werke AG

1 - collector; 2 - emulsion pump; 3 - heat exchanger; 4 - reactor vessel; 5 - separator; 6 - dosing container; 7 - solution container; 8 - dissociating agent pump; 9 - fan; 10 - oil reservoir; 11 - oil pump; 12 - flowmeter; 13 - stirrer; 14 - mixing vessel; 15 - vapor; 16 - condensation water; 17 - to waste water; 18 - to transportation container

The waste waters are neutralized by a treatment with acids or alkalis. This step is finalized by the sedimentation of the precipitated metal hydroxides.

The pickling baths are purged with hydrochloric acid or sulfuric acid, and contain dissolved divalent iron. In the neutralization step, first iron (II) hydroxide, $Fe(OH)_2$, is formed, which has a low solubility only from pH 9 and above. Since this substance shows a very quick reaction with oxygen in the alkali range, with the formation of iron (III) hydroxide, $Fe(OH)_3$, the initially set pH value decreases very quickly, with the iron being dissolved again. To prevent such redissolution, the neutralization is combined with the oxidation of the divalent into the trivalent iron when the purging waters are treated. For this reason, air is introduced into the neutralization stage that operates either intermittently or continuously. The pH value is kept around 8.

The oxidation is very quick, the pH value need no longer be reduced by the addition of an acid when the water is discharged. This would be required when higher pH value would have been set.

The flow-through or duration time in neutralization or oxidation basins must not be rated too short. If possible, it should be determined empirically. The use of hydrate of lime as the neutralizing agent is common practice. This also provides for a reduction of the sulfate concentrations down to the residual concentrations determined by the calcium sulfate solution.

The solubility of the compound $CaSO_4 \cdot 2\,H_2O$ (gypsum) in water at 20° C is 2038 mg $CaSO_4$ per liter, which corresponds to 1438 mg SO_4^{2-}/l.

Pure iron (II) hydroxide is white. When it is contaminated with small amounts of iron (III) hydroxide (which is mostly the case), it is greenish. A further oxidation produces a dark green to nearly black color, and when the conversion to iron (III) hydroxide is completed, an intensive brown color can be noted. Then the iron (III) hydroxide is sedimented and dehydrated in specific equipment. In batch operation, neutralization basins are used for this purpose, while, in continuous operation, final sedimentation basins are provided.

The sedimentation process can be accelerated by the addition of an organic anion-active polymer as an auxiliary flocculating agent. The addition of such an agent is also advisble to improve dehydration.

The used pickling baths in small and medium size pickling plants may also be processed in an operation including the stages of neutralization, oxidation, sedimentation and sludge dehydration, possibly together with the purging waters.

Various methods are used to <u>recover</u> the acids and metal salts contained in the pickling baths. The pertinent technical literature (e.g. MEINCK, SIERP, and others) contains detailed descriptions of these techniques.

6.6.1.2.5 Waste Waters from Anodizing Plants

Waste waters discharged by anodizing plants are characterized by their high fluoride concentrations.

The analysis of the discharge from a neutralization and sedimentation plant for mixed domestic and commercial waste waters with a very high percentage of fluoride-carrying waste waters from anodizing works, furnished fluoride concentrations as high as 7.8 to 11.7 mg/l F.

The neutralization of used acid concentrates, semiconcentrates and purging waters is best achieved when hydrate of lime is used. This process also precipitates calcium sulfate:

$$Ca^{2+} + SO_4^{2-} \longrightarrow CaSO_4$$

As has been discussed above in relation to the sulfate-carrying iron pickling baths, the solubility of $CaSO_4$ at a water temperature of 20 °C is 2,038 mg $CaSO_4$ or 1,438 mg SO_4^{2-}/l. Consequently, the corrosive effects of sulfate-carrying waste waters on the concrete in sewerage systems and structures cannot be eliminated by the precipitation using lime, since the usual requirement of 400 mg/l of SO_4^{2-} as the permissible upper limit for discharge into the sewerage system cannot be met. When the concentrates and semiconcentrates from anodizing plants are neutralized, very large volumes of sludge are produced. The sludges predominantly consist of calcium fluoride, calcium sulfate, and aluminum hydroxide. They are easy to dewater when a cell filter press is used. It is possible to obtain total solids values of 40 to 45 % by weight.

6.6.1.2.6 Other Waste Waters from Metal-Processing Industries

The process steps
- neutralization with an acid or an alkali,
- separation and thickening of the sludges produced in the preceding step (hydroxides, fluorides, sulfates, phosphates)
- dewatering of the concentrated sludges,

are also used for the other waste waters discharged from metal-processing industries, similar to the treatment of waste waters from iron-pickling and anodizing plants.

Chromate-carrying and alkali-cyanide containing waste waters must undergo a special treatment before neutralization. The waste waters from the metal-processing plants contain many substances that reduce the efficiency of biologically operating waste water treatment plants and of sludge treatment systems (methanogenic fermentation). In practice, the biological processes are entirely prevented in many cases. The retention of the various constituents of such waste waters is therefore of a predominant importance for both the unimpaired efficiency of biological sewage plants and the organisms living in the water. A health risk for man, which results from the use of surface water to produce drinking water or from the consumption of damaged water fauna, cannot be precluded.

Owing to the toxicity of the waste waters, the treatment systems are termed detoxication and neutralization systems. It is the objective with such systems to remove environmentally harmful substances from the waste water by way of oxidation, reduction, neutralization, flocculation, and precipitation, or to convert them to substances that are less harmful.

Typical examples of such processes are:
- the oxidation of unbound and bound cyanide,
- the conversion of hexavalent chromium (chromic acid, chromate) by way of reduction to trivalent chromium which can be precipitated in the form of a hydroxide,
- the detoxication of nitrite by way of oxidation or reduction,
- the precipitation of dissolved metals by using an acid or base to set the appropriate pH range,
- the precipitation of anions, using hydrate of lime or calcium salts ($CaCl_2$) to convert them to almost insoluble calcium compounds (fluoride, phosphate, sulfate).

The risks involved in many substances, including those to human health, demand that the personnel entrusted with the operation and maintenance of the equipment and systems be thoroughly trained. Necessary safety measures must be considered in the design and construction of the plants. For instance, collector basins for dangerous liquids are required in zones where such liquids may leak or escape due to overflow, spillage, through leaks in piping or containers, and possibly occurring leaks in pumps, fittings, or measuring instruments. Specifically, critical zones that involve the risk of released unpleasant or noxious gases must be provided with exhausters or air scrubbers. Shower facilities with rapid access must be installed for the personnel.

Wearing appropriate protective clothing, gloves and goggles must be mandatory for dangerous works. Special training for accidents and for first aid is required.

The establishment of detoxication and neutralization centers has proven to be useful in towns and communities with many small and medium size metal-processing plants. The individual plants pay a certain fee and deliver then the used concentrates and semiconcentrates to such a center for detoxication and neutralization. Such centers are far better to operate from both the economic and engineering viewpoints than are the plant-internal systems. The individual companies are then only required to detoxicate and neutralize the purging waters.

The quantities of additives required for detoxication and neutralization can be quickly and reliably determined by laboratory analyses and tests. For this reason, the plants must have an appropriately equipped laboratory with skilled labor. The operating facilities must be well equipped with instruments to permit the exact weighing and metering of volumes of the solutions to be treated and the agents to be added. This allows for a suitable transfer of the laboratory results to the operational environment.

6.6.1.2.6.1 Chromate Detoxication

Waste waters carrying chromic acid and chromates are produced when metals are pickled and when metal baths are mixed, when chromate-laden solutions are used to anodize aluminum, when bright and hard chrome platings are electrolytically applied, in electrolytic brightening and polishing, and when metals (even synthetic materials) are pickled and etched.

Even though chromate forms relatively insoluble salts (e.g. barium chromate) with a small number of cations, it has become general good chromate removal practice in waste water treatment to reduce the chromium present in a hexavalent form in chromate by chemicals to trivalent chromium, which can then be precipitated by caustic soda solution or hydrate of lime in the pH range from 7.5 to 8.5.

The chemical reduction may take place
- in an acidic solution (pH value ca. 2 to 2.5) with gaseous sulfur dioxide (SO_2 gas) or sodium hydrogen sulfite ($NaHSO_3$),
- in a neutral or slightly alkali medium, using sodium dithionite ($Na_2S_2O_4$),
- in an alkali solution (pH value 8.5 to 12), using iron (II) hydroxide

Equations

For the reduction with hydrogensulfite

$$Cr_2O_7^{2-} + 3\ HSO_3^- + 8\ H^+ \longrightarrow 2\ Cr^{3+} + HSO_4^- + 4\ H_2O$$

The equation shows that the addition of an acid involves a shift of the equilibrium to the right-hand side. During the reduction process, the pH value must be regularly checked and kept in the pH range below 2.5 by the addition of acid. Otherwise it will constantly rise.

For the reduction with dithionite

$$2\ CrO_4^{2-} + 3\ S_2O_4^{2-} + 4\ H^+ \longrightarrow 2\ Cr^{3+} + 6\ SO_3^{2-} + 2\ H_2O$$

This equation applies to the chromium reduction in the neutral to weakly alkali range. This process oxidizes dithionite only up to the sulfite level.

An application in the acidic range is also possible. Here, the reduction potential of this substance is even essentially higher than in the neutral or alkali medium, since the oxidation is carried out up to the sulfate level:

$$Cr_2O_7^{2-} + S_2O_4^{2-} + 6\ H^+ \longrightarrow 2\ Cr^{3+} + 2\ SO_4^{2-} + 3\ H_2O$$

For the reduction with iron (II) hydroxide

$$CrO_4^{2-} + 3\ Fe^{2+} + 4\ OH^- + 4\ H_2O \longrightarrow Cr(OH)_3 + 3\ Fe(OH)_3$$

This process is advantageous when alkali regenerates from anion exchangers are conditioned and processed. The acidification required for a reduction to take place, which is combined with a high consumption of chemicals and considerable salt production, is omitted.

Directions for the detoxication of concentrates and semiconcentrates containing chromate and chromic acid, using sodium hydrogen sulfite solution $NaHSO_3$:

Sodium hydrogen sulfite in a solid state is not known; it occurs only in solution. The salt formed when this solution is concentrated is sodium disulfite, $Na_2S_2O_5$. Sodium hydrogen sulfite is commercially available in an aqueous form with a concentration between 37 and 40 % $NaHSO_3$, density 1.34 to 1.36 (kg/l). It can also be obtained when the solid salt sodium disulfite is dissolved in water:

$$Na_2S_2O_5 + H_2O \longrightarrow 2\ NaHSO_3.$$

Chromium and concentrated chromating baths should be diluted at least in a ratio 1 : 5, whereas the dilution ratio should be 1 : 2 for diluted chromating baths. The pH value must be lower than 2.5. Hydrochloric acid or diluted sulfuric acid can be used for acidification. Hydrochloric acid is expensive and, when handled inappropriately, may give rise to corrosion due to the release of HCl into the atmosphere. It is preferable, however, to the application of sulfuric acid when the sulfate concentration in the subsequent discharges must be low in consideration of the aggressive effects on concrete. When reduction takes place in batches it should be noted that the vessel does not contain excessive concentrate quantities, so that provisions are made for the accommodation of additional diluting water and detoxication chemicals. A 10 to 20 % sodium hydrogen sulfite solution is then slowly introduced while stirring well, until an excess of the reduction agent can be determined. The heat of reaction may result in a temperature increase above 60 °C. This should be definitely avoided (for the reason that larger sulfur dioxide (SO_2) quantities escape into the air and that the chemicals have a corrosive effect on the container and the stirring equipment). The mixture should possibly be cooled before further chemicals are added. This process entails an SO_2 escape into the air. The reduction vessels should therefore be connected to exhausters and exhaust air scrubbers.

The detoxication containers should be clad with PVC, polyethylene, or ceramic material resistant to chromic acid and heat. The respective stages of reduction can be clearly noted by the color changing from yellow/orange (chromate color) to an intensive green (color of the trivalent dissolved chromium). The end of reaction can be determined when the redox potential is measured with gold or platinum electrodes. Since the reaction is finished when an excess quantity of the reduction agent is available, provided that the prescribed pH range of 2 has been observed, this condition can also be checked as follows.

river water, whereas only ca. 10 % of the cyanide were released in a depth 500 mm below the water surface.

Instead of forming salts with alkalis, the cyanidic iron complexes can also form salts with di- and trivalent iron ions in the following forms:
(a) $Fe_2(Fe(CN)_6)$ iron (II) hexacyanoferrat-2, Prussian white
(b) $Fe_4(Fe(CN)_6)_3$ iron (III) hexacyanoferrate-2, Prussian blue
(c) $Fe_3(Fe(CN)_6)_2$ iron (II) hexacyanoferrate-3, Turnbull's blue
(d) $Fe(Fe(CN)_6)$ iron (III) hexacyanoferrate-3

The compounds (a), (b), and (c) are almost insoluble in water. The compound (d) dissolves in water with a dark brown color, so that it is not suited for cyanide removal by precipitation. The precipitation producing Prussian blue is successful only in a pH range from 3 to 5. With the pH value increasing, however, the Prussian blue re-dissociates into ferrous potassium cyanide and precipitated iron (III) hydroxide. With pH values of 8 and higher, this reaction is quantitative:

$$Fe_4(Fe(CN)_6)_3 + 12\ OH \longrightarrow 3\ (Fe(CN)_6)^{4-} + Fe(OH)_3$$

This means that, when Prussian blue is combined with alkali waste waters, this substance is soluble again when both constituents are sedimented or dehydrated together. The same applies to landfills. The precipitation in the form of Prussian blue is therefore not a satisfactory solution either for cyanide removal.

When only iron (II) salts are used for precipitation and when neither iron (III) compounds nor oxidation agents are present, Prussian white is precipitated, which is insoluble even in weak alkalis.
The precipitation of cyanides in the form of Prussian white is well suited for the treatment of concentrates, even concentrates including cyanidic zinc and cadmium complexes. In these cases, the metals are precipitated in the form of hydroxides when iron (II) salts are used for cyanide precipitation. When the cyanide is precipitated as Prussian white, it is only removed from the waste water but it is not detoxicated. It continues to be a potential risk in landfills. Only the thermal destruction of the Prussian white with cyanide destruction leads to depositable residues that are harmless to the environment.

Nickel. The difficulty of destroying the cyanidic nickel complex with hypochlorite results from the extremely low rate of dissociation. With a 20% excess of hypochlorite, a reaction time of 40 minutes is required to detoxicate the cyanidic nickel complex, whereas slight excesses of chlorine demand reaction times of ca. 12 hours.

The detoxication of concentrates and semiconcentrates in batch operations does not involve any difficulties in determining the conditions for the destruction of the cyanidic nickel complex.

Fig. 6.6.-6 illustrates a flow diagram of a flow-through type of detoxication system. All the reactor basins are subdivided to improve the duration behavior. Stirrers are provided in the compartments 1-1, 2-2, and 4-4. Compressed air is blown into the compartments 3 and 3 so as to produce turbulence. The basin 5 is a clarification basin of common design, e.g. a funnel-type basin.

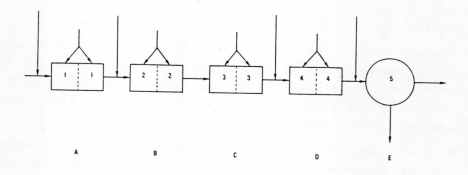

Fig. 6.6.-6: Flow diagram of a flow-through detoxication system

The following processes apply to the individual stages:

A. CYANIDE DETOXICATION

    Purpose: destruction of cyanide and metal cyanide complexes

    Conditions: pH value ca. 11, possibly to be set with NaOH, NaOCl addition, ca. 40 minutes duration at peak inflows

    Added chemicals: NaOH liquid, controlled through pH value measuring electrode;

NaOCl liquid, controlled through cyanide measurement or measurement of free chlorine.

B. CHROMATE DETOXICATION

Purpose: reduction of Cr^{6+} to Cr^{3+}, which is then precipitated as $Cr(OH)_3$

Conditions: pH value higher than 9, possibly to be set with NaOH, addition of divalent iron, ca. 15 minutes duration time with peak inflows

Added Chemicals: NaOH liquid, controlled through pH value measuring electrode.
Iron (II) sulfate solution, possibly used iron pickling baths provided they contain sufficient quantities of divalent iron. Controlled through redox potential measurement

C. OXIDATION OF $Fe(OH)_2$ TO $Fe(OH)_3$

Purpose: conversion of the iron (III) hydroxide, which is largely insoluble in the neutral range

Conditions: whenever possible, the pH value should be left as it is. Oxygen supply by blowing in compressed air, ca. 15 minutes duration time with peak inflows.

D. NEUTRALIZATION AND METAL PRECIPITATION

Purpose: precipitation of the metals by setting the pH value at the optimum for precipitation. This value is usually within a range that permits the waste water discharge to stage E without any additional measures. If this is not the case, the discharge from E must be readjusted to the permissible range by the addition of an acid or a base. Then a further neutralization stage is required in the outflow from stage E.

Added chemicals: Depending on the requirements, acids or bases are added. Control through pH value measurement.

Conditions: duration time 30 minutes with peak inflows.

E. SLUDGE SEPARATION BY SEDIMENTATION

Purpose: separation of the solids from the clear water

Added chemicals: an auxiliary polymer flocculating agent is added in dosed quantities upstream of the sedimentation basin to improve the separation of solids (sludge separation). This agent should be introduced into the clarification zone at moderate turbulence and with a duration time of 2 minutes or more before discharge into the clarification zone. The type and quantity of the polymer must be determined by laboratory tests.

The essential dimensions of basin 5 (such as final clarification surface, volume of the sludge deposition and sludge collecting zones, and total volume) are defined by the quantity and the properties of the sludge to be separated. They must also be determined by preliminary laboratory testing.

6.6.1.2.6.3 Neutralization

The neutralization of the waste waters discharged from metal-processing plants, including the precipitates of the dissolved metals or the anions (fluoride, phosphate, sulfate with certain restrictions) accessible to precipitation with calcium, is normally the last treatment stage following the various detoxication processes. When concentrates are neutralized, the resulting development of heat must be considered. Fig. 6.6.-7 illustrates the dependence of the solubility of various metal hydroxides on the pH value, according to WEDEKIND and REUTER /107/. In practical operation, the metals are precipitated not only in the form of pure hydroxides but also in the form of phosphates, carbonates, borates and other nearly insoluble compounds (to the extent by which they can be bound accordingly), depending on the composition of the waste waters.

To avoid excessive salinity in the waste waters, it is advisable to use alkali or acidic concentrates and semiconcentrates as neutralizing agents, after a previous detoxication.

The separate collection and treatment of various waste waters before the final neutralization is a prerequisite for a thorough removal of the metals by way of precipitation, as seen from the foregoing sections. For instance, cyanidic waste waters must be kept separate particularly from nickel- and iron-carrying waste waters, before they are detoxicated with hypochlorite. The reason is that the complex cyanides of these two metals are difficult or impossible to destroy with hypochlorite.

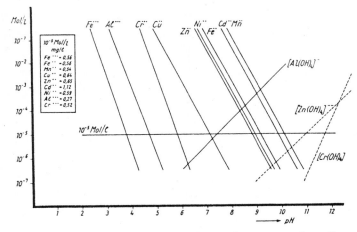

Fig. 6.6.-7: Solubility of different metal ions as a function of the pH value /107/

10^{-5} mole/l of different metals correspond to the following mg of metal/l values:

10^{-5} metal	mg of metal/l
Fe^{3+}, Fe^{2+}	0.56
Mn^{2+}	0.55
Cu^{2+}	0.64
Zn^{2+}	0.65
Cd^{2+}	1.12
Ni^{2+}	0.59
Al^{3+}	0.27
Cr^{3+}	0.52

Silver, copper, zinc, cadmium and nickel, when combined with ammonia, form <u>amine complexes</u>.

$$Me^{x+} + y\ NH_3 \longrightarrow (Me(NH_3)_y)^{x+}$$

In this equation, <u>me</u> represents a metal, <u>x</u> represents the number of positive charges of the metal ion, and <u>y</u> represents the number of NH_3 molecules. With monovalent silver, $y = 2$, with the divalent metal ions copper, zinc, and cadmium, $y = 4$, with divalent nickel, $y = 6$. The metal ions form the amine complexes only with NH_3 but not with NH_4^+. Since the ratio between NH_4^+ and NH_3 mainly depends on the pH value in aqueous solutions (as is illustrated in Fig. 6.6.-8), amine complexes are not stable in the acidic range. When metal ions are precipitated in the form of hydroxides that are able to form amine complexes, and when ammonium is simultaneously present, a thorough removal may therefore take place in a pH range slightly below the value common for an optimum hydroxide precipitation. The precipitation of the aforementioned metals, however, by way of neutralization should be carried out in ammonium-free solutions whenever possible.

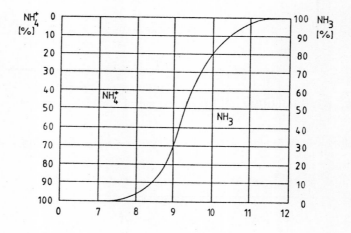

Fig. 6.6.-8: Dependence of the NH_4^+/NH_3 equilibrium on the pH value of the aqueous solution at ca. 20° C water temperature

Complexing or chelating agents other than cyanide are also added to the baths or rinsing liquids when metals are finished. These chelating agents prevent a metal precipitation in the form of hydroxides or dissolve already precipitated metal hydroxides by the formation of complexes. Such agents include condensed phosphates, triethanolamine, nitrilotriacetic acid, ethylene diamine tetraacetic acid (EDTA), tartaric acid, citric acid, gluconic acid, and other organic compounds. HARTINGER /65/ describes a testing method to detect chelating agents in waste water, where copper is used as the testing substance since this metal forms complexes in most instances.

Copper sulfate is dissolved in 1 liter of waste water, or is added to the waste water in a dissolved form, in an amount sufficient to achieve a copper concentration of 100 mg/l. Then sodium hydroxide solution NaOH is added for neutralization to a pH value of 8.5. Two hours later, the mixture is filtered through a very tight filter. The filtrate is then analyzed for copper (wet chemistry tests after destruction of the complex or, even better, by means of flame atomic absorption spectrometry). When more than 1 mg/l of copper is found in the filtrate, chelating substances are present.

Generally applicable methods of treating waste waters carrying metal complex compounds, other than the cyanide methods, are not available. Therefore, in practical operation, the methods of destroying the chelating agents must be found experimentally or empirically. The following methods come under consideration:
- precipitation of the metal with a large hydroxide excess after the addition of calcium chloride,
- precipitation of the metal in the form of a sulfide,
- reduction of the metal bound in the complex,
- oxidation of the chelating agent with chemicals,
- electrolytic methods with deposition of the metals at the cathode and anodic oxidation of organic chelating agents,
- treatment with ion exchangers.

The application of the majority of these treatment methods is much more successful when concentrates and semiconcentrates are treated rather than rinsing liquids.

After detoxication, neutralization, and sludge separation, the waste waters discharged from metal-processing industry still often contain organic compounds accessible to biological decomposition so that a final treatment in biological clarification methods is useful.

6.6.1.3 Waste Waters from Mines and Ore-Dressing Plants

Five important types of waste waters must be mentioned in relation to pit coal mines (see Chap. 2.4.1.3), which are distinguished by their quantities and compositions.

Since these waste waters may cause damage to receiving waters, they need a special treatment before they are introduced into the natural waters.

6.6.1.3.1 Mine Waters

Mine waters from pit coal mines are mostly so clear that they could be considered to be suited for direct introduction into the receiving water. Since they may be acidic, however, they undergo a specific treatment. Otherwise, considerable volumes of such waste waters would result in problems regarding water usage in fishing, agriculture, and drinking.

The dissolved salts can be removed from the mine waters only at considerable cost. Less strongly salty mine waters are softened and then used as boiler feedwater or as cooling water for surface condensers.

6.6.1.3.2 Waters from Pithead Baths

The same demands as on the methods for the treatment of domestic waste waters must be made on the purification of waters discharged from pithead baths. For this reason, these waters are also introduced into the public sewerage system in many cases, or they are treated in suitable sewage plants.

6.6.1.3.3 Waste Waters from Coal Washing

The waste waters from coal washing operations need a thorough mechanical purification. Whenever possible, this purification is combined with methods recovering and reusing the separated coal particles.

The clarified water is fed into the production cycle again and can be reused for coal washing purposes.

6.6.1.3.4 Waste Waters from Coal Coking Plants

Most of organic pollutants in waste waters discharged from coking plants are phenols. These can be removed by extraction methods. This measure considerably reduces the load of the waste waters with organic pollutants.

The best-known method with a wide range of applications for different phenolic substances and their absolute concentrations in waste waters is the Phenosolvan process, which used butyl acetate initially and is now using isopropylether as the solvent and extracting agent. This process is also used in many operations on a large industrial scale to dephenolize waste waters from pit coal coking plants.

In addition, many coking plants use the activation method to purify the phenol-laden waste waters. The waste waters are detarred and deoiled. Then they are continuously supplied through a volumetric equalizing container to an activation system. Phenols, other degradable organic compounds, sulfides, cyanides, and thiocyanates are thoroughly oxidized. GROSS and DRECHSEL report the operating results of such systems.

It is also possible to treat these waste waters together with domestic waste waters.

JACOBS* isolated phenol-decomposing bacteria strains from the activated sludge of a biological dephenolizing system. These bacteria belong to the varieties Brevibacterium, Comamonas, Pseudomonas, Cellulomonas, Mycoplana, Nocardia, and Alcaligenes. Moreover, a phenol-utilizing Mycelium sterilium was found. He could detect and cultivate bacteria

---

* M. Jacobs, <u>Microbiological Studies on a Dephenolizing Sewage Plant</u>. Doctoral thesis. 1974

that are able to utilize various phenols, thiocyanates, cyanates, cyanides, sulfide, and ammonia.

The thorough removal of substances that, in a high concentration, have toxic effects, by a biological process is always possible in the activation method when these substances, in a low concentration, are accessible to a biological decomposition, and when the operating conditions are known and observed that are important for the success of the method on a large industrial scale.

An important step is the start-up. The biological processes cannot be successfully initiated with undiluted phenol-laden waste waters. The activation system is therefore charged with a mixture of river water and phenol-laden waste water. The phenol concentration should be set to a value below 80 mg/l. The waste waters are practically free of phosphates, so that phosphates must be added in quantities of ca. 0.3 kg P per 100 kg BOD_5 (e.g. in the form of industrial phosphoric acid).

A phenol concentration of 2.2 to 2.8 BOD_5 (in mg O_2/l) applies to the BOD_5 in waste waters from pit coal coking plants. Degradable substances accompanying the phenols are included in this value. The start-up and the operation of the system can be facilitated by the addition of dissolved iron salts into the activation basin in quantities of 20 to 100 g $FeSO_4 \cdot 7\ H_2O$ per m³ of phenol-laden waste water, for example. After these preparatory steps, the contents in the activation basin are aerated without any inflow or outflow. The phenol decomposition is monitored by frequent determination of the phenol concentration. When the values come into the "zero" range, the continuous inflow of phenol-laden waste water commences. It should be noted that the inflow be increased only to the extent by which the growing biological matter is able to keep the phenol concentrations in the basin close to "zero". The system can be loaded to full capacity in most cases after a few days. The water contained in the activation compartment of a system in full operation has the same quality as the water discharged from the sewage plant. It contains practically no pollutants. The operation of biological dephenolization plants, therefore, does not require any bacteria adapted specifically to high phenol concentrations.

The activation basins should be basins with total intermixture. The inflowing raw water should be distributed to the entire basin volume as quickly as possible so that local harmful concentration cannot arise.

Basins with a unidirectional flow (drop flow) are therefore not appropriate.

Moreover, the recirculated sludge from the final clarification basin should be introduced into the activation basin entirely separately from the raw water. The usual introduction of the recirculation sludge into the raw water flowing into the activation basin would be totally inappropriate in this case owing to the high toxicity of the raw water.

A thorough phenol decomposition is achieved in practice with volumetric loads up to ca. 2 kg BOD_5/m^3 of aeration volume per day. The less the volumetric load, the more thoroughly sulfides, cyanides and thiocyanates are eliminated. With volumetric loads less than 0.5 kg BOD_5/m^3 of aeration volume per day nitrification occurs to an ever-increasing extent.

Owing to the changes in the economical situation, coal coking plants are frequently operated with a highly varying utilization of capacity. The raw materials supplied may also be of different quality. As a result, the concentrations and pollution loads in phenol-laden waste waters are varying. When biological sewage plants are constructed, methods should therefore be selected which, from both a technical and an economical viewpoint, ensure an appropriate adaptation to these variations. This includes the following measures:
- addition of phosphates and iron salts,
- temperature control by indirect cooling or heating, as required.

6.6.1.3.5 Waste Waters from Coal Gasification

From an economic viewpoint, the phenol recovery by extraction as the first stage in the treatment of waste waters from coal gasification operations, after detarring and deoiling, is much more favorable than phenol recovery from the waste waters from high-temperature coking plants, which carry less phenol, since the operating expenditure (energy for mixing and distillation of the extracting agent, solvent losses) per m^3 of waste water, which depend on the throughput, permit the recovery of a substantially higher amount of phenols. A subsequent purification can then be carried out with biological methods.

The waters from lignite coal gasification are very polluted. Their total phenol concentration is up to 30 g/l, they contain up to 16 g of volatile organic acids per liter, and their BOD_5 is as high as up to 50 g/l. The first purification stage is the recovery of the phenols by extraction methods. Then the waste waters may undergo a biological treatment in multistage systems.

6.6.1.3.6 Waste Waters from Charcoal Production

Waste waters result from the purification of wood gas with water (scrubber waters), which cannot be condensed.

Scrubber waters that are practically free from sulfur and nitrogen compounds, show acidic reactions. They contain acetic acid, creosote (phenols), acetone, methyl alcohol, etc.

The waste water treatment depends on the composition of the waste waters. In any case, however, thorough cooling, detarring, and deoiling are required, also, if necessary, neutralization of the free acids with limewash. The natural (in land irrigation or ponds) and biological methods on an industrial scale are available to continue the treatment of the waste waters.

Dephenolization can also be achieved by exchange adsorption using synthetic resins.

6.6.1.4 Waste Waters from Chemical Industry

In accordance with Section 2.6.1.4, the following discusses briefly the treatment of the waste waters from mineral acid production, potash industry, soda works, fertilizer production, and antimony red production.

6.6.1.4.1 Waste Waters from Mineral Acid Production

Depending on the method used for the preliminary treatment of the exhaust gases, alkali and partly phenol-laden waste waters are discharged from the <u>production of sulfuric acid</u>. On the other hand, acid-containing cooling waters possibly also arise as a result of the acid condensation. Both types of waste waters are mixed in equalizing tanks for neutralizing purposes.

<u>Nitric acid production</u> gives rise to waste waters containing nitric acid of ca. 2 % concentration and cooling water. If the slightly acidic waste waters cannot be neutralized directly in the receiving water, they must be treated before their introduction.

The <u>production of hydrochloric acid</u> involves cooling waters, weakly hydrochloric discharges from exhaust gas scrubbing, as well as wastes and arsenic sludges containing mineral oils. The polluted waste waters must undergo a specific purification before they are introduced into the receiving water.

6.6.1.4.2 Waste Waters from Potash Industry

These waste waters contain salts that practically resist removal with both the usual treatment methods and biological decomposition. Attempts have therefore been made to utilize the salts contained in these waters. A great success, however, has not been achieved in this field, so that the residual waste waters are either filtered into the soil or diluted and then introduced into the natural waters.

MEINCK /107/ suggests the following procedures for salt utilization:
- recovery of magnesium chloride for further processing or for use as sizing agent in textile industry,
- production of magnesia cement, Sorel, or wood cement,
- dust binder for road treatment,
- herbicides for garden walks,
- precipitating agents for the purification of certain industrial waste waters,
- production of magnesite and magnesium oxide,
- production of a spreadable fertilizer, the so-called foots lime.

The effect of waste water surge loads on the salt concentration of the water must be observed when the discharges are introduced into the receiving water. If necessary, equalizing basins should be provided. Basins with a capacity of 3,000 to 7,500 m^3 are generally deemed sufficient for crude salt quantities of 100 tons processed per day.

6.6.1.4.3 Waste Waters from Soda Works

Strictly speaking, the purification of calcium chloride final liquors involves only a thorough retention of the sludge and, together with it, the remaining suspended matter, whereas the chlorides remain dissolved. In small and medium size companies, sedimentation tanks (for eight hours) are provided for purification. Large companies use basins in the ground or ponds.

When the discharges are introduced into the receiving water a thorough distribution of the salt-containing solutions must be observed.

6.6.1.4.4 Waste Waters from Fertilizer Production

Regarding the treatment of the waste waters arising in the production of potash salts reference is made to Section 6.6.1.4.2.

The <u>ammonia synthesis</u> (if combined also with a gas production system) involves waste waters from gas scrubbing, which contain carbon and ash particles, hydrogen sulfide, phenols, bicarbonates of the ammonium and the hardening constituents, etc.

Sedimentation basins (for the retention of the undissolved substances) with 90 minutes duration time, equipped with aerators (to oxidize the hydrogen sulfide) are provided for waste water purification.

The oxidation of the hydrogen sulfide (in tower aerators where the waste water is introduced at the top) demands appropriate measures to avoid the pollution of the atmosphere, e.g. closed aeration systems, gas collection and incineration, etc.

In the <u>production of superphosphates</u>, a final waste water treatment is generally not required. In critical cases, therefore pollutants that are possibly present can be rendered harmless.

Occasionally, however, the addition of limewash alone is not sufficient. Since the deposition is based on a flocculation reaction, a slow stirrer must keep the liquid moving so that a maximum flocculation will permit an optimum separation.

6.6.1.4.5   Waste Waters from Antimony Red Production

The waste waters that arise from soda production (e.g. in the Solvay process) still contain small amounts of antimony red and sulfur. These waste waters must undergo a sedimentation process (with a duration time of five hours at minimum) before they can be introduced into the receiving reservoir.
The sedimentation basins should also have the function of equalizing basins, specifically when the waters are small (e.g. less than 30 m^3/s).

6.6.2 Organic Industrial Waste Water

The manufacture of a great variety of products in one plant is common. This means that the waste waters discharged from large-scale operations may contain hundreds or even thousands of different inorganic and organic compounds. They are the residues of the raw materials, of the intermediate and final products, and of the auxiliaries utilized in the individual steps of production, such as solvents, acids, bases, and salts. There is a strong tendency toward a combined mechanical-biological purification of these waste waters. This means the least expensive method for a foreseeable period. It offers the additional advantage that there are no negative affects on the biological conditions in he waters when the biological purification causes no problems.
The purification in mechanical-biological sewage plants alone or together with municipal waste waters is in fact successfully carried out by a great number of companies in the large-scale chemical industry. Depending on their type, composition, concentration, and harmfulness, the waste waters discharged from production must be preliminarily treated before the biological purification, or must even be kept away from the biological treatment stages. Such preliminary treatment includes specifically the fat elimination and deoiling, neutralization, and the removal of substances that pass through mechanical-biological sewage plants without being changed, which, however, cause changes in the waters by color changes, turbidity, subsequent precipitations or an increase of the fertilizing effect, even though their own behavior is neutral. Substances with a definitely harmful effect must be initially detoxified at their places of occurrence, or must be kept away from normal waste water purification by suitable measures. Such measures include, for

instance, evaporation and subsequent incineration of the residues, direct incineration without previous concentration by evaporation, extraction methods, or methods of steaming, and also the treatment with adsorbing or precipitating agents. The application of these methods is advantageous in cases where the pollutants arise in quantities almost accessible to quantative collection, which is true for the batch production methods, which are mostly discontinuous.

Another important prerequisite for the successful mechanical-biological treatment of the waste waters discharged from production operations is their equalization in view of quantities, concentration, and temperature. It can be noted in many cases of practical operation that waste waters discharged from such plants contain slight quantities of colorants or substances causing turbidity even after a thorough mechanical biological treatment. Even though these negligible quantities are harmless in these concentrations, they may considerably change the color of the receiving waters so that outsiders may get the impression of strong pollution. This is often the cause for citizens' initiatives with protests and charges against the companies. The retention of these colorants or turbidity-causing substances frequently demands investments beyond proportion for additional installations and operating equipment. In many cases, however, this is actually inevitable. The situation may be different in each case. Therefore, a study should be made as to the location in the plant where the retention of the colorants and turbidity-causing matter is most effective at the lowest expenditure.

Owing to the special constituents in the waste waters discharged from the manufacture of certain products, similar problems often occur in the treatment of waste waters discharged from companies purchasing these products and employing them in wet processes. This applies, for instance, to the production and dyeing of textiles and papers, wood-fiber board production, tanneries, and leather production.

6.6.2.1 Waste Waters from Pharmaceutical and Cosmetics Industry

6.6.2.1.1 Waste Waters from Pharmaceutical Industry

The waste water treatment methods must be adapted to the specific conditions. If necessary, detoxication, deoiling, neutralization, treatment with precipitating agents, decolorization, sedimentation, biological purification, and equalizing basins must be provided.

The production of antibiotics also demands that the waste waters be almost sterile and that they must therefore be mixed with domestic (municipal) waste waters so that a biological treatment can be carried out.

Wherever possible, the different waste waters should be definitely separated and undergo the specifically optimum waste water treatment process. This is particularly true for a separate handling of cooling water and rainwater.

6.6.2.1.2 Waste Waters from Cosmetics Industry

It has become common for small cosmetics producers to introduce their waste waters into the municipal sewerage system. Larger companies must provide the corresponding measures of treatment that are comparable to those required in the pharmaceutical industry.

6.6.2.2 Waste Waters from Dye and Colorant Production

Waste waters discharged from plants producing dyes and colorants must definitely undergo an appropriate treatment (see also Section 6.6.2.1) before they are introduced into the sewerage system or the receiving reservoir.

6.6.2.2.1 Inorganic Dye and Colorant Waste Waters

Waste waters containing toxic substances, e.g. lead or arsenic compounds, must be detoxicated.

Detoxication is carried out with chemicals solutions that form insoluble compounds with these toxic substances. Examples:

for lead-containing waste waters	soda (sodium carbonate, synthesis of insoluble lead carbonate
for arsenic-containing waste waters	sodium sulfide (synthesis of arsenic sulfide) or iron (III) in an alkali solution (synthesis of iron (III) arsenate, or adsorptive co-precipitation

for barium-containing waste waters	soda (synthesis of barium carbonate)
for chromate-containing waste waters	iron vitriol (reduction to chromium (III) salt which can be precipitated with limewash)
for final liquors from titanium white	crystallization of the iron sulfate and its processing to iron oxide waste and sulfuric acid, or carrying off the diluted acid for submersion into the sea

It is most advisable for detoxication to use treatment basins directly at the location where the substances arise. After detoxication, the waste waters are supplied into sedimentation basins.

6.6.2.2.2 Organic Dye and Colorant Waste Waters

The waste waters are discharged in considerable quantities. They are composed of numerous small quantities of different waste waters with harmful pollutants, which are highly concentrated for certain periods. As far as these pollutants render the overall purification difficult, they must be neutralized directly at the place where such waste waters are produced.

It is most important that the waste waters are carried off in separate sewers and that they are individually treated in consideration of their composition, the necessity of treatment, and their specific type. Waste waters containing toxic substances (e.g. cyan compounds, arsenic salts, free chlorine, fluorine, bromine, etc.) are detoxicated in chemical processes of waste water treatment.

Waste waters containing oil and fat are retained in specifically provided oil and fat separators (see Section 6.1.2.4).

Highly concentrated waste waters with predominantly organic substances must be treated by extraction, evaporating concentration, etc.

Waste waters developing strong odors that could become a nuisance to the environment must be neutralized, e.g. by aeration in blowing columns, evaporation, treatment with oxidizing agents (e.g. chlorine), etc.

When chemical neutralization is required, this step is provided only after a collection of all types of waste waters, using the possibilities of mutual interreaction. Limewash, lime waste, or other alkali agents are added for neutralization. When an additional chemical precipitation is required to separate substances in colloidal dissolution or solids causing turbidity, this stage is best located upstream of the neutralization stage. Since the chemical precipitation is based on a particularly thorough adsorption, the waste waters, mixed with the precipitating chemicals and set to the appropriate pH value, are treated in a flocculation basin with a built-in slowly rotating stirrer for 15 to 30 minutes before the coarse sludge flakes are sedimented in a sedimentation basin with a continuously operating sludge removal equipment.

Various methods are available to eliminate the color (see also the statements relating to the textile industry). Some examples:
- chemical _precipitation_ by flocculating additives such as iron or aluminum salts, if necessary in combination with caustic lime. Owing to the adsorptive effect, dyes and colorants in pseudo-solution are enveloped by the formed flakes. But in this process the precipitating agents do not have any effect on the dyes and colorants in genuine solution. In most cases, they must first be destroyed by a reducing agent such as iron (II) chloride bases, sodium hydrogen sulfide. Azo dyes are cracked by reducing agents.
- _Chlorination_ may have a destructive effect on some dyes and colorants, whereas it is ineffective with others. The suitability of a particular method should be determined by tests.
- The _land filtration_ can often be an appropriate method of decolorization.
- _Biological purification_ may result in decolorization unless toxic substances are present in the waste water.
- In many cases, the _combination of chemical precipitation with biological purification_ (NIERS process) has lead to a satisfactory result.

6.6.2.3 Waste Waters from Soap and Synthetic Detergent Production

6.6.2.3.1 Waste Waters from Soap Production

The waste waters from plants that may discharge them into the public sewerage system must first be cooled to temperatures below 30 °C.

Waste waters with a strong odor, which result from crude fat refining and the injection condensers, can be deodorized by chlorination.

Fatty waste waters must be principally treated in fat separators at the place of their origin. The intermediate suds from saponification need a separate treatment. Cooling may effect an additional separation of the substances. The residues produced in this treatment stage might hardly be reusable. The remaining waste waters that arise from fat cracking should also be neutralized by lime. The channels must be provided with airtight covers to avoid odor nuisance.

Before the waste waters are introduced into the receiving water, first the zinc salts must be reclaimed from the acidic waters. The glycerine, oil, and fat residues contained in the condensation and cooling waters can be removed by a fat collector equipped with a coke or activated charcoal filter, before they are discharged into the receiving water.

The introduction of the waste waters through an equilibrating and sedimentation system, which should be as large as possible, is recommended. When a predominantly acidic reaction is involved, the addition of limewash may result in both neutralization and precipitation of some dissolved organic pollutants. Such equalizing sedimentation system should be designed for a capacity corresponding at least to the waste water volume per day.

The biological treatment is considered to be the final purification stage. The waste waters from soap production plants, which contain glycerin, protein, and fatty acids (except the zinc-rich scrubbing waters from the fat cracking stage) can be decomposed in a crude and in a diluted condition. The cooling waters are a suitable diluting medium. The nitrogen and phosphoric acid quantities, which are required for biological decomposition, may be partly introduced by the inclusion of the plant's toilet and operational waste waters into the biological purification system. If necessary, nutritive salts containing nitrogen or phosphorus must be added.

6.6.2.3.2 Waste Waters from Synthetic Detergent Production

When synthetic detergents are introduced into surface waters and the groundwater through waste waters, they cause serious damage owing to strong foaming and the high oil and fat concentration. They may also disturb water processing.

The amounts of cooling and condensation waters are mostly a multiple of the quantity of genuine sewage. They should therefore be separated from the polluted operational discharges. All discharges from operation that are polluted by fat and oil can be treated by good fat and oil separators at their origin. This prevents clogging of the outlet lines and allows for recovery of valuable products. The actual purification is then best carried out after the combination with other polluted discharges from operation, possibly by means of chemical precipitating agents in a sufficiently dimensioned sewage plant which, at the same time, has an equalizing function for the waste water composition. The result is a mutual neutralization of the alkali and acidic waste waters, a dissociation of any fat and oil emulsions present, and thus a more effective separation of the lightweight substances, an equalization of the waste water surge loads, and a more uniform composition of the final discharge. Iron (III) or aluminum salts are suitable precipitating agents, even though only a 50 % success of the purification step would result. Such treatment is particularly suited to remove turbidity and any extremely fine oil and fat suspensions.

The suitability of biological methods for final waste water purification must be tested. This is the normal road leading to success, after elimination of the interfering effects of toxic or biologically harmful constituents by preliminary treatment of certain discharges from operation in situ.

After a corresponding preliminary treatment, the waste waters from paraffin oxidation are well suited for a biological purification. The low fatty acids, alcohols and ketones, which are contained in these waste waters, are easily accessible to biological decomposition.

When the purified waste water is introduced into the receiving reservoir, the foaming tendency must be considered and correspondingly counteracted (e.g. by floating scum bars, by spray water, with antifoam agents, etc.).

6.6.2.4 Waste Waters from Plastics Industries

6.6.2.4.1 Waste Waters from Plastics Industry Producing Cellulose

The waste waters arising from the production of transparent films (e.g. cellophane) by precipitation in sulfuric salt baths can be compared to those resulting from the production of viscose, so they need a similar treatment.

The production of vulcanized fibers involves extraction or washing waters, which need neutralization and clarification. Waste acids may be concentrated by evaporation and reused after the addition of freshly prepared acid. The organic (slimy) and inorganic contaminations are removed from the used zinc chloride lyes and the first extraction waters, which may contain 1 g/l or more of zinc, in various processes. Then the residues are vaporized in vacuum equipment. Before zinc-containing waste waters are introduced, the zinc must be precipitated. This is achieved by alkalinization and flocculation of the waste waters at pH values of 9.5 to 10.5, with final clarification of the resulting turbid liquid in sedimentation basins.

6.6.2.4.2 Waste Waters from Plastics Industry Producing Condensed Products

The condensation of phenoplastics and aminoplastics furnishes large quantities of cooling waters and so-called reaction waters.

The <u>reaction waters</u> (see also Section 2.4.2.2.2) may, for instance, be dephenolized by extraction, reduced by evaporation and incineration, or decomposed by biological methods.

After a corresponding preliminary treatment, they can be purified in one month in activated sludge basins or biological filters. Domestic waste waters should provide for a sufficient supply of nutritive salts.

Following neutralization and dilution with cooling waters or domestic waste waters, the waste waters may be introduced into the municipal sewerage system. A preliminary treatment by phenol extraction is also advisable in such cases.

After their neutralization and dilution with cooling waters or mixing with domestic waste waters, these waste waters can be introduced into municipal drainage systems. In this case, too, a preliminary treatment by phenol extraction is expedient, since such a treatment means a considerable reduction of the waste water load and thus of the purification cost. Possibly present formaldehyde does not have any effect on the biological decomposition of sufficiently diluted waste waters.

The <u>wood and wood-fiber board production</u> in wet processes involves waste waters that are characterized by their acidic reaction, their high $KMnO_4$ consumption, and a relatively low BOD_5. The water is as turbid as an emulsion.

Limewash is added to flocculate the waste water and purify it biologically. Domestic waste waters or nutritive salts must be added.
The sludge quantities produced are, however, considerable (ca. 100 l of sludge per m³ of waste water), with a low cellulose concentration (2 - 3 %).

6.6.2.4.3 Waste Waters from Plastics Industry Producing Polymerized Products

In view of their considerable pollution, these waste waters must be treated where they originate to reduce the pollution level. Apart from corresponding internal measures (recirculation and reuse of waste waters), the incinerator is the appropriate method of eliminating the organic concentrates, solvent residues, etc.

6.6.2.5 Waste Waters from Tanneries and Leather-Producing Plants

6.6.2.5.1 Waste Waters from Tanneries

The waste waters from tanneries contain substances that have been used for tanning:
- digestable organic substances (hides, gelatine),
- nondigestable substances (calcium oxide, calcium sulfide, calcium carbonate, sand),
- dissolved chemicals,
- dissolved organic substances (protein-containing substances, salts of organic acids, detergents),
- floating substances (hair, fat).

The waste waters from tanneries are characterized by specific constituents and conditions such as
- the sulfide concentration,
- the high pH value in combination with a high alkali concentration,
- the high $KMnO_4$ consumption.

Formerly, the discharges from tanneries were pretreated only to an extent that the objective was a quantitative equalization over time, on the one hand, and a reasonable duration in a correctly sized basin allowed for sedimentation, on the other hand. In isolated cases, flocculating agents were also added to increase the separation.

The modern techniques of waste water purification involve the following processes:
- neutralization of the alkalinity,
- binding or destruction of the sulfides,
- precipitation of the organic constituents with their very strong reducing effect,
- sedimentation of the solids.

An additional precipitation stage is recommended to support the sedimentation behavior.

Biological filters or activated sludge basins are used for the biological treatment. Small tanneries may also use land filtration or sewage lands.

The sludge volume produced in tanneries is considerable, while the percentage of water is high (even higher than 97 %).

Dehydration on drying beds is inefficient and slow. And yet, this method of dehydration is applied in most cases. Rotary vacuum filters and filter presses are better suited for this purpose.

After dehydration on drying beds, by vacuum filters or filter presses, the sludge from tanneries is useful matter to improve the soil, particularly after composting with other cellulose-rich wastes that have low water content.

When mixed with municipal waste waters, the purification of discharges from tanneries is easier and more economical. For this reason, the connection to the local drainage network, if there is any, is a solution of the waste water problem.

Waste waters from tanneries need preliminary treatment before they are introduced into the local sewerage system. The undissolved substances must be removed by sedimentation from the lime liquors. If they contain sodium sulfide this substance must be removed as well. Two processes have proven useful for this purpose in practical operation: Treatment with flue gases and the chemical treatment with iron sulfate and subsequent oxidation by aeration (see above).

Hides may be infected with milzbrand pathogens. In these cases, disinfection is necessary, e.g. using chloramine (1 hour of duration at a 20 mg/l concentration of chloramine).

6.6.2.5.2 Waste Waters from Leather Production

Regarding the decolorization of the waste waters, reference is made to Section 6.6.2.2.

6.6.2.6 Waste Waters from Textile Industry

The waste waters arising in the textile industry have a very different composition depending on whether the plant processes raw materials, or whether semifinished products are processed and improved to produce finished goods.

Most textile companies comprise various combinations of different branches rather than individual branches (e.g. dyeing, printing, spinning, etc.). Each of these branches results in a great number of different types of waste waters, which can be treated only after a corresponding study and testing.

6.6.2.6.1 Waste Waters from Spun Fabric Production

6.6.2.6.1.1 Waste Waters from Flax Retting and Hemp Roasting

The aerobic <u>flax retting operations</u> involve a BOD_5 reduction by at least 50 %. Owing to the high concentration of vegetable nutrients in these waste waters, mainly agricultural utilization (spray irrigation or land irrigation) or intermittent land filtration are to be suggested as methods of treatment.

A preliminary neutralization of the organic acids with limewash, alkali waste lyes, or the admixture of domestic waste waters with subsequent clarification in sedimentation basins is suggested.

Experience so far gathered in practical operation has shown that artificial biological techniques (treatment on biological filters or in activated sludge systems or in oxidation ponds with the addition of ammonium phosphate) are also applicable in cases where sufficient areas are not available for agricultural utilization of the waste waters. Such techniques demand a preliminary thorough neutralization and a dilution with a double or triple quantity of pure water when the biological system has a single stage only. A fish pond may be provided for final treatment, complete utilization of the nutrients still present, and simultaneous check of the biological purification efficiency.

Waste waters from flax retting are highly concentrated, while their temperature is increased, so that they are not generally suited for introduction into municipal sewerage systems. A partly biological preliminary treatment of the waste waters, if applicable, before their introduction into the sewerage system, would be conceivable.

Hemp roasting involves acidic (KORTE process) or alkali (lye process) waste waters with organic compounds in different concentrations. These waste waters are generally neutralized with lime.

The digestable and fermentable waste waters, which result from hemp and flax processing, belong to the group of waste waters that have the highest concentrations of pollutants known. In addition to the decomposing agents, they contain various extracted acids, they quickly start to decompose and rot, and cause severe oxygen loss or fish-killing with enormous fungus growth immediately upon their introduction into the receiving reservoir.

A simple mechanical purification must be deemed insufficient.

6.6.2.6.1.2 Waste Waters from Silk Cooking

The total waste water discharged from silk cooking operations is a viscous brown liquor with a considerable concentration of dissolved substances (due to soap, wetting agents and other detergents, silk glue, caterpillar bodies, damaged cocoons) in a highly decomposed state.

The anaerobic decomposition in digestion tanks (at 30° C) with 3 to 4 days of digestion and subsequent treatment (on land filters, by agricultural utilization, or in artificial biological installations) is appropriate for the purification of these waste waters.

The digested waste water, however, has an annoying odor, which must be taken into consideration.

6.6.2.6.1.3 Waste Waters from Rayon and Staple Fiber Production

Before the discharge of the diluted waste water, the valuable substances are recovered in all rayon and staple fiber producing plants, mainly from the used liquors and acidic precipitation baths.

The alkalinic liquors are regenerated. Sulfates are recovered from the acidic precipitation baths by evaporation in vacuum coolers and by crystallization of the sodium sulfate decahydrate (Glauber's salt).

Moreover, screens are installed to retain the coarser undissolved fibrous matter while sedimentation facilities (3 to 4 hours) are used to eliminate the finer undissolved and colloidal matter.

A further purification step is normally provided to reduce the contents of zinc, hydrogen sulfide, and organic pollutants. After the elimination of zinc (e.g. in an ion exchanger process, with precipitating agents, etc.), of the heavy metals and the hydrogen sulfide, the water undergoes biological purification. The prerequisite, however, is a preliminary mixture of these waters with domestic waste waters or the addition of nutritive salts.

The reaction of the waste waters in the <u>copper oxide ammonia process</u> may be partly acidic, partly alkaline. These waste waters contain copper, sulfates, ammonia, soap residues, etc. The waste water treatment aims at a recovery of the chemicals (in ion exchangers, by vacuum distillation, etc.). They need a treatment in accordance with the statutory regulation of waste waters.

The <u>cellulose acetate process</u> furnishes strongly acidic waste waters with small quantities of suspended matter (below 100 mg/l) but with a great variety of dissolved constituents such as sulfuric acid, acetic acid, sulfates, chlorides, perchlorates, aldehydes, etc.

Before their introduction into the receiving reservoir, the waste waters are passed through equalizing basins (e.g. with a capacity of 3 to 4 times the waste water quantity per hour). After sufficient neutralization, a biological purification is carried out. These waste waters require the addition of nutritive salts containing nitrogen or phosphoric acid, or mixing with domestic waste waters.

Neutralization as a preliminary treatment stage is required when these waste waters are introduced into the municipal sewerage system.

6.6.2.6.2 Waste Waters from Cotton Bleaching

The concentration of chemicals in these waste waters can be considerably reduced by the reuse of the cooking liquors and the recovery of the alkali in the used mercerization liquors by evaporation for condensation or by dialysis. In certain cases, more than 90 % of the caustic soda contained in the waste liquor, even though in a diluted form, can be recovered.

For purification, the waste waters are introduced into equalizing basins (sized to accommodate the waste waters flowing in for 12 or more hours). The outflow from such systems still shows a distinct alkaline reaction. Regarding its biochemical oxygen demand, it corresponds to municipal waste water of low concentration so that it should be correspondingly treated alone or, even preferably, in combination with domestic waste waters.

6.6.2.6.3 Waste Waters from Wool Washing

Owing to the widely varying composition of the waste waters from wool washing, they must be purified in separate systems, particularly when the wool grease is to be recovered, which is usually done. This substance is a raw material for the pharmaceutical and the cosmetics industries (lanolin), since its cholesterol concentration is particularly high.

The treatment begins with the retention in rotary sieves of the wool fibers carried in the waste waters. Then suitable chemicals are added to remove the wool grease. Centrifuges, however, are preferable for this purpose.

The washings discharged from the preliminary treatment are either processed or clarified in sedimentation basins for further neutralization. A biological treatment then follows.

Wool washing plants that do not recover wool grease may consider quick digestion a partial purification step. Such quick digestion involves an anaerobic decomposition of the waste water with digested sludge. The sludge is digested in heated tanks for twenty days.

The waters discharged from preliminary purification and washing may be introduced into the municipal sewerage systems without preliminary treatment, whereas the wool washing lyes may be discharged only after thorough deoiling. If the waste waters are pretreated with limewash and calcium chloride where they originate, they may be introduced after a 1 : 10 mixture with municipal waste water.

6.6.2.6.4 Waste Waters from Cloth Production

Waste waters from cloth production result form different operations. It is recommended that the waste waters from those different operations be channeled into different networks and that the required purification methods are selected depending on their pollution.
Further measures applied should be the recovery of reusable substances (e.g. spinning oil from milling liquor, fertilizers, fibrous matter, dyes or dyeing baths, etc.).

The following techniques and operations are used in the purification of waste waters from textile industry: fiber separation with rotary sieves, quantitative and qualitative equalization in mixing and equalizing basins, neutralization, cooling, elimination of dissolved sulfides by the addition of iron salts with proper aeration, possibly combined with preliminary flocculation or precipitation, biological purification with activation or biological filter systems, possibly subsequent treatment in combined flocculation/precipitation stages (iron or aluminium salts plus polyelectrolytes), and adsorption agents.

Natural or artificial biological processes are available for further treatment.

With a large sized activation system (rated better than the biological filters), activated charcoal powder or other chemicals common in waste water treatment (hydrate of lime, iron or aluminum salts, polyelectrolytes) may be added to achieve a discharge quality that is considerably higher than can be achieved with a biological treatment alone. The locations of addition and the doses of activated charcoal and chemicals must be determined in preliminary tests.

The introduction of the waste waters into the municipal sewerage system may be an economically interesting solution after a corresponding preliminary treatment (temperature and color must be noted!) under the previously described conditions.

6.6.2.7 Waste Waters from Cellulose, Paper, and Cardboard Production

6.6.2.7.1 Waste Waters from Pulp Factories

The waste waters from pulp factories are characterized by their high concentration of fine and ultrafine wood fiber particles (higher than 1 g/l). They are also laden with dissolved organic substances.

Depending on the respective grinding operation, the waste waters produced are distinguished as follows:

	Waste waters from	
	white pulp production	brown pulp production
reaction	neutral	weakly acidic
color	colorless	brownish
odor	slightly resinous	strongly resinous
$KMnO_4$ consumption	64 mg/l	841 mg/l
ammonia	not det.	distinct reaction
population correction factor per ton of wood	100	300

The treatment of the waste waters from <u>white pulp production</u> is less extensive due to the introduction of the hot-grinding process (water temperature 60° C) and recirculation. Since the required water supply can be restricted to an extent that the same water volume may be used for grinding for several weeks, only a slight replenishment is necessary and, accordingly, the waste water removal is simple. The waste water arising when the water volume is exchanged is clarified and then best used for agricultural purposes.

Plants still using the older processes (cold grinding) must provide for a mechanical clarification of the waste waters (e.g. with flotation collectors) and possibly also for biological treatment, e.g. in an activated sludge system.

Biological treatment techniques are suited for the <u>steam condensation products arising in brown pulp production</u>.

6.6.2.7.2 Waste Waters from Cellulose Production

Cellulose is mostly produced according to the sulfite and sulfate method (cf. also Section 2.4.2.7.2).

The waste waters from all cellulose-manufacturing plants contain considerable quantities of cellulose fibers. It is economically advisable to recover them. The best effect is achieved by waste water recirculation so as to avoid fiber losses entirely. This also permits a waste water reduction by 50 % at maximum.

The fibers may be separated by screening (e.g. filtration), sedimentation with and without flocculating agents, flotation or collector funnels (see Fig. 6.6.-9).

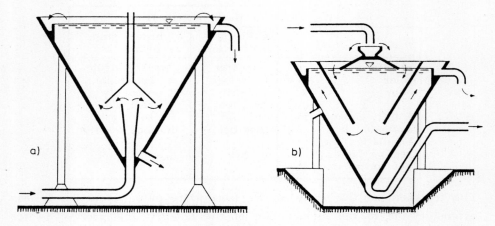

Fig. 6.6.-9: Collector funnels
a - fiber collector (Füllner company); b - KROPP fiber collector system

Moreover, the waste waters, which have partly been treated with aluminum sulfate and sulfuric acid, are passed through sedimentation basins or sedimentation ponds (2 hours minimum duration).

A sufficient liquor treatment remains one of the most important requirements.

6.6.2.7.2.1 Waste Waters from Sulfite Cellulose Production

The undiluted sulfite liquor is a dark brown to black liquid with a slightly acidic odor and varying composition. A satisfactory technique of liquor purification is not yet available. Since this liquid must not be introduced into the receiving water, incineration or utilization are the only possibilities of disposal.

Incineration involves various disadvantages (expansion and incomplete incineration of the concentrated liquor, incrustation of the furnace walls, flue dust as a nuisance to the neighborhood). These disadvantages can partly be eliminated when the concentrated liquor is injected through jets under highly superheated steam into the cylindrical incinerator chamber with tangential introduction of the incinerating air. This technique permits complete incineration without any additional fuel. The flue dust nuisance is avoided when an incineration temperature is selected that is so high that the ash melts into a utilizable slag with a high specific weight.

Sulfite liquors can be used in different ways:

- for alcohol production by fermentation of the sugar contained therein. The undiluted sulfite liquor of
  spruce contains 9 - 12 g/l pentoses, and 29 - 33 g/l hexoses,
  beechwood contains 48 - 56 g/l pentoses, and 12 - 14 g/l hexoses.
  10 l of alcohol are produced from 1 m^3 of sulfite liquor in a fermentation process.

- for yeast production since certain yeast fungi are able to utilize even those liquor pentoses that cannot be fermented into alcohol to build up cell substance. Approximately 40 t of pressed feed yeast can be produced from 1000 m^3 of liquor.

- use of the concentrated liquor as a raw material for glue, dust binders for roads, substitutes or fillers for tanning agents, cementing agent for lime briquetting (cellular pitch) and for the molding sand in foundries, as an additive for Portland cement, for vanillin production, as a soil upgrading agent, etc.

When mixed with domestic waste waters (75 %), pretreated sulfite liquor (25 %) may also be purified by land filtration.

6.6.2.7.2.2 Waste Waters from Sulfate Cellulose Production

As in the case of sulfite pulp, the chemicals in liquors discharged from the sulfate cellulose production plants can be recovered. To this end, the liquors are passed through multistage condensers, then thickened in disk condensers, and finally incinerated in a rotary furnace.

The remaining washings have a dark brown color, are strongly alkaline, and contain large quantities of fiber matter and dissolved organic substances.

To purify the waste waters, the fibrous matter is first retained either in a fiber collector or by preliminary hydrolysis.
The organic substances that negatively affect receiving water (up to BOD_5 values of 2,500 mg/l) can be eliminated in activated sludge basins with the addition of nutrients.

The coloring lignin substances can be reduced by flocculation with aluminum sulfate.

High lignin concentrations may considerably impede the sludge dehydration. Up to 4 - 5 years may be required for a natural dehydration. Recently, a mixed bacteria culture (Thermobacter mix) has been developed that is suited for application in the treatment of waste waters from the paper and cellulose industry.

6.6.2.7.2.3 Waste Waters from Chemical Straw Pulp Production

The cooking lye used in chemical straw pulp production is also regenerated to recover the caustic soda fraction. For this reason, the most thorough lye collection is deemed important. Therefore, only ca. 50 % of the volume of caustic soda and organic pollutions, in some cases even less, are introduced into the waste water to be discharged. The waste water concentration mainly depends on the extent by which the waste water is recirculated. The total waste water quantity is in the range between 500 and 1000 m^3 per ton of finished product.

6.6.2.7.2.4 Waste Waters from Paper Mills

On average, 200 m³ of water are required per ton of paper. This demand may even be as high as 500 m³. As a rule, the water flowing from the paper presses is sedimented or flotated to remove the solids and fibrous matter. Then it is circulated into an open installation cycle so that the quantities of freshwater and waste water are considerably smaller than the previously mentioned 200 or 500 m³/ton of paper. Modern paper mills produce 12 to 45 m³ of waste water per ton of paper, or 12 to 20 m³ of waste water per ton of cardboard. Various collectors are used such as flotation collectors, disk collectors, suspension filter, vacuum drum filters, centrifuges, etc. (see Section 6.2).

After preliminary mechanical clarification or chemical precipitation by flocculation with a preliminary mechanical clarification, the BOD_5 is 2 to 7 kg/ton of paper or 4 to 6 kg/ton of cardboard after this first purification stage.

For the low-pollution waste waters from fine paper production, a chemical-mechanical treatment is usually sufficient. Fibers and fillers are flocculated by the addition of chemicals, hydrate of lime, and aluminum sulfate in most cases. Auxiliary flocculating agents are used to enlarge the thus created flakes and to render them sedimentary. Then the flakes are separated in common sedimentation basins, even though specific basins are preferable (such as Cyclator, Koagulator), with a final mechanical dehydration.

When the pollution load in waste waters is stronger so that the BOD_5 is distinctly higher than 50 mg O_2/l after this purification stage, a complete mechanical-biological purification is required. The complete purification in activation systems should not involve a volumetric load higher than 0.7 to 0.9 kg $BOD_5/(m^3 \cdot d)$. Nitrogen and phosphorus deficits can be balanced with the addition of appropriate compounds in the appropriate amounts. In such a case, a preliminary mechanical purification should be provided to eliminate any interfering sludge growth due to inert solids in the activation system and to protect the equipment such as aerators and pumps against clogging by coarse matter. If, however, the concentration of solids in the water is low due to internal mill measures, the preliminary purification preceding the activation stage may be omitted.

It could be noted in some cases that various organic substances added during paper production (such as starch degradation products) inhibit the flocculation of mineral opacifying substances in a preliminary chemical-mechanical stage so that these substances become destable and able to flocculate only after the decomposition of the organic substances in the activation system. When the mineral opacifiers are not bound into the activated sludge, the final discharge remains more or less turbid. A secondary purification by the addition of main and auxiliary flocculating agents in a final chemical-mechanical stage is appropriate in such cases.

When circulation systems are installed in paper and cardboard mills, the following aspects should be considered, since they may negatively affect waste water purification.

- Within the circulation systems, suspended matter may be converted to dissolved and semidissolved matter under various conditions, so that a load increase in kg COD or kg BOD_5 per ton of finished product may arise in the waste water.

- The waste waters discharged from the circulation system to the purification stages are often very warm so that the rate of biological processes is considerably higher. The oxygen consumption in the individual stages of clarification is accelerated. Where free oxygen is not available, sulfate respiration commences, which is accelerated by the heat. The final product of sulfate respiration, the malodorous and toxic hydrogen sulfide, is well known.

The production of hydrogen sulfide may begin in the preliminary and secondary clarification basins of the mechanical-biological systems, so that the activated sludge is impaired.

Even when the treatment of the sludges from the individual clarification stages is continued in thickeners and dehydrators, putrefaction and the production of hydrogen sulfide occur, naturally, much more quickly and intensively when the water temperatures increase. Circulation systems should therefore be closed only to the extent by which they facilitate the final waste water purification. It must be noted that the installation of highly self-contained circulation systems requires an increased use of chemicals to counteract foaming and slime formation. Particularly, the slime-preventing chemicals, which are disinfectants, may cause problems in the purification of mill waste waters.

6.6.2.7.2.5 Waste Waters from Straw-Board Production

The purification of the waste waters from straw-board production has so far created considerable difficulties.
The waste waters discharged from small and medium size mills were passed to extensive land areas.
In large mills, however, the waste water is mixed with hot water and then passed into large compartments for fermentation (ca. 8 days). Then the waste waters are completely fermented by other operational waters, are diluted, and left in large reservoirs in the soil for some weeks for final fermentation. This method, however, involves considerable odor emission.
Another method used is the anaerobic decomposition with digested sludge in sludge digestion compartments.
The waste water from straw-board production can also be purified in activation basins with 8 hours of aeration and with the addition of nutrients.

6.6.2.7.2.6 Waste Waters from Wood-Fiber Board Production in a Wet Process

The example of the waste water situation in a wood-fiber board plant using a wet process is suited to demonstrate that a restriction of the production water circulation system considerably increases the product-specific BOD_5 loads and thus causes difficulties in the purification of the waste water. The reason is the prolonged thermo-chemical treatment of the wood extracts in a sulfuric environment. DEWES* quotes the following values for an operation processing ca. 90 % beechwood and 10 % spruce.

---

* E. DEWES: Method of complete purification of the waste waters from the Renitex wood-fiber board company in Niederlosheim. Schriftenreihe des Ministeriums für Arbeit, Gesundheit und Sozialordnung des Saarlandes (Publications by the Ministry for Labour, Health and Social Affairs of the Saarland), Volume 11, 1978

Plant Discharges m³ per ton of board	BOD5 kg O_2 per ton of board
50	33
25	37
15	50
10	66
5	100

After its neutralization and the addition of nitrogen and phosphate-containing compounds, the waste water can be easily purified in an activated sludge process. With volumetric loads of ca. 2 kg BOD_5/ (m³ d), 90 to 95 % of BOD_5 and 85 to 90 % of COD can be eliminated. Such a system has been operated successfully for some years in a plant discharging a BOD_5 of ca. 6 to 8 tons O_2/d.

6.6.2.8 Waste Waters from Petroleum Processing Industry

6.6.2.8.1 Waste Waters from Crude Oil Production in Oil Fields

Salt brines (well waters) and oil are produced in oil fields; they are discharged from pipes and from various processes, and account for the majority of the waste waters. Moreover, rainwater is also polluted with oil, so that it must be deemed polluted waste water.

A complete separation of the brines and the oil by sedimentation and flotation is possible only in rare cases. For this reason, disemulsifiers are used to achieve a chemical separation.

The brines are pumped back into the dry wells for removal (after an appropriate treatment).

The oily waters are passed through open channels installed in the oil fields to a treatment plant on the site. There, they are treated with the processes known to be effective for oil waste waters.

6.6.2.8.2 Waste Waters from Mineral Oil Refineries

Several (in some cases even 6 to 7) different networks are installed in mineral oil refineries to drain the various types of waste waters. These networks serve to collect and drain the waste waters to the various treatment areas.

In general, the refinery sewage networks are subdivided into separate systems, i.e. in systems for oil-carrying and non-oily precipitation waters, oily process water, sulfide- and phenol-laden or sulfur-containing process waters, as well as domestic waste waters.

Cooling waters from surface coolers and oil-free rainwaters may sometimes be introduced into the receiving reservoirs directly or after a short-term oil separation stage.

All the other waste waters from a mineral oil refinery must undergo a purification process before they are introduced into the receiving water.

The process waste waters are pretreated by way of evaporation for concentration or by stripping in order to remove strongly smelling components that have negative effects on the subsequent purification stage. In this step, pollutants such as sulfides, mercaptans, and phenols must be incinerated.

The floating oil is separated at present in various types of oil separators, such as recommended by the American Petroleum Institute (see Fig. 6.6.-10), SHELL parallel separators (Fig. 6.2.-20), laminar oil collectors, etc.

The oil recovered in the oil separators, which may contain several percent of solids and more than 50 % of water, must be purified before its treatment is continued.

The oil which cannot be included into the separation under gravity effects is removed from the waste water by a chemical process involving flocculation. Usually aluminum sulfates, iron (II) sulfate, and iron (III) sulfate are used as flocculating agents, partly in combination with activated silicic acid or synthetic precipitating agents.

Further purification can be achieved in an attached biological activated-sludge system.

When the demands are extremely high, a second biological stage or filtration through an adsorptive material can be added.

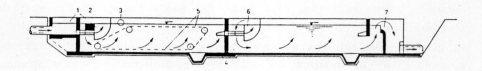

Fig. 6.6.-10: Oil separator according to the recommendations of the American Petroleum Institute

1 - distribution chamber; 2 - filter; 3 - oil collector tube; 4 - compartments; 5 - sludge scraper and scum stripper; 6 - distribution compartment; 7 - overflow

6.6.2.8.3 Waste Waters from Petrochemical Plants

Methods are employed for the preliminary waste water treatment that eliminate the disturbing components. Examples of these methods are steaming, gassing, extraction, chlorination.

Solid, semisolid, and highly viscous residues from the production operation, which are difficult to decompose, must be incinerated.

Waste waters carrying inorganic pollutants must undergo a preliminary treatment that is appropriate to their chemical composition before they may be discharged. Such pretreatment depends on the requirements in each case, and may include neutralization, detoxication, separation of heavy metals, mechanical clarification, or flocculation, or simply a stage balancing the concentration. The salt concentration of the waste water may be reduced at most by internal measures. The various waste water treatment techniques alone are not suited for this purpose.

The combined treatment of the mixtures of the organically polluted operational waste waters is initiated, in some cases, with a chemical flocculation and sedimentation. Then, depending on the individual case, a biological purification stage may be provided with single or multistage operation.

6.6.2.8.4 Waste Waters from Gasoline Stations

Gasoline stations and company car pools must be equipped with gasoline separators (see Section 6.2.2.1.3.).

6.6.2.9 Waste Waters from Foodstuff Production

The waste water from the foodstuff industry usually does not cause problems, so that it can be biologically purified together with municipal waste water provided that the discharges do not contain excessive quantities of pollutants. A thorough recovery of reusable substances and savings in freshwater may essentially reduce the pollution volume in the waste water.

6.6.2.9.1 Waste Waters from Sugar Factories

The waste waters discharged from the sugar industry are characterized by their high level of decomposition. They begin to rot very quickly so that various specific measures are required for their treatment:
- strict separation of the waste waters by the types of processing or according to the circulation systems,
- reutilization of the water so that the smallest quantities possible of waste water need treatment,
- provision of dump ponds relieving the receiving water.

Modern factories produce very little waste water, ca. 25 % to 30 % of the industrial water volume supplied.

It is common to operate three circulation systems:
- the system circulating the beet fluming and washing waters (see Fig. 6.6.-11) includes sedimentation basins and a chlorination and/or liming stage for the waste water. The chlorination must <u>follow</u> the sedimentation, while liming must <u>precede</u> this treatment.
- the system circulating the discharges from diffusion and chip presses,
- the system circulating the condenser waters in cooling towers and spraying systems (Fig. 6.6.-12) includes subsequent chlorination to prevent fungus growth.

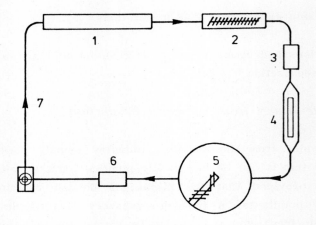

Fig. 6.6.-11: Beet fluming water cycle /107/

1 - beet fluming stage; 2 - beet washer; 3 - beet-leaf catcher; 4 - sand trap; 5 - sedimentation basin; 6 - chlorination stage; 7 - recirculation

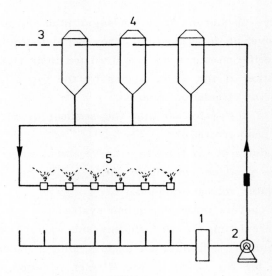

Fig. 6.6.-12: Condenser water cooling /107/

1 - chlorination; 2 - recirculating pump; 3 - possible freshwater supply; 4 - condensation; 5 - spraying system

When specific requirements apply to the discharges, the decomposition of the dump ponds may be increased by aeration, or the water leaving the pond may go through another natural or artificial biological treatment.

6.6.2.9.2 Waste Waters from Milk Processing Plants

The waste waters result from washing and flushing tank vehicles, containers, pipes, pumps, apparatus, and other production equipment, as well as from cleaning the rooms and yard areas. The waste water volume per 1000 liters of milk supplied amounts to 0.5 to 3 m³, with a BOD_5 between 0.3 and 4 kg.

Milk involves a BOD_5 of ca. 100,000 to 120,000 mg O_2/l, while the BOD_5 in whey is ca. 30,000 to 36,000 mg O_2/l.

Both natural and technological methods are used for waste water treatment.

Examples of natural methods are agricultural utilization by irrigation or spray irrigation, fish ponds, and, in rare cases, land filtration.

Low-load and two-stage biological filters, multistage submersed biological filters, activated sludge systems, oxidation basins, or aerated ground basins are proven technological treatment systems.

Presumably the best method of waste water removal in a dairy is the connection to the local sewerage system provided that such a connection is available. The predominant factor in this solution, however, is the ratio of dilution with the domestic waste water. When the dilution ratio is lower than 1 : 10, specific measures of preliminary treatment of the dairy discharge are not necessary. When a dairy is located in the final branch of the sewerage system, it is recommended that cooling water volume be discharged at the end of each working day in order to clean the piping and protect the channels against acid fermentation products. When a dilution higher than 1 : 10 is not allowed in the municipal or communal sewerage system, specific measures are recommended to avoid a surge load on the municipal sewage plant due to the dairy discharges. In such a case, a collector basin on the dairy site could be one solution. The outflow of the waste waters should be delayed, while a simultaneous balancing is achieved. Another example of such measures is the purification of the mixed waste waters in systems with a higher buffering capability, e.g. oxidation ditches or oxidation ponds sized for several days of waste water duration.

6.6.2.9.3 Waste Waters from Production of Nonalcoholic Beverages

Waste waters from juice and mineral water production should not be introduced into the natural waters without previous treatment (except into very large rivers, seas, oceans).

In a company's own sewage plants, the waste waters are treated in a biological process including screens and activation systems or oxidation ponds. Owing to their lack of nitrogen and phosphorus, all unfermented fruit juices are not sufficient as nutrients for the mineralizing bacteria. Therefore, the addition of nutrients (BOD : N : P = 100 : 6 : 1) to the waste water is required, or the waste water must be mixed with domestic waste water.

The pH value should be regularly checked and corrected or diluted in order to keep it within the limits of 6.5 and 9.

The installation of equalizing basins upstream of the treatment stages is recommended so that variations in quantity and composition can be balanced.

The clarified waste water can be reused to wash bottles, as process water, or for general floor, car, and other cleaning purposes. In such a case, however, it is necessary that the clarified waste waters undergo a further purification by

- a second biological treatment stage, or
- sand filtration and
- disinfection (with chlorine) and a subsequent
- dechlorination by means of activated charcoal or sulfur dioxide.

6.6.2.9.4 Waste Waters from Breweries

Unnecessary pollution of the waste waters and disturbances of the operation in the waste water purification system should be avoided. Therefore, malt and hop residues, yeast particles, protein-containing sediment or sludge, and other solids should be retained by screening and sedimentation as early as possible in the various operational stages. Economically it is better to make them usable as feedstuffs and fertilizers than loading the waste water with these substances, which also involves additional investments for waste water purification. When such uses are not possible, these substances can be anaerobically decomposed in digesters. Their organic substance is decomposed into carbonic acid and gaseous methane with a small portion of low-molecular-weight intermediary compounds. The remaining solid residues can then be accommodated on land areas. The sludge water is best mixed with the waste water to be purified.

Waste waters are mechanically treated (screening is sufficient) and are then passed through a biological process in land treatment, biological filters (in two stages with a preceding collector basin), biological disk filters, activation basins, oxidation basins. A preceding equalizing basin may be economically advisable when artificial biological methods are applied.

When the waste waters from malt production are treated in a company's biological system, the addition of nitrogen compounds may be necessary.

Waste waters may also be discharged into the public sewerage system. In such a case, however, equalizing basins are necessary.

6.6.2.9.5 Waste Waters from Margarine, Edible Fat, and Oil Production

The main function of waste water treatment in this field is the separation of fatty from fat-free waste waters.

The fat substances are retained in fat separators and reused. As far as fat emulsions are concerned, flocculating agents (lime, chloride of lime, aluminum or iron salts) must also be added. A subsequent flotation is preferable. The resulting sludge is concentrated and dehydrated either naturally or artificially. The waste water treatment is continued in agricultural uses or in activation basins or two-stage low-load biological filters. The dosed addition of nutrients or mixing with domestic waste waters is necessary.

The hot gas scrubbing waters discharged from fat hardening and the condensation waters from vacuum distillation must be cooled in graduators, cooling towers, or ponds, etc., before their discharge. The gas scrubbing waters need a preliminary clarification in sedimentation systems. If they cannot be recirculated into the production operation after such treatment, they may contribute to the dilution of the purified concentrated waste waters. The same applies to warm cooling waters of a higher purity.

When waste waters from margarine production are introduced into municipal sewerage systems, they may cause destruction of cement lines and joints and gas production (hydrogen sulfide) when they are combined with domestic waste waters. These effects can be avoided when the waste waters are neutralized and cooled before they are discharged into the public sewerage system.

6.6.2.9.6 Waste Waters from Slaughterhouses and Meat Processing Plants

When the utilizable substances and those substances that might clog the piping (e.g. claws, horns, bristles, etc.) are removed, the waste waters from slaughterhouses can be introduced into the municipal sewerage system.

The rumen contents often can be introduced into the sewerage system. When this is not possible, they can be pressed in centrifuges or worm presses and then incinerated.

Parts of the intestines should be comminuted before their introduction into the sewerage system. Hot waste waters (> 30° C) must be cooled before they are discharged.

When engineering or economic aspects prevent the connection to the sewerage system, the waste waters must be purified in separate sewage plants.

The prerequisite of any such sewage treatment is the separation of fats. Solid offals may be digested in a biogas installation at 30 °C for 25 days, together with the stable dung and manure, the sludge from the fat separators, and the sludge from the waste water clarification facilities. Composting is another possibility.

The waste waters are first anaerobically decomposed with digested sludge in heated facilities for 2 to 3 days. Then the pretreated waste water can undergo thorough biological purification in waste water ponds, by land filtration, by agricultural utilization, in biological

filters, and in activated sludge basins (aeration time 54 hours, final clarification basin 8 hours) or in oxidation trenches.
The waste waters are first disinfected by chlorine addition.
The sludge resulting from the purification of the waste waters is exposed to further treatment by anaerobic decomposition in heated digesters for stabilization. This process prevents negative environmental effects.

6.6.2.9.7 Waste Waters from Fish Curing Plants

Waste waters discharged from fish curing plants fluctuate widely. Moreover, they contain considerable fat. A simple treatment is a thorough defatting stage, e.g. in a SCHREIBER-GANSLOSER fat separator (see Fig. 6.6.-13).
The discharge of these waste waters into the public sewerage system involves primary unpleasant odors. Chlorination of the waste waters is therefore recommended, while the piping length up to the sewage plant should be as short as possible.

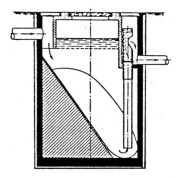

Fig. 6.6.-13: Fat separator for fermentable waste waters /107/

When engineering or economic considerations do not allow discharge into the sewerage system, the waste water treatment becomes difficult due to the protein compounds and the high salt concentration. The waste waters should first be neutralized with lime or other alkaline agents. If necessary, chlorine should be added before the mixture is left in ground basins for several days. Dilution would be another method, possibly using domestic waste waters.

6.6.2.9.8 Waste Waters from Fruit and Vegetable Canneries

A thorough separation of the strongly polluted washing and blanching waters from the cooling and condensation waters is required. The undissolved constituents should be separated from the waste water as quickly as possible and used as feedstuff in fresh condition. Another possibility is composting with lime and peat or other water-absorbent material, to convert these substances into a fertilizer.

After the previously described treatment, the waste waters from fruit and vegetable canneries can be discharged into the public sewerage system, provided that a dilution of ca. 1 : 20 is observed.

Such waste waters can be treated in combined mechanical and biological sewage plants. A preliminary treatment with chemical flocculating agents may reduce the BOD_5 by 40 to 50 %.

Waste water ponds with ca. 5 to 6 months duration (note the development of odors), land treatment by irrigation and spray irrigation, activated sludge techniques, digestion processes (96 % BOD_5 reduction in 9 days of digestion) come into question as natural methods of treatment.

The sludges may be colored, e.g. red from red beet, green to light brown in the vegetable harvesting season, black in cold seasons. Unless biological treatment is successful, decolorization attempts should be made with chloride of lime or directly with chlorination.

The occurrence of bulking sludge in the clarification basins could be prevented when iron sulfate was added.

6.6.2.9.9 Waste Waters from Starch Production and Potato Chip Processing Plants

Considering of the relatively large quantities, the high concentration, and the varying composition of the waste waters discharged from starch production, their introduction into municipal drainage systems is in most cases not advisable. In cases where it is advisable, however, a preliminary treatment of the waste waters in a digester system or an activated sludge basin (if necessary with a preceding neutralization stage) is recommended so that the waste water load can be reduced to a level which is easier to manage in the municipal sewage plant. Potato starch companies with seasonal operation cannot be considered for such a set-up.

First of all, the waste water quantity should be reduced in such companies. The gluten-laden waste waters and the starch washing waters can be utilized for soaking and swelling, and, owing to the subsequent steaming for concentration, as a nutrient for the production of antibiotics. The protein substances and gluten can be collected so that the BOD_5 level is reduced to ca. 80 %.

Mechanical and biological methods can be used to purify the waste waters.

Depending on the types of raw materials and on the processing methods, the actual clarification stage should be relieved by preceding leaf catchers, screens, pulp collectors, etc., upstream of the locations where such matter arises.

The waste waters are neutralized with lime and introduced into sedimentation basins.

Spray irrigation, biological filters, and activated sludge basins are used as biological methods. If, however, a final BOD_5 of less than 50 mg/l has to be achieved, the biological treatment will have to be carried out in two stages. The decomposition efficiency that can be reached is in the range of 98 - 99 %.

Waste waters from potato chip processing operations are passed through vibrating screens or centrifuges, and can then be discharged into the municipal sewerage system. Another possibility of purification is the agricultural utilization or the treatment in oxidation ditches and similar biological systems.

6.6.2.9.10 Waste Waters from Wine and Champagne Production

These waste waters carry the following pollutants:
- grapes and tendrils, which are collected when the equipment and the floors are cleaned during and after wine pressing,

- unfermented juice slime (slime lees) during decanting and cleaning of the barrels,
- yeast residues (yeast lees) during the first and second racking and after the subsequent barrel cleaning,
- residues of clarification agents such as bentonite or blueing,
- wine and dirt residues when used bottles are cleaned,
- residues from liquor and sulfuric water discharged from bottle rinsing, even when new bottles are cleaned,
- wine residues spilled during barrel decanting and bottling.

The solid wastes such as tendrils or shoots, juice and yeast lees, and any other filter residues, including diatomaceous earth, are retained and composted.

Aerated waste water ponds and biological trickle filters are the suitable methods of waste water purification.

The clarification of waste waters from wine-growing operations, however, definitely demands the selection of a method that is sufficiently flexible to manage the considerably fluctuating waste water peak loads even outside the harvesting and racking seasons. The design of a sewage plant accommodating only the peak load, which lasts for about two months, is no longer economically acceptable. For this reason, a solution must be found that will manage a sufficiently sized basic load and is suited to absorb peak loads.

When the flow rate in the receiving water exceeds 1 m³/s, without any other load, the waste waters may be introduced after a preliminary mechanical treatment.

6.6.2.10 Waste Waters from Rendering Plants

Various types of sewage plants are available for waste water purification based on the activation techniques. The first purification stage should always be preceded by a thorough fat removal by precipitation and flotation. It is also advisable to install a buffering basin upstream of the activation system, to equalize the loads discharged per day and during the weekends. Activation systems should be operated in the range between 0.2 and 0.4 (at maximum) kg BOD_5/m^3 of aeration volume per day. The addition of phosphates is required when a biological purification is carried out.

Owing to the frequently very high ammonium concentrations, the low-load ranges involve certain problems:
- Ammonium causes toxic inhibiting effects due to free ammonia, specifically in the alkaline range.
- If nitrification commences, it often goes up to the nitrite level, so that the discharges may contain up to several hundreds of mg nitrite per liter. In such cases, the pH value may drop to 5 and even below, which may inhibit the decomposition of carbonaceous compounds.
- The production of nitrite and nitrate may result in floating denitrification sludge in the final clarification basins.

Provisions must be made in activation systems for the addition of neutralizing agents to readjust the pH value from the alkaline to the neutral range, using mineral acids, or from the acidic to the neutral range, using hydrate of lime.

Depending on the operational results at hand, activation systems are suited for nitrogen elimination from the waste water; these systems are operated with a volumetric load of 0.4 kg BOD_5/m^3 of aeration space per day at maximum. These systems offer an opportunity of realizing nitrification and denitrification processes simultaneously due to the creation of oxygen-rich and low-oxygen intermixing zones.

The formation of floating sludge in final clarification basins can be avoided when their inflow structures are designed as a compartment with stirring equipment, without aeration, for ca. 40 minutes time. It is also advisable to provide the final clarification basins with scum remover equipment.

6.6.2.11 Waste Waters from Fish Meal Production

The waste waters discharged from presses, centrifuges, and fish transportation and spawning waters are first treated to recover train oil and proteins. Fat and train oil separators such as those of the SCHREIBER-GANZLOSER type, rotojectors, and the like are proven equipment for this purpose.

The introduction of a chlorine/air mixture (80 g chlorine per m³ of air) has been successful to improve the fat separation and to retard digestion in the total waste water mixture.

The total waste water is treated in mechanical-biological sewage plants. The treatment is facilitated by a preliminary treatment and chemical precipitation with iron (III) chloride, lime, and clay.

Specific measures for deodorizing the total waste waters are not required when the water used for vapor condensation is chlorinated before application.

The waste waters must be passed through fat separators (see above), with a preceding emulsion cracking stage, before they may be introduced into the municipal sewerage system. A sufficient dilution with domestic waste waters must be ensured at the discharge location. Especially, unacceptable conditions arise when the waste water discharged from fish meal production is mixed with municipal waste water already in the putrefying stage in the sewerage system, since this combination will highly accelerate putrefaction. A practical solution to this problem consists in the addition of odor-masking, antiputrefactive chlorinated carbohydrates to the municipal waste water before putrefaction starts. Another successful method is the chlorination of the waste water from the fish meal plant with chlorine quantities of 50 to 150 g/m³. Odors must also be expected when other companies discharge acidic waste waters at the same location into the municipal main sewer.

6.7 Handling Residues from Waste Water Treatment

Since the waste water treatment system must be maintained in an operable condition, it is extremely important to remove and process the residues. This residual matter is nearly always hygienically critical or hazardous so that its treatment and elimination is an important problem. The costs of the facilities to be installed for this purpose in sewage plants often amount to 30 % and more of the total investment.

The essential locations where such residues incur are

ca. 130° C for dehydration. After a treatment with sulfuric acid they are neutralized, then bleaching clay is added for filtration. Following distillation and deodorization, they are mixed with various additives, depending on their application as motor oil, lubricating oil, hydraulic oil, etc. Fig. 6.7.-1 is a schematic of an installation to process used oil /189/. Oil and fat residues that can no longer be used are incinerated in specific incinerators. Concerning its thermal value, good used oil is practically equal to light fuel oil.

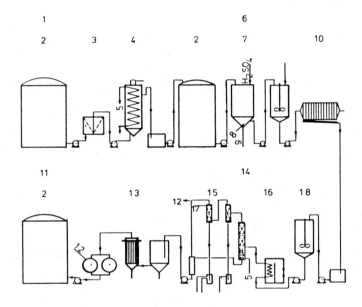

Fig. 6.7.-1: Schematic of a used-oil separator /189/

1 - used oil; 2 - tank; 3 - purification; 4 - drying; 5 - steam; 6 - refining stage; 7 - agitators; 8 - air; 9 - acid resin; 10 - filter press; 11 - processed oil; 12 - vacuum; 13 - filter cartridge; 14 - distillation; 15 - columns; 16 - furnace; 17 - water; 18 - bleaching clay

6.7.4 Sludge Treatment

The sludge quantities retained in screens, sand traps, and oil or fat collectors are very small, compared to the sludge volume arising in the sedimentation basins of the mechanical, biological, and chemical purification stages. To achieve a constantly good purification efficiency in a sewage plant, it is therefore very important to select and sufficiently size the best sludge treatment and removal processes possible in each case.

6.7.4.1 Sludge Volume and Composition

The sludge types occurring in the sewage plant are classified as follows:
- <u>Sludge from preliminary clarification (primary sludge)</u>, which is produced when the sedimentary substances sediment in the raw waste water. The remover pushes it into the sludge funnel from where it can be discharged with ca. 90 - 95 % water content.
- <u>Sludge from the biological purification</u>, which arises from the biological activities of the microorganisms used in the purification process. The sludge has a flaky structure and is separated in the secondary sedimentation basin. In the biological filter process, it is removed from the final clarification basin in the form of a so-called <u>biological filter sludge</u>, whereas, in the activation process, the predominant portion is recycled into the activation basin as recirculation sludge. Only the respective sludge increase is removed as <u>excess sludge</u>. In most cases, biological filter sludge and excess sludge are reintroduced into the inflow of the sewage plant. They are then separated, together with the primary sludge, in the preliminary clarification basin. This technique reduces the high water content of biological sludges.
- The term <u>chemical sludge</u> relates to the sludge arising in the chemical-physical purification. It is sedimented in downstream sedimentation basins.
- The sludge from the preliminary clarification is generally also called <u>fresh or crude sludge</u>. When it also contains the sludge from the biological purification stage, roughly 95 - 96 % water content can be expected when the sludge is removed from the funnel in the preliminary clarification basin. The crude sludge is generally also mixed with the <u>floating sludge</u> provided that the latter is not particularly oily in special cases so that it needs separate elimination.

- **Digested sludge** is the term applying to the crude sludge digested in the digester compartments. The digestion process reduces its volume to roughly two thirds of the initial solids. Depending on the thickening extent, the water content is 90 - 96 %.

6.7.4.1.1 Sludge Volume

Table 6.7.-1 /117/ gives a survey of the sludge volumes. For plants without a large share of commercial waste waters, the sludge quantity is best calculated from the number of inhabitants connected to the sewage plant. The sludge volume produced per person per day is subject to certain variations that depend on the living standard and the habits of the population (water consumption, nutrition, working place within or outside the catchment area of the same plant, etc.).

Owing to the considerably varying water contents of the sludge, which depend on the clarification process selected, on the mode of operation (e.g. duration in the sludge funnels), on the sludge properties, and on other factors, the variations in the unit volume data must be normally expected to be greater than the variations arising for solids. Therefore, the statements on the solids percentages in sludges and thus on the volumetric sludge quantities, e.g. in liter per person per day, are inaccurate. This is the reason why the volumetric information given in Table 6.7.-1 should be handled with caution.

The influence of the water content in a particular sludge at a constant solids concentration on the sludge volume is obtained from the following formula and Fig. 6.7.-2.

$$\frac{V_1}{V_2} = \frac{TS_2}{TS_1} = \frac{100 - WG_2}{100 - WG_1}$$

where:

V_1, V_2 = sludge volumes, e.g. before and after thickening
TS_1, TS_2 = solids content, in % by weight
WG_1, WG_2 = water content, in % by weight

This formula shows the importance that must be given to the drainage of water from the sludge to reduce the sludge volume and thus facilitate the sludge utilization or elimination.

Tab. 6.7.-1: Sludge list, quantities and compositions of sludges from the mechanical and biological purification of domestic waste waters for different purification purposes /117/

	a	b	c	d	e
	mean specific solids quantity	organic solids content therein (measured as ignition loss)	solids content (TS)	water content	mean specific sludge volume $\frac{a}{c} \cdot \frac{100}{1000}$
	g TS/(P·d)	g oTS/(P·d)	%	%	l/(P·d)
crude sludge from mechanical waste water purification	54	38	5-10	95-90	av. 0.72
from biological waste water purification with biol. filters					
acc. to minimum requirements	32	14-16	4-8 av. 6	96-92 av. 94	av. 0.54
with thorough nitrification	24	9-11	4-8 av. 6	96-92 av. 94	av. 0.4
with activation process					
acc. to minimum requirements	40	20.4	0.5-2.5 av. 1.5	99.5-97.5 av. 98.5	av. 2.67
with thorough nitrification	36	16.2	0.5-2.5 av. 1.5	99.5-97.5 av. 98.5	av. 2.4
aerobically stabilized sludge (primary and secondary sludge)	60-70	25-45	2-6 av. 4	98-94 av. 96	av. 1.5-1.75
digested sludge (anaerobically stabilized sludge) from mechanical waste water purification	34	15-12	5-8	95-92	av. 0.52
from mech./biol. waste water purification (biological filter)	61-55	27-22	4-6	96-94	av. 1.16
from mech./biol. waste water purification (activation process)	66-62	29-25	4-6	96-94	av. 1.28

The relationship, which is given in the graph and the formula, between the sludge water content (WC) and its volume (V) at a constant solids percentage (TS) (in the graph, 5 on the ordinate axis corresponds to 5 kg) strictly speaking applies only when the solids density is selected to be 1. In reality, it is ca. 1.3 to 1.4 for sludges from domestic waste waters, but this difference may be deemed negligible in practice with the water contents of 75 % and more, which are relevant from an engineering viewpoint. The statement of the solids content is generally preferable to the statement of the water content. One of the reasons for this is the fact that it indicates directly the reduction of the sludge volume, e.g. in dehydration operations.

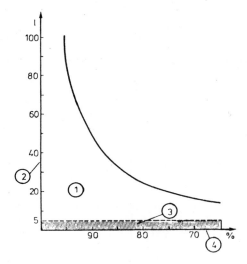

Fig. 6.7.-2: Sludge volume as a function of the water content /3/
1 - water; 2 - sludge volume; 3 - solids content; 4 - water content of the sludge

An increase of the solids content (e.g. by dehydration) e.g. from 5 to 10 % thus corresponds to a volume reduction by 50 % from 5 to 20 % to a reduction by 25 %.

Since the sludge volume produced per person per day also depends on the specific waste water purification process, Table 6.7.-1 also gives the figures rated for the most frequently used methods.

The figures relate to the most frequent case of a drainage network that must also accommodate the stormwater (combined operation). When the rainwater is separately drained (separate operation), substantially smaller sludge volumes cannot usually be rated. When, in the combined operation method, the sludge carried by the rainwater is supplied into the sewage plant, i.e. when well functioning rainwater basins are operated, the figures must be increased by 20 %.

Certain additions to the sludge volumes given in Table 6.7.-1 must be made when the catchment area of the sewage plant includes small and medium industrial areas and establishments such as laundries, dairies, slaughterhouses, and similar installations (Table 6.7.-2).

When these establishments are supraregionally important or when other major industrial establishments are connected to the network, it is difficult to indicate the quantity of sewage sludge arising. Specific studies are generally required in these cases. Table 6.7.-3 is a guideline for such calculations.

Tab. 6.7.-2: Population values for some commercial establishments for local servicing /3/

Commercial operation	Production or raw materials unit (output per day)	Correction value per unit of production
dairy, without cheese production	1000 l of milk	20 - 50
dairy, with cheese production	1000 l of milk	100 - 200
slaughterhouse	1 hd. of cattle ≙ 2.5 pigs	20 - 100
brewery	1 hl of beer	5 - 15
laundry	1000 kg of laundry	100 - 1000

Tab. 6.7.-3: Mean total solids load of suspended matter in the waste waters from various types of industrial operation /3/

Waste water discharging plant	unit of production or the like	total load of suspended matter, kg/unit
brewery	1000 l of beer	2.5 - 5
malting plant	1000 kg of malt	1 - 3
starch production	1000 kg of cereals	2 - 4
fruit juice production	1000 l of finished product	1 - 2
yeast production	1000 kg of molasses	10
dairy	1000 l of milk	0.5 - 1
dairy without whey utilization	1000 l of milk	2
slaughterhouse	1000 kg of meat	2.5 - 10
meat processing	1000 kg of finished product	1 - 2
laundry	1000 kg of laundry	3 - 5
margarine production with refinery	1000 kg of finished product	4
vegetable cannery	1000 kg of finished product	3 - 5
tannery	1000 kg of hide	3 - 60
paper production from cellulose	1000 kg of paper	5 - 20
paper production from waste paper	1000 kg of paper	30 - 50

About two-thirds of the total suspended matter can be expected to sediment on average and to separate by ca. 90 - 95 % after 2 hours of sedimentation.

In chemical precipitation, a precipitating agent always increases the sludge volume considerably, since more pollutants are sedimented due to the fact that the conversion products of the precipitating agent are added. The main reason, however, is that the water content is much higher. The volume increases to two to three times the sludge volume that would be obtained from sedimentation alone. When lime is also used, the increase may be even higher. Paper or coal not only considerably increase the volume but also the dry residue /73/.

6.7.4.1.2 Sludge Composition

The composition of the sludge produced in sewage plants depends on the origin and the digestion stage of the sludge. The primary sludge from the preliminary clarification basins is yellowish gray. Coarser constituents (feces, vegetable and fruit residues, corks, paper, etc.) can still be distinguished. In warmer weather this sludge quickly begins to putrefy. The release of the great quantity of turbid and odorous sludge water becomes more and more difficult. The sludge from the biological purification stages is more homogeneous. In the fresh condition, its color varies from gray to brown. It begins to putrefy even more quickly, because it contains a high volume of organic matter (60 - 75 %), which is easily attacked by the saprophyte bacteria. The dehydration properties of such putrefying sludge are very poor.

Completely digested sludge is black (iron sulfide), its odor reminds one of tar, and it is relatively easy to dewater. The contents of organic matter is still ca. 50 %, while the solids are reduced to approximately two-thirds.

Aerobically stabilized sludge has a brownish color and an earthy odor. The engineering limit in aerobic stabilization is also ca. 50 % of organic solids in the stabilized sludge. When spread in a thin layer, it dries on beds within two weeks in the summer without creating an odor nuisance.

The composition and the characteristics of sewage sludges are given in Table 6.7.-4.

- pH value. The pH value of a particular sewage sludge is generally around 7. Completely digested sludge or sludge distinctly in the methanogenic fermentation stage displays a slightly alkaline reaction (7.0 - 7.5), whereas the reaction of crude sludge or sludge under "acidic fermentation" is in the slightly acidic range (up to 6.0 or even less). The pH value of a sewage sludge is therefore a useful indicator of its state of digestion unless an industrial load is involved.

  The pH limits where a troublefree methanogenic fermentation takes place are narrower and the pH values are higher, the higher the load of the digester compartment is. Excessively high pH values, e.g. values that may arise from excessive lime addition, distinctly negatively affect the digestion process, be it by direct toxic effects or by a shift of the ammonium ion/ammonia equilibrium toward high toxic ammonia values.

- Water content, solids content. The water content of the sludge is the most important property in terms of engineering. It is determined from the loss in weight due to the sludge completely drying in the water bath or drying oven (at 105° C). When, for instance, the loss is 90 %, the water content is 90 % and the dry residue or solids content is 10 %.

- Ignition loss. The ignition loss is even more important than the solids content, especially when the control of the digestion process is involved. The predried material is heated red hot (ca. 30 minutes at 550° C) and then the percentage of organic substance, i.e. decomposable or gas-producing matter, is determined. Crude sludge contains ca. 70 % organic and 30 % mineral matter.

- Volatile acids. In the course of the digestion process, organic, low-molecular-weight acids always occur as intermediate products. Since they can be distilled off they are also termed volatile acids. Their quantity is generally given in mg/CH_3COOH (acetic acid) or in mmole/l. Their quantity is very important for the assessment of the digestion process. A sudden increase of organic acids beyond the

Tab. 6.7.-4: Magnitude ranges of the composition and characteristics of sewage sludges /3/

Cons. No.	sludge characteristics		dimension	crude sludge from mechanical systems	crude sludge of a biological nature (biological filters or activated sludge)	sludge type			
	constituent or type of characteristic of a particular type of sludge					digested sludge insufficiently digested ("acidic fermentation")	digested sludge moderately digested	digested sludge well digested	digested sludge very well digested
1.	pH			5.0-7.0	6.0-7.0	6.5-7.0	6.8-7.3	7.2-7.5	7.4-7.8
2.	solids (TS)		% by weight	5-10	4-8	4-12	4-12	4-12	4-12
3.	loss at red heat		% by weight rel. to TS	60-75	0.5-3.0 or 55-80	55-70	50-60	45-55	30-45
5.	acid consumption		mg/l CaCO$_3$ or mmole/l	500-1000	500-1000 in cases <500	1000-2500	2000-3500	3000-4500	4000-5500
				20-40	20-40 in cases <20	40-100	80-140	120-180	160-220
6.	volatile acids		mg/l of acetic acid or mmole/l	1800-3600	1800-3600	2500-4000 and above	1000-2500	100-1000	<100
				30-60	30-60	40-70 and above	15-40	2-15	2
9.	ether extract		% by weight rel. to TS	10-35	5-10	2-15	2-8	1-6	1-4
10.	total nitrogen		same, as N	2-7	1.5-5.0 or 3-10	1-5	1-3.5	0.5-3.0	0.5-2.5
11.	total phosphorus		same, as P	0.4-3	0.9-1.5	0.3-0.8	0.3-0.8	0.3-0.8	0.3-0.8
	potassium		same, as K	0.1-0.7	0.1-0.8	0.1-0.3	0.1-0.3	0.1-0.3	0.1-0.3
13.	spec. filtration resistance		m/kg	$10^{11}-10^{13}$	$10^{12}-10^{13}$	$5 \cdot 10^{11}-5 \cdot 10^{12}$	$10^{11}-10^{12}$	$5 \cdot 10^{10}-5 \cdot 10^{11}$	$10^{10}-10^{11}$
16.	caloric value		kJ/g TS	16-20	15-21	15-18	12.5-16	10.5-15	6.3-10.5

value normal in a particular sewage plant is an indicator of organic matter overload or of damage to the methanogenic bacteria (intoxication). Such disturbed conditions can often be recognized by the increase of the organic acids rather than by the reduction of the pH value, the gas yield, or the increase of the carbon dioxide percentage of the gas.

When only the volatile acids are used for assessment, a proper odorfree methanogenic fermentation can be expected at volatile acid values up to 1000 mg/l, related to acetic acid. In the range from 1000 to 2000 mg/l, the methanogenic fermentation is predominant, however, a slight hazard is involved. With values higher than 2000 mg/l, a transition to acidic fermentation and thus odors must be expected, whereas values higher than 3,500 mg/l may occur only in acidic fermentation due to the overburden of or damage to the methanogenic bacteria by toxic agents.

- Organic substance, producer gas. The thermal value and the methane percentage are normally determined for producer gas. Substantially depending on the fat percentage, an average of 850 - 1000 cm^3 gas/g of decomposed organic substance can be expected as the gas yield, for instance, from municipal crude sludge, with methane percentages of ca. 65 to 70.

- Nutrients. Nutrients are required for the bacteria to become biologically active. A wide range of nutrients is available in domestic waste water sludge. Industrial waste sludges, e.g. sludges containing mainly organic carbon, other nutrients are often lacking. According to BUSWELL /3/, 7 mg of nitrogen are required per 1 g of sludge to be decomposed. When 1 l of crude sludge contains 60 g of solids of which 15 g can be decomposed, 7 x 15 = 105 mg of nitrogen must be available. More recent studies on the carbon/nitrogen relationship have shown that a C/N ratio of 10 to 16 is particularly favorable. The nitrogen/phosphorus relationship should be 7.

Even with agricultural utilization, the nutrient content is highly relevant. The organic percentage in completely digested sludge is ca. 45 % - 50 %, while ca. 50 % - 55 % of minerals are present.

- Heavy metals, toxic matter. The sludge must be analyzed for toxic matter in order to avoid disturbing the digestion process. When

certain concentration limits are exceeded, a reduction of the gas yield must be expected, as can be noted from Table 6.7.-5.
The general statement applies that the ability of a well operating digester compartment to absorb toxic surge loads is higher, the higher the pH value and the acid consumption are, and the higher the load of substance is which the digester can still process properly. The agricultural utilization of such a sludge, however, demands special attention (sewage sludge regulation). This applies also to the presence of heavy metals (see Section 6.7.4.4.1.6).

- <u>Dehydratability</u>. The dehydration properties are determined on a paper filter or, preferably, on a sand bed. This analysis determines the time elapsing until the sludge becomes solid. The sludge behavior on the filter paper also allows conclusions as to whether the sludge can be dehydrated in fresh, completely digested, washed condition or mechanically after the addition of precipitating agents.

- <u>Caloric value</u>. Depending on their contents of organic substance, sewage sludges have a more or less high caloric value. Domestic and municipal sewage sludges show a close relationship between the loss at red heat as a measure of the organic substance and the caloric value. When the caloric value is indicated as an upper limit H_o (i.e. cooling of the incineration products to the initial temperature, all the water occurring in incineration is present in a liquid form), a caloric value of ca. H_o = 26 kJ can be expected per 1 g of loss at red heat according to NIEMITZ /3/. The values are higher by ca. 5 - 10 % for fresh sludges.

Table 6.7.-6 is a survey of the caloric values of different sewage sludges.

Tab. 6.7.-6: Caloric value of sewage sludges /3/

sludge type	solids %	ignition loss % of dried matter	H_o kJ per g of solids	
			empirical	calculated
crude sludge	7.7	63.3	17.4	16.5
moderately digested sludge	4.5	52.2	13.4	13.6
well digested sludge	9.2	40.8	11.1	10.6

Tab. 6.7.-5: Concentration limits of the most important constituents in commercial waste waters, having toxic effects on the digestion process /3/

substance	concentration limits above which reduction of gas yield and thus disturbance of sludge digestion begins		remarks
	mg/l in w. water	% rel. to sludge solids	
ammonia	-	2	depends on the pH value of the sludge. The higher the value, the higher is toxicity
arsenate	4	-	-
gasoline	-	1	-
benzene, toluene, etc.	-	2	partly subst. lower data (above 0.1 %)
lead	-	0.2	
cadmium	1-5	0.2	
chlorinated hydrocarbons	10	0.01	depends on the structure of the compounds, owing to their high density accumulation in the sludge
chromium, trivalent	10	1-2	owing to precipitation only intermed., certain acclimatization is possible
chromium, hexavalent	1-2	0.05-0.4	same as above
cyanide	2	0.01-0.02	same as above
formaldehyde	-	0.2	-
copper	1	0.1-0.5	accumulation in the sludge at values as low as 1.0 mg/l in the waste water is possible
solvents, e.g. alcohols etc.	-	0.5-1	varying, depends on the solvent, minor quantities have a stimulating effect
nickel	1-10	0.2-1	-
oils, lubricating oil etc.	-	-	more mechanical disturbance, scum layer
phenols	-	0.2-0.4	-
thiocyanate	-	1	partly lower data (above 0,3 %)
salts	-	10	depends on the cation and other factors, toxicity increases in the sequence Ca, Mg, Na, K, NH_4
sulfates		1	strong hydrogen sulfide formation
sulfides, organic S-compounds		0.1	varying data
surfactants	30-40	0.5-1.0	foaming disturbed according to surfactant type
zinc	3-10	0.3-0.5	-

- Pathogenic germs, parasite eggs, etc. Fresh sludge may contain all kinds of pathogenic organisms, parasite eggs, and the like, which subsist in human or animal excrement. In view of epidemic control and hygiene, this sludge is therefore an extremely hazardous material. The conditions in a well operating digester compartment are so ideal for the methanogenic bacteria that pathogenic germs are killed or that their virulence is weakened. With the digestion process advancing, parasite eggs are also generally destroyed or they lose their ability to develop. Owing to the content of pathogenic germs, however, the direct land treatment with completely digested sewage sludge always demands particular precautionary measures. Proper composting, e.g. in combination with solid wastes, offers a higher reliability. Only pasteurization (at 70° C) or hot drying (at 100° C and above) provides definite safety.

6.7.4.2 Crude Sludge Stabilization

Crude sludge from mechanical-biological sewage plants has a high water content. Unless it is processed as quickly as possible, acidic fermentation begins. Dehydratability rapidly decreases while mephitic substances are formed, which have frequently been the cause of complaints and court charges.

Moreover, crude sludge is rich in pathogenic germs and worm eggs and, therefore, hygienically hazardous.

Crude sludge is usually processed in anaerobic or aerobic biological processes with a thorough decomposition of the organically digestable substance in both methods. Odors are thus prevented, while the sludges are easier to dehydrate.

MUDRACK /118/ gives the following classification in terms of decomposability:

a) inorganic matter that is not decomposable, e.g. sand, metal oxides, carbonates,

b) organic matter that is not digestible or is difficult to decompose:
   from the crude water: plastics, coal, partly cellulose and wool fibers,

from the excess sludge: wall and envelope substances of the microorganisms,

c) decomposable organic matter:

from the crude water: parts of excrements, vegetable and meat residues, fat, carbohydrates (starch grains),

from the excess sludge: decomposable suspended matter and colloids (e.g. proteins) adsorbed to the flakes, reserve matter stored in the microorganisms (glycogen, lipoids, volutin) and the protoplasma (protein) of the microorganisms themselves.

The organic matter contained in the crude sludge and in other highly concentrated substances are excellent nutritive solutions for heterotrophic bacteria, yeasts, and fungi. This knowledge has long been applied in industry, in the food sector, as well as in waste water treatment and other fields.

The first step in the chain of biological conversions of the sludge solids into a soluble substrate that the bacteria cells can absorb is the same under anaerobic and aerobic conditions. It is the hydrolytic disintegration of the vegetable supporting structure substance (paper, vegetable wastes, polymer carbohydrates) and of fats (glycerol ester) and proteins of animal and vegetable origin. The only difference is that the bacteria reproduction rate in an aerobic environment is higher than under anaerobic conditions.

With hydrolytic enzymes, which are secreted by the cell (exoenzymes) and become effective outside the cell, cooperating, polymer carbohydrates are converted to sugar, fats are converted to fatty acid and glycerin, and protein is converted to peptides. The hydrolytic disintegration of solids very quickly increases their ability to swell. For instance, paper becomes slimy in water, cellulose is disintegrated and absorbs water. The dehydratability decreases. This is one explanation for the rapid reduction of the dehydratability of crude sludge.

After this conversion, which is equal under both aerobic and anaerobic conditions, the different aerobic or anaerobic decomposition processes start, which are different also in view of the final decomposition products. In a single metabolic step, CO_2 and H_2O are produced as final products under aerobic conditions. Fig. 6.7.-3 is a schematic of the metabolic processes in anaerobic stabilization according to MUDRACK and SCHOBERT /118/.

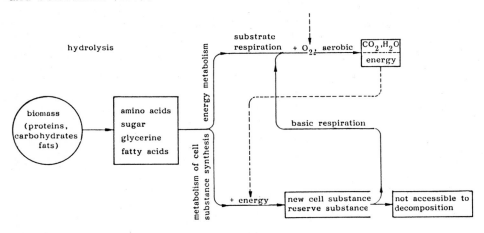

Fig. 6.7.-3:   Schematic of the metabolic processes in aerobic stabilization /118/

Fig. 6.7.-4 is a schematic of the metabolic processes inanaerobic decomposition (digestion) according to MUDRACK and SCHOBERT /118/.

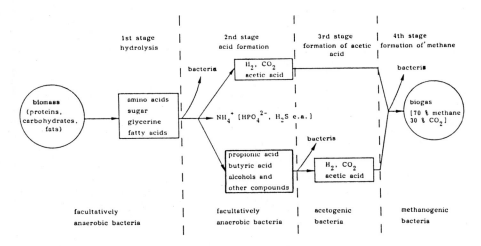

Fig. 6.7.-4:   Schematic of the metabolic processes in anaerobic decomposition /118/

1. <u>Hydrolytic stage</u>. Enzymes convert macromolecular often undissolved substances (polymers) to dissolved fragments of molecules.

2. <u>Acidification stage</u>. In the acid formation stage, short-chain organic acids, acetic acid, alcohols, H_2, and CO_2 are formed. From these intermediate products, however, the methanogenic bacteria are able to convert only acetic acid, H_2 and CO_2 into methane directly. Numerous species and strains of fermentative bacteria are responsible for the two first stages.

3. <u>Acetogenic stage</u>. All the fermentation products that are not accessible to the attack of the methanogenic bacteria must be converted into H_2, CO_2, and acetic acid by the bacteria of the acetogenic stage. This stage is the bottleneck in the methanogenic fermentation process. Acetogenic bacteria display their activities only in a biogenergetic symbiosis with methanogenic bacteria or other H_2-consuming organisms. More than 50 % of the metabolic conversion processes between fermentative bacteria and methane producers pass through the acetogenic stage.

4. <u>Methanogenic stage</u>. During the methanogenic stage, methane is produced predominantly from acetic acid, H_2 and CO_2.
From a morphological viewpoint, the methanogenic bacteria are heterogeneous, from a biochemical viewpoint, however, they are homogeneous, strictly specific as to their substrates and extremely sensitive to oxygen. In well operating digesters, the individual stages of decomposition take place simultaneously, so that there is no accumulation of intermediate products.

The digestion process is highly responsive to any variation in the environment. A temperature decrease of as little as 2 - 3° C results in a restriction of the bacterial activities. The pH value and the organic loading must also be carefully monitored in order to ensure a stable digestion. This sensitivity reduces the reliability of the anaerobic process.

The decisive advantage over the aerobic stabilization is the energy produced in the form of gaseous methane. Even though a considerable portion of the producer gas produced is required to heat the sludge, it is normally sufficient to cover the system's energy demand.

Table 6.7.-7 indicates the gas volumes for various commercial waste waters. The last column indicates the number of days (according to REINHOLD /73/) that had to lapse in the bottle test until half of the gasifiable matter had been gasified. This half life period is a reference value for properly sizing a digestion compartment.

Tab. 6.7.-7: Gas volumes for commercial waste waters /73/

Substance	gas volume at 30° C in l/kg of the		methane concentration %	half life period (days)
	total dry residues	organ.		
A. Municipal Operations				
waste water sludge	431	607	78	8
refuse without ashes	281	305	66	10
including paper	227	259	63	8
including vegetable scraps	608	643	62	6
slaughterhouse offals				
contents of cattle intestines	461	524	74	13
intestinal parts	87	89	42	2
cattle blood	158	159	51	2
B. Industrial Operations				
dairy waste water	975	1025	75	4
whey (67.3 g/l dry res.)	670	–	50	–
waste water from pressed yeast	486	796	85	–
waste water from paper prod.	250	–	60	–
beet and turnip chips	400	423	75	4
apple residues	313	322	75	4
grape residues (rotten)	137	200	79	27
brewery wastes (hops)	426	445	76	2
orange residues	482	500	72	5
banana leaves (dry)	413	450	78	18
C. Agricultural Operations				
manure with straw	286	342	75	19
horse excrement	391	430	76	16
cattle excrement	237	315	80	20
pig excrement	257	415	81	13
wheat straw	348	367	78	12
potato leaves	526	606	75	3
maize leaves	485	514	83	5
sugar beet leaves	456	501	85	2
grass	490	557	84	4
chaff from the thresher (wheat chaff)	338	386	73	10
broom	434	446	76	7
reed	285	314	79	18

The use of the gas to heat the digestion compartments and buildings is general practice in communities where the mean annual temperature is not higher than 10° C (heating is not provided in the conventional Emscher tank). The excess gas is normally flared. Depending on the size of the digestion tanks, between 30 and 45 % of the gas volume normally occurring are required to heat the crude sludge. The excess gas can be utilized in various ways.

- Sale of the gas. Gas sales to the municipal gasworks will be the exception since a special arrangement regarding the caloric value is necessary (standard gas 18,000 kJ/m^3). A better opportunity are the sales to certain municipal operations (slaughterhouse, swimming pool) or industrial establishments.
- Utilization for heating purposes in final sludge treatment. Various thermal methods are available for sludge conditioning. Thermal energy is equally used in methods of sludge drying and incineration.
- Power generation for internal supply. 0.5 - 0.6 m^3 of gas are necessary to generate 1 kWh of electricity. Approximately 5,200 kJ/kWh of waste heat can be recovered from the exhaust gases and the cooling water of the gas engines. This energy can be used to heat the digestion tank. The question of whether a gas-operated power station is economically advisable depends on the local electricity costs and on the size of the plant. The power supply of a sewage plant can be generally operated as an autonomous unit when ca. 100,000 or more inhabitants are connected to the network. This means that the sewage plant can cover its entire power demand from the producer gas. A gas power station should be established only after a thorough feasibility study.

Digestion water. A telescopic tube or three tubes of different lengths are vertically immersed into the sludge. They terminate in a shaft with a vertical slide valve in its discharge. This slide valve is used to control the sludge level in the digestion compartment. The immersed tubes have the effect of an overflow. The digestion water is returned into the inflow into the sewage plant, if necessary through a thickener.

However, the withdrawal of the turbid water is not always a definite advantage. The surge loads of concentrated turbid water on the sewage plant mostly result in a deterioration of the biological purification efficiency, and certainly lead to a wash-out of substances resulting in eutrophication of the receiving water.

Digestion waters contain the following substances /3/:

solids	0.5 -	10 g TS/l
ammonium nitrogen	400	- 1100 mg N/l
BOD_5	400	- 2000 mg O_2/l
potassium permanganate consumption	500	- 2000 mg $KMnO_4$/l
phosphate	200	- 2500 mg PO_4^{3-}/l

A continuous recirculation of the contents in the digestion compartment lead to better results in the digestion water. The following values have been obtained in such a system: COD 290 - 774 mg O_2/l, BOD_5 20 - 125 mg O_2/l.

The digested sludge. The actual aim in sludge digestion is a sufficient stabilization of the sewage sludge. This aim is normally achieved when the organic matter is reduced by ca. 50 % (engineering digestion limit). With 67 % organic dried matter (oTS) in the crude sludge this means a one-third reduction of the total solids quantity. The dehydratability is improved at the same time so that the crude sludge volume is reduced to 40 - 50 %. Water, the so-called digestion or turbid water, is released to a corresponding extent.

This calculation is now explained in detail.

A crude sludge containing 100 kg of solids and 95 % of water is a volume of 100/0.05 = 2000 l; 67 % of the solids, i.e. 67 kg, are assumed to be organic, the remaining 33 kg mineral.

Half of the organic solids disappear during digestion, i.e. 33 kg. Thus, 34 kg oTS + 33 kg mTS = 67 kg TS remain in the digested sludge. This sludge, however, contains 50 % of organic matter in the solids (34 out of 67 kg). The water contents are assumed to be reduced to 92 %, for instance. Then the volume is 67/0.08 = 840 l, which corresponds to 42 % of the crude sludge volume. The released quantity of turbid water is 2000 - 840 = 1160 l.

Table 6.7.-8 is a survey of the values that KEEFER /26/ recommended for the calculation of the decomposition of organic matter at the engineering digestion limit.

Tab. 6.7.-8: Required % reduction of the organic substance at the engineering digestion limit /26/

Mineral solids in the crude sludge	%	20	25	30	35	40	45
organic solids in the crude sludge	%	80	75	70	65	60	55
decomposition, related to organic solids	%	85	75	65.5	55	46	36.5
decomposition, related to total solids	%	68.0	56.2	45.8	36.1	27.6	20.1
organic solids residue, rel. to the total solids	%	12.0	18.8	24.2	28.9	32.4	34.9
organic solids in the digested sludge	%	37.5	43.0	44.6	45.2	44.8	43.7

The completely digested sludge is withdrawn from the tip of the funnel, since it is thickest there. It is usually black. The digested sludge is passed on to further sludge treatment stages (thickening, dehydration, etc.).

<u>Charging of the digestion compartments</u>. Crude sludge is pumped into the digestion compartment usually once or several times a day in small and medium size plants. Then the sludge is recirculated for some time provided facilities are available for well mixing and inoculating the crude sludge.

Larger tanks should be continuously charged, if possible, with continuous recirculation. It is then necessary to arrange open or closed thickeners (closed thickeners as unheated closed final digestion tanks, for instance) downstream of the heated first digestion tank. This arrangement provides for the separation of the sludge from the digestion water. The functional relationship between the digestion tank and the downstream thickener is then comparable to that between the activation basin and the final clarification basin in activation systems.

The digestion tank functions as a reactor. The sedimented sludge can be returned selectively from the downstream thickener back into the reactor, or passed on to further processing stages (pasteurization, wet transfer to agricultural utilization, dehydration, etc.).

This mode of operation equalizes the load on the aerobic purification stages of the sewage plant, which is due to the surge discharge of the very harmful digestion water. Another advantage is a more uniform gas production (80 - 90 % of the gas volume produced in the first stage) so that the gas containers can be made smaller. Another advantage is the reduced load of the digestion waters from containers with continuous recirculation.

When a two-stage digestion or a digestion compartment and a thickener are provided, it is recommended to design both units with the same size and the same installations so that a changeover to single-stage operation and, if necessary, parallel operation is possible in the event of a problem or repair.

Load and dimensioning. Digestion is a biological process following an asymptotic course. Its last stages extend over a very long period so that they eventually become irrelevant from an engineering viewpoint. The process is therefore stopped at a certain point, which is termed the engineering digestion limit. This limit is reached when ca. 90 % of the gas volume arising at 15° C is produced. FRIES and ROEDIGER /141/ have determined that at this point roughly half of the charged organic matter is decomposed.

The required size of the digestion compartment can be approximated from the l/P data resulting from experience.

The sizes of digestion tanks per person connected are indicated in Table 6.7.-9 according to IMHOFF /73/.

Tab. 6.7.-9: Dimensioning of the digestion compartment sizes, in l/P /73/

	Emscher tank	heated digestion compartiment (30°C)	unheated digestion compartment
sedimentation system	50	20	150
biological filter system			
low-load	75	25	180
high-load	100	30	220
activation system			
low-load	150	40	320
high-load	100	35	220

When a heated digester is to be installed for 50,000 persons connected to the network, its useful volume must be 50,000 · 0.04 = 2,000 m³ when the sewage plant comprises a preliminary clarification basin and a low-load activation system. Additions up to 20 % must be made for the operation of rainwater clarification basins.

The gas volume produced per day amounts to 50,000 · 0.026 = 1,300 m³/d. Expressed in another way, the gas yield per m³ of digester volume and day is then

$$1,300 : 2,000 = 0.65 \text{ m}^3/(\text{m}^3 \cdot \text{d}).$$

Apart from the digestion time and the crude sludge volume, the organic volumetric load (kg oTS/(m³·d)) is a dimensioning parameter. The duration times then substantially depend on the extent by which the water contents of the sludge can be reduced by preliminary thickening. Table 6.7.-10 contains some standard values of the volumetric load and the digestion time for the dimensioning of heated digesters.

Tab. 6.7.-10: Digester dimensioning values /31/

	persons	volumetric load	digestion time
load for heated digesters 30 - 33° C	< 50 000 50 000 - 100 000 >100 000	2 kg oTS/(m³·d) 3 kg oTS/(m³·d) 4 kg oTS/(m³·d)	20 - 30 d 15 - 20 d 10 - 15 d

Even though operations with volumetric loads up to 5 kg oTS/(m³ d) have been successful, the full utilization of these high values is not recommended since the digester operation becomes very sensitive to variations. Especially when the occasional occurrence of toxic substances in the crude sludge must be expected the dimensioning should include proper reserves. Values of 2.0 kg oTS/(m³·d) will normally not be substantially exceeded.

Digesters may also be installed in a two-stage arrangement. The first stage is provided with a gas cover, heating, and scum breakers. Two-thirds of the theoretically possible gas output is produced in such digester at 30° C in 5 days only. The overflowing water from the first stage is passed to the second stage, where it is finally separated. The second stage has a most simple design, either in the form of an open reinforced concrete container or a ground basin. Heating and scum breakers are no longer required, the gas cover can also be saved in view of the low gas yield. When a two-stage arrangement is planned, the volumetric load is calculated with the useful volume of both digester stages. The volume of the first stage amounts to roughly one third of the total volume. Digestion compartments provided to accommodate sludges laden with toxic agents must be larger. According to BUCK-STEEG /38/, for instance, 0.4 % copper hydroxide causes an inhibition of the gas production by ca. 13 %. The required digestion time should be increased to 37 days in such a case.

6.7.4.2.2 Aerobic Stabilization

The lack of decomposable compounds in activation systems causes the activated sludge to change over to the basic respiration phase. In this phase, the microorganisms utilize the stored reserve substances and the body's own matter for survival. The growth rate in this phase is very low. The production of excess sludge is therefore correspondingly low.

Two methods of aerobic sludge stabilization have been developed in practice on the basis of these metabolic processes.

The combined stabilization (sludge stabilization in activation basins) is preferred in small plants, since it is a very simple method in terms of both structural design and operation. In larger plants (for 5,000 to 10,000 persons + correction factor) the separate stabilization is frequently applied. In addition to the conventionally designed sewage plant, a separate basin is installed for sludge stabilization. This method can be considered competitive to the sludge digestion method for a catchment of 100,000 persons + correction factor.

When stabilization is carried out directly in the activation basin, the arrangement of a preliminary clarification stage is nonsense, since the primary sludge also needs stabilization. The systems must be dimensioned for a sludge load BTS $\leq$ 0.05 kg BOD_5/kg $(TS \cdot d)$ and total solids TS of ca. 4 g/l.

When 60 g BOD_5 are rated for the crude waste water per person per day, an activation volume of 300 liters per person connected must be available for an aerobic sludge stabilization in combination with waste

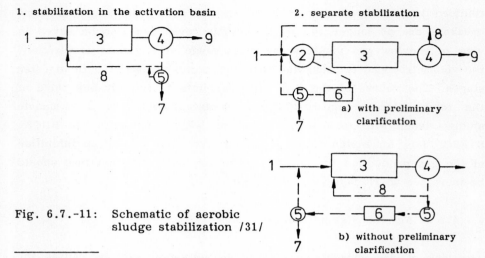

Fig. 6.7.-11: Schematic of aerobic sludge stabilization /31/

1 - inflow; 2 - preliminary clarification; 3 - activation basin; 4 - final clarification; 5 - thickener; 6 - stabilization basin; 7 - stabilized sludge; 8 - recirculation; 9 - outflow

water purification. On the other hand, the preliminary clarification volume of ca. 30 liter per person connected is not required. When an activation system downstream of a preliminary clarification is operated with nitrification, 80 liter activation volume plus 30 liter preliminary clarification volume = 110 liter per person connected are required for BR = 0.5 kg $BOD_5/(m^3 \cdot d)$ and 40 g BOD_5 per person per day. Another 40 liter volume per person connected is required for the heated digester.

The above considerations show that the combined stabilization is acceptable only for small numbers of persons connected. Moreover, inexpensive structural designs, e.g. trench systems or circulation basins, must be selected. The so-called Carrousel system according to PASVEER /2/ (Fig. 6.7.-12) has been developed for industrial and larger sewage plants.

In systems with a separate stabilization stage, the sewage sludge arising in the purification of waste water is separately aerated subsequently, if necessary after a preceding concentration in a thickener. Aeration basins for the separate aerobic sludge stabilization have the same design as aeration basins. The sludge is charged continuously or in batches.

Figs. 6.7.-13 and 6.7.-14 illustrate a lightweight construction and a solid construction of stabilization basins.

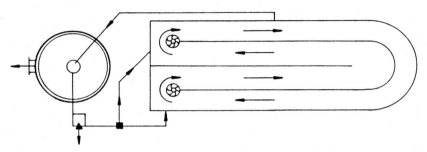

Fig. 6.7.-12: Carrousel system /2/

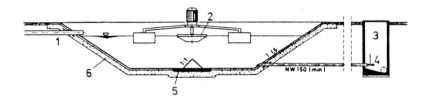

Fig. 6.7.-13: Lightweight construction of a stabilization basin /3/

1 - inflow; 2 - floating rotary aerator; 3 - sludge shaft; 4 - outflow; 5 - bottom reinforcement; 6 - concrete foundation

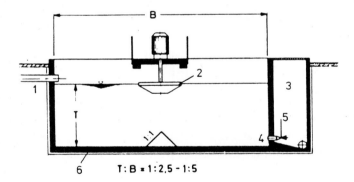

Fig. 6.7.-14: Solid construction of a stabilization basin /3/

1 - inflow; 2 - stationarily mounted rotary aerator; 3 - sludge shaft; 4 - outflow valve; 5 - outflow pipe; 6 - concrete foundation
(B - width; T - depth)

The stabilized sludge can be discharged directly on drying beds. In general, the installation of an intermediate small storage tank in the form of a thickener is recommended for a further reduction of the water contents in the sludge. This is particularly valid when a subsequent artificial dehydration stage is provided.

An essential dimensioning parameter for the stabilization systems is the stabilization time or the sludge age (see Section 6.4.4.5.2).

An unambiguous definition of the "engineering stabilization level" is the prerequisite for safe data in relation to the stabilization time to be selected.

Even though many different criteria can be enumerated, such as
- fat contents $\leq$ 65 mg/g TS
- digestability, odor test with lead acetate paper - no coloring after 6 weeks,
- BOD_5/COD ratio
- reduction of the organic percentage of the dry substance (oTS), engineering limit ca. 50 % oTS residue

the value of 0.1 kg O_2/kg (oTS·d), as mentioned by MUDRACK /118/, for the respiration intensity of a sufficiently stabilized sludge is an assessment measure suited for practical operation.

When the aeration of a normal activated sludge is continued without the supply of additional waste water, its respiration activity is reduced over time. After ca. 6 days a substantial further reduction can no longer be observed. This time is deemed a sufficient minimum for stabilization (comparable to the engineering digestion limit).

The sludge changes to a biologically stable condition in this stage. After discharge from the aerobic process stage, it may then be left unaerated for a while in thickeners, sludge silos, and dehydrators, without odor-generating anaerobic processes (desulfurization, fermentation) occurring.

The oxygen volume consumed up to that stage amounts to 0.8 - 0.9 kg O_2/kg oTS (supplied). The stabilization basin may be loaded up to ca. 5 kg oTS/(m³·d). However, the duration time of the sludge solids is decisive for the stabilizing effect.

Design recommendations for a separate aerobic sludge stabilization involve 20 days to be rated as the stabilization time for the crude sludge. In a mechanical-biological sewage plant with a sludge volume of ca. 2 liters per person per day, this means 40 liters of basin volume. The installed aerator should be designed for an efficiency of 50 to 70 W/m³ of basin volume.

The aerobic sludge stabilization, like any biological metabolic process, strongly depends on temperature, as is illustrated in Fig. 6.7.-15 according to LOLL /3/.

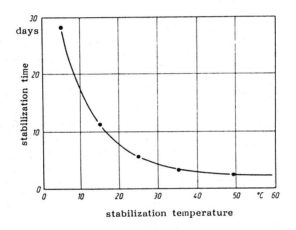

Fig. 6.7.-15: Interdependence between the stabilization time and the stabilization temperature /3/

According to RUEFFER /145/ and LOLL, the stabilization temperature can be maintained at a level higher than 45° C even at very low outside temperatures, owing to exothermal autonomous decomposition processes in a separate aerobic stabilization system (Fig. 6.7.-15). An aeration time of 3 to 7 days is normally sufficient for stabilization with the aerobic thermophilic sludge treatment method.

The aerator should be selected considering the climatic conditions and the respective situation. Systems that do not clog and that ensure proper adaptation to the occasionally strongly varying oxygen demand should be preferred. Surface aerators have proven successful.

The aerobic stabilization always demands an energy supply beyond the normal requirements of an activation system.

Compared to the anaerobic digestion process, which may be seriously disturbed by toxic agents owing to the sensitivity of the methanogenic bacteria, the aerobic stabilization process can be adapted more quickly to toxic agents due to the abounding number of bacteria species and strains involved. The continuous inoculation with waste water and crude sludge after surge loads with toxic agents provides for a rapid and unproblematic resumption of the process.

The method is therefore recommended not only for small and medium size plants but also for sewage plants treating a high percentage of industrial waste water or containing toxic substances.

The aerobic stabilization slightly reduces the dehydration ability of the sludge. The sludge conditioning expenditure for an artificial dehydration thus increases accordingly. From a hygienic viewpoint, the aerobically stabilized sludges must be classified as very harmful when an appropriate final treatment or an aerobic thermophilic treatment is not provided.

Even though the pathogenic germs disappear to a satisfactory extent (90 - 98 %) by the digestion method and likewise in aerobic stabilization, the worm eggs in the sludge are not destroyed. When hygienic objections arise (e.g. when the sludge is spread on pastures and used in horticulture) pasteurization, composting (with sufficient inherent heating), or several months of deposition in anaerobic dumping ponds are recommended.

The economical feasibility of the methods must always be critically examined when prolonged stabilization times are involved.

6.7.4.3 Sludge Dehydration

The high water content of the sewage sludges renders their treatment complicated and bothersome, so that a substantial concern in the sludge treatment methods should be the reduction of the water content. When a method is successful, for instance, in thickening a sewage sludge from 95 % to 85 %, it reduces its volume down to one-third of the initial volume.

Fig. 6.7.-16, according to MOELLER /116/, is a schematic illustrating the volume reduction of a completely digested sludge from domestic waste waters. The sludge water is subdivided, according to its combination with sludge particles, into intermedium, connate and capillary water, internal and adsorption water. With the bonding intensity increasing, the energy expenditure for water separation becomes higher.

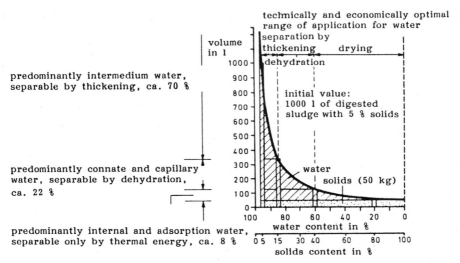

Fig. 6.7.-16: Volume reduction of a sewage sludge by dehydration /116/

For the separation of the intermedium or void water, gravity is sufficient, i.e., technically, the process of thickening.

The separation of the connate or adhesion water and the capillary water demands considerable mechanical force, such as

- negative or superpressure in filters, or

- artificially enforced gravity fields in centrifuges or decanters.

When biological processes are not used, only thermal forces are left to separate the adsorption or interlinking water.

The separation of the internal water finally demands the destruction of the cell walls, e.g. due to their biological decomposition by aerobic or anaerobic operations (aerobic or anaerobic sludge stabilization), by strong heating or freezing.

The reduction of the water contents also changes the condition of the sludge (Table 6.7.-11).

Tab. 6.7.-11: Composition of sewage sludges with decreasing water contents /3/

Water contents	Properties
> 85 %	liquid and suited for pumping
75 - 65 %	generally properly solid, still plastic, pasty and slimy
< 65 - 60 %	crumbly-solid, no longer pasty
< 40 - 35 %	spreadable, permanently solid
< 15 - 10 %	dusty

When thickening under gravity is applied, the predominant part of the water contained in the sludge can be separated. The arrangement of thickeners is therefore an economic solution in almost all cases.

The dehydration methods can be subdivided into natural and artificial methods (Table 6.7.-12).

Tab. 6.7.-12: Correlation and applications for the individual stages of method of separating the sludge water, and their efficiency characterized by the respectively achievable final water percentage /3/

Stage	process	effective due to	individual method	final water content achievable (without conditioning)	associated properties of the sludges by their water-binding capacity or their thickening and dehydration abilities	
NATURAL PROCESSES						
1. stage	thickening	gravity	continuous or discontinuous thickeners	90-85% 75% 99-97%	moderate good poor	dehydration dehydration dehydration
2. stage	dehydration	gravity (natural gravity field) + thermal forces (evaporation)	sludge beds sludge drying beds sludge dumps sludge ponds up to	70-60% 50-30% 85-75% 50% 80-75% 50% 50% 85-80%	moderate good poor moderate moderate with very long storage good poor	dehydration dehydration dehydration dehydration dehydration dehydration dehydration
3. stage	drying	thermal forces	sludge drying beds		only in warm, arid zones, in other zones artificial methods	
ARTIFICIAL METHODS						
1. stage	thickening	gravity	continuous or discontinuous thickeners	90-85% 75% 99-97%	moderate good poor	dehydration dehydration dehydration
2. stage	dehydration	in static methods: pressure difference from atmospheric pressure, produced by negative or excess pressure in the machinery	static methods: vacuum filtration band filter presses (excess) pressure filtration	80-70% 70-50% 85-80% 65-60% 40-30% 75-70%	moderate good poor moderate good poor	dehydration dehydration dehydration dehydration dehydration dehydration
		in dynamic methods: gravity (artificial gravity fields)	dynamic methods: centrifuges/decanters (with 95 % separation levels)	80-70% 70-50% 85-80%	moderate good poor	dehydration dehydration dehydration
3. stage	drying	thermal forces	drying by any method	2-1%		

Particularly in the artificial dehydration methods, the water contents achievable depend on the composition and properties of the sludge and on the conditioning process selected.

Sludge drying is generally considered only a preliminary stage to sludge incineration or a stage applicable when thermal energy is relatively inexpensive (excess producer gas, waste heat from power plants).

6.7.4.3.1 Thickening

The static thickening process involves a separation of the sludges into an upper liquid phase and a lower solid phase under the influence of gravity.

The sludge thickening effect in the sludge funnels of the sedimentation basins (crude sludge) or in the digester (digested sludge) is frequently unsatisfactory. A dilution with waste water can hardly be avoided in discharge. It is therefore recommended to thicken the sludge in specifically designed thickeners, which also permit a better control of the thickening process.

Thus, thickeners can be used for the following sludge types.

a) For crude sludge: The sludge from the preliminary clarification stage with a relatively high water content is pumped into a thickener in the freshest condition possible. The sludge funnel in the preliminary clarification stage can therefore be made considerably smaller. The sludge from the biological stage (biological filter sludge or excess sludge from the activation system) in the secondary clarification stage need not be definitely passed through the preliminary clarification stage. It may be transferred directly to the thickener. The thickened sludge is then passed on to the following sludge treatment stages (e.g. digester).

b) For digested sludge: Particularly in cases where only a single digestion stage is provided, the installation of a thickener is advisable so that not only the water contents in the digested sludge are further reduced but also the turbid water is clarified.

c) For all sewage sludges (fresh or stabilized): Thickening is a stage of preliminary treatment and storage for artificial dehydration. It thus joins the thermal conditioning stages.

The static thickeners are usually circular containers with a funnel-shaped bottom, for up to a 5 m sludge depth.

The installation of a rabbling system has proven successful in larger thickeners. These systems comprise vertical rods, which are slowly drawn through the sludge to open channels for the water so that the water can rise more easily. Such a system accelerates the thickening process. The rabbler is simultaneously used as a remover for the basin bottom.

Since most thickeners are installed above ground, pumps are mounted for charging. The sludge is delivered centrally through an inflow structure with facilities to slow down and uniformly spread the sludge over the thickening compartment. The sludge water flows off over a gutter with spillway weirs on the inside of the thickener. The provision of a submerged partition opposite the discharge gutter prevents floating sludge from drifting off. It is recommended to provide the rotating rabbler with a floating-sludge remover.

The following modes of operation are possible.
1) Continous charging:
   The thickener is operated just like a sedimentation basin (see Fig. 6.7.-17). The thickened sludge is withdrawn from the lowermost point of the basin (funnel), the separated water flows off over a spill crest. It has become common to recycle the water into the inflow into the sewage plant.
2) Discontinuous charging:
   When a thickener is discontinuously charged (static thickening) (Fig. 6.7.-18), the separated water must be withdrawn over a overflow or over several overflows in staggered arrangement on the various levels. The water may also be displaced over a spill crest during the charging operation.

The thickener is dimensioned on the basis of the surface solids load. For mixed domestic sludges it ranges between 40 and 100 kg TS/$(m^2 \cdot d)$. The lower values should be selected for sludges with a substantial percentage of biological matter.

The hydraulic charge rate in the delivery of crude sludge should not exceed 0.75 m/$(m^2 \cdot h)$.

The solids concentrations achievable for the different sludge types can be taken from Table 6.7.-13.

Tab. 6.7.-13: Solids concentrations achievable by thickening /3/

Sludge type	Solids concentration achievable by thickening without conditioning
sludge from the preliminary clarification stage and heavy industrial sludge	10 - 30 %

sludge from the preliminary clarification stage
  loss at red heat above 65 %
  loss at red heat below 65 %
	
	5 - 7 %
	7 - 12 %

sludge from the preliminary clarification stage and activated sludge
  index above 100 mg/l
  index below 100 mg/l
	
	4 - 6 %
	6 - 11 %

sludge from stabilization systems	3 - 5 %
sludge from preliminary clarification and biological filters	7 - 10 %
digested sludge - preliminary clarification	8 - 14 %
digested sludge - activation system	6 - 9 %
thermally conditioned activated sludge	10 - 15 %

When a thickener is charged every hour with 10 m³ of crude sludge with a solids content of 25 kg TS/m³, at a load of 40 kg TS/(m²·d), a surface of

$$F = \frac{24 \text{ (h/d)} \cdot 10 \text{ (m}^3\text{/h)} \cdot 25 \text{ (kg TS/m}^3\text{)}}{40 \text{ (kg TS/(m}^2\cdot \text{d))}} = 150 \text{ m}^2$$

is required.

The thickness of the sludge layer determines the duration of the sludge solids in the thickener. When a high concentration of very active biological sludges (e.g. from high-load sewage plant) is involved, attention should be paid to the fact that digestion does not begin during thickening. Such beginning digestion negatively affects the thickening effect. Particularly in summer, the duration of the solids must be restricted to 1 - 2 days.

Static thickening systems are dimensioned for a useful volume that corresponds to the crude sludge quantity produced in one day. Advantages in terms of operation can be achieved by a subdivision of the required thickening volume into several units, which may also be operated in tandem. With such an arrangement, crude sludge, digested sludge and turbid water are alternately treated.

The turbid water in digested sludge is estimated to have a BOD_5 of 1,500 mg/l. It should therefore be recycled into the sewage plant during night hours and as uniformly as possible.

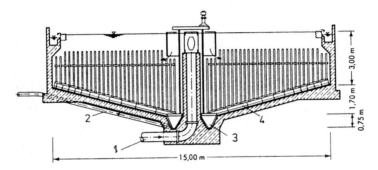

Fig. 6.7.-17: Schematic of a continuously operated sludge thickener with a lattice rabbler /3/

1 - inflow; 2 - sludge discharge; 3 - outflow gutter; 4 - remover

Sludge thickening by flotation has also become a common method. The flotation of sludges is a method where the sludge matter is caused to rise to the surface by adhering fine air bubbles. The solids concentrations achievable are higher than those reached in static thickening by 1 to 2 %. Moreover, the flotation process is substantially quicker than static thickening, so that the sludge are processed in fresh condition with mostly better dehydration properties. Preferably, the sludge flotation technique has been applied for thickening and recovery of industrial sludges.

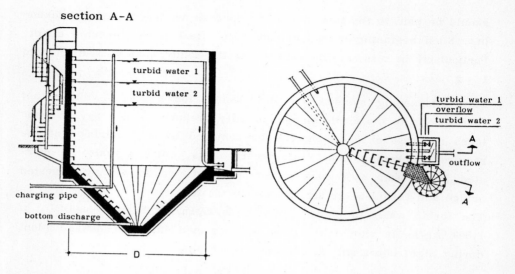

Fig. 6.7.-18: Schematic of a dicontinuously operated sludge thickener (static thickening) with stage outflow for turbid water /3/

6.7.4.3.2 Natural Dehydration

The term of natural dehydration (or dewatering) is a general definition of all methods of removing water under the influence of natural forces, e.g. gravity or thermal forces. Such methods include:
- sludge drying beds,
- sludge ponds.

<u>Sludge drying beds</u>. Sludge dehydration in sludge drying beds must be deemed the oldest dehydration method.

The bottom of the sludge beds (Fig. 6.7.-19, from the German federal regulations on waste water treatment, ATV /3/) comprises drainage lines for recirculation to the inflow into the sewage plant. Gravel and sand layers (200 - 400 mm) of different grain size and thickness are arranged as a filter medium above the drainage bottom. The side walls are now generally made of prefabricated concrete elements. The bed width is 4 m for manual removal, 7 m for mechanical removal in medium size plants, and 12 - 25 m in larger installations. The sludge is spread in thin layers (150 - 300 mm).

Practical operation has shown that the solids thickened on the bottom prevent the sludge water from seeping. This is the reason why the sludge beds are operated as thickeners in many sewage plants, even though the discharge of the accumulating water is not entirely without problems and dehydration up to a proper solidity is not achieved.

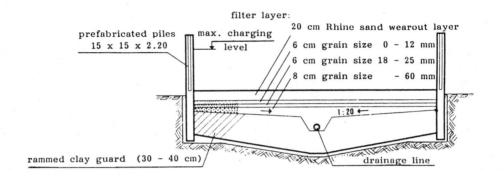

Fig. 6.7.-19: Sludge bed /3/

In warmer climates, evaporation might be highly relevant. In moderate climates, dehydration is difficult because of frequent precipitations.

The indispensable sludge removal renders the operation of such beds highly expensive in terms of labor. Meanwhile, systems have been developed to facilitate bed clearing. In these systems, machinery (crawlers, tractors, caterpillars, trucks, etc.) spread the wet sludge and remove the dehydrated sludge from the beds. Specially designed sludge scoops are a great improvement over earlier methods; they travel along the partition between the beds and are suited to scoop the dried sludge in layers. They permit roughly the double loading of the beds. Their application, however, becomes economical only with plant sizes beyond 30,000 P + correction factor. Another possibility consists in roughening the sludge surface so as to accelerate evaporation.

An annual charge of 1.0 - 2.0 m^3 of wet sludge/($m^2 \cdot a$), corresponding to 100 - 150 kg TS/($m^2 \cdot a$), can be permitted on the basis of operational experience. This corresponds to an approximate load of 5 - 8 P/m^2. The efficiency can be increased when sludge scoops are used for removal, when flocculating agents are added, when the charges are spread in thin layers. Only experiments in situ can determine the optimum bed efficiency achievable.

A well dried sludge has solids concentration of 35 to 45 %. From a hygienic viewpoint, such sludge must be classified hazardous.

Drying beds can be provided in small and medium sized plants if sufficient land is available. Moreover, drying beds are useful in warmer climatic zones.

Sludge ponds. Sludge ponds or lagoons have also been known since the beginnings of waste water engineering. Initially, they were operated as uncovered digesters (cf. Section 6.7.4.2.1.1).

Now they are mostly used to equalize the dicontinuous discharge of sludge to agriculturally utilized areas and also partly as interim dumps prior to final elimination.

After charging, a rapid sedimentation of the solids on the bottom can be observed, similar to the sedimentation on drying beds. Therefore, the main aspect in dehydration is the removal of the accumulated water. The ponds are provided with shafts to withdraw the water accumulating on top of the solids. Other facilities are floating withdrawal systems connected through flexible piping.

MILLER /111/ recently published data on sludge dump pond dimensioning. He considers a dumping period of roughly 1 year necessary in dimensioning when agricultural sludge utilization is intended, with interim dumping during the periods without sludge distribution to the farmers. This leads to a specific volume of 150 l/P as a rule of thumb. With a dumping height of 3 - 4 m, an inner bank grade of 1 : 1 and a bank grade of 1 : 2 at the air side, a specific area requirement of ca. 0.15 m^2/P is calculated. The solids concentration of the dumped sludge doubles over the duration time.

When the sludge pond is envisaged as a dehydration method independent of agricultural utilization, longer durations are recommended, e.g. 2 to 5 years. The charging height should not exceed 1 - 1.5 m/a. Depending on the charging period, the storage time and the intensity of turbid water withdrawal, and also in consideration of the climatic conditions, solids concentrations between 10 and 25 % can be achieved.

The basin volume should not exceed 5,000 m³. Otherwise, the charging and the sludge removal operations will become difficult.

The sludge pond cost substantially depends on the construction work involved. When only earthwork is to be carried out, the costs are relatively low. The lowest costs are involved when the ponds are fairly close to the sewage plant and when the sludge may be left there for final deposition.

The general statement, however, applies that the sludge ponds, in contrast to the drying beds, are suitable also for larger plants when the required land is available and when interference with the environment can be avoided.

Root dehydration. The natural methods also include the dehydration by reeds. When reed is planted on drying beds, the natural dehydration can thus be distinctly improved. A sewage plant in the Augsburg district, in operation since 1974, is reported to be charged every fortnight with ca. 10 m³. After 4 years, a residual sludge volume of 2 %, related to the spread volume, in a 35 cm thick layer, was measured. With operation for four or more years, the method appears rather interesting, mainly in cases where sludge drying beds are already available in a sewage plant. The technical literature quotes 0.06 - 0.1 m² of bed area per person /66/ as the dimensioning value.

6.7.4.3.3 Sludge Conditioning

In almost no cases can sewage sludge be dehydrated in the condition in which it occurs in the treatment stages of the sewage plant without additional mechanical treatment. Such additional treatment precedes the dehydration stage and is termed sludge conditioning.

A fundamental distinction is made between
- chemical sludge conditioning and
- thermal sludge conditioning.

Chemical conditioning. Iron chloride, iron sulfate, and lime are the usual chemical additives. Their addition changes the chemicophysical condition of the sludge (Fig. 6.7.-20).

Iron or aluminum salts as precipitating agents, mostly in combination with lime, flocculate the colloids, reduce the adhesive force of the sludge, and thus facilitate the release of water. Such addition may be interpreted as an energy supply. As a standard rule, 10 kg iron (II) sulfate (in the form of a heptahydrate) or 2 - 3 kg of iron (III) chloride ($FeCl_3$, anhydrous) + 10 kg hydrate of lime are used per m³ of wet sludge.

Organic auxiliary flocculating agents or polyelectrolytes may offer more advantages than do the iron salts, when used as filtering auxiliaries since they are easier to dose and because they avoid aggressive sludge behavior. They are effective either through their electric charge or through cross links formed between the sludge particles. Approximately 100 to 200 g of active matter are added per m³ of sludge. It is important that the flocculating point be exactly found and that any unnecessary load on the flocculated sludge be avoided.

The suitability of flocculating agents and their respectively required quantities must be empirically determined in laboratory tests in each case.

Compared to iron salts, the added quantities are substantially lower, whereas the costs are accordingly higher.

The addition of ash, wood meal, or the like changes or loosens the grain structure of the sludge structure so that the build-up of a filter layer in filtration equipment is facilitated; 1 to 2.5 kg of ash per kg of sludge solids is the standard value.

The conditioning success depends on the best method of mixing and adding chemcials, and on the subsequent reaction. An addition of the acid metal salts separately from the basic hydrate of lime is particularly important, separately in terms of both location (continuous reactor) or time (static reactor). The agent added first must be completely distributed over the volume before the next agent is added.

The effects of volume and sequence of the chemicals to be added on the filtering behavior can be tested in the laboratory. The simplest test involves the conditioning of 300 ml of sludge, for instance, which is then charged into a BUECHNER funnel, diameter 13.5 cm, with inserted black band filter, diameter 12.5 cm. A graduated cylinder is placed underneath to determine the increase of the filtrate quantity as a function of time.

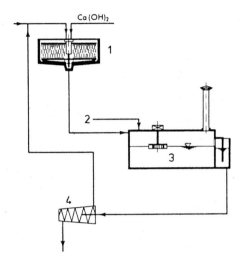

Fig. 6.7.-20: Schematic of sludge conditioning before dehydration, using a centrifuge /3/

1 - thickener; 2 - CO_2-containing gas; 3 - neutralization; 4 - centrifuge

This test may also be used for simple preliminary experiments to compare the various metal salts. It should be noted, however, that the application of this technique in large-scale plants may involve entirely different assessment criteria, e.g. the influence of gypsum formation on filter cloths, lines, and containers.

Thermal conditioning. Sewage sludge can be thermally conditioned when heat is withdrawn or supplied. The different temperature ranges are associated with the following methods.

freezing conditioning: freezing to $-20°$ C

chemical and thermal conditioning: heating to $+60°$ C
utilization of flue gases or addition of precipitating agents

hot thermal conditioning: heating to $+180°$ C up to $+220°$ C
Porteous and Zimmermann method (partial oxidation)

For reasons of completeness, the complete wet oxidation of the sludge at temperatures above $250°$ C with simultaneous oxygen supply must be mentioned (Zimpro method).

The PORTEOUS method (Fig. 6.7.-21) has gained special importance in Germany in the last 10 years. In this method, the sewage sludge is heated to a temperature between 180 and $220°$ C at a pressure of $(10$ to $25) \cdot 10^5$ Pa, and maintained at this temperature for 30 minutes to 1 hour, which means that it is boiled. In a process comparable to boiling a piece of meat, a considerable percentage of the organic substances dissolves.

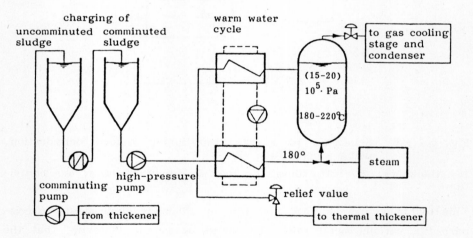

Fig. 6.7.-21: PORTEOUS method with indirect heat exchange and direct heating

When the sludge has been heated in the reactor, it is passed through a heat exchanger where it gives off heat to the crude sludge to be treated.

The treated sludge has very good dehydration properties, without any further additives, even though the filtered water is highly polluted. This is shown in Table 6.7.-14 /66/.

Tab. 6.7.-14: Characteristics of filtered waters from thermally treated sludges /66/

dissolved matter	mg/l	5,000 to 20,000
BOD_5	mg/l	3,000 to 20,000
potassium permanganate consumption	mg/l	7,000 to 30,000
COD (as O_2)	mg/l	4,000 to 30,000
organic carbon	mg/l	3,000 to 8,000
NH_4-bound N	mg/l	400 to 1,000
organic N	mg/l	500 to 1,600
P (total)	mg/l	10 to 80

The filtered water resembles a highly concentrated waste water. The BOD_5 can be reduced by ca. 70 % after 6 days digestion time at 37° C. Another possibility is the treatment of the filtered water together with the waste water in the activation basin when sufficient aeration time is provided. The inflow load is then increased by ca. 25 %.

The thermal treatment ensures not only a favorable sludge conditioning for dehydration but also a thorough sludge disinfection (pasteurization).

The thermal treatment involves a considerable expenditure in equipment. On the other hand, the operating costs are very low. This method is applied only in very large sewage plants.

From the viewpoint of industrial processing engineering, the freezing conditioning technique is more difficult to realize so that it is applied in rare cases only.

6.7.4.3.4 Technical Dewatering

Technical (mechanical) methods are employed to reduce the dehydration process in terms of space and time. The achievable efficiency depends on the energy spent in the process.

The conventional methods of technical sludge dewatering can be subdivided into two groups:

- the methods based on water separation due to increased gravity (centrifuges, separators),
- the methods based on sludge water separation due to external pressure (band filter, filter press, vacuum filter) by means of a water-permeable separating membrane.

From among these methods, centrifuges, band filters, and filter presses are generally more common.

Apart from the procedural differences, the three most important dewatering units, i.e. centrifuge, band press, and filter press, are fundamentally distinguished from each other by
- the ability to retain solids,
- the achievable solids concentration in the thickened sludge or filter cake,
- the mode of conditioning.

Even in modern times, the decision to adopt one of these methods mainly depends on the final sludge elimination method and the associated demands on the nature of the dehydrated sludge /66/.

6.7.4.3.4.1 Dewatering in Centrifuges

Centrifugation is an operation in which a tube rotating at a high speed separates the solids from the water owing to their different sedimentation rates. As a result of their higher sedimentation rate, solids are deposited on the wall of the drum with increasingly finer solids being deposited toward the center. The aqueous phase is accumulated on top of the solids. The water flows out from the centrifuge through a sill disk and is drained off as centrifuged liquid (Fig. 6.7.-22). The solids are generally discharged from the centrifuge by a worm conveyor corotating with the centrifuge at a low differential speed /66/.

One feature of all centrifuges, i.e. the fact that a distinct separation beween solids and liquid, like in filtering processes, cannot be achieved, causes problems. A lower water concentration in the filter cake is at the expense of a relatively high concentration of solids of the finer fractions in the filtered liquid (poor solids yield).

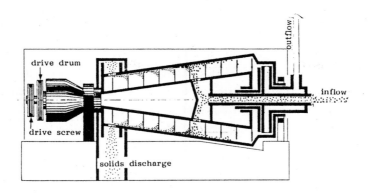

Fig. 6.7.-22: Schematic of a centrifugal decanter /107/

The solids yield can be improved when organic flocculating agents (predominantly cation-active polyelectrolytes) are added in dosed quantities directly into the clarification zone inside the centrifuge.

In contrast to the use of inorganic flocculation agents, the dosed addition of polyelectrolytes has the advantage of avoiding an artificial solids supply. Owing to the relatively high cost of organic flocculating agents, a most balanced adaptation to the demand should be attempted, while excess doses are to be avoided. The main reason lies in the fact that excess doses result in a negative impact on the dehydration properties.

It has not yet been possible to size centrifuges on the basis of characteristics determined on a laboratory scale. Data such as the specific filtration resistance do not furnish a distinct correlation with the level of separation or the concentration achievable in the thickened sludge.

Centrifuges with a hydraulic charge are commonly available. Good separation levels are achieved when the centrifuge is charged with 50 to 60 % of the hydraulic charge rate, depending on the solids concentration in the sludge to be dehydrated.

The centrifuge cost is relatively low. The systems wear out, however, very quickly unless a good sandtrap is provided.

The operating cost substantially depends on the conditioning method required. The centrifuges are characterized by their continuous operation. It is possible to automate the addition of the flocculation agent, e.g. as a function of the composition of the centrifuged liquid which is determined by a turbidity measurement. The thick sludge concentration, however, is limited. But this restriction may be overcome when quicklime is subsequently added in measured quantities. Depending on the added quantities, the thus treated centrifuged sludge can be consolidated to a degree that a deposition in landfills becomes possible.

6.7.4.3.4.2 Dewatering with Band Filters

Various designs and types of band filters are available. The best known filter is presumably the screen belt press.
A thin layer of sludge is spread on a circulating screen belt. After a preliminary dehydration under gravity, the sludge is pressed through the pressing band above the belt, and raked off at the end of the machine. This arrangement is based on the principle of filtration with a screen as separating membrane.

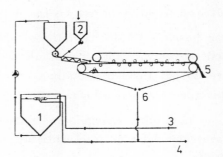

Fig. 6.7.-23: Schematic of sludge dehydration with a screen belt press
/107/
1 - thickener; 2 - reservoir with flocculating agents; 3 - inflow; 4 - outflow; 5 - filter cake; 6 - filtrate and rinsing waters

Mobile supporting and press rollers can be mounted in such a way in the last belt section that they force the belts to reroute several times so that the pressing forces are augmented by additional shear forces acting on the sludge.

The belts can be guided both horizontally and vertically. The screening effect can be enhanced by belt vibration.

The returning screen belt section is passed along washing jets.

The total recirculated water volume (filtrate and washings) normally corresponds to 1.5 to 1.7 times the charged volume. The solids they contain amount to ca. 0.5 to 2 % by weight of the charged sludge dry substance, provided that the sludge has good dehydration properties.

The recycled water is normally strongly polluted in terms of BOD_5 and COD. It is therefore passed through a mechanical clarification stage in a downstream small sedimentation basin, with recirculation of the sedimented sludge into the press.

The addition of organic flocculating agents is a proven successful method of sludge conditioning, as used centrifuges. The required polyelectrolyte quantities are in the range from 75 to 250 g/m^3. Polyelectrolytes are commercially available in the form of dry powder or viscous solutions with 30 % of active substance. Owing to the considerable work involved when the 0.05 to 1 % solutions, which are required for addition, are prepared from the dry powders, the easily dilutable viscous solutions are more and more coming into use. A so-called activator, which is supplied together with the solution, must be added.

In a way comparable to the centrifuge dimensioning, the band filters cannot be sized on the basis of specific or characteristic values even though the specific filtration resistance ought to be a suitable theoretic parameter. Therefore, tests with industrial-scale equipment are recommended to determine the filter efficiency, the solids concentration in the filter cake, and the composition of the filtrate. At a constant filter zone length, the throughput is related to the width of the press. Ratings between 2.0 and 5.0 $m^3/(m \cdot h)$ are common. Depending on the sludge type, filter cakes with 25 to 30 % solids concentration can be achieved.

Since band filters are constructed with 2 m maximum band width, their practical application is restricted to small and medium size sewage plants /66/.

The costs involved are similar to those of centrifuges. The operating costs are slightly higher than those incurred by centrifugation, but, in return, the thick sludges have a higher concentration. The band filter operation is also continuous. These filter systems are suitable for the dehydration of crude and digested sludges.

6.7.4.3.4.3 Dewatering with Filter Presses

Filter presses are distinguished from the other two dewatering methods in view of essential characteristics, especially by the composition of the filtrate, the achievable filter cake concentration, and their operation. They are composed of a number of plates suspended in a supporting and guide structure. The edges of these plates are reinforced, and the plates are covered with a filter cloth acting as a separating membrane. When the plates are pressed together, a cavity is formed between two plates, into which the sludge is charged for dehydration. Pressure ($12 - 16 \cdot 10^5$ Pa) is applied to press the sludge water through the filter cloth while the solids slowly accumulate as a filter cake on the inner side of the compartment, until the compartment is completely filled with the solids. The filtrate flows off through grooves on the filter plates.

The plates are then moved apart (most commonly by an automatic feed mechanism) for filter cake removal.

When all compartments are empty, the press is closed again for another charge. The dehydrated sludge is thus discontinuously produced, in contrast to centrifugation and dehydration in band filters. The filter cake should come off and drop out, if possible without any mechanical means, immediately on opening of the compartment. If necessary, different filter cloths can be tried to find the best filter cake solution. Since fine solids or conditioning chemicals stay in the filter cloths after each filtration operation, which increases the cloth resistance, the cloths must be flushed from time to time under high pressure. Acidification may occasionally also be necessary to dissolve gypsum or calcium carbonate deposits. When clogging occurs too frequently, other filter cloth types should be used to improve the situation.

Sludges with poor filtration properties can be dehydrated in the so-called precoat method which involves the application of a thin layer of wood meal or ash (precoat) in the press.

The sludge to be dehydrated in filter presses is normally conditioned with inorganic flocculating agents (iron and aluminum salts, lime). Moreover, the thermal conditioning technique has proven successful. The specific conditioning operation should aim at a specific filtration resistance of ca. 10^{12} cm^{-2}.

The addition of inorganic chemicals means the supply of additional solids, which also have to be treated in the subsequent steps of operation. When the sludge is dumped in landfills, these additional solids increase the volume, and in the event of sludge incineration they involve a furnace load, since they are nonflammable. The technique of conditioning with organic polyelectrolytes has so far not proven successful.

The use of membrane filter presses offers new possibilities for dewatering. This improvement of the known compartment filter presses includes two plates with a membrane plate between them that expands when filled with water or air, thus pressing against the adjacent volumes. In contrast to the normal filter press using a high charging pressure, this membrane press is charged with sludge at normal pressure. The result is a more uniform pressure distribution, while the charging and pressing operation is less forceful. This method has been developed to improve the filtering properties of particularly poorly dehydrating sludges. This process is expected to provide for conditioning with organic flocculating agents.

Filter presses are sized taking into consideration the filtering efficiency, related to the solids concentration, or the volumetric charge, at a defined filtration period. Filter ratings of 5 - 15 $kg/(m^2 \cdot h)$ are common at pressing periods between 2 and 3 hours. Chamber or volume depths of 20 - 30 mm have been successful in operation.

In contrast to the aforementioned dewatering methods, the dehydration in filter presses may end up with 40 - 60 % solids concentrations in the dewatered sludge. The filter cake is suitable for dumping. Another feature distinguishing the filter press from the other dehydration methods is the degree of solids separation, which is 100 %. This means that the filtered liquid is free of solids unless a plate fracture or cracks in the filter cloths transfer the solids from a particular compartment into the filtrate. Structural measures such as webs and knubs at the plates, or the separation of coarse material from the sludge before press charging may avoid plate failure. The filter cloths must be regularly checked for signs of wear, e.g. when they are sprayed for cleaning. If necessary, they must be replaced.

The filter presses and their buildings require investments much higher than the costs of centrifuges and band filters. On the other hand, this method permits the highest solids concentrations in the dehydrated sludge and thus the smallest residual volumes. This aspect may be very important particularly for subsequent dumping and in cases where a restricted dumping volume is available.

There are practically no restrictions as to the size of such filter press installations. The largest filter presses manufactured at present have a filter plate edge length of 2 x 2 m, with 150 plates per unit.

Filter presses are suited for the dehydration of crude and digested sludge.

6.7.4.3.4.4 Dewatering by Vacuum Filtration

This method is applied almost exclusively in the treatment of commercial and industrial waste waters.

A vacuum of ca. $1.6 \cdot 10^5$ Pa is generated inside a slowly rotating drum (cf. Fig. 6.2.-11). The lower part of this drum is submersed into a basin filled with sludge. When the drum moves through the basin, sludge is attached to the filter cloth on the drum by aspiration, and is then dehydrated when moving through the air. Shortly before the filter cake arrives at the point of drum immersion it is removed. Compressed air improves this sludge removal. Mechanical means such as scrapers, rolls, and continuously passing threads or wires can be used to lift off the sludge. In certain drum types the entire filter cloth is lifted off and passed along washing jets. The precoat layer can also be applied in the vacuum drum filters.

Solids concentrations of ca. 30 % can be achieved with digested sludge. Aspiration filters are normally operated with iron chloride as the flocculating agent. The filter load is 10 - 30 kg dried matter/(m² of filter area per hour), the dehydration efficiency is ca. 200 l/(m²·h).

The final selection of the respective method depends on the type of the sludge to be treated and on the final composition of the dehydrated sludge.

When crude sludge is dehydrated, attention should be given to the fact that the sludge must stay fresh, since sludge in which digestion has begun has very poor dehydration properties in spite of the higher quantities of conditioning agents.

The efficiency of the dehydration method depends on the flocculating agents selected and on where the flocculating agents are added.

6.7.4.4 Sludge Ulitization and Removal

Sewage sludge is a waste product arising in affluent societies. It contains undesirable substances in varying amounts, e.g. heavy metals. The utililization or removal of the sludge therefore always involves certain difficulties.

6.7.4.4.1 Sewage Sludge Utilization in Agriculture

When the sludge can be utilized in agriculture, this method should always be considered, if only for economic reasons. The sludge can be spread either in wet condition or after a preceding dehydration and drying stage, if necessary. Agricultural use, however, also makes certain demands on the sludge.
- The sludge must not involve any hygienic or health hazard to man or animals.
- The sludge must not result in a harmful enrichment of undesirable substances in plants or in the soil.
- The sludge must not impair the yield and the quality of plant products.

- The sewage sludge should contain sufficient humus and/or nutrient effects.

The digestion (with the digestion times now common) and the aerobic stabilization are not sufficient to achieve a total destruction of infecting germs. It is therefore often indispensable to disinfect the sewage sludge so as to permit the agricultural sludge utilization.

The following methods of sewage sludge disinfection are available.
- with liquid sludge: pasteurization, aerobic-thermophilic stabilization (circulating aeration, liquid composting, hot aeration), the consolidation with quicklime, composting in pits or silos, exposure to irradiation from cobalt sources or electron accelerator tubes;
- with dehydrated sludge: composting in aerated stationary or mobile rotting cells with or without the addition of substances such as straw, sawdust, bark, garbage, etc. Lime is added as a dehydration auxiliary in compartment filter presses.

6.7.4.4.1.1 Pasteurization

The heating temperature and the reaction time are the factors decisive when sewage sludges from mechanical-biological plants are pasteurized. Experiments with a pasteurization installation with batch operation have demonstrated that salmonella and worm eggs are destroyed within 30 minutes at temperatures between 60 and 70° C. A continuous pasteurization operation is also possible.

Operations in Switzerland have shown, however, that digested sludge after pasteurization almost regularly resulted in sludge recontamination with enterobacteriaceae and salmonella (BREER and collaborators /37/).

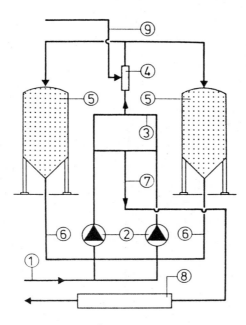

Fig. 6.7.-24: Schematic of a sludge pasteurizing installation (ROE-DIGER system)

1 - sludge for pasteurization; 2 - sludge pumps; 3 - sludge line to steam nozzle; 4 - steam nozzle; 5 - reaction container, alternately charged, with intermittent operation; 6 - discharge lines for pasteurized sludge; 7 - pasteurized sludge to sludge cooler; 8 - sludge cooler; 9 - low-pressure steam

Pertinent studies in Switzerland have shown that crude sludge pasteurization before digestion (preliminary pasteurization) is a reliable method of avoiding recontamination.

6.7.4.4.1.2 Aerobic Thermophilic Sludge Treatment

The aerobic thermophilic sludge treatment aims at a decomposition of the organic sludge constituents to inorganic final products in an exothermic automatic decomposition process. The prerequisite for this process is an intensive aeration with atmospheric oxygen or pure oxygen.

Thermophilic bacteria develop in this process. They increase the temperature and accelerate the reaction (40° C - 70° C, depending on the container insulation).

Studies have shown that this method is based on a coordination of temperature and pH value to destroy infecting germs.

Fundamentally, within the temperature range from 40° C to 55° C, the temperature must be the higher the lower the pH value is, and that it may be the lower the more alkaline the pH value is. Only this condition ensures a safe destruction of salmonella and viruses. Worm eggs are destroyed only when a temperature level of 50° C is maintained for 24 or more hours. If the necessary parameters are observed, aerobic thermophilic stabilization is a suitable method to destroy salmonella together with viruses and worm eggs.

6.7.4.4.1.3 Sludge Consolidation with Quicklime

When quicklime is used for sludge consolidation, this material is added to the previously dehydrated sludge (20 % TS).

A heap of the sludge/lime mixture is piled up. Temperatures in the range of 67 to 70° C arise in 24 hours. The final pH value is ca. 13.5. Salmonella and worm eggs introduced together with the sludge were destroyed after only 2 hours of exposure to quicklime.

6.7.4.4.1.4 Silo Composting

When digested sludge is composted together with straw in silos or pits, the sludge is spread together with bales of straw, and then well mixed with reversing equipment. The rotting sludge/straw mixture is turned over for aeration once per week.

6.7.4.4.1.5 Composting in Bio-Reactors

The bio-reactor sludge composting technique is a continuous composting method. The sludge is aerated from the reactor bottom through special distributors. The dehydrated sludge is composted with sawdust and discharged through the bottom, in most cases using cutters, conical worms, or bit worms. Experiments have shown that this process produces a hygienically unobjectionable compost. When, however, the operation is disturbed, e.g. due to aerator failure, such disturbance has negative effects on the disinfecting effect.

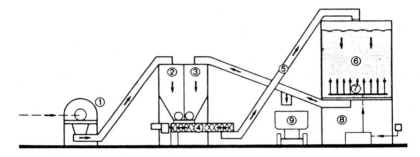

Fig. 6.7.-25: Schematic of a continuous reactor without level splitting, for sewage sludge composting (KNEER bioreactor system)

1 - mechanical sludge dehydration; 2 - silo for dehydrated sludge; 3 - silo for recirculated matter; 4 - mixing worm; 5 - vertical conveyor system; 6 - bioreactor; 7 - aerator; 8 - switching center and blower compartment; 9 - humus removal

The thermal conditioning methods provide for an additional disinfection.

6.7.4.4.1.6 Problems with Heavy Metals

It is not only the hygienic harmlessness but also the heavy metals concentration that is important in the agricultural sewage sludge utilization.

The tolerable heavy metal concentrations were determined on the basis of a maximum sludge quantity of 5 tons TS/ha per year. The "final" limits given in the regulation on sewage sludge spreading in the Federal Republic of Germany (sewage sludge regulation) or the "lower recommended values" (Table 6.7.-15) presently applying in Baden-Württemberg permit the heavy metal loads given in Table 6.7.-16.

Tab. 6.7.-15: Limits or recommended values of heavy metal concentrations in sewage sludges, in mg/kg of dry substance

	limits according to the sewage sludge regulation		recommended values in Baden-Württemberg	
	for transitional 8 years	final limits	lower value recommended	upper value recommended
lead	1200	600	400	800
cadmium	30	10	10	30
chromium	1200	600	800	1200
copper	1200	800	400	600
nickel	200	100	100	200
mercury	25	10	10	25
zinc	3000	2000	2000	3000

These limits or recommended values and the thus permitted heavy metal supplies are so dimensioned that the tolerable soil concentrations (Table 6.7.-16) will not be exceeded within an application period of 50 - 100 years. Thus no restrictions in plant production or even damage need be expected on the basis of today's knowledge.

Tab. 6.7.-16: Tolerable heavy metal quantities supplied with sewage sludge, and tolerable concentrations in the soil

	tolerable supply in kg/(ha·a)	tolerable soil concentrations mg/kg
lead	2.0	100
cadmium	0.05	3
chromium	4.0	100
copper	2.0	100
nickel	0.5	50
mercury	0.05	2
zinc	10.0	300

The heavy metal quantities given in Table 6.7.-16 are opposed by much smaller quantities <u>consumed by plants</u>, as becomes apparent from Table 6.7.-17.

Tab. 6.7.-17: Mean heavy metal quantities consumed by agricultural cultivations per crop rotation cycle, from DIEZ and ROSOPULO

	unloaded soils		polluted soils*
	consumed g/(ha·a)	% of the tolerable supply	g/(ha·a)
lead	10-20	0.5-1	15-30
cadmium	0.5-2	1-4	10-30
chromium	5-30	0.2-1	50-150
copper	20-50	1-3	30-70
nickel	10-20	2-4	20-60
zinc	100-350	1-3	600-1200

* 400 t dried matter/ha, strongly polluted sewage sludge for many years, tolerable concentrations in the soil partly exceeded several times

The actual heavy metal quantities transferred into the soil together with the sewage sludge are far below the permitted maximum values. And yet the <u>heavy metal balance</u> in sewage sludge application shows a marked positive trend, particularly since sewage sludge is not the only source of heavy metal transfer into the soil (fertilizers, agricultural pesticides).

Hence, it may be assumed that the heavy metals transferred into the arable soil together with the sewage sludge remain predominantly in the topsoil where they are correspondingly enriched by repeated application.

Sewage sludge contains, apart from the undesirable constituents, valuable nutrients for the plants, mainly phosphate and nitrogen (Table 6.7.-18).

Tab. 6.7.-18: Mean nutrient concentrations in sewage sludge (based on studies made by LUFA, Augustenberg, unpublished)

	1978	1979	1980
number of samples studied	83	298	237
dried matter (TS) (%)	20.4	17.6	13.7
total N (% in TS)	3.5	3.5	4.5
phosphate (% P_2O_5 in TS)	2.4	2.2	2.3
potassium (% K_2O in TS)	0.2	0.2	0.2
magnesium (% Mg in TS)	0.05	0.05	0.06
organic matter (% in TS)	39.2	40.6	43.1

Different sewage plants, however, may produce distinctly different concentrations.

Dehydrated or composted sewage sludge has a lower nutrient but a higher humus effect than liquid sludge, since it contains substantially larger quantities of organic matter.

6.7.4.4.2 Sewage Sludge Deposition in Sanitary Landfills

With insufficient area available or with hygienic considerations (heavy metals, etc.) not permitting, sludge reuse is not possible, so that the sludge is frequently deposited or dumped.

The methods of sewage sludge deposition can be classified in two groups:
- mono-deposition, which means that sludge alone is deposited,
- combined deposition, meaning that the sludge is deposited together with solid waste.

Further process varieties can be specified in either group, as becomes apparent from Table 6.7.-19 from KOEHLHOFF /84/. The deposition, in each form, must be carried out in a way that the requirements of groundwater protection are met (as defined in the German law on water economy), while also complying with the requirements of the law on waste disposal. Sludge deposition, also in mono-deposition, falls under the regulations on waste disposal.

Tab. 6.7.-19: Possibilities of sewage sludge deposition /84/

Sewage sludge deposition	
Mono-deposition	combined deposition

- liquid deposition (pond or lagoon)
 	
- deposition of the dehydrated sludge
	- consolidation deposition
	  with preliminary comminution
	  without preliminary comminution
	
	- rotting landfill
	
	- deposition of pressed garbage bales

A mono-deposition of the sludge should be envisaged only in exceptional cases, since the total area required for a separate dump is considerably larger. Therefore, the trend should be toward combined deposition together with solid domestic waste.

A further sludge dehydration is mostly desirable to maintain the useful lifetime of the landfill, which depends on relevant parameters such as water contents, cohesion, and friction resistance.

The effects of the conversion processes in the landfill on the long-term sludge stability are still largely unknown. This is why appropriate safety reserves must be considered in sizing the landfill.

A controlled combined deposition demands the establishment of delivery and dumping schedules.

6.7.4.4.3 Sewage Sludge Composting

It is best to compost the sludge in combination with pre-comminuted garbage. With 25 - 30 % of water, the sludge contains the moisture that the garbage needs for composting. On the other hand, the dehydrated garbage provides for the necessary air cavities. The sewage sludge/garbage mixture should have a water concentration of 40 - 60 %. This demands a preliminary sewage sludge dehydration to 70 %. Moreover, the mixture has a carbon/nitrogen ratio of ca. 15 : 1, which is favorable for composting since the lack of carbon in the sludge is compensated by the addition of garbage.

6.7.4.4.4 Sludge Discharge in the Open Sea

The transfer of the sludge by ship and its discharge or dumping into the open sea is a solution available in coastal areas, even though this solution is doubtful since the oceans and the sea do not have an unlimited capacity of absorbing pollutants.

6.7.4.4.5 Sludge Drying and Incineration

The sludge water arising when the sludge is dried is vaporized or evaporated. These processes consume thermal energy. The separation of the sludge water from the solids in mechanical dehydration processes is less expensive than the thermal methods. Since, however, the water concentration after dehydration is still as high as 50 to 80 %, drying methods may be applied to vaporize the residual water.

The dried sludge should contain less than 10 % of water. In such a dry condition it can be milled and used as spreadable fertilizer. The high temperatures ensure disinfection.

Two groups of equipment, i.e. flue gas driers and contact driers, are available.

safety measures are required. The person taking the sample must be secured with a rope by a second operator. Explicit reference is made to the relevance of such safety measures.

7.1.2 Methods of Waste Water Sampling

Waste water should be sampled in a way that the results of the analyses of the waste water samples will provide representative information about the composition of the waste water, or for particular problems only, if necessary. Any secondary contamination of the sample, be it by the sampling method, by the sampling containers, etc., must be avoided. Physico-chemical, chemical, or biological changes in the composition of the waste waters should be avoided as far as this is possible. Unless such a secondary change cannot be precluded, the samples must be preserved. Cooling of the water upon sampling and the quickest transfer possible into the analytical laboratory are often sufficient. Such immediate cooling is particularly important for the microbiological waste water analyses or for the determination of the biochemical oxygen demand. Waste water sampling particularly involves cooling or freezing or a chemical sample preservation as particularly important steps. Waste waters usually contain many microorganisms whose microbiological reactions continue in the waste water sample taken. Cooling reduces such microbiological activities. The water samples should be kept at + 4° C in cooling boxes containing ice or in a refrigerator in the sampling vehicle. Then they must be transferred to the analytical laboratory as quickly as possible. In the laboratory, the waste water samples should never be allowed to stand for an unnecessarily long time.

It is possible to freeze the samples in order to fix the waste water for subsequent analyses in the laboratory. In this case, the water should be frozen at − 20° C as quickly as possible, and should then be maintained in this frozen condition until the sample is analyzed.

Microbiological reactions may also cause secondary chemical changes in the waste water, e.g. desulfo-vibrio bacteria may cause a sulfate reduction to produce hydrogen sulfide with secondary precipitations of sulfides of heavy metals. Microbiological reactions may cause rearrangements of the nitrogen compounds, e.g. due to nitrification and denitrification processes. Not only cooling but also the addition of 1 ml of trichloromethane per liter of water is recommended to inhibit these processes. Such a preservation may also be applied, for instance, for the phenol determination.

It is also common to use a sodium hydroxide solution to alkalinize water samples to a pH value of 10 to 11 for the determination of cyanides or phenol.

Another method of inhibiting secondary microbiological reactions is the addition of mercury chloride (ca. 50 mg/l) instead of trichloromethane. Waste water samples taken to determine traces of heavy metals must be preserved by acidification with hydrochloric acid, sulfuric acid, or nitric acid upon sampling. Any preliminary treatment of the waste water must be noted in the sampling record. When chemical additives are used to preserve samples, it is also recommended to note on the label of the particular sample that "blind tests are required".

It is possible to fix varying substances in the waste water, e.g. divalent iron, hydrogen sulfide or sulfides, carbon dioxide, etc. Such fixing must also be indicated in the sampling record (see also Section 7.2 - Site Inspection).

Waste water samples can be taken in random sampling, repeated random sampling, time-dependent sampling, or quantity-related sampling. Manual sampling is possible, or semiautomatic or fully automatic equipment can be used (see Sections 7.1.3 through 7.1.5.3).

Sample quantities of 1 to 2 liters of water are normally sufficient to perform conventional waste water analyses. When specialized analyses are required, e.g. to determine heavy metals or certain organic substances or checks for toxicity to fish, the person taking the sample must have the pertinent information so as to take the appropriate number of additional water samples.

It may be desirable to fill several bottles, each preserved with a different acid, to check for inorganic trace elements. It may also be necessary to bottle water samples for the determination of organic substances, e.g. by gas chromatographic steam space analysis, by extraction or infrared analyses, or for the measurement of radioactivity. In these cases, the sampler must also be informed by the supervising expert before the samples are taken. If necessary, both have to discuss and agree on the optimum sampling techniques and bottling methods, as well as the best preservation of the samples.

When a phosphate determination is required or when the contents of surfactants (substances active as detergents) are to be determined in the waste water samples, bottles cleaned with phosphate-containing detergents in a bottle washing machine must not be used. The use of new glass or plastic flasks is recommended for such analyses (see also Section 7.1.6.3 - Bottles).

When waste water samples undergo a special treatment, this fact must be noted on the label, with an explanation in the sampling record.

Depending on the data processing methods, it may also be necessary to code the sample designation and the sampling information for the electronic data processing system as early as the sample taking. The waste water samples taken must be unambiguously labeled directly upon their taking. Either tags or stickers can be used for labeling. The labels may be designed as forms and should include information about the sampling place, site, date, and time; the possible preservation of the sample; and the sampling technique.

7.1.3 Random Sampling

Random sampling, i.e. the taking of individual samples at discrete times, furnishes the characteristics only of the particular water flowing at the time of sampling. A simple random sample is advisable only when a very slow change in the composition of the water to be analyzed is expected. A random sample can only provide information about the substances in the water and the composition of the waste water at the time of sampling. Random sampling cannot be applied to gather information about contamination levels or purification efficiency.

When, however, several random samples are taken within a specified period, e.g. two hours, the analyses of the individual random samples can reveal inadmissible emission exceeding certain thresholds by a particular enterprise, e.g. by waste water surge loads or waste water discharge in batches into the sewerage system.

The information that can be gathered from simple waste water random sample is normally insufficient. Multiple waste water sampling within a specified period is already a step toward the time-dependent sampling technique.

7.1.4 Time-Dependent Sampling

The time-dependent or proportional time-related sampling, which may also be termed quasi-continuous sampling, involves the sampling of equal volumes of water at specified intervals within a certain period, and their combination to form composite samples. The sampler may carry out such time-proportional sampling by hand, or semiautomatic or fully automatic equipment may be used.

The proportional time-related sampling method involves, for instance, the combination of composite samples from the individual samples taken during 2, 8, or 24 hours. Such composite samples are combined from partial samples taken every 5 or 15 minutes or hourly, for example, which are then combined to form the composite samples with respectively equal partial volumes.

When proportional time-related samples are taken during periods of more than 2 hours, it is recommended that two samples at a time be taken. One sample is used to produce the time-dependent composite sample, whereas the second partial sample contributes to a 2-hour composite sample used for certain local measurements of parameters such as the pH value, the redox potential, etc., or for the determination of the sedimentary matter.

Such a determination of the sedimentary matter in time-dependent partial samples is necessary because secondary changes with flocculation in the waste water cannot be precluded with prolonged storage of the composite waste water samples, e.g. when a composite sample per day is established. Such secondary changes, however, would lead to faulty results when the sedimentary matter is measured, and possibly also furnish faulty results in the pH measurement due to secondary conversion processes.

Time-dependant sampling may also demand the immediate analysis of each sample for certain parameters, e.g. temperature and pH value, so as to detect, for instance, short term exceeding of the statutory pH range in the waste water.

In this case, the composite sample proper is always combined from successive equal waste water quantities, parallel to the waste water sampled for immediate measurement.

A time-dependent sampling technique can also be used to determine pollution in combination with a separate volumetric measurement. Pollution can be determined by time-dependent sampling, in combination with a quantitative analysis, when

a) the waste water volume varies, with the concentrations of the contents constantly remaining unchanged, or
b) the waste water quantity remains approximately constant, with the concentrations of the contents varying.

When variations in both the waste water volume and the concentrations of the contents are noted, a volume-dependent sampling technique must be used (see Section 7.1.5). The time-dependent sampling techniques should aim at selecting the shortest time intervals possible between the individual sampling operations so as to achieve an approximately representative composite sample. This method, however, always requires a compromise due to the local conditions and the efficiency of the person taking the samples. It has therefore become possible to use automatic equipment with selectable intervals for time-dependent sampling (see Section 7.1.5.3).

When the waste water samples are to be taken by hand, be it in random or time-dependent sampling, porcelain, plastic, or high-quality steel pans have proven useful. Appropriate plastic or alloy steel buckets are suited to combine composite samples. Sampling ladles with extension shafts up to several meters in length have become a common tool avoiding the climbing into shafts. Bottles or pans for manual operation are attached with clamps to bamboo poles. Waste water sampling using suitable equipment is continually advancing (see Section 7.1.5.3).

7.1.5 Volume-Dependent Sampling

The volume-dependent sampling technique is suited for the determination of pollution loads, especially when the quantities and concentrations of the waste water vary. This technique demands that the rate of water outflow per unit time be known (see Section 7.1.5.1 - Volumetric Measurement).

With the waste water quantity discharged per unit time known, partial water samples are taken in correlation to the outflow, through a water flow dividing network at certain intervals over a constant time period. These partial samples are then combined into volume-proportional composite samples.

When, for instance, the water volume measuring instrument indicates 10 l/sec at the time of sampling, 0.1 l water are taken for the composite sample. With 20 l/sec, the corresponding volume of 0.2 l is taken. The respective samples are combined into a composite sample as a function of the indicated water volume. In this field, however, automated sampling equipment is used increasingly more often (cf. Section 7.1.5.3).

The volume-dependent sampling also requires that the information on measurements on the site, e.g. temperature and pH value, and on the determination of sedimentary matter in separate composite partial samples must always be noted.

7.1.5.1 Volumetric Measurement

A manual measurement of waste water quantities is possible only with small inflow volumes and only when the waste water can be collected in vessels or containers with a specified volume. In such a case the time up to the filling of a cylinder with a specified volume, e.g. a 10 l bucket, must be taken, for example. This results in an instantaneous value that can be repeated several times. This value permits the approximate estimation of the outflow per hour and per day only when the outflow variations are small.

This volume measuring technique is applicable in particular cases. It furnishes information that is quite reliable, particularly with very low waste water discharges.

Waste water volume measurements are normally of interest only when they are taken in a long-term series of measurements, with continuous recording of the readings if possible. Such series permit the recognition of the actual conditions with maximum and minimum values, and their correlation with the responsible events such as precipitations or discharges from different enterprises, etc.

<u>Inductive measurement</u>. The application of this method is possible in closed filled tubes even with surge loads and up to very high flow rates. The complete tube section is maintained so that there is no increase in pressure. The relatively short part is mounted into the piping, with flanged couplings attached at both sides. To ensure a reliable measurement, the tube must be completely filled with the waste water. The flow movement must be uniform. It is therefore advisable to mount the equipment in a vertical pipe or, with installation in a horizontal pipe, upstream of an extension leading toward a vertical section. In proportion to the rate of waste water flow, according to FARADAY's law of induction, an AC signal is generated in the inductive flowmeter (Fig. 7.1.-1). Indicators and recording instruments are provided to record the reading of this AC signal. Pollutants carried in the waste water do not interfere with the measuring operation.

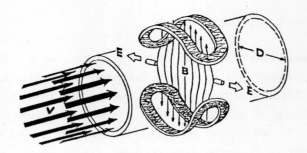

Fig. 7.1.-1: Principle of measurement in the inductive flowmeter

<u>Actual operating diagram of the sensor</u>. The inductive flowmeter measurement is based on FARADAY's principle of induction.

An alternating magnetic field is established in a direction normal to the direction of flow. An electrically conductive liquid passing through this magnetic field induces a voltage at the electrodes. This measuring voltage is proportional to the magnetic induction B, to the rate of flow v, and the electrode spacing D (Fig. 7.1.-1).

$$E = B \cdot D \cdot v$$

Considering the fact that the magnetic induction B and the electrode spacing D are constant, the following relationship is obtained:

$$E \sim v$$

Through integration of the cross section, the generated measuring voltage is proportional to the volume of flow Q:

$$Q = \frac{D^2}{4} \cdot \pi \cdot v$$

hence $E \sim Q$,
where:

    E = measuring voltage, in V
    D = diameter (electrode spacing), in m
    v = flow rate, in m/s
    B = magnetic induction constant, in V
    Q = volume of flow, in m³/s

separately into the laboratory for analysis. A volume-dependent analysis of this partial sample, however, is not possible in such a case.

7.1.5.3 Volume-Dependent Sampling Using Instruments

The sampling instruments used for volume-dependent sampling include a built-in flowmeter controlling the automatic sampling device. Such a sampling device, for instance, continuously takes a partial flow from the waste water flow using a pump. This partial flow is then passed along a magnetic switch. The magnetic switch receives the impulses from the volumetric measuring unit to fill, on each pulse, a specified sampling volume, e.g. 5 or 10 ml, into the bottle storing the composite waste water sample. The sampling intervals are then increased or reduced to an extent corresponding to the waste water volumes that have passed. The illustrations given below show examples of such automatic sampling instruments.

The diagrammatic sketch (Fig. 7.1.-5) from STEINECKE-DIETRICH is a proven layout for a sampling device.

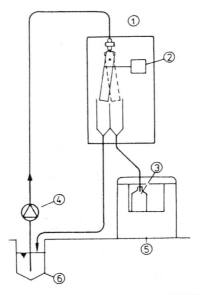

Fig. 7.1.-5: Sampling device according to STEINECKE-DIETRICH
1 - water switch; 2 - driving element; 3 - collector; 4 - Mohno pump; 5 - freezer; 6 - sampling point

Fig. 7.1.-6 illustrates an electro-pneumatic sampler with four filling stations.

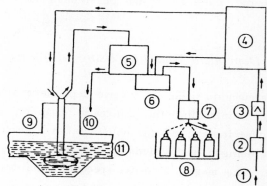

Fig. 7.1.-6: Electro-pneumatic sampler

1 - compressed air; 2 - air filter; 3 - pressure reducer; 4 - controller; 5 - collector; 6 - dosing pump; 7 - distributor; 8 - filler unit; 9 - waste water pump; 10 - to sewer; 11 - sewer

The following function layout (Fig. 7.1.-7) illustrates the operational flow of an automatic sampler.

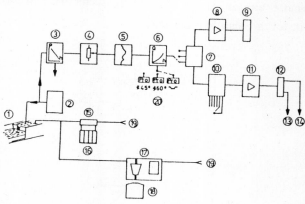

Fig. 7.1.-7: Flow diagram of an automatic sampler system

1 - weir; 2 - pump; 3 - differential pressure gauge; 4 - remote transmitter; 5 - recorder 0. 100 %; 6 - linearizer; 7 - integrator; 8 - amplifier; 9 - m^3 counter; 10 - interval; 11 - amplifier; 12 - impulse, 0.1 m^3; 13 - ext. for sampler; 14 - cont. for sampler; 15 - sampling pump 12 B; 16 - sampler; 17 - sampling vacuum pump; 18 - composite sample; 19 - volume-proportional connector; 20 - weir, for changeover

Depending on the type of such automatic sampler system, the following additional installations are possible:

- cooling or heating system

- automatic sample changer, e.g. 12 bottles for 12 2h-composite samples per day,
- explosion protection system when the installation in channels or other confined spaces is intended, which are liable to contain explosive mixtures,
- a system for continuously measuring and recording various parameters such as pH value, conductivity, and temperature
- time setter to specify the beginning and the end of the sampling operation

When automatic devices are used for waste water sampling over a prolonged period, all parts and components in the device, which are in contact with the waste water, e.g. lines, pipes, containers, etc., must be checked for the growth of microorganisms (biological slime). When such slime occurs the sampling operation must be interrupted for a short time for thorough cleaning of the device, using mechanical means whenever possible. Transparent lines should be kept as dark as possible in order to avoid the growth of algae on the waste water side. This aspect must be considered when samples are taken from treated waste water over prolonged periods, with the water being delivered through transparent piping. As a result of such growth in sampling devices and piping, the waste water to be sampled may have already been exposed to a more or less strong biochemical influence so that insufficient maintenance will result in incorrect results to be expected from the analyses.

Waste water delivery in intermittent operation (which means that the pump is restarted for each sample filling) causes less problems due to piping contamination. When the pump operation is resumed at a certain interval before the actual sampling step, such operation ensures sampling from a continuous cycle.

The respective technique of volumetric waste water measurement and the employed automatic sampler are to be noted in the sampling record.

The automatic devices always require thorough maintenance. Particular attention must be paid to the risk of clogging and to the question of whether the waste water to be sampled carries an oil film. These oil films are frequently not included into the sample in proportionate quantities.

7.1.6 Waste Water Sampling in Practice

Waste water sampling is a difficult task as a result of the usually heterogeneous composition of the waste water as well as the occurrence of surge loads with pollutants. Moreover, secondary changes are to be expected in the waste water samples, e.g. resulting from the admission of air or microbiological reactions. When waste water samples contain sedimentary matter, the composition of composite samples may require the use of the excess liquid, i.e. the water separated from the sedimentary matter, for the production of the composite samples, depending on the respective sampling task. When other precipitations are separated from the composite sample on the transport to the laboratory, the samples must be homogenized or dissolved so that these deposits will be included into the analysis. The sedimentary matter separated at the sampling point may be separately kept in bottles and analyzed in the laboratory independently. The labels should bear a remark as to whether the sedimentary matter was collected after one or two hours of settlement, together with an indication of the temperatures involved.

As much information as possible should generally be gathered when the local conditions are inspected. A maximum of measurements possible, particularly of variable parameters, should be made.

A sampling team should be provided not only with sufficient glass and plastic bottles, but also units for electrometric pH measurement, the measurement of the specific conductivity, the redox potential and oxygen, together with spare electrodes and the associated standard or calibration solutions.

The common utensils such as thermometer, pycnometer (to determine the specific gravity), barometers, stop watches, bottle labels, IMHOFF funnels to determine the sedimentary matter, water ladles, soldering lamps, field test filtration set, cooling boxes, optical sounders, and pumps must be taken along. If available, water flowmeters and automatic samplers with the associated equipment for automatic sampling and for the measurement of the pH value, oxygen, conductivity, temperature, etc. should also be included in the kit. It may be necessary to provide gas collecting vessels and equipment for simple in situ analyses, e.g. pipettes, burets, ERLENMAYER flasks, etc.

The kit must also include equipment for a manual measurement of the gas composition in shafts, e.g. ex meters or gas detectors, explosion-proof flashlights, and protective clothing.

All equipment and bottles used for sampling must be clean and dry, or, before use, should be rinsed with the waste water to be sampled several times.

The bottles are normally not completely filled but only to a level that leaves an air space of ca. 1 cm^3 on top. This air space is to prevent a bottle from breaking as a result of the expansion of the water due to temperature increases. Water samples for microbiological analyses must be sampled under sterile conditions in sterile glass receptacles.

7.1.6.1 Sample Preservation

Certain analytical techniques require the preservation of the waste water samples in order to prevent secondary changes in the concentration of these variable substances on transport. It should be emphasized again that the best preservation of the samples is the quickest analysis possible of the sampled water quantity. The water sample should be cooled to ca. + 4° C (cooling box or refrigerator in the vehicle) until the analysis is performed.

The cooled waste water samples should be delivered to the laboratory on the next day or within 2 days after sampling at the latest. The waste water samples should not be allowed to stand in the laboratory for an unnecessary length of time. The laboratory manager will have to decide, on the basis of technical considerations, which analyses are to be made immediately and which work can be carried out later. In such a case, the waste water must also be stored in cooled condition.

The waste water samples may also be quickly frozen in plastic bottles down to – 20° C to allow fixing of the samples. The waste water samples must be maintained in their deep-frozen condition until the analyses begin. Deep-freezing cannot be recommended, however, for sewage sludges, since frequently the solids are then separated from the sludge water.

When waste water samples are preserved, particular attention must be given to the fact that chemical and mainly microbiological reactions continue after sampling. The rates of reaction are kept low by cooling. Microbiological reactions may lead to a further conversion or decomposition of organic substances in the water so that total or combined parameters such as the biochemical oxygen demand may become unreliable. The biological activities of desulfo-vibrionic organisms in the anaerobic range may also result in a sulfate reduction with the formation of hydrogen sulfide. This hydrogen sulfide production in its turn initiates other secondary chemical processes, e.g. iron precipitation in the form of iron sulfide and the co-precipitation of other heavy metals.

When the sampled waste water contains disinfectants, e.g. chlorine compounds, sodium thiosulfate must be added, after sampling, to the samples taken for a bacteriological analysis, in order to eliminate these agents. Since silicate and borate may be released, glass bottles should be avoided and plastic bottles are to be used for the quantitative determination of these substances.

A transport in cooled condition and the quickest water analysis possible are important for the quantitative determination of the phenols and the various nitrogen compounds such as ammonium, nitrite, nitrate, organically bound nitrogen. If local conditions are not favorable, the water samples may be preserved by the addition of 1 to 5 ml of chloroform per liter water, for instance.

Waste water samples taken for a determination of the total cyanides and the easily releasable cyanides are best alkalinized and set to a pH value of ca. 10 to 11 with caustic soda solution for transport.

Sulfides or hydrogen sulfide are to be bottled in special flasks with ca. 250 ml volume, with the addition of an aqueous zinc acetate or cadmium acetate solution to fix the hydrogen sulfide.

Waste water samples for the determination of the biochemical oxygen demand (BOD_5) or the chemical oxygen demand (COD) must be deep-frozen and delivered to the laboratory in a deep-frozen state unless the analyses can begin within one day in a chilled condition at + 4° C. With such a deferred analysis, the tests must be carried out as quickly as possible.

Specific purposes may demand analyses to determine specified toxic substances that may interfere with a biological clarification process. These applications involve the acidification of the waste water samples with a suitable acid (hydrochloric acid, sulfuric acid, or nitric acid), for instance, for the determination of heavy metals in the laboratory. This acidification is carried out at the sampling site to an extent that secondary precipitations will not occur on transport.

When waste water samples are taken to carry out a fish test or a bacteriological test for toxicity, larger waste water volumes, e.g. 10 to 20 liters, must be sampled in one operation. A chemical preservation of these samples is not allowed. Moreover, a chemical preservation is out of the question for the determination of the biochemical oxygen demand. The only useful method involves cooling of the waste water sample and rapid delivery to the laboratory.

When in exceptional cases the samples must be chemically preserved for other analyses, the following agents are used:
- bases for alkalinization,
- acids for acidification,
- addition of mercury chloride (50 mg/l) to inhibit microbiological reactions,
- addition of trichloromethane (1 to 5 ml/l) to inhibit microbiological reactions.

Whenever a waste sample is preserved, the type and volume of the fixing agent must be noted on the respective bottle.

Special analyses, e.g. the determination of the extractable matter or tests for pesticides, may require that a specified volume of the solvent be added to the unfixed waste water in situ, which will be used later on for extraction and enrichment, e.g. petroleum ether, n-hexane, or a chlorinated organic compound. Such addition must also be noted by type and quantity on the bottles. When such organic solvents are added, glass bottles should be used.

Further information about the fixing of samples or measures to preserve waste water samples is to be found in the "Collection of methods of waste water analysis" /55/.

7.1.6.2 Waste Water Volumes for Analysis

Waste water quantities of 1 to 2 liters are sufficient for a simple waste water analysis, whereas 0.25 to 1 l are necessary for a bacteriological water analysis. The trace analyses often require several liters of water filled in different bottles each containing 1 or 2 liters.

When extractable organic material is to be determined, the required sampling receptacles must be preliminarily cleaned in the laboratory, using the extracting agent. This applies also to the bottle neck and the cap. Quantities as large as up to 5 liters must often be sampled for these purposes. When the analyses also include studies on the toxicity to fish, ca. 10 to 20 liters of waste water must be sampled, depending on the toxicity level.

It is generally recommended that the analyzing authority or facility or the expert informs the person taking the waste water samples about the scheduled volume of the analyses before sampling so that the sampler may then be able to sample the respectively required quantities and keep them under the appropriate conditions, using his experience. The importance of such a cooperation between the analyzing laboratory and the sampling department can never be overestimated.

7.1.6.3 Bottles

Samplers frequently like to use plastic bottles to take waste water samples. Polyethylene has largely been accepted among the plastic materials. Bottles with both a wide and a narrow neck with screw caps may be used. Plastic bottles made of polyvinyl chloride or polystyrene are less common, e.g. as a result of the risk of foreign matter being released to the waste water.

Disposable plastic bottles have been successful when the labor cost of cleaning the used bottles exceeds the price of new bottles.

Specific demands on the so-called hydrolytic quality of the material are generally not involved when glass bottles are used. When waste water samples are taken in glass bottles, it is common to use narrow-neck bottles with ground glass plugs.

7.1.6.3.1 Glass Bottles

Glass bottles are generally used in waste water sampling to take waste water samples for microbiological analyses or for the determination of the oxygen demand, but also for the determination of oily and fat-type substances by way of extraction. The steam space gas chromatographic analyses demand that the samples be filled into the bottles in a gas-tight form. To this end, special gas flasks are used that have a tube at the side and contain a septum or a tightly closing plug with a septum so that gas samples can be taken out of the gas chamber for the gas chromatographic analysis in the laboratory, simply using a suitable syringe without having to open the bottles.

Glass bottles are to be used as water receptacles especially when secondary reactions are to be avoided between the plastics and the water. In this respect, the release of organic substances into the water may be relevant, but also the partly selective adsorption of substances from the water to the plastic material. As an example the heavy metal, e.g.

silver, adsorptions should be mentioned, and also the collection of pesticides at the plastic surface.

Reactions at the walls of glass bottles must, of course, also be noted. It is necessary, for instance, to add a suitable acid to the water sample taken, to set the sample to the acidic range immediately upon sampling in order to determine traces of heavy metals. The reason is that such wall adsorptions for heavy metals are considerably decelerated in the acidic range.

Glass bottles rinsed with detergents containing phosphates and surfactants must not be used for waste water sampling destined for a quantitative analysis of phosphates or detergent substances.

Before their first use, new glass bottles must be cleaned with a diluted acid and rinsed with distilled water.

The risk of breaking must be considered when glass bottles are used. This is the reason why they are not to be filled completely. Rather a gas space with a volume of 1 cm^3 or more should remain on top after closing.

It is recommended that prepared transportation cases with padded cutouts in foamed material be provided, corresponding to the size of the glass bottles containing the waste water samples. This avoids the risk of glass breakage on transport.

7.1.6.3.2   Plastic Flasks

The use of plastic flasks or bottles is frequently possible in waste water sampling. These plastic receptacles have the advantage of not bursting when they are filled to the top. But a plastic material must be used that is as inert as possible in relation to the filled water. Polyethylene flasks have proven successful. It is possible to use plastic receptacles with the usual round bottle form, and also plastic receptacles in square form that are particularly easy to stack. A screw-on lid or top is generally preferable. The advantage of using disposable plastic bottles has already been emphasized. Whenever the water sample must be deep-frozen for preservation, plastic bottles are to be used.

The plastic bottles may be identified with adhesive stickers or label tags. When the labeling is written directly onto the plastic bottles, e.g. with felt-tip pens, attention should be paid to the fact that the labeling must be waterproof and that a mechanical abrasion of the labeling on transport must be avoided.

7.6.1.6.3.3 Special Bottling

As stated above, it may be necessary to use either larger bottles or special flasks made of a specific material, for special samples. Some examples:

Taking more voluminous waste water samples, e.g. 50 or 100 liters in plastic cans or containers or in plastic drums for studies on the possibilities of waste water purification.

Taking water samples to carry out certain analyses in metal containers, e.g. for the determination of the polycyclic aromatic hydrocarbons.

Special waste water bottling may also be required to determine gas releases. In this application, the waste water samples are taken in gas collecting glass vessels of 1 l volume, with a lower and an upper ground tap. One liter glass flasks with a special Teflon plug and a septum for gas sample removal from the gas compartment have generally proven successful when steam space gas chromatographic analyses are carried out. For a pycnometric determination of the waste water density, the specific gravity glass bottles are also to be filled at the sampling site directly.

Special bottling in measuring cylinders or flat bottom flasks may also be required for the determination of divalent iron, for the photometric determination of silicic acid, or to fix hydrogen sulfide with zinc acetate or cadmium acetate at the sampling site directly. It may finally be required to take waste water samples in prepared small flasks filled with NaOH or $Ca(OH)_2$ for carbonic acid analyses in the laboratory.

When waste water sampling with special flasks is required, it is indispensable that the expert or the laboratory manager discuss these aspects before sampling.

7.1.6.4 Sludge Sampling

It may become necessary to take samples not only of sludge-containing waste waters but also of the sewage sludges themselves. Examples of sludges arising in the waste water sector:

- waste water sludges,
- sewage sludges (stabilized and unstabilized),
- fully digested sludges,
- floating sludge,
- neutralizing sludges,
- industrial sludges.

A particular aspect in sludge sampling is the homogenization of the average sample taken. The reason is that sludges often have a heterogeneous composition. Such average samples are to be taken from sludge in a manner that, depending on the consistency of the sludges at various locations of the sludge deposit, partial samples can be taken, using a suitable sampling vessel or sampling pipette. These partial samples are then combined to form a composite sample. Because the materials are inhomogeneous, ample sample volumes of ca. 5 to 10 kg are desirable. Sufficiently solid sludges will only in rare cases separate into sludge matter and water. Such a separation of sludge and water must, however, be expected in diluted sludges. It may thus be necessary to homogenize the sludge with the water before the analysis, or separate analyses of the sludge and the water will be necessary after a mechanical separation. When the water concentration in the sludges or the concentrations of solids are to be determined, the sludge samples must be delivered to the laboratory in watertight receptacles for water determination, to avoid losses of water. Heating of the samples should be avoided as far as possible.

The sludge samples taken may be used not only to determine the solids concentrations but also for digestion tests. As early as during transport, the release or formation of certain substances must be expected, e.g. the release of hydrogen sulfide or the production of a gas (methane or carbon dioxide).

The assessment of the hygienic condition of the sludges also requires sterile sludge sampling in sterile receptacles for microbiological analyses. Wide-neck glass bottles (volume 500 or 1,000 g) are used for this purpose.

Specific sludge samples, each 1 kg or more, must also be taken when not only the sludges are to be analyzed but also the rainwater percolated through sludge or seepage containing carbonic acid, e.g. water occurring in refuse landfills.

It is very important in sludge sampling that the person taking the sample furnishes the analyst with information about the local conditions and on the types of sludges involved, as well as on possible preliminary treatment, e.g. mechanical, thermal, chemical, or microbiological treatment.

Nonstabilized sludges must be transported in a cooled casing and should be delivered to the laboratory as quickly as possible. Local pH value and redox potential measurements, if necessary with puncture electrodes, may be necessary.

7.2 Testing on the Site

The importance of the sampler recording the local conditions in a sampling record should be reemphasized. This information is indispensable in addition to the results of the local tests listed below. This local determination of the waste water occurrence and the sampling technique used, as well as on the method of fixing, is relevant in addition to information about whether quantitative waste water measurements have been taken or about the sampling operation and composition.

The following parameters are relevant:
- atmospheric pressure, in hPa
- altitude of the sampling site above mean sea level elevation

- sensory testing in situ
- water temperature
- air temperature
- transparency and turbidity (sighting depth)
- flow rate
- sedimentary matter
- pH value, measured electromagnetically in situ
- electrical conductivity, measured with an indication of the temperature
- redox potential, measured electrically
- oxygen, electrometrical measurement or fixing for subsequent analysis in the laboratory
- density (aerometer or filling of specific gravity tubes)
- fixing of variable substances, e.g. carbonic acid, divalent iron, hydrogen sulfide, silicic acid, cyanides
- nitrogen compounds
- heavy metals
- pesticides, oils and fats, etc.

In exceptional cases and when the person taking the sample is sufficiently qualified and the suitable measuring equipment is available, it is also possible to determine transient radionuclides immediately on the sampling site.

The following sections contain explanations on the methods of local waste water analyses.

7.2.1 Temperature

<u>Air temperature</u>

Equipment:
calibrated mercury thermometer, 0.5° C graduation.

Method:
The temperature is measured with a thoroughly dried thermometer (otherwise the cold due to evaporation will cause inaccurately low readings) near the sampling site, 1 m above the water or the ground. When the sun is shining, the measurement must be taken in the shade or with screening against reflected radiation (light walls of houses or rock).

Water temperature:

Equipment:

calibrated mercury thermometer, 0.1° C graduation. Common measuring ranges: − 5 to + 30° C, − 5 to + 60° C, − 5 to + 100° C.

Special thermometer:

maximum thermometer (especially for hot waste waters).

Method:

The thermometer is submersed into the water up to the reading level. The tester then waits until the reading remains constant (ca. 1 min.). When a direct measurement is impossible, the water sample is taken in a receptacle with a volume of 1 l or more. This receptacle must have been set to the water temperature. Then proceed as usual.

The water temperature can also be measured using an electrical resistance thermometer or a thermocouple element.

7.2.2 Sensory Testing in Situ

Appearance: indicate e.g. clear or turbid, color, undissolved matter, gas release, odor, etc.

The odor should be characterized as exactly as possible as to its intensity and its kind. The following five grades are common to define the odor intensity:

"very weak"	can be percepted only by an experienced analyst
"weak"	compared against a reference sample
"distinct"	perceptible by every water consumer
"strong"	unpleasant
"very strong"	nauseating

The following designations are used to describe the kind of odor:

General odors:

"metallic"	iron-carrying groundwater
"earthy"	occasionally caused by blue-green algae
"moldy-musty"	impure water
"peaty"	boggy water

"stale-foul"	impure water
"ichorous"	heavily polluted water
"fecal"	heavily polluted water
"fishy"	occasionally caused by blue-green algae
"like train oil"	caused by flagellata
"like seaweed"	caused by flagellata

Chemical odors:

odor reminding of	hydrogen sulfide	tar substances
	mineral oils	phenols
	chlorophenols	chlorine
	ammonia	soap
	acetic acid	butyric acid

<u>Hydrogen sulfide and free chlorine</u> may camouflage other odors. A sample can be dechlorinated by the addition of approximately stoichiometric quantities of sodium thiosulfate. When hydrogen sulfide is present, a small amount of cadmium acetate is added to the water sample. Then another sensory test is made.

The relevance of the odor thresholds in waste water is normally not too high. If the sensory threshold is to be determined, reference is made to the "Collection of methods of water analysis" /55/.

When the visual description of the water must also include the <u>color</u>, this test must be made in situ. Colorimeter tubes may be used for this purpose. The coloring can be determined by visual inspection, and can be described as follows:

"colorless"

"very weak coloring"

"weak coloring"

"strong coloring".

Each indication is completed by the indication of the shade, e.g. yellowish, yellowish-brown, brownish, yellowish-green, red, black-brown, black, etc. Regarding the color measurement, reference is made to the "Collection of methods of water analysis" /55/.

There is no <u>testing of the taste</u> of waste water analyses.

Transparency and turbidity. The inspection of waters for transparency and turbidity results in a mere description for orientation. It does not allow the conclusion of certain qualities and substances contained in the water. The transparency results from the color and possible turbidity of the water. The turbidity is caused by inorganic and/or organic matter in suspension or colloidal solution. Sludge particles, silicic acid, iron and manganese hydroxide, as well as organic colloids, bacteria and plankton may be involved. This aspect should be determined immediately, but not later than 24 hours after sampling.

The turbidity of water can be determined

a) after taking of a sample in a calibrated colorimeter cylinder with a plane-parallel bottom disk (25 mm inside diameter, cm graduation) in a manner that the readability of a standard lettering (black letter, 3.5 mm high, lines 0.5 mm thick) is determined by vertically viewing through the tube in diffuse daylight. The value is indicated in cm (mean value from several "readings").

b) Another method is the determination of the sighting depth of the waste water. The measure of sighting depth is the water depth in which a white disk of 20 cm edge length or diameter, submersed into the water and attached to a chain or a rod, can still be recognized. Up to 1 m, the values are indicated in cm, for depths exceeding 1 m the sighting depth is indicated in 0.1 m graduations (rounded).

For a simple turbidity test, a colorless clear glass bottle (volume 1 l) is filled with the water sample by two thirds. The bottle is shaken well. The sampler looks through it, first against a black and then against a white background (cf. TYNDALL effect). The following grades of clarity are distinguished:

"clear"

"opalescent"

"weakly turbid"

"strongly turbid"

"opaque".

A comparison of the sample turbidity with the turbidity in a graded series of diluted standard suspensions of diatomaceous earth (1.00 g and 0.10 g SiO_2/l) furnishes more exact definitions. The values can be interpreted both visually and photometrically (nephelometric analysis).

7.2.3 Density

The ratio between the mass "m" and the volume "V" of a substance is defined to be the density "d" of this substance:

$$d = \frac{m}{V}$$

The density is indicated in g/ml in water analysis.

When the density of waste water samples is to be measured, either an aerometer or a pycnometer (specific gravity bottle) is used for this measurement.

<u>Pycnometric method</u>:

Equipment:

pycnometer, volume 200 - 250 ml, with a flask neck restricted to an inside width of 5 - 6 mm, with an etched mm-graduation over a length of 50 mm. Thermostat, precision balance

Method:

The inside of the pycnometer must be totally dry. When the sample is taken in situ, any gas loss should be avoided while filling the measuring bottle with the sample to a level that is in the middle of the scaled pycnometer neck. At temperatures lower than + 10° C a correspondingly lower quantity must be filled, whereas more water must be filled into the bottle at temperatures of ca. + 30° C and even higher. The reason for such different filling levels is the fact that the warmer water body contracts and the colder one expands after a temperature standardization to + 20° C. When the bottle has been filled, it is tightly sealed immediately with a rubber plug. It is then kept for safe transport.

In the laboratory, the filled pycnometer is tempered at + 20° C ($\pm$ 0.2° C) in a thermostat for 1 hour. Then the water level is exactly read at the scale on the pycnometer neck. The liquid should not contain any air bubbles and the wall of the bottle neck should be free of any water droplets.

When the water level has been noted, the pycnometer is exactly weighed (precision 1 mg), emptied, cleaned, and filled with distilled water of + 20° C to the previously read water level. The receptacle is closed with the rubber plug and exactly weighed, emptied, dried, and weight at + 20° C in empty condition (with plug).

Calculation of the density:

The density is calculated by the formula

$$d_{20°C} = \frac{(G_1 - G_L) \cdot 0.99823}{G_2 - G_L} \quad [g/ml]$$

where:

G_1 = weight of the pycnometer filled with the water to be analyzed, in g

G_2 = weight of the pycnometer filled with distilled water, in g

G_L = weight of the empty dry pycnometer, including its plug, in g

0.99823 = density of water at + 20° C

With densities below 1, the value is indicated to four decimal places. Density values higher than 1 are given with three decimal places.

Example:

$$d_{20°C} \text{ (pycnom.det.)} = 1.006 \text{ g/ml}.$$

When the density is determined at environmental temperatures, the value must be converted to the density $d_{20°C}$. In such a case, the factor 0.998230 (density of water at + 20° C) must be replaced by the factor "F" which is indicated in Table 7.2.-1 giving the values for temperatures from 0 to + 30° C.

Tab. 7.2.-1: Density of water (g/ml) at the indicated temperature (in °C) at atmospheric pressure, from KUESTER-THIEL-FISCHBECK

Temperature °C	Factor F	Temperature °C	Factor F
0	0.999 868	16	0.998 970
1	0.999 927	17	0.998 801
2	0.999 968	18	0.998 622
3	0.999 992	19	0.998 432
4	1.000 000	20	0.998 230
5	0.999 992	21	0.998 019
6	0.999 968	22	0.997 797
7	0.999 929	23	0.997 565
8	0.999 876	24	0.997 323
9	0.999 808	25	0.997 071
10	0.999 727	26	0.996 810
11	0.999 632	27	0.996 539
12	0.999 525	28	0.996 259
13	0.999 404	29	0.995 971
14	0.999 271	30	0.995 673
15	0.999 126		

7.2.4 pH Value

The theoretical fundamentals of pH value measurement can be taken from the "Collections of methods of water analysis" /55/. The regulations for actual measurement are as follows.

Equipment:

pH meter

single-rod measuring chain (measuring + reference electrode)

The principle of measurement and the application of the advanced glass electrodes are based on the production of phase boundary potentials between the glass and an electrolyte, i.e. on an ionogenic interaction between the membrane glass and the solution. It is assumed that the contact with aqueous solutions on the glass membrane results in ion exchange and swelling processes that produce a silica gel layer (these processes have not yet been entirely clarified). An ionic equilibrium between this silica gel layer and the solution is produced to the effect that hydrogen ions are adsorbed or desorbed at the gel, depending on whether the adjacent liquid contains an hydrogen ion excess or lack, compared to the gel. The electric double layer provided at the phase

Regarding calibration and temperature compensation, reference is also made to the "Collection of methods of water analysis" /55/. The measured value may be compensated in accordance with the measured temperature, if necessary.

7.2.7 Oxygen

The oxygen concentration of water that is in solubility equilibrium with the atmosphere above is called <u>oxygen saturation</u> (oxygen saturation concentration) at a specified temperature and pressure (which depends on the altitude above mean sea level and other factors). When the concentrations are below this value, this condition is termed <u>oxygen deficit</u>, while values above the saturation value denote the condition of oxygen oversaturation or supersaturation.

The <u>actual oxygen concentration</u> defines the oxygen concentration in the water, which results from an analysis of a sample.

The oxygen concentration in the water is vital for both animal and vegetable organisms. This applies particularly to the metabolism of the microorganisms (bacteria) that effect the decomposition of the pollutants in the water. Such aerobic decomposition of pollutants in the water consumes dissolved oxygen. A reduction of the normal oxygen concentration is thus an indicator of pollution or of the presence of oxygen-consuming substances.

The "Collection of methods of water analysis" /55/ contains further theoretical information and processes.

7.2.8 Electrometric Determination of Dissolved Oxygen

The membrane-covered polarographic oxygen analyzer operates on the basis of the so-called CLARK principle. The Clark cell comprises two electrodes arranged on an insulator, which are interconnected through a liquid or pasty electrolyte. The electrode gap is separated from the medium to be measured (water, waste water) by a membrane permeable to O_2.

With a constant electrode area, a constant thickness of the membrane and a constant concentration of the electrolyte, the diffusional current generated in the analyzer substantially depends only on the partial oxygen pressure and on temperature. The diffusion of the oxygen through the membrane in the analyzer depends on the effect that the indicated internal current increases (decreases) the more the temperature increases (decreases) at the same O_2 concentration or O_2 partial pressure. Moreover, the O_2 solubility (e.g. in water) is also a function of temperature. Temperature compensation is therefore important for an oxygen measuring electrode. This compensation can either be produced in the analyzer unit using a thermistor associated with the cathode, or it is carried out by a simple calculation when less complex units are used.

Equipment and chemicals:

Oxygen membrane electrode (single-rod measuring chain). The design of such an electrode differs from one manufacturer to the other. The essential part of such an electrode is the measuring head (Fig. 7.2.-1).

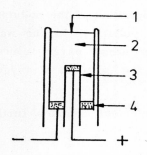

Fig. 7.2.-1: Schematic of a measuring head

1 - membrane; 2 - electrolyte volume; 3 - cathode; 4 - anode (circular)

The cathode is usually made of fine gold, while the anode consists of a metal more electronegative than gold. Silver is frequently used. Polyethylene, polypropylene, or Teflon are suitable membrane materials. A pasty substance is used as the electrolyte, which has recently replaced the KOH/KCl electrolyte that was formerly used. The other electrode parameters such as flow rate, polarization DC voltage, zero (residual) current, temperature compensation, time used to set the measuring value, steepness or sensitivity, and measuring accuracy are given in the particular instructions for use which accompany each unit.

Oxygen zero solution: a 3-5 % sodium sulfite solution prepared with warm water at ca. 60° C, is filled into a flask with a long narrow neck and is allowed to stand, closed, at room temperature for 24 hours.

Oxygen saturation solution: ice water set to a temperature of approximately 0° C is stirred with an agitator so that it is well mixed with air (10 - 15 min). The saturation value remains constant at the ice water temperature.

Calibration: The oxygen zero or neutral point is calibrated by dipping the measuring head into the oxygen zero solution, which must be entirely free of any air bubbles. After ca. 15 minutes, the zero or neutral value has been set, which is fixed with a control on the unit to "zero mg O_2/l".

The measuring head is dipped into the oxygen saturation solution with thorough agitation, in order to determine the standard saturation point. After ca. 3 minutes, a constant measuring value must be indicated, which is adjusted to the air saturation value using a control on the equipment ("steepness"). (The partial pressure of a gas in a liquid saturated with this gas corresponds to the partial pressure of the gas above this liquid (air saturation value = water saturation value)).

Standardization can also be carried out in air. To this end, the membrane and its vicinity are dried. The electrode is exposed to air for 5 - 10 minutes. Then the calibration procedure is the same as that described before. This method, however, involves an unavoidable calibration error of $\pm$ 3 %.

Determination. The single-rod measuring chain is dipped into the sample in the measuring cylinder to determine the oxygen concentration. Then a stirrer is started to provide the required flow rate. In units without a temperature compensation feature, the temperature of the sample is measured simultaneously. Air bubbles adhering to the electrode are removed by a short period of shaking. The measuring value is read when the indication has remained constant for 1 minute or longer. The value is read from the unit directly in mg/l O_2. The results are rounded to 0.1 mg/l.

Regarding the storage of the electrode, its preparation for calibration and measurement, as well as measurement checks to determine failure, or indications of wear, reference is made to the directions which are delivered together with each unit.

7.2.9 Sedimentary Matter

For a determination of the volume of the sedimentary matter, the application of the method is recommended that is described in the "Deutsche Einheitsverfahren zur Wasser-, Abwasser- und Schlammuntersuchung" (DEV), Section H 9 (DIN Standard 38409) [German Standard Methods of Water, Waste Water and Sludge Analysis]. This method is applicable for the determination of sedimentary matter in a volume above 0.1 ml/l.

The sedimentary matter should be determined as early as possible after sampling so as to avoid errors resulting from the subsequently occurring flocculation processes. Shocks or vibrations during sedimentation and thermal flows (convection) may result in interference and disturbance of the operation.

Equipment:
Sedimentation flask according to IMHOFF, clamping device

Method (simplified procedure):
2 x 1 l each of the well agitated sample are filled into the IMHOFF sedimentation flasks immediately after sampling, if possible. The flasks are rotated in jerks about the vertical axis during the sedimentation time after 50 and 110 minutes, to allow the matter adhering to the glass wall to settle. The volume of the sedimentary matter is read after 1 and 2 h.

Indication of the result:
Depending on the volume or concentration of the sedimentary matter, the values are rounded. The respective ranges are as follows:

volume: 2 ml/l, rounding to 0.1 ml/l
volume: from 2 to 10 ml/l, rounding to 0.5 ml/l
volume: from 10 to 40 ml/l, rounding to 1   ml/l
volume: 40 ml/l, rounding to 2   ml/l

Example: "volume of sedimentary matter 9.5 ml/l".

sodium hydroxide solution, 1 N

activated charcoal, nitrite-free

nitrite stock solution: 0.150 g sodium nitrite p.a. are dried at 105° C
for 1 h before weighing, dissolved in nitrite-free water, and topped up to 1 l with nitrite-free water after the addition of 1 ml trichloromethane. The solution is adjusted, using 0.1 N potassium permanganate solution. Its titration standard must be checked once per week. 1 ml $\triangleq$ 0.1 mg NO_2.

nitrite reference solution I: 10 ml of nitrite stock solution are diluted to 100 ml with nitrite-free water; 1 ml of this solution contains ~ 0.01 mg NO_2. This reference solution must be prepared before each application.

nitrite reference solution II: 10 ml of nitrite stock solution are diluted to 1 l with nitrite-free water; 1 ml of this solution contains ~ 0.001 mg NO_2. This reference solution must be prepared before each application.

nitrite-free water (according to the Standard Methods): 1 ml of conc. sulfuric acid (d = 1.84 g/ml) and 0.2 ml manganese (II) sulfate solution (36.4 g $MnSO_4 \cdot H_2O$ in 100 ml of aqueous solution) are added to 1 ml distilled water, then 1 - 3 ml potassium permanganate solution (400 mg $KMnO_4$ in 1 l aqueous solution) are added until a remaining pink color is achieved. The preparation is allowed to stand for 15 min. Then ammonium oxalate solution with 900 mg $(NH_4)_2C_2O_4 \cdot H_2O$ in 1 l aqueous solution is added to decolor the solution.

Procedure

<u>Standardization</u>. Increasing volumes of the nitrite reference solutions I and II with NO_2^- concentrations from 0.005 to 0.05 mg are transferred into 100 ml measuring cylinders with a pipette and treated according to the directions of procedure given below (Determination). The calibration plot is established from the extinction values.

<u>Determination</u>. The water sample volume required to determine the nitrite concentration of the aliquot from 0.0005 to 0.05 mg to be analyzed must be determined in a preliminary test. To this end, 10 - 50 ml of the water sample are transferred into a 100 ml measuring cylinder with a pipette and mixed with 4 ml of a freshly prepared mixture of the same volumetric parts of the sulfanilic acid solution and the 1-naphthylamine

solution. After topping up to the mark at 20° C, the preparation is mixed and the solution is allowed to stand in a dark space for 2 hours. After this reaction time the extinction is measured at 530 nm.

On the basis of the result obtained from this preliminary test, an appropriate volume of the water sample is transferred into a 100 ml measuring cylinder with a pipette, with 0.05 mg NO_2^- at maximum. The volume to be used must not exceed 95 ml; 4 ml of the mixture comprising the same volume proportion of sulfanilic acid and 1-naphthylamine solution are added to this volume, as described above, and the solution is topped with distilled water up to the mark and well mixed. After standing in the dark for 2 hours, the extinction is measured at 530 nm. A blind test is carried out in an analogous way, using distilled water.

7.3.2.3 Nitrate

Waste water samples may contain higher concentrations of nitrate ions. Other substances in the waste water may interfere with the direct spectrophotometric determination, e.g. in the sodium salicylate process. Therefore, the reductive distillation with a titrimetric final determination is described below as an optional method. The determination is based on the reduction and distillation process.

When a DEVARDA alloy (= alloy constituted by 50 % copper, 45 % aluminum, 5 % zinc) in an alkaline solution is added, nascent hydrogen reduces nitrate ions to ammonia:

$$NO_3 + 8\ H \longrightarrow NH_3 + OH + 2\ H_2O$$

The ammonia is redistilled and is determined in the distillation product by an acidimetric or spectrophotometric method.

The procedure is suited to determine nitrate in turbid or colored waters or waste waters as well as in samples with relatively high salt concentrations (salt waters). With concentrations higher than 5 mg/l NO_3 titration is applied for final determination; whereas with low concentrations, the spectrophotometric method using indophenol is applied.

Since ammonium ions cause problems, they are distilled off from the alkaline sodium hydroxide solution before the DEVARDA alloy is added. Nitrite ions are analyzed together with the nitrate ions. They must be determined separately and given due consideration in the calculation of the final result. The analysis must be carried out in an ammonia-free atmosphere. If filtration is necessary, it should be noted that only nitrogen-free filter paper is used. In spite of these precautionary measures, blind determinations are definitely required.

Equipment and chemicals

distillation apparatus: 1 l ERLENMEYER flask with WAGNER tube adapter with drop collector in the bulb; vertically descending cooler with extended tip.

DEVARDA alloy (50 % Cu, 45 % Al, 5 % Zn) p.a. as a powder

sodium hydroxide solution, 30 %, p.a.

boric acid solution: 40 g H_3BO_3 p.a. are dissolved with ammonium-free water to obtain 1 l (1 ml of this solution absorbs ca. 2 mg NH_4^+).

reagents for the acidimetric titration: see "ammonium" (acidimetric determination after distillation)

Procedure

Standardization. Increasing volumes of the nitrate reference solution with 1 to 100 mg NO_3^-/l are distilled according to the directions for procedure below (Determination). The ammonium ions are acidimetrically determined in aliquot parts of the obtained distillation products. An NO_3^--related calibration plot is established. Individual points on this curve are selected to check the plot by a determination from the ammonium reference solution.

Determination. The water sample volume required to carry out the determination is established in a preliminary test. This volume is transferred into the distillation flask in the main analysis, where it is mixed with 30 ml of the 30 % sodium hydroxide solution, whereupon approximately half of the volume is concentrated by evaporation. When the volume determined in the preliminary test is less than 250 ml it must be topped with ammonium-free water to 250 ml before transfer into the flask.

Upon removal of the ammonium ions, the solution is allowed to cool in the distillation flask, then 1 g DEVARDA alloy is added to the solution and the distillation apparatus is immediately closed. The foremost end of the cooler is immersed into the adsorption solution (50 ml boric acid solution). At room temperature, the nitrate ions (and also possibly present nitrite ions) are reduced in an alkaline solution by the DEVARDA alloy.

After a complete dissolution of the DEVARDA alloy, 100 ml of the reaction solution are redistilled. The distillate is transferred into a 200 ml measuring cylinder and topped with ammonium-free water up to the mark at 20° C. The ammonium ions in aliquot parts are subjected to an acidimetric determination (see "Ammonium"). A blind test must be carried out in an analogous way.

When nitrite ions are present simultaneously, the separately determined amount of nitrite ions must be converted to that of nitrate ions, using the factor 1.3478, and deducted from the nitrate amount determined.

7.3.2.4 Phosphate

In natural waters and waste waters, orthophosphates, condensed phosphates (polyphosphates and metaphosphates), and organic phosphorus compounds may be found. Where natural conditions, i.e. conditions not influenced by man, prevail, the concentrations of phosphorus compounds, calculated as hydrogen phosphate ions, are generally below 0.1 mg/l. Higher concentrations in ground and surface waters (rivers and lakes) are very often to be ascribed to the influences of civilization. They contribute substantially to the eutrophication of stagnant waters.

The pH value of a water determines whether or not the dissolved orthophosphates occurs in the form of

$$PO_4^{3-}, \text{ or } HPO_4^{2-}, H_2PO_4^- \text{ or } H_3PO_4$$

This is illustrated in the diagram below:

7. Evaluation

The phosphate (P) mass concentration of the water sample is calculated according to the equation

$$\beta = \frac{(A_s - A_{50}) \cdot f_s \cdot V_m \cdot k}{V_p}$$

where

β = phosphorus (P) mass concentration of the sample, in mg/l
A_s = extinction of the measuring solution
A_{50} = calculated extinction of the blank solution
f_s = factor established according to Section 6, in mg/l
s = index number as an indication of the chosen thickness of the layer
V_m = maximum volume of the water sample used in ml (V_m = 40 ml)
V_p = volume of the water sample applied, in ml
k = factor allowing for the acid added when taking the sample (1.005)

8. Indication of the result

When giving mass concentration of total phosphorus of a water sample, the values are rounded as follows:

        0.1 mg/l to 0.001 mg/l
  0.1 to 10   mg/l to 0.01 mg/l
      10   mg/l to 0.1 mg/l

not considering, however, more than 3 significant places.

7.3.2.5 *Sulfide Sulfur*

Sulfide sulfur may be present in waste waters in the form of dissolved sulfuretted hydrogen, hydrogen sulfide, or sulfide, depending on the pH value. The determination method described here covers the total sulfur present in oxidation stage 2.

<u>Determination as methylene blue</u>

Combined with dimethyl p-phenylene diamine, hydrogen sulfide forms an

intermediary compound (intermediate stage) containing sulfur, which changes to leucomethylene blue. The leucomethylene blue is oxidized to methylene blue by iron (III) ions. The methylene blue is photometrically determined at 670 nm. This method can be applied either for a direct sulfide sulfur determination in the sample solution or after distillation.
The sulfide sulfur, fixed in the form of zinc sulfide, is distilled over as sulfuretted hydrogen after the addition of phosphoric acid, in a distillation apparatus. The recipient vessel in the apparatus, which collects the sulfur, contains a zinc acetate solution.

The method is suitable to determine sulfide sulfur in the concentration range from 0.005 to 5 mg/l. The extinctions come under the LAMBERT-BEER law up to 70 g. The mean error ranges at $\pm$ 1 % in the sulfide range from 10 to 60 µg.

The method is also suitable for a sulfide sulfur determination in more heavily polluted surface waters and waste waters, as well as in mineral waters and other waters having a mineral concentration higher than 2000 mg/l, since high mineral concentrations in the test solution result in a reduction of the extinction and thus in insufficient sulfide values.

More than 1 mg iodide in the water sample aliquot used for the analysis causes a problem. This iodide concentration, however, is achieved only in exceptional cases.
Nitrite ions in quantities below 10 µg in the solution to be analyzed are not disturbing. The disturbance by even higher quantities can be eliminated by the addition of 3 drops of a 5 % urea solution.
Iron (II) ions in concentrations up to 25 mg/l and sulfite ion concentrations up to 10 mg/l do not affect the determination.
The presence of considerable quantities of hydrogen carbonate or carbonate results in the formation of CO_2 when the strongly acidic dimethyl p-phenylene diamine reagent solution is added. This CO_2, when escaping, may carry with it H_2S. The sample is therefore placed on top of the reagent, the flask is closed, and then it is agitated. When the flask is opened, only CO_2 escapes, since the amine has already bound H_2S.

All the disturbances and the disturbance by pollutants and opacifying substances or excessively high mineral contents (see Application) may be eliminated by the application of the distillation method.

Equipment and chemicals

Spectrophotometer (670 nm) or photometer (dark red filter)
cuvettes, 1, 2 and 5 cm
measuring cylinders with ground plug, 100 ml and 250 ml
distillation apparatus according to QUENTIN:
A 100 ml measuring cylinder is used as the recipient (Fig. 7.3.-2) whose diameter corresponds to an equal diameter in the ground section of the introduction pipe.

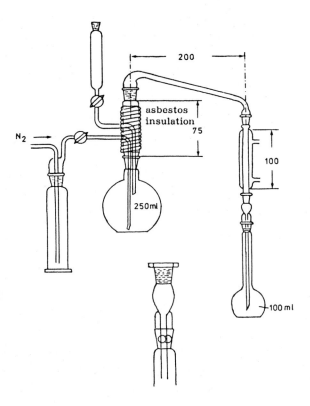

Fig. 7.3.-2: Distillation apparatus

steel cylinder containing nitrogen

zinc acetate solution, 2 %: 2 g zinc acetate dihydrate p.a. are dissolved in water and topped with water to 100 ml while some drops of 1 N acetic acid are added.

nitrogen cleaning solution: 2 % potassium permanganate solution, containing 5 g mercury (II) chloride in 100 ml.

amine reagent solution: 2 g dimethyl p-phenylene diamine hydrochloride are suspended with 200 ml dist. water in a 1 l measuring cylinder. Then 200 ml conc. sulfuric acid (d = 1.84 g/ml) are carefully added, and after cooling to 20° C the solution is topped up to the mark.

iron (III) reagent solution: 10 ml of conc. sulfuric acid (d = 1.84 g/ml) are poured over 50 g iron (III) ammonium sulfate, $NH_4Fe(SO_4)_2 \cdot 12\ H_2O$. Then the solution is topped with dist. water up to the mark.

phosphoric acid, d = 1.70 g/ml

sulfide reference solution: 500 ml of fresh gelatine solution (150 mg in 500 ml) are added to 250 ml of the 2 % zinc acetate solution. A brown glass flask containing this solution is placed as a recipient under the distillation apparatus. The discharge pipe of the cooler should be deeply immersed into this solution (if necessary, the pipe must be extended). A moderate nitrogen flow is passed through the washing bottle with the nitrogen cleaning solution and then circulated in the distillation flask. In this flow, 150 l of boric acid solution (6 g H_3BO_3 p.a. dissolved in 150 ml warm water) are heated to boiling. While the solution is heated, a sodium sulfide solution (dissolve a washed sodium sulfide crystal in a small water volume, ca. 40 mg $Na_2S \cdot 9\ H_2O$ cryst. p.a., corresponding to ca. 5 mg S) is added into the separating funnel. When the boric acid solution boils and water is already distilling over, the boiling process is interrupted briefly and the sodium sulfide solution is slowly added in drops. The released hydrogen sulfide is distilled off within 15 minutes, and is bound as colloid ZnS in the collecting recipient. When the recipient is topped with water in a 1 l measuring cylinder up to the mark, the sulfide concentration in this solution is iodometrically determined. (Use only boiled water to prepare all solutions! Stock the water in brown glass flasks!)

Procedure

Standardization. Depending on the result of the iodometric sulfide determination in the sulfide reference solution, varying volumes of this solution, which contain between 10 and 70 µg sulfide, are filled into 100 ml measuring cylinders and mixed with water until the liquid level is slightly below the shoulder of the bottle neck. Then the procedure continues with methylene blue coloring as described below (Determination). A calibration plot is thus established which shows a straight curve up to 70 µg sulfide. The measurement is taken at 670 nm.

Annotation:

Dark blue solutions with sulfide concentrations higher than 70 µg cannot be measured, not even after dilution. The established calibration plot does not apply to such dilutions.

Determination. 250 ml measuring cylinders (distillation flasks) are charged with 5 ml zinc acetate solution (sufficient to bind 14 mg sulfide). Depending on the sulfide sulfur concentration to be expected in the water, up to 160 ml of the analytic water are introduced into the sample with a pipette at the sampling site. In the laboratory, the ground plugs are removed from the flasks and the flasks are connected to the distillation apparatus. 10 ml zinc acetate solution are charged into the recipient (100 ml measuring flask with bore, see Fig. 7.3.-2). The ground introduction pipe with a bore is immersed into this zinc acetate solution. The bores at the introduction pipe and the measuring flasks are aligned by rotation.

With a moderate nitrogen flow (see above, preparation of the sulfide reference solution, Equipment and chemicals) is heated to boiling. When the solution is boiling and water is already distilling over, the boiling process is interrupted and 5 ml phosphoric acid (d = 1.70 g/ml) are added through the separating funnel. H_2S then passes into the recipient together with the steam, and is bound there again to the zinc.

After ca. 15 minutes the liquid volume in the recipient is ca. 60 ml (10 ml zinc acetate solution + 50 ml distillate) and the distillation process is finished.

The measuring flask, including the introduction pipe, is removed from the apparatus. The amine reagent solution and then the iron (III) reagent solution are added through the introduction tube. The bores of the two ground sections are separated so that the sample solution can be agitated. The flask with the introduction pipe is then closed with a ground plug (see Fig. 7.3.-2). The introduction pipe is drawn out and rinsed with water only when the iron solution has been added and the solution has been agitated. The 100 ml measuring flask is now in "normal condition" again. It is allowed to stand for 10 minutes at room temperature, then the solution is topped with distilled water up to the mark, and the extinction of the colored solution is measured by comparison to the blind test solution which has been prepared simultaneously.

7.3.2.6 Sulfite

Since the iodometric methods to determine sulfite in waste water are often susceptible to disturbances, the gravimetric determination of the sulfite ions after distillation as barium sulfate is described here as an example of another method.

The procedure explained below is termed <u>gravimetric determination as barium sulfate after distillation</u>.

The sulfite ions are distilled over from a phosphate solution as sulfur dioxide into a recipient containing an iodine/potassium iodide solution, with oxidation of the sulfite ions to sulfate ions in the solution in the recipient:

$$SO_3^{2-} + J_2 + H_2O \longrightarrow SO_4^{2-} + 2 H^+ + 2 J^-$$

$$SO_4^{2-} + Ba^{2+} \longrightarrow BaSO_4$$

A barium chloride solution is added to precipitate the sulfate ions as barium sulfate, then the precipitate is weighed.

The method is suitable to determine sulfite ions in the concentration range from 5 to 30 mg/l and in the presence of considerable quantities of disturbing organic substances.

The distillation method largely eliminates any disturbance.

Equipment and chemicals

distillation apparatus with a long cooler end

phosphoric acid, d = 1.15 g/ml: 12.5 ml phosphoric acid, d = 1.70
g/ml, are diluted with 100 ml distilled water

iodine/potassium iodide solution: first 7.5 g potassium iodide and then
5 g iodine are dissolved in 500 ml distilled water, the solution
is then diluted to 1 l.

barium chloride solution: 12.5 g $BaCl_2$ 2 H_2O are dissolved with
distilled water and topped to 500 ml.

hydrochloric acid, d = 1.05 g/ml: 75 ml hydrochloric acid, d = 1.19
g/ml, are diluted with 200 ml distilled water.

carbon dioxide gas

Procedure

Determination. Depending on the sulfite ion concentration, which is determined in a preliminary test, 300 to 500 ml of the water sample are distilled after the addition of 20 ml phosphoric acid (d = 1.15 g/ml). The complete distillation apparatus must be filled with carbon dioxide before the sample is charged, in order to prevent an oxidation of the sulfur dioxide by atmospheric oxygen. The cooler end is immersed into the iodine/potassium iodide solution in the recipient, which must be present in an excess quantity.

When the distillation process is finished, the sulfate formed is precipitated with an excess of hot barium chloride solution in boiling heat (the barium chloride solution is added in one batch), after the addition of 1 ml hydrochloric acid (d = 1.05 g/ml), in the iodine/potassium iodide solution in the recipient. The collected precipitate is allowed to stand overnight, then it is filtered through a dense filter, washed with hot distilled water. The filter is incinerated, after drying, in a dish; the ash is calcined at 800° C until the weight remains constant; and then the residue is weighed.

The factor applying to the conversion of barium sulfate to SO_3^{2-} is 0.3430. Using this information, the sulfite concentration in the sample is calculated according to the formula

$$\frac{g \cdot 343}{V} = G$$

where:

    g = $BaSO_4$ weight, in mg
    V = used water volume, in ml
    G = SO_3^{2-} concentration in the water sample, in mg/l

7.3.2.7 Sulfate

Polluted waters and waste waters contain considerable quantities of chloride, nitrate, and phosphate, and mostly they have an increased sulfate concentration, which may range from ca. 100 mg/l up to and even beyond several hundreds of mg/l. When such high sulfate concentrations occur within the territory of drinking water test installations, specific analyses are required for a hygienic rating of the drinking water. These high concentrations might result from urine, manure, and outflows from sewage lands, waste waters from metallurgical plants and chemical plants.

<u>Gravimetric determination as barium sulfate</u>

The sample is first acidified with 5 ml hydrochloric acid (d = 1.125 g/ml) and evaporated until dry, so that disturbing silicates are removed. After the addition of 5 ml of the same hydrochloric acid, the residue is heated, diluted with 50 ml dist. water, and the mixture is filtered in hot condition. The filtrate is washed out with hydrochloric acid in a 1 : 50 dilution.

Silicates and organic substances may also be removed from the sample solution together by evaporating the sample in a platinum or quartz dish in a water bath until almost dry. After the addition of some drops of hydrochloric acid (d = 1.125 g/ml) and 2 - 5 drops of 10 % NaCl solution, the dish is inclined so as to contact with the hydrochloric acid the substance adhering to the rims, then the dish contents are evaporated until dry, and incinerated at weakly red heat. The residue is heated with 3 ml distilled water and 1 ml of hydrochloric acid, taken up with hot water, and filtered in hot condition. The filtrate contains the silicic acid, which has become insoluble. It is washed with hot distilled water until the filtrate is free of chloride ions. Both the filtrate and the washing waters are used to determine the sulfate ions.

solution, if possible. The absorption vessel is charged with 10 ml 1 N sodium hydroxide solution. The flowmeter is applied to set the air stream (scooped by the water jet pump and passed through a washing bottle charged with 1 N sodium hydroxide solution, before its introduction into the flask) at a flow rate of 20 l/h.

After the addition of 10 ml 25 % hydrochloric acid per 100 ml of the sample, the heater is started to heat the charge in the flask up to boiling temperature for ca. 45 minutes (backflow ca. 1 to 2 drops per second). After this, all decomposable cyanides are transferred in the form of hydrocyanic acid through the carrier gas flow into the absorption vessel.

When the distillation process stops, the absorptive solution is turbid or when the transfer of any substances that disturb the determination process must be assumed, the absorptive solution undergoes a second distillation step. To this end, the contents of the absorption vessel are introduced into a second distillation flask previously charged with 10 ml cadmium acetate solution and 40 ml buffer solution. The absorption vessel is rinsed with 40 ml distilled water. This washing liquid is also introduced into the flask. Then the procedure is the same as that described in relation to the first distillation process.

For the cyanide ion determination, the contents of the absorption vessel (after the first or the second distillation) are transferred into the 25 ml measuring cylinder, with subsequent triple washing with 3 ml of distilled water each, whereupon the measuring cylinder is topped up to the mark with distilled water.

7.3.2.8.1.1 Spectrophotometric Determination Utilizing Barbituric Acid Pyridine

Cyanide ions react with the active chlorine in chloramine T, forming chlorine cyanide:

$$CN^- + Cl_2 \rightarrow CNCl + Cl^-$$

When an aqueous solution of pyridine and barbituric acid is added, the chlorine cyanide is accumulated at the pyridine ring while separating it. The resulting product is an intermediary glutacondialdehyde, which when combined with 2 molecules of barbituric acid ($C_4H_4O_3N_2$) is condensed to produce a reddish violet polymethin dye accessible to photometry at 573 nm:

$$C_5H_5N + CNCl + 2\ C_4H_4O_3N_2 \longrightarrow NH_2CN + HCl +$$

$$\begin{array}{c}HN-CO\\|\quad\;|\\OC\quad C\\|\quad\;|\\HN-CO\end{array}=CH-CH=CH-CH=CH-CH\begin{array}{c}OC-NH\\\diagdown\quad\diagdown\\\quad\;OC\\\diagup\quad\diagup\\OC-NH\end{array}$$

The method can be applied to determine cyanide ions in surface and waste waters with concentrations from 0.005 to 0.05 mg CN^-/l after a preliminary distillation. This range corresponds to a cyanide concentration of 2.5 to 25 µg in 10 ml absorptive solution, assuming that 500 ml water sample have been employed, and comes under LAMBERT-BEER's law.

Almost all disturbing substances, nitrogen/oxygen compounds, hydrogen sulfide, sulfur dioxide, or sulfide, are practically eliminated before or in the second distillation (see "Distillation process" above). Distillation products that are then still dyed or contain organic compounds which form dyes in combination with the used reagents, are not accessible to this determination method.

Equipment and chemicals

photometer
cuvettes, thickness of layer 1 cm to 5 cm
measuring cylinders, 25 ml, 50 ml, 250 ml
hydrochloric acid, d = 1.19 g/ml
hydrochloric acid, 1 N
sodium hydroxide solution, 0.4 N
buffer solution, pH 5.4: 60 g sodium hydroxide p.a. are dissolved in
  ca. 500 ml distilled water; 118 g succinic acid are stirred into
  the still warm sodium hydroxide solution. After cooling, the
  solution is topped with distilled water up to 1 l.
chloramine T solution: 1 g chloramine T (p-toluene sulfochloramide
  sodium) p.a. is dissolved in distilled water to obtain 100 ml.
  The solution is stable for ca. 1 week. It should be checked
  for efficiency, however, by iodometric methods.

barbituric acid pyridine reagent: 3 g barbituric acid p.a. are weighed into a 50 ml measuring cylinder, soaked with a small quantity of distilled water, and then mixed with 15 ml of freshly distilled pyridine (boiling point 115 - 118° C). The solution is then diluted with distilled water while agitating, until the barbituric acid is almost completely dissolved. Then 2.5 ml hydrochloric acid p.a. (d = 1.19 g/ml) are added. After cooling at 20° C, the solution is topped with distilled water up to the mark. The reagent is stable in a brown bottle for 1 day and in the refrigerator up to 1 week.

cyanide reference solution: 25 mg potassium cyanide p.a. are dissolved in 1 l of 0.4 N sodium hydroxide solution. 1 ml of this solution contains ca. 0.01 mg cyanide ions. The factor of this solution is determined by titration using 0.001 N silver nitrate solution (see "titrimetric determination with silver nitrate").

Procedure

Standardization. 2, 5, 10, 20 and 25 ml each of the cyanide reference solution are introduced into a 250 ml measuring cylinder with a pipette. Each measuring cylinder is topped with 0.4 N sodium hydroxide solution up to the mark. Ten milliliters each are removed from these measuring cylinders, introduced into small 25 ml measuring flasks, and mixed with 1 N hydrochloric acid, buffer solution, chloramine T solution, and barbituric acid pyridine reagent, as indicated under the heading "Determination". Then the photometric analysis is carried out. The measured values must be located on a straight line. The calibration plot must be verified from time to time.

Determination. 10 ml are taken from the 25 ml measuring flask containing the absorptive solution after distillation, and this quantity is transferred into another 25 ml measuring flask. Then, while agitating, exactly 2 ml buffer solution, 4 ml 1 N hydrochloric acid, and 1 ml chloramine T solution are added successively, and the mixture is allowed to stand in the sealed flask for at least 1 minute but not longer than 5 minutes. Then 3 ml barbituric acid pyridine reagent are added, the solution is topped with distilled water up to the mark, and the contents in the flask are well shaken. After 20 minutes, the photometric measurement is carried out at 578 nm against a reference solution that is to be produced as follows:

10 ml 0.4 N sodium hydroxide solution, 2 ml buffer solution, 4 ml 1 N hydrochloric acid, 1 ml chloramine T solution, and 3 ml barbituric acid pyridine reagent are introduced into a 25 ml measuring flask while agitating, and topped with distilled water up to the mark.

A blind value must be considered for the evaluation, which comprises a complete analysis including distillation. Instead of the water sample, distilled water is used. The blind sample is also subjected to photometric measurement against the reference solution.

The blind value B (mg CN^-) is subtracted from the cyanide concentration P of the analyzed sample (mg CN^-), which is read from the calibration plot, the difference is multiplied by 1000; and the product is divided by 0.4 V. The result is the total cyanide concentration G in the water sample, calculated as cyanide ion concentration. In this calculation, V corresponds to the used sample volume (ml). The factor 0.4 must be used since only 10 ml, i.e. 40 %, of the absorptive solution topped up to 25 ml, were used in the photometric measurement. The formula for this calculation is:

$$G = \frac{(P - B) \cdot 1000}{0.4 \cdot V}$$

7.3.2.8.1.2 Titrimetric Determination Utilizing Silver Nitrate

A complex silver cyanide ion

$$2\ CN^- + Ag^+ \longrightarrow (Ag(CN)_2)^-$$

is obtained from the titration of a cyanide solution with a silver nitrate solution. After the conversion of all cyanide ions, the excess silver ions react with p-dimethylamino benzylidene rhodamine (FEIGL's reagent). The resulting lax-red coloring of the titration solution indicates the end of the titration process.

The method is applicable to all distillation products (cf. "Distillation process") containing more than 0.05 mg CN^-. This amount corresponds to a cyanide concentration in the water sample of more than 0.1 mg/l when 500 ml of the sample were distilled.

A preceding single or double distillation eliminates practically all disturbing factors.

Equipment and chemicals

beakers, magnetic stirrer with magnetic bar

buret, 10 ml, or flask buret

silver nitrate solution, 0.001: it is most advantageous to prepare the solution before use from 0.01 N silver nitrate solution, and to check it for its concentration.

indicator solution: 20 mg p-dimethylamino benzylidene rhodamine, $C_{12}H_{12}N_2OS_2$, are dissolved in 100 ml acetone p.a. The solution is stable in a dark bottle for at least 1 week.

sodium hydroxide solution, 1 N.

Procedure

Determination. First the blind value is determined, since the color shade obtained from the blind value determination is the reference for the final product of titration from the cyanide-containing sample solution; 10 ml of 1 N sodium hydroxide solution are introduced into a beaker, together with ca. 20 ml distilled water and 0.1 ml indicator solution. Then the magnetic stirrer is started while the silver nitrate solution is allowed to flow into the solution from the buret whose tip is immersed in the liquid. The silver nitrate solution is introduced until the color of the solution changes from yellow to yellowish red to lax red. The color remains constant only for a short time.

Then the cyanide-containing sample is analyzed. The content of a 25 ml measuring flask containing the adsorptive solution of the distillation process is transferred into another beaker, 0.1 ml of the indicator solution is added, and titration takes place as for the determination of the blind value. The indicator solution is added until the reference shade obtained with the blind value is reached with the sample.

One milliliter of 0.001 N silver nitrate solution corresponds to 0.052 mg cyanide ions. Thus the total cyanide concentration in the sample, G, results from the amount of 0.001 N silver nitrate solution consumed by the sample solution (absorption solution), calculated as cyanide ions in mg/l, A, reduced by the amount of 0.001 N silver nitrate solution, B, consumed by the blind sample. This value is multiplied by 0.052 (conversion factor $AgNO_3/CN^-$), multiplied again by 1000 (conversion of ml into l), divided by the obtained value of the volume V of the water sample used for distillation:

$$G = \frac{(A - B) \cdot 0.052 \cdot 1000}{V}$$

7.3.2.8.2 Easily Releasable Cyanide

The term easily releasable cyanides in the meaning of the present definition is to be understood to apply to those cyanides or cyanide compounds whose cyanide contents are covered by the conditions of analysis described here. This cyanide group includes not only hydrocyanic acid, alkali and earth alkali cyanides, but also cyanides in complex compounds, except iron and cobalt cyanide complexes. This term does not cover either nitriles, cyanates, thiocyanates, or chlorine cyanide. The quantitative determination is determined by spectrophotometry utilizing barbituric acid pyridine after separation ("cold distillation") in an air stream from zinc and cadmium sulfate containing solution at pH 4.

Separation of easily releasable cyanide (distillation process)

The separation process expels the cyanide contained in the water sample, in an air stream, in the form of hydrogen cyanide to the extent by which it is releasable in a slightly acidic environment (pH 3.9 - 4.0) at room temperature, and of collecting the hydrogen cyanide in an alkali absorptive solution. The pH value is maintained by the addition of potassium hydrogen phthalate buffer solution. The inclusion of iron cyanide complexes in the analysis is prevented by the addition of zinc sulfate, whereas the zinc powder, which is also added, results in an accelerated decomposition of the copper (II) cyanide complex. EDTA is added as a chelating agent to largely inhibit the catalytically decomposing effect of heavy metal ions, especially the copper ions. Cadmium sulfate buffers the disturbance by sulfides or hydrogen sulfide.

The method is applicable to all types of water whose concentration of easily releasable cyanide does not exceed 50 mg CN^-/l. With higher cyanide concentrations, the water sample must be diluted.

Practically all aspects in the explanations and descriptions given in relation to total cyanide (distillation process) apply also to the process of separating easily releasable cyanide.

Equipment and chemicals

Distillation apparatus, flowmeter, washing bottle, and 25 ml measuring cylinder are the same as listed in the chapter on "total cyanide" (distillation process).

hydrochloric acid, conc. and 25 %

sodium hydroxide solution, 5 N and 1 N

stannic (II) chloride solution: 500 g $SnCl_2 \cdot 2\ H_2O$ and 200 ml concentrated hydrochloric acid are dissolved with distilled water to obtain 1 l.

trichloromethane phenolphthalein solution: 0.3 g phenolphthalein are dissolved in 900 ml ethanol, and 100 ml $CHCl_3$ are added.

zinc sulfate/cadmium sulfate solution: 100 g $ZnSO_4 \cdot 7\ H_2O$ and 100 g 3 $CdSO_4 \cdot 8\ H_2O$ are dissolved in distilled water to achieve 1 l.

buffer solution, pH 4: 80 g potassium hydrogen phthalate, $C_8H_5KO_4$, p.a. are dissolved in 920 ml warm distilled water.

EDTA solution: 100 g di-sodium salt of the ethylene diamine tetraacetic acid p.a. are dissolved in 940 ml of slightly warmed distilled water.

zinc powder p.a.

Procedure

Preparation of the sample. The preparation of the water sample for the determination of easily releasable cyanide is particularly important from the sampling time onward. This is necessary to fix the exact cyanide concentration.

To this end, 5 ml of a 5 N sodium hydroxide solution, 10 ml trichloromethane phenolphthalein solution, and 5 ml stannic (II) chloride solution are added per liter of the water sample (after a preliminary dilution when waste waters with a known high cyanide concentration are involved). When the solution becomes red, the coloring is removed by the dropwise addition of 1 N hydrochloric acid. When no red coloring occurs, the sample is mixed with 1 N sodium hydroxide solution until a slight shade of red appears. Colored solutions are adjusted to a pH of ca. 8 either electrometrically or with an indicator paper.

Then 10 ml zinc sulfate/cadmium sulfate solution are added. The thus prepared sample is kept cool in a dark space. The analysis should be carried out as soon as possible.

When the absorption vessel filled with 10 ml of 1 N sodium hydroxide solution has been connected to the reflux condenser, and when the air stream has been set to an initial rate of 30 to 60 l/h, the three-necked flask is charged with 10 ml zinc sulfate/cadmium sulfate solution, 10 ml EDTA solution, 50 ml buffer solution, and 100 ml of the prepared water sample. When a concentration lower than 0.1 mg/l of easily releasable cyanide must be expected, a larger sample volume (up to 500 ml) can be introduced. This requires, however, a corresponding increase of the quantities of zinc, zinc sulfate/cadmium sulfate solution, and buffer solution.

A pH electrode is introduced through the lateral adapter into the solution, while hydrochloric acid or sodium hydroxide solution is introduced dropwise through the charging adapter until the pH value is 3.9 ± 0.1. After removal of the electrode, ca. 0.3 g zinc powder are introduced through the lateral adapter, the opening is closed with a glass plug, the charging adapter is connected to the washing bottle filled with 100 ml of 1 N sodium hydroxide solution, and the air flowrate is increased to 60 l/h.

This process is stopped after 4 hours. Then the absorptive solution is transferred into the 25 ml measuring flask, washed three times, each time with 3 ml dist. water. These washing solutions are transferred into the measuring flask and topped with distilled water up to the mark.

Spectrophotometric determination utilizing barbituric acid pyridine

The procedure of the spectrophotometric determination of the "free cyanide" is identical in each aspect with the procedure described in the chapter on "total cyanide".

7.3.3 Organic Substances

The concentrations of oxidizable organic substances in waters is highly important in water assessment. They are determined indirectly through the quantity of oxygen consumed for a decomposition of such chemically oxidizable organic substances as complete as possible. The total oxygen quantity required for this process is called "chemical oxygen demand" (COD).

The COD is determined by the introduction of a known oxygen quantity in the form of a chemically effective oxidant into a measured volume of the water sample. Then the oxidation is chemically carried out, with a determination of the oxidant quantity consumed, calculated as mg O_2/l of water. It should be noted that this process also covers oxidizable inorganic substances. The oxygen consumed by these substances must be taken into account when the analytical results are evaluated.

The ideal case is involved when a certain oxidant quantitatively decomposes the total of the present organic carbon compounds down to the final products CO_2 and H_2O. This ideal case, defined as "total oxygen demand" (TOD), is generally not achieved with the oxidants common in water analysis, i.e. potassium permanganate and potassium dichromate. According to W. LEITHE, the average $KMnO_4$ consumption in the oxidation of domestic waste waters is as low as 25 % of the TOD value corresponding to total oxidation. When the oxidation is carried out with $K_2Cr_2O_7$ (with addition of Ag_2SO_4 as the oxidation catalyst and $HgSO_4$ to mask the chlorides) up to 95 - 98 % at maximum of the absolute (100 %) TOD value can be achieved. Disturbances may occur as a result of the different oxidation processes with the different nitrogen compounds, as well as the nonvolatile and hardly oxidizable organic substances.

Potassium permanganate and potassium dichromate are mainly used as oxidants to determine the chemical oxygen demand. The chemical oxygen demand is defined to be the $KMnO_4$ or $K_2Cr_2O_7$ amount, converted to oxygen, which is required to oxidize the organic substances in the water under specified conditions.

The addition "converted to oxygen" is important. The formerly common indication of the volume of oxidant consumed should be avoided since only the standardized indication in mg O_2/l allows comparability of the results obtained in the various processes, and a comparison to the results obtained from the determination of the biochemical oxygen demand (see the section on "biochemical oxygen demand"). The fact should never be overlooked that this comparability is always only very limited since there are organic substances which are oxidized by a particular oxidant to a considerably higher level than others, and that this condition may be the exact opposite with other organic substances. Depending on the concentrations of such substances in a given water sample, totally different correlations between the oxidants to be compared may occur.

When potassium permanganate is used as an oxidant, it is still frequently common in practice to define the $KMnO_4$ consumption as "oxidizability" and indicate it in mg $KMnO_4$/l water. In such a case, however, the oxygen equivalent of the $KMnO_4$ consumption must also be indicated.

It has also become common to restrict the application of the term "chemical oxygen demand" (COD) essentially to the potassium dichromate consumption, which is indicated in mg/l. This is, however, only a restriction based on nothing but practical usage. It would be principally appropriate to indicate the results of both procedures as COD-Mn and COD-Cr (each expressed in mg/l water).

It should also be noted that even though the COD is expressed with the same measuring unit as that applying to a volume concentration, since the oxygen quantity consumed is given, it must not be interpreted as a concentration of mass. The reference to the element oxygen merely takes the character of a symbol of the type of the chemical effect. It is therefore not recommended to indicate the COD in mg O_2/l, particularly since the molecular oxygen does not have anything to do with this parameter (as contrasted to BOD). The German process standards therefore provide for an indication of the total parameters as COD or BOD in mg/l.

In conclusion, the following definitions apply:

<u>Oxidizability</u>: The potassium permanganate consumption determined from the chemical oxidation of organic substances contained in the water, to be given as "mg $KMnO_4$/l water", and principally also as the corresponding oxygen equivalent in "mg/l water".

<u>Chemical oxygen demand (COD)</u>: the potassium dichromate consumption determined from the chemical oxidation of organic substances contained in the water, to be given as the oxygen equivalent in "mg/l water".

Just like the biochemical oxygen demand discussed in a separate section, the oxidizability and the chemical oxygen demand are not an exact measure of the total of the organic substances present in the water. But, in practice, they are used as proven indices in the assessment of the total of organic pollutants in a particular water.

Depending on the purpose of the analysis, the oxidizability or chemical oxygen demand is determined in untreated, sedimented, or filtered water. The result of the analysis should include a remark on the particular condition applied in the analysis.

Waters are analyzed in an untreated condition when a statement must be made on the total pollution by organic substances, e.g. the loading of a natural water by waste waters.

When a statement must be made on the chemical oxygen demand still present in a particular water after a mechanical treatment, the oxidizability or COD value is determined with a sedimented sample. The dissolved oxidizable substances contained in water are determined with a sample filtered through a soft folded filter or a glass fiber filter.

Moreover, the determination of phenols, oils, and fats is described.

7.3.3.1 Oxidizability

Potassium permanganate in an acidic, neutral, and alkaline solution has an oxidizing effect on many organic and a number of inorganic substances. The oxidation level depends on the composition, type, and concentration of the organic substances (according to W. LEITHE, substances such as acetic acid, acetone, benzoic acid, phthalic acid, and many amino acids are not at all accessible to an attack by $KMnO_4$) as well as on the concentration of the potassium permanganate solution and the utilized sulfuric acid, and the reaction temperature and reaction time. For this reason, it is definitely required that the conditions of analysis be strictly observed.

An excess of potassium permanganate is always applied for the determination. In an acidic solution, the permanganate ion is reduced to a manganese (II) ion:

$$MnO_4^- + 5\ e^- + 8\ H^+ \longrightarrow Mn^{2+} + 4\ H_2O$$

The KMnO$_4$ excess is determined by titration with oxalic acid:

$$2\ MnO_4^- + 5\ C_2O_4^{2-} + 16\ H^+ \longrightarrow 2\ Mn^{2+} + 10\ CO_2 + 8\ H_2O$$

In an alkaline solution, the permanganate ion is reduced only to the level of the tetravalent manganese, which is precipitated as brown manganese (IV) oxide:

$$MnO_4^- + 3\ e^- + 2\ H_2O \longrightarrow MnO_2 + 4\ OH^-$$

Since after acidification both the excess permanganate ions and the manganese (IV) oxide are reduced by the oxalic acid to manganese (II) ions:

$$MnO_2 + C_2O_4^{2-} + 4\ H^+ \longrightarrow Mn^{2+} + 2\ CO_2 + 2\ H_2O,$$

the type of the medium (acidic or alkaline) is irrelevant for the calculation of the oxygen consumption.

The method is appropriate for the determination of the oxidizability in any water with a potassium permanganate consumption of at least 1 mg/l (corresponding to an oxygen equivalent of 0.25 mg/l or higher).

The potassium permanganate process is less suitable in the analysis of industrial waste waters since the oxidation effect is lower and varies according to the composition of the water. In these cases, the oxidation with potassium dichromate (COD) is recommended. It is more complete and is better suited to furnish comparable values for waters with a strongly differing organic loading.

The samples should be analyzed as soon as possible after sampling. If the samples are filtered and paper filters are used for this purpose, they must be washed with distilled water until the filtrate no longer takes up any potassium permanganate.

When oxidizable inorganic matter is present, the following corrections must be considered.

1 mg/l Fe^{2+}: 0.57 mg/l $KMnO_4$ or 0.14 mg/l (oxygen equivalent)
1 mg/l NO_2^-: 1.37 mg/l $KMnO_4$ or 0.35 mg/l (oxygen equivalent)
1 mg/l S^{2-} : 1.97 mg/l $KMnO_4$ or 0.50 mg/l (oxygen equivalent)

must be deducted from the result.

Rounding of the results:

$KMnO_4$ consumption:
    to   0.1 mg/l up to 10 mg/l
    to   1   mg/l for 10 to 100 mg/l
    to   10   mg/l for 100 to 1000 mg/l
    to   100 mg/l for more than 1000 mg/l

Oxygen:
    to   0.1 mg/l up to 10 mg/l
    to   1   mg/l for 10 to 100 mg/l
    to   10   mg/l for 100 to 1000 mg/l
    to   100 mg/l for more than 1000 mg/l

Example: oxidizability of the filtered sample
         in an alkaline solution: 18.6 mg/l $KMnO_4$,
         corresponding to: 4.6 mg/l O_2.

7.3.3.2 Chemical Oxygen Demand (COD)

7.3.3.2.1 Method Utilizing Potassium Dichromate

Potassium dichromate in an acidic solution has an oxidizing effect on almost all organic substances and on a number of inorganic compounds or ions. The level of oxidization depends on the type and concentration of the organic substances, on the concentration of potassium dichromate and sulfuric acid, on the reaction temperature, and on the reaction time. It is therefore always necessary to observe the conditions of analysis.

The determination is always made with an excess of potassium dichromate, with one part of the dichromate being reduced to the chromium (III) ion:

$$Cr_2O_7^{2-} + 6\ e^- + 14\ H^+ \longrightarrow 2\ Cr^{3+} + 7\ H_2O.$$

The excess of potassium dichromate ions is titrated back with iron (II) solution while ferroin is used as the redox indicator:

$$Cr_2O_7^{2-} + 6\ Fe^{2+} + 14\ H^+ \longrightarrow 2\ Cr^{3+} + 6\ Fe^{3+} + 7\ H_2O.$$

The method is suited to determine the chemical oxygen demand in any type of water and waste water absorbing more than 40 mg/l potassium dichromate when the 0.25 N potassium dichromate solution is used, or consuming 10 - 40 mg/l when 0.05 N potassium dichromate solution is used. In the latter case, the results are generally lower by 10 %, compared to the results obtained with 0.25 N potassium dichromate solution. This is an indicator of the fact that the values obtained with different concentrations of the oxidant cannot simply be compared to each other.

Some organic substances in the water, e.g. benzene and pyridine, are only incompletely oxidized by the indicated method. The COD value therefore reaches the 100 % level of the TOD (total oxygen demand) in particularly favorable cases only, whereas it may remain considerably lower in water samples of an unfavorable composition.

The addition of mercury (II) sulfate is an attempt to reduce the disturbance by the chloride ion oxidation. The occurring mercury (II) chloride (sublimate) is highly toxic and must be detoxicated in a special process (see Section 7.3.3.2.3 "Removal of Hg and Ag from residue solutions for COD determination") in order to avoid subsequent environmental pollution.

Sulfur compounds containing the sulfur in a lower oxidation stage than +6 (sulfate sulfur) are practically completely oxidized to the sulfate stage. Their influence can be considered in the calculation provided that it is possible to apply a selective method of determining these compounds.

Reduced nitrogen (ammonium ion, amino compounds, amides, nitriles) normally retains its oxidation level of -3. Exceptions are possible.

Oxidized nitrogen (nitrite, nitroso, and nitro compounds) is practically completely oxidized to nitrate.

Iron (II) ions are completely oxidized to iron (III) ions. Their influence on the result can be covered in a separate iron (II) determination and in the calculation.

In spite of these remarks, the fact remains that the influence of disturbances causes problems in the COD determination. In most cases where waters without an extreme loading are involved, however, silver sulfate can be added as the oxidation catalyst, so that an average oxidation level of 95 - 98 % TOD can be achieved.

Equipment and chemicals

transfer pipettes, volume 2 ml, 10 ml, 20 ml, 30 ml,
cylindrical ground reaction vessels (standard grinding 29/32), as shown in Fig. 7.3.-4.

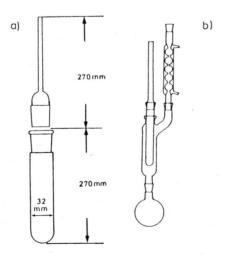

Fig. 7.3.-4: Reaction vessel

a - cylindrical reaction vessel; b - apparatus to heat the preparations for COD determination, set up on the principle of the boiling thermostat

thermostat bath with circulation pump, operating temperature at 180 °C,
  temperature constancy ± 0.1° C or better
buret, volume 25 ml, 0.05 ml graduation
ERLENMEYER flask, volume 300 ml
measuring cylinders, volume 100 ml, 1000 ml
ammonium iron (II) sulfate solution, ca. 0.1 N: 40 g $(NH_4)_2Fe(SO_4)_2 \cdot 6 H_2O$ p.a. are dissolved in bidistilled water. The solution is mixed with 20 ml sulfuric acid p.a. (d = 1.84 g/ml) cooled, and topped to 1000 ml with bidistilled water.
ammonium iron (II) sulfate solution, ca. 0.02 N: 8 g $(NH_4)_2Fe(SO_4)_2 \cdot 6 H_2O$ p.a. are dissolved in bidistilled water. The solution is mixed with 20 ml sulfuric acid p.a. (d = 1.84 g/ml) cooled, and topped to 1000 ml with bidistilled water.
potassium dichromate solution, 0.25 N, sulfuric: 390 ml sulfuric acid p.a. (d = 1.84 g/ml) are diluted in 400 ml bidistilled water in a 1000 ml beaker and cooled to 20° C; 12.259 g potassium dichromate are dissolved in the thus prepared sulfuric acid in a 1000 ml measuring cylinder, and topped to the mark with bidistilled water at 20° C. The potassium dichromate is dried at 105° C for 2 hours before weighing.
sulfuric acid, containing silver: 15 g silver sulfate, Ag_2SO_4, cryst., are dissolved in 1 l sulfuric acid p.a. (d = 1.84 g/ml).
mercury (II) sulfate solution, containing sulfuric acid: solution A (to mask chloride concentrations up to 500 mg/l). 10 ml sulfuric acid (d = 1.84 g/ml) are diluted to ca. 80 ml with bidistilled water in a 100 ml measuring cylinder. Five grams of mercury (II) sulfate are dissolved in the thus prepared sulfuric acid, and topped to 100 ml with bidistilled water.
mercury (II) sulfate solution, containing sulfuric acid: solution B (to mask chloride concentrations up to 1500 mg/l). 10 ml sulfuric acid are prepared as described for solution A. Instead of 5 g, however, 15 g mercury (II) sulfate are added. The solution is also topped then to 100 ml.

ferroin solution: 1.485 g of 1.10-phenanthroline and 0.695 g iron (II) sulfate are dissolved in starting water and topped to 100 ml. The solution is stable.

0.025 N iron (II) solution: ca. 1 l starting water (amply measured) is boiled for 5 minutes and then cooled to remove the oxygen as far as possible; 500 ml of this solution are mixed with 20 ml sulfuric acid 95 to 97 %, and cooled. Then 10 g ammonium iron (II) sulfate are dissolved in the solution, and the complete solution is transferred into a measuring cylinder, 1000 ml. The solution is then topped to 1 l with the boiled starting water.

adjustment of the iron (II) solution: 10 ± 0.02 ml 0.1 N potassium dichromate solution are transferred as a recipient into an Erlenmeyer flask. Then 100 ml starting water and 40 ml sulfuric acid 95 to 97 % are added, and the solution is cooled to room temperature. After the addition of 2 to 3 drops of ferroin solution, the titration is carried out with iron (II) solution. After the addition of ca. 2 drops, the color changes from blue-green to reddish-brown.

Annotation: the final point can be determined also by an appropriate electrochemical method.

Calculation of the factor:

$$f_{Fe} = \frac{A \cdot N_{Cr}}{B \cdot N_{Fe}}$$

where:

f_{Fe} = factor of the iron (II) solution (must be determined at least once a day)

A = volume of potassium dichromate solution used (10 ml)

N_{Cr} = normality of the potassium dichromate solution, 0.1 N

B = volume of iron (II) solution used, in ml

N_{Fe} = normality of the iron (II) solution, 0.025 N

This results in a simplified formula of evaluation:

$$f_{Fe} = \frac{40}{B}$$

Procedure

Sample

- measure 5 ± 0.1 ml mercury sulfate solution into the reaction flask
- measure 20 ± 0.03 ml of the sample to be analyzed, add this quantity into the reaction flask, and mix well
- measure 10 ± 0.02 ml 0.1 N potassium dichromate solution, and add this quantity into the flask
- add 2 - 3 boiling beads with a COD-free instrument
- connect the reaction flask to the cooler (do not grease the ground sections!), add 40 ± 0.5 ml sulfuric acid 95 to 97 % through the cooler, mix thoroughly (observe the safety regulations!)
- heat up to boiling temperature, start stop watch when the solution begins to boil
- after 5 minutes boiling, add 5 ± 0.1 ml silver sulfate solution through the cooler into the reaction flask
- stop boiling after another 10 minutes
- allow to cool in the open air for 5 minutes
- carefully add 50 ml starting water through the cooler while agitating (avoid delay in boiling!)
- remove flask and rinse the ground section with starting water
- let cool to room temperature in water bath
- add 2 - 3 drops of ferroin solution
- titrate with 0.025 N iron (II) solution to achieve a color change from blue-green to reddish-brown.

The final point can also be determined by an appropriate electrochemical method.

Blind value

Instead of the water sample to be analyzed, the procedure is carried out using 20 ml of starting water.

Interpretation

The COD value is calculated as follows:

$$COD = \frac{(C - D) \cdot N_{Fe} \cdot f_{Fe} \cdot \frac{O}{2} \cdot 1000}{E} \quad [mg/l]$$

where:

C = volume of iron (II) solution consumed to titrate the blind value, in ml

The value to be used here is the arithmetic mean value of the two or more blind values determined.

D = volume of iron (II) solution consumed to titrate the sample, in ml

E = sample volume used (20 ml)

$\frac{O}{2}$ = equivalent weight of oxygen (= 8)

This furnishes the simplified formula of evaluation:

$$COD = (C - D) \cdot f_{Fe} \cdot 10 \qquad [mg/l]$$

Calculation of the blind value:

$$b = 100 - \frac{C \cdot N_{Fe} \cdot f_{Fe} \cdot 100}{F \cdot N_{Cr}} \qquad [\%]$$

where:

b = blind value (per cent) of the oxidant quantity used

F = volume of potassium dichromate solution used (10 ml)

This results in a simplified formula of evaluation:

$$b = 100 \, (1 - 0.025 \cdot C \cdot f_{Fe}) \qquad [\%]$$

The blind value b in per cent of the oxidant quantity used permits a simple check of the cleanness of the equipment, reagents, and starting water used. 10 % are permissible at maximum.

7.3.3.2.3 Removal of Hg and Ag from Residue Solutions for COD Determination

The method of largely detoxicating Hg and Ag residues from the COD determination is based on the fact that alkaline thiosulfate solutions in the acidic solutions containing mercury, silver, and chromium react to produce the sulfides of mercury and silver as well as chromium (III) hydroxide. The aforementioned compounds are separated by way of filtration.

Procedure

The reaction solutions containing mercury, silver, and chromium ions are collected when the determination of the chemical oxygen demand (COD) has been completed. They are first mixed with 30 g sodium thiosulfate per 500 ml, and then with 300 ml 30 % sodium hydroxide solution. This reaction removes mercury and silver from the solution in the form of sulfides, while chromium is removed in the form of chromium (III) hydroxide. Hg^{2+} ions can no longer be detected in the filtrate of the excessive colorless liquid (use cold vapor atomic absorption spectrophotometry, detection limit at 0.001 mg/l). The residual silver concentration was below 0.1 mg Ag^+/l while the chromium (VI) concentration was below 0.5 mg/l.

7.3.3.2.4 Total Organically Bound Carbon (TOC) and Dissolved Organic Carbon (DOC)

The presence of organic carbon is common to all organic substances. Depending on the respective problem, it is therefore possible to determine the biochemical oxygen demand (BOD), the chemical oxygen demand, or the TOC and DOC values for a total assessment. The two methods mentioned last also oxidize the substances that are not accessible to attack by other processes.

The equipment has been so improved that after a thorough homogenization of the suspended matter or after their correct removal, the determination of higher and mean TOC/DOC concentrations occurring in waste waters is no longer a problem. The main application of this method, however, is the concentration range below 10 mg/l.

Attention should be drawn to certain difficulties in the methodology since the equipment manufacturers use different oxidation processes. The carbon is converted to CO_2 by chemical wet oxidation, by exposure to UV light, or by incineration at higher temperatures, with CO_2 measurement, e.g. by means of an IR photometer. It is also possible to use hydrogen and precious metal catalysts to reduce the CO_2 produced to methane with final measurement at an ionization flame detector. All these methods are not absolute, so that the equipment must be calibrated with standard solutions. The C concentration in the water used to prepare the standardization solutions must be as low as possible. Bidistilled water is therefore used. Deionized water is less suitable for this purpose.

The following description of the methods is based on the German federal draft standard of July 2, 1981.

Application

The methods are applicable to any kind of water with TOC and DOC contents of ca. 0.1 mg/l up to more than 1 g/l.

Waste water with a high TOC/DOC is generally analyzed in other methods and equipment than those applied to analyze water with a low TOC/DOC.

It may be necessary to dilute the water to adapt it to the measuring range of the respective method or equipment.

Equipment and chemicals

Homogenizing equipment, e.g. ultrasonic equipment with a sufficient rating to homogenize water samples containing suspended matter

measuring cylinders with a suitable nominal volume, e.g. measuring cylinder DIN Standard 19664 - MS A 100

transfer pipettes with an appropriate nominal volume, e.g. pipette DIN Standard 12691 - VP AS 10

measuring pipettes with an appropriate nominal volume, e.g. pipette DIN Standard 12697 - MP AS 10-01

Only chemicals may be used whose purity is defined as "suitable for analysis". Only those chemicals are mentioned that are largely used in all the methods. Further chemicals must be used and pretreated, if necessary, according to the directions given by the equipment manufacturer.

phthalate stock solution: (C) = 1000 mg/l

2.125 g potassium hydrogen phthalate are dissolved in ca. 700 ml water in a measuring cylinder, nominal volume 1000 ml. The solution is then topped with water to 1000 ml. When well sealed, it can be stored in a refrigerator and remains stable for ca. 4 weeks.

phthalate standard solution: (C) = 100 mg/l

100 l of the phthalate stock solution are transferred with a pipette into a measuring cylinder, nominal volume 1000 ml, and topped with water up to the calibration mark. When sealed well, the solution is stable in a refrigerator for ca. 1 week.

reference solution for the determination of the inorganically bound carbon: (C) = 1000 mg/l

4.404 g sodium carbonate (dried at 255° C for 1 hour) are dissolved in ca. 500 ml water in a measuring cylinder, nominal volume 1000 ml. Then 3.497 g sodium hydrogen carbonate (dried over silica gel) are added. The solution is topped with water to 1000 ml. When sealed well, the solution remains stable in a refrigerator for ca. 4 weeks.

gases: air, nitrogen, oxygen, free of CO_2 and organic pollutants.

Pretreatment

The samples are filled into clean glass bottles. When the analyses cannot be made immediately, the samples must be kept in the refrigerator.

When only the dissolved organic carbon (DOC) is to be determined, all the undissolved matter must be removed. The filtration through a diaphragm has not proven satisfactory in practice since the filter material may release carbon.

Glass fiber filters are suitable filter materials provided they are previously washed with bidistilled water. A sufficiently long centrifugation is also an appropriate method of separating solid particles.

Procedure

Standardization

Comparative methods definitely require standardization, whereas such standardization serves to check the systems used in absolute methods, e.g. acidimetric or coulometric methods.

The calibration plots are established by the preparation of standardization solutions from the phthalate stock solution or phthalate standard solution. These standardization solutions have concentrations of TOC mass so high that the expected measuring range is amply covered. The extrapolation of one calibration plot to another measuring range is not permitted. Regarding the TOC/DOC mass concentration range from 10 to 100 mg/l, for example, the procedure is as follows:

1.0 ml, 2.5 ml, 5.0 ml, 7.5 ml, and 10.0 ml of the phthalate stock solution are each transferred with a pipette to a measuring cylinder, nominal volume 100 ml. The cylinders are topped with water up to the calibration mark. A sixth cylinder contains only water as a reference or blind sample. All solutions, including the blind solution, are analyzed at least three times, in correspondence with the operating directions of the equipment manufacturer.

The TOC mass concentrations of the individual standardization samples are marked on the abscissa in a coordinate system. The values are calculated as follows:

$$K = \frac{V_K \cdot S}{V_o} \qquad [mg/l]$$

where:
- K = TOC mass concentration of the respective standardization sample, in mg/l
- V_K = volume of phthalate stock solution (or phthalate standard solution) used, in ml
- V_o = maximum volume of the standardization sample, here V_o = 100 ml
- S = TOC mass concentration of the phthalate stock solution (or phthalate standard solution), in mg/l

The value to be indicated on the ordinate depends on the type of equipment. It is to be taken from the operating instructions.

This series of measured values is used to establish the corrective straight line. The reciprocal of the inclination of the straight lines furnishes the factor f in the unit which depends on the method selected.

Procedure

The operating instructions of the equipment manufacturers must be observed. If necessary, the water sample must be diluted to an extent that the achieved concentration of organically bound carbon comes under the measuring range of the equipment used.

Before the TOC/DOC determination, it should be noted that the functional checks have been made for the equipment at the intervals prescribed by the manfuacturers. The total system should be regularly checked for tightness.

Depending on the TOC measuring instrumentation used, different quantities are obtained that are to be used to calculate the TOC concentration of the analyzed sample. Such quantities may be the peak altitudes occurring in discontinuous measurement (only with a peak shape independent of the substance), peak areas or the lye volume required for carbon dioxide titration, or other factors. An example of a continuous TOC measurement involves the plotting of a line on a recorder that corresponds to a certain CO_2 concentration in the carrier gas flow, which is obtained with a specific water sample. The spacing of this line from the zero line is proportional to the TOC concentration. The TOC mass concentration in the sample proceeds from the calibration plot independently of the TOC or CO_2 measuring instruments.

Rounded values are given, with three significant digits at maximum. The rounding depends on the TOC mass concentration:

        TOC below 1 mg/l rounding to 0.01 mg/l
        TOC from 1 to 10 mg/l rounding to 0.1 mg/l
        TOC above 10 mg/l rounding to 1 mg/l

Examples:
organically bound carbon (TOC)   0.76   mg/l
organically bound carbon (TOC)   530   mg/l
organically bound carbon (TOC) $6.32 \cdot 10^3$ mg/l.

7.3.3.3 Biochemical Oxygen Demand (BOD)

The biochemical oxygen demand (BOD_n) denotes the mass of dissolved molecular oxygen that microorganisms require for the oxidizing decomposition (or rearrangement) of organic substances contained in the water, under defined conditions within a defined period (the subscript "n" defines days or hours).

To determine the BOD, the bacterial decomposition is allowed to take place under controlled conditions in specimen bottles. Then the oxygen quantity consumed is determined. In BOD determination, the consumption of dissolved molecular oxygen for purely chemical oxidation processes (without microorganisms being involved, and mainly without the oxidation of inorganic substances) must be considered. It is generally assumed that these purely chemical oxidation processes have come to an end within 2 hours from water sampling.

The biochemical oxygen demand is to be indicated in mg/l water. It is frequently also termed "oxygen consumption" within the period of analysis. It is a measure to assess the influence of organic pollution on the oxygen balance of a particular water. It must be noted, however, that the biochemical decomposition or rearrangement of the organic substances occurs in two stages that are not distinctly separated from each other. In the first stage, organic compounds are largely decomposed into inorganic compounds. In the second stage, also termed nitrification, the ammonium produced from organic nitrogen-containing compounds is mainly oxidized to nitrite and nitrate. This second stage of decomposition, however, begins only after a certain interval, so that it is mostly irrelevant, e.g., when the BOD_5 is determined. It can be determined separately in special cases.

The oxygen demand in the first stage of decomposition is only required to determine the BOD, with nitrification being regarded as a disturbance. This situation clearly shows that the BOD in the defined and strict sense is not a measure of the total of the organic substances contained in the water. It cannot be directly compared to the chemical oxygen demand (COD) using potassium dichromate or to oxidizability determined with potassium permanganate. But there are empirical values at hand which indicate that in waters with mean pollution, the oxidizability analysis with potassium permanganate indicates 25 %, the BOD_5

analysis indicates 70 %, and the COD analysis under optimum conditions indicates up to 98 % of the oxygen required for complete oxidation of the organic substances (W. LEITHE).

The magnitude of the BOD depends on many factors: the composition and concentration of the organic substances to be decomposed in the water; the type, number, and adaptation level of the microorganisms; the nutrients available to the microorganisms; the incubation time (consumption period); the temperature; the effects of light; and the influence of toxic substances on the biological or biochemical processes.

A period of 5 days (BOD_5) has proven a successful incubation time. Since the starting period of the bacterial processes varies, a reduction of this time could result in incorrect or insufficiently reproducible values. On the other hand, reproducibility is also affected by an excessively long time of BOD determination, as well as by the fact that the bacterial flora effecting the decomposition, which is the most important "reagent", cannot be determined in terms of quality and efficiency. According to LEITHE, reproducibility (range of variation for several analyses with the same sample) ranges between 20 and 50 % of the values found, especially when several people or several laboratories are involved.

The BOD_5 is a term denoting the quantity of dissolved molecular oxygen that is consumed by microorganisms during the incubation time of 5 days, for the oxidizing decomposition of organic substances contained in the water, at 20° C. The determination of BOD_2 is also frequent for an analysis of surface waters. Incubation times of 24 hours and 10 or 20 days may also be selected for special purposes.

Depending on the pollutants, the BOD_5 is determined in a water sample that is not pretreated, sedimented, or filtered. The respective condition must be indicated in the results of the analysis.

Oxygen cylinders are used for the BOD_5 determination. When a sample is analyzed, it should be noted that the oxygen and nutrient quantities required for an optimum decomposition, as well as suitable microorganisms, are available in sufficient quantities.

Two different principles are fundamentally applicable in BOD determination:

1. The consumption of a certain quantity of dissolved oxygen is measured in a sealed calibrated receptacle, from the difference of the oxygen concentration in the sample at the beginning and after the incubation time. The value is volumetrically or electrometrically determined.
2. The water sample is kept in closed equipment with continuous O_2 supply from the gas reservoir. The O_2 consumption during the incubation time is followed up with a pressure gauge on the basis of the changing oxygen concentration in the gas reservoir (e.g. according to WARBURG).

The method used in this prescribed process corresponds to the principle explained in item 1.

7.3.3.3.1 BOD_5 Determination after Dilution

This method is suitable for BOD5 determination in more strongly polluted waters and waste waters. These are waters where the oxygen concentration is insufficient, even after shaking the sample with air (oxygen saturation), to cover the biochemical oxygen demand. They are diluted with an oxygen-rich "consumed" diluting water to an extent that at least 2 mg oxygen per liter are still present after the attrition time.

Substances that can be oxidized by oxygen without cooperating microorganisms cause problems, which can be eliminated when the beginning of the incubation time for BOD_5 determination is set at 2 hours after the dilution of the sample.

Biologically inhibiting or toxic substances must be made ineffective. When the removal of such substances entails a substantial change of the water sample composition, the BOD_5 determination gives false results.

Free chlorine is removed by the addition of an equivalent quantity of sodium thiosulfate solution (1 ml 0.01 N solution = 1 mg Cl_2).
Strongly acidic or alkaline waters are set to a pH value of 7 to 8.

The sodium hydroxide solution contains sodium azide to remove the disturbance caused by nitrite which is partly formed only during the incubation time.

In waters in which the decomposition of the organic matter into inorganic compounds (first stage of decomposition) progresses very rapidly, the second stage of decomposition, which is called nitrification, may occur as early as during the first 5 days, i.e. during the BOD_5 incubation time. When 0.5 mg n-allyl thiourea per liter of diluting water are added, the nitrification process can be largely suppressed, since this additive inhibits the NH_4^+ oxidation.

Waters with a low microorganism count demand additional inoculation. This is achieved by the addition of sedimented domestic waste water, biologically purified waste water or considerably polluted river water, all in consideration of these additives' own BOD.

The dilution of the water sample may cause a disturbance by lack of nutrient salts. To prevent this disturbance, nutritive salt solutions are added, whose compositions are given below in the section on equipment and chemicals.

Equipment and chemicals

Oxygen cylinders: volume 110 - 130 ml or 250 - 300 ml with the same numbering of the cylinders and plugs

transfer pipettes: volume 2 ml, graduation in 0.1 ml

diluting water: the diluting water is warmed to 20° C and must be saturated with oxygen at this temperature. Its BOD_5 must not exceed 1 mg/l ("consumed" diluting water). It is either taken from the water serving as receiving water, or it is prepared from distilled water to which nutritive salts are added. Preparation: see below.

nutritive salt solutions:

solution 1: 8.5 g KH_2PO_4, 21.75 g K_2HPO_4, 33.4 g Na_2HPO_4, 2 H_2O and 1.7 g NH_4Cl are dissolved in ca. 500 ml distilled water and topped to 1 l. pH = 7.2.

Solution 2: 22.5 g $MgSO_4 \cdot 7\ H_2O$ are dissolved in distilled water to prepare 1 l.

Solution 3: 27.5 g $CaCl_2$ are dissolved in distilled water to prepare 1 l.

Solution 4: 0.25 g $FeCl_3 \cdot 6\ H_2O$ are dissolved in distilled water to prepare 1 l.

preparation of diluting water: The water used for this purpose must be free of silver and copper ions and effective chlorine. 1 ml each of solutions 1, 2, 3, and 4 is added to 1 l distilled water. The thus prepared diluting water must be repeatedly aerated until it is saturated with oxygen. It is kept in a dark space.

manganese (II) chloride solution: 800 g $MnCl_2 \cdot 4\ H_2O$ are dissolved in 1 l distilled water.

sodium hydroxide solution, containing KJ/NaN_3: 360 g NaOH (containing no nitrite), 200 g KJ (free of iodate), and 5 g NaN_3 are carefully (protective goggles) dissolved with distilled water to prepare 1 l. The solution is then filtered through glass wool.

phosphoric acid, d = 1.70 g/ml

sodium thiosulfate solution, 0.01 N

zinc iodide starch solution: 4 g starch are rubbed with a small quantity of distilled water. This slurry is added into a boiling solution of 20 g zinc (II) chloride in 100 ml distilled water. The evaporating water is supplemented and boiling continues until the solution is clear, then it is diluted, mixed with 2 g zinc iodide, topped with distilled water to prepare 1 l, filtered, and kept in a brown bottle. After dilution with a water volume 50 times its own volume, it must not take a blue color when acidified with diluted sulfuric acid.

Procedure

<u>Preparation of the samples</u>. When coarsely dispersed substances in the sample are to be included in the determination, the sample must be homogenized in a mixer for 3 minutes.

When the BOD of the sedimentary matter is to be omitted from the determination, the water is sedimented in the Imhoff funnel for 2 hours, decanted, and analyzed.

When the BOD of the dissolved matter only is to be determined, the filtered sample is used. The first 100 ml of the filtrate are discarded since they might give rise to errors caused by soluble substances of the filter that consume the oxygen.

Determination of the BOD_5. Waters with a low microorganism count are inoculated with 0.3 ml of sedimented waste water per liter of the diluted sample, or with 5 - 10 ml of river water. The inoculating liquid is added after the dilution of the sample with approximately half of the diluting water quantity prepared.

Depending on the BOD to be expected, <u>two or more dilutions</u>, possibly with addition of inoculating liquid, must be prepared of the water sample and the diluting water, possibly after a preliminary water sample treatment. When domestic and municipal waste waters are analyzed, whose potassium permanganate consumption (oxidizability) is known, the following table can be applied:

Tab. 7.3-2: BOD_5 determination by dilution

oxidizability (O_2)	BOD_5 to be expected (O_2)	ml water sample to be diluted to prepare 1000 ml
mg/l	mg/l	ml
up to 4	up to 10	250 and 150
4 to 10	10 to 30	100 and 75
10 to 15	20 to 50	50 and 40
15 to 30	40 to 100	30 and 20
30 to 60	80 to 200	15 and 10
60 to 90	160 to 300	10

When the BOD_5 to be expected is higher than 300 mg/l, the sample is diluted with diluting water in a ratio of 1 : 9, and then the table values are applied.

The dilution is made in a 500 ml or 1 l measuring cylinder in a way that the cylinder is partly filled with diluting water. After the addition of the water sample, the volume is topped up to the mark and is then mixed. Before each dilution, three oxygen cylinders are filled, without producing air bubbles. One of the three samples (1) is immediately mixed with manganese (II) chloride solution and sodium hydroxide solution containing KJ/NaN_3. When the smaller oxygen cylinders (110 - 130 ml) are used, these reagent quantities are added per 0.5 ml, whereas these quantities are added per 1.0 ml when the larger cylinders (250 - 300 ml) are used. All three bottles are then immediately sealed, avoiding air bubbles.

The sample (1) mixed with reagents is well shaken. The second and third samples are kept in a dark space at 20°C for the incubation time (attrition period). After 5 days, the waters in bottles 2 and 3 are also mixed with the two reagent solutions in the manner described above. The treatment of the volume in each of the three bottles is continued as follows (oxygen determination according to WINKLER):

The sedimented precipitate is dissolved by the addition of 2 ml phosphoric acid. The released iodine is titrated with 0.01 N sodium thiosulfate solution. Toward the end of titration, 1 ml of zinc iodide starch solution is added to the light yellow solution, and then the titration of the solution is continued until it becomes colorless. At the same time, the BOD of the diluting water is determined and considered in the evaluation.

Only those samples are evaluated in which the attrition (O_2 consumption), without the attrition of the diluting water, is higher than 1 mg/l and whose residual oxygen content is 2 mg/l or higher.

One milliliter of 0.01 N sodium thiosulfate solution corresponds to 0.08 mg O_2. On the basis of this relationship, the oxygen content in the individual samples is calculated according to the formula:

$$G = \frac{a \cdot F \cdot 80}{V - V_R} \qquad [mg\ O_2/l]$$

where:
- a = volume of 0.01 N sodium thoisulfate solution consumed, in ml
- F = factor of the 0.01 N sodium thiosulfate solution
- V = volume of the oxygen cylinder, in ml
- V_R = reagent solutions added, in ml
- G = oxygen concentration, in mg O_2/l

The O_2 concentrations before and after the standing time may also be electrometrically determined.

The oxygen consumption (attrition) in the dilutions is calculated as follows:

$$G_1 - \frac{G_2 + G_3}{2} = Z \quad [mg/l]$$

where:
- G_1 = O_2 concentration of sample 1, in mg/l
- G_2 = O_2 concentration of sample 2, in mg/l
- G_3 = O_2 concentration of sample 3, in mg/l
- Z = O_2 consumption (attrition) in mg O_2/l

$$BOD_5 = \frac{V_a}{V_b} \cdot (Z_p - Z_v) + Z_v \quad [mg/l]$$

where:
- V_a = total volume after dilution, in ml
- V_b = volume of the undiluted sample, in ml
- Z_p = oxygen volume consumed in the dilution in 5 days, given in mg O_2/l
- Z_v = oxygen volume consumed in the diluting water (or the inoculated diluting water) in 5 days, given in mg O_2/l
- BOD_5 = biochemical oxygen demand of the water in 5 days, given in mg O_2/l

7.3.3.3.2 BOD_5 Determination with Dilution as a Field Test Method

This method is applied to determine the BOD_5 in considerably polluted waters and waste waters when, because of special conditions, the time between sampling and the laboratory analysis would be so long that false results ought to be expected.

This method is fundamentally the same as the dilution method. The sampled (and possibly pretreated) water, however, is diluted with diluting water on the site directly in the oxygen cylinder. This method is more easily accessible to errors than the diluting method.

7.3.3.3.3 General Information and Special Directions

The BOD_5 determination is a biological chemical analysis provided that the diluting method is applied. The conditions in terms of temperature, pH value, inoculation, addition of mineral salts, and storage in a dark space must be strictly observed. It is also important that the diluting water temperature roughly corresponds to the incubation temperature of 20 °C. Otherwise major errors could arise. The maintenance of the glass bottles is particularly important, since biological growth may form in them when they are frequently used. Such growth may affect the results. It is best to clean the bottles thoroughly with brushes immediately after the analysis. Then the bottles are rinsed several times with clear water free of chemicals. It is also advisable to store the bottles filled up to their rim with clear water until they are used again.

The use of detergents for cleaning always involves the risk of detergent residues remaining in the bottles, which may impair the result. When it is necessary for any reason whatsoever to clean the bottles with a chromate-containing concentrated sulfuric acid, they must be rinsed repeatedly with clean water.

Apart from the correct and exact performance of the tasks required to prepare and store the samples, the determination of the oxygen contents in the bottles has an important influence on the result. When the titrimetric determination is carried out according to the modified WINKLER method, reliable results can be obtained only when the persons entrusted with this work are technically qualified and are very well acquainted with the associated problems, such as secondary reactions, titer settings, indicator errors, or the like.

A quick and exact determination of the oxygen concentration is possible with modern oxygen meters provided with a digital indication of the results.

It is known that ca. 30 to 70 % of the oxygen concentration should be present after 5 days attrition in a BOD_5 determination, related to the initial values present at the beginning of the attrition. When it is assumed, for instance, that the oxygen concentration in freshly prepared samples amounts to 9 mg O_2/l, the residual concentrations after 5 days are within the evaluation range from 2.1 to 6.3 mg O_2/l, which means that the sample need not be diluted in the BOD_5 range between 2.7 and 6.9 mg $O_2 l$.

The following is a list of waste water with diluting water as a function of BOD_5:

1 + 9 - range	27 to	69 mg O_2/l
1 + 99 - range	270 to	690 mg O_2/l
1 + 999 - range	2700 to	6900 mg O_2/l

The extremely strong dilutions in loaded waste waters give rise to a considerable multiplication of errors.

In addition to the dilution method, manometric methods (quick tests that can be applied also by trained operators) are also used to determine the BOD_5. The general procedure involves the following steps: filling the waste water sample to be analyzed into a flask having an airtight internal air volume, moving the waste water by stirring, shaking, or agitating so that a permanent oxygen supply throughout the sample is ensured, since the oxgen is absorbed from the air volume above. To be able to measure the oxygen volume consumed by a determination of the pressure drop in the flask, the CO_2 produced thereby, unless dissolved or bound in the sample, must be removed from the airtight internal air volume. The CO_2 is absorbed by bases, mostly potassium hydroxide. Since a considerably greater quantity of oxygen is available, with the airtight internal air volume, than in the diluting water, the range of application is much wider than the application of equally diluted or undiluted preparations made according to the dilution method. In the last analysis, however, it is restricted.

Special equipment can be used to shift this limit upward. Such equipment takes measurements according to the same principle, but the pressure drop is merely used to control the supply of pure oxygen into the airtight internal air volume. This oxygen quantity is then measured and recorded.

Compared to the BOD_5 dilution method, such instrumentation has the advantage that the processes of decomposition can be continuously followed up and recorded so that there are various possibilities of determining and detecting different parameters, e.g. temperature, toxicity of substances, lack of mineral salts, effects of dilution, and adaptation of biocoenoses to substrates.

Remarks resulting from practical operation:

The BOD_5 is generally understood to cover only the decomposition of the organic compounds. This decomposition is also defined as carbon decomposition or as the decomposition of the first stage as a delimitation from the decomposition of the second stage which involves the ammonium nitrification, also consuming oxygen, and begins only after the thorough carbon decomposition. All indications as to the BOD_5 volumetric and sludge loads for the purification method relate to the decomposition in the first stage.

There are also some other biochemical oxygen-consuming reactions. In terms of quantity, above all the biochemical conversion of ammonium nitrogen to complete nitrification is important. It requires nearly as much oxygen as the BOD_5, for instance, for the biological purification of domestic waste waters up to a complete nitrification.

Nitrification does not occur in a BOD_5 test with unpurified waste water, or it occurs only to an almost unnoticeable extent. But other conditions occur for the discharges from mechanical-biological sewage plants: The further the decomposition of organic compounds progresses, the more likely is the possibility of nitrification taking place simultaneously with the BOD_5 test, thus increasing the oxygen consumption. Many efforts have therefore been made to suppress nitrification in the BOD_5 test in order to obtain reproducible and comparable BOD_5 values for the first stage of decomposition. It can be noted that no differences occur in BOD_5 determination with or without elimination of nitrification in the raw water when both methods are applied. But at the discharge of an activation system with a very weak nitrification, the values are substantially higher without elimination of nitrification (up to ca. 30 % on average) than in discharges where nitrification was eliminated.

It is interesting that the frequently applied BOD_5 requirement of 25 mg O_2/l at 15 % was exceeded in all tests when nitrification was not eliminated. This requirement is met, however, when nitrification processes are inhibited.

The following diagrams (Figs. 7.3.-5 and 7.3.-6) according to THERIAULT /73/ illustrate this condition.

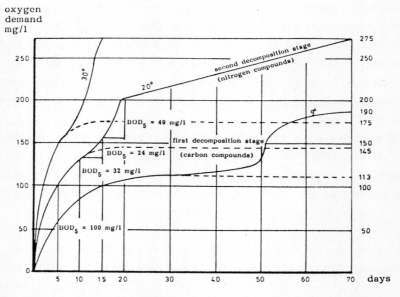

Fig. 7.3.-5: Oxygen demand reduction in fresh waste water when carbon and nitrogen compounds are decomposed /73/

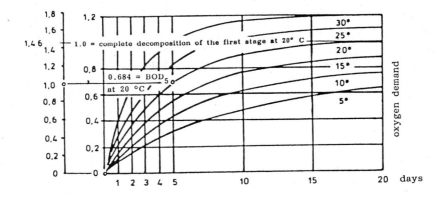

Fig. 7.3.-6: Decomposition of the carbon compounds in aerated water, as a function of temperature /73/

When the results of the BOD analysis are evaluated, nitrification must be considered for the BOD_5 determination in order to ensure the extent by which the BOD_5 is responsible for the decomposition of the carbon compounds or for nitrification. This question has a considerable relevance particularly when contracts are concluded that involve warranties in terms of BOD_5 values in the discharge for the construction and operation of sewage plants.

It should be emphasized that it is an important aspect in the BOD_5 determination in waste waters that many organic compounds in high concentrations have toxic effects on the microorganisms whereas these substances can be decomposed by the same microorganisms when they are present in low concentrations. Phenolic substances are an example of this phenomenon. With undiluted phenol-containing waste waters, e.g. those discharged from coal coking plants, manometric measuring methods cannot be applied to determine BOD_5 values, whereas the dilution method is successful.

When the biochemical oxygen demand of the sedimentary matter must be determined, it is possible to calculate this value from the loss at red heat when primary sludge is involved. According to IMHOFF, 30 g of organic matter in the primary sludge cause 20 g BOD_5. Particular attention must be paid to an appropriate thorough homogenization when BOD_5 is determined in the presence of sedimentary matter or sludges.

7.3.3.4 Phenols

7.3.3.4.1 Phenols Volatile in Steam

The property of the phenols to form strongly colored red azo dyes as a result of coupling reactions with diazo compounds is the basis of the photometric method described here.

$$O_2N\text{-}C_6H_4\text{-}NH_2 + 2\,HCl + NaNO_2 \longrightarrow O_2N\text{-}C_6H_4\text{-}N\!=\!NCl + NaCl + 2\,H_2O$$

$$O_2N\text{-}C_6H_4\text{-}N\!=\!NCl + C_6H_5\text{-}OH \longrightarrow O_2N\text{-}C_6H_4\text{-}N\!=\!N\text{-}C_6H_4\text{-}OH + HCl$$

The individual phenols, however, have different color intensities at the same mg/l concentrations. The phenols distilled with steam from a mineral acid solution therefore furnish the values given below, when they are determined with p-nitraniline, related to an equivalent phenol volume = 100 % (according to DEV):

o-cresol:	147 %,	o-xylenol:	16 %,	guaiacol:	165 %,
m-cresol:	120 %,	m-xylenol:	52 %,	pyrocatechol:	29 %,
p-cresol:	21 %,	p-xylenol:	92 %,	2-naphthol:	23 %.

The percentages given in this list show the uncertainty that must be expected in an evaluation of the photometrically measured results.

The method is suited to determine phenols volatile in steam over a wide concentration range. With phenol concentrations of more than some mg/l, the steam distillation is carried out in two stages. For the extraction with n-butanol, a distillate volume is used that is so large that the concentration of phenols volatile in steam, contained in the butanol extract, corresponds to a phenol concentration of 0.001 to 0.4 mg/l in the water sample.

Disturbances by sulfide or cyanide ions are eliminated by the addition of $CuSO_4$ or $CoSO_4$ solutions.

Equipment and chemicals

distillation apparatus comprising a 500 ml round flask, a spherical attachment and a descending LIEBIG condenser
measuring cylinders, 100 ml
phosphoric acid, d = 1.70 g/ml
copper (II) sulfate solution: 10 g $CuSO_4 \cdot 5 H_2O$ + 100 ml H_2O
cobalt (II) sulfate solution: 10 g $CoSO_4 \cdot H_2O$ + 100 ml H_2O
p-nitraniline solution: 1.38 g p-nitraniline are dissolved in 310 mg/l of 1 N hydrochloric acid and topped with distilled water up to 2 l.
sodium nitrite solution, saturated
sodium carbonate solution, 1 N
n-butanol
phenol stock solution: 1.000 g phenol p.a. (freshly distilled) is dissolved with distilled water to achieve 1 l.
photometer with cuvettes

Procedure

<u>Standardization</u>. Suitable dilutions of the phenol stock solution with phenol concentrations of 0.001 - 0.4 mg/l are prepared, transferred into the distillation flask, and each topped with distilled water to achieve 200 ml, to establish the calibration plot. The calibration solutions are then subjected to the same series of analyses as the water sample.

<u>Determination</u>. 200 ml of the water sample (or a correspondingly lower volume that must then be topped to 200 ml with distilled water) are mixed with 1 ml $CuSO_4$ solution and/or $CoSO_4$ solution, if necessary (binding sulfide and/or cyanide) in the distillation apparatus. After the addition of 10 ml phosphoric acid (d = 1.70 g/ml), the solution is distilled into a recipient charged with 30 ml of 1 N sodium carbonate solution, until the distillation flask contains a residue of 20 ml of the water sample used.
The distillation product is transferred into a 500 ml separating funnel and set to ca. pH 11.5 by dropwise addition of 30 % NaOH (electrometric check), if necessary, and mixed with 20 ml of diazotized p-nitrani-

line solution (preparation by the addition of some drops of saturated sodium nitrite solution). The solution is allowed to stand for 20 minutes, then the resulting dye is extracted by thorough shaking with 50 ml n-butanol. After another 10 minutes, the aqueous phase is separated and the extinction of the butanol extract is measured by comparison to a parallel analyzed blind sample at 530 nm.

<u>Calculation</u>. On the basis of the extinctions measured, the associated concentration of phenols volatile in steam is taken from the calibration plot. The value is converted to mg/l in consideration of the water sample volume used for the determination. Reference must be made here, however, to the above survey, which gives an indication of the possible assessment of the result from the photometric measurement in view of the qualitative composition of the phenols in the water sample.

7.3.3.4.2 Total Phenols

The phenols are separated by extraction with a 4:1 mixture of benzene and quinolin. The sodium phenolates, which are produced when the mixture is agitated with sodium hydroxide solution, are reconverted into the free phenols with sulfuric acid, they are then bromated with a known bromium volume, and the excess bromium is iodometrically titrated back:

$$C_6H_5OH + NaOH \rightarrow C_6H_5ONa + H_2O$$
$$2\ C_6H_5ONa + H_2SO_4 \rightarrow 2\ C_6H_5OH + Na_2SO_4$$
$$BrO_3^- + 5\ Br^- + 6\ H^+ \rightarrow 3\ Br_2 + 3\ H_2O$$
$$C_6H_5OH + 3\ Br_2 \rightarrow 3\ HBr + C_6H_2Br_3(OH)$$

Excess bromium is iodometrically titrated back:

$$Br_2 + 2\ J^- \rightarrow J_2 + 2\ Br^-$$
$$J_2 + 2\ S_2O_3^{2-} \rightarrow 2\ J^- + S_4O_6^{2-}$$

The method is suitable to determine the total phenol concentrations higher than 100 mg/l. When the method is correspondingly varied, it is possible to cover concentrations up to 10 mg/l.

As a result of the phenol trend to oxidize when exposed to air, and to polymerize, and in consideration of the possibility of bacterial decomposition, it is recommended that NaOH be used to render the water samples alkaline after sampling. Then the samples can be kept in a dark space at 4° C for some days.

Equipment and chemicals

Buret, 25 ml

separating funnel, volume 1 l

ERLENMEYER flask with ground section and attached drip funnel, volume 500 ml

measuring cylinder, volume 500 ml

benzol/quinoline solution: 800 ml benzol are mixed with 200 ml quinoline (each of highest quality)

sodium hydroxide solution, ca. 4 N (d = 1.15 g/ml): 160 g NaOH are dissolved in 1 l distilled water

carbon tetrachloride, CCl_4, highest purity

KBr/$KBrO_3$ solution, 0.1 N

sulfuric acid, 30 % (d = 1.22 g/ml): 230 ml conc. sulfuric acid (d = 1.84 g/ml) are dissolved in 1 l of distilled water.

KJ solution: 110 g potassium iodide are dissolved in 1 l distilled water.

sodium thiosulfate solution, 0.1 N

starch solution: 1 g soluble starch and 25 g sodium chloride p.a. are dissolved in 100 ml distilled water.

Procedure:

500 of the neutral or slightly acidic (pH = 5) water samples are extracted three times with 100 ml each of the benzol/quinoline solution. The combined organic phases are washed once with 100 ml distilled water and extracted twice with 100 ml sodium hydroxide solution (d = 1.15 g/ml) each for 5 minutes after separation (and discarding) of the water. The sodium phenolate extract is clarified by short evaporation, and is extracted, after cooling, with 100 ml carbon tetrachloride. The aqueous phenolate solution is quantitatively transferred into a 500 ml measuring cylinder and topped up to the calibration mark with distilled water.

50 ml of this solution (V) are mixed with exactly 15.0 ml 0.1 N KBr/$KBrO_3$ solution in a ground ERLENMEYER flask with the separating

funnel attached, and with 10 ml sulfuric acid (d = 1.22 g/ml) through the funnel. The closed flask is allowed to stand in complete darkness for 1 hour, then 10 ml of the 10 % KJ solution are added through the separating funnel, and after 10 minutes the released iodine is titrated with 0.1 N sodium thiosulfate solution.

An analogous blind test is carried out with 500 ml distilled water. When the difference between the consumption in the blind sample and the analyzed sample is higher than 10 ml sodium thiosulfate solution, 50 ml of the sodium phenolate solution are diluted with distilled water to 500 ml before bromation; 50 ml of this diluted solution are mixed with KBr/$KBrO_3$ solution and the procedure is continued as described above.

1 ml of 0.1 N sodium thiosulfate solution corresponds to 1.70 mg total phenols when a mean gram-molecular weight of the phenols of 102 is taken as a basis.

$$\text{mg/l total phenols} = (a - b) \cdot 1.70 \; \frac{1000}{V} \quad [\text{mg/l}]$$

where:
- a = volume of 0.1 N $Na_2S_2O_3$ solution consumed in the blind test (ml)
- b = volume of 0.1 N $Na_2S_2O_3$ solution consumed by the sample (ml)
- V = volume of the aliquot used for bromation

7.3.3.5 *Oils and Fats*

Waters and particularly surface waters and waste waters may be polluted by mineral, vegetable, and animal oils and fats. These substances may occur in very low quantities as a solution or dispersion in the water, or they may be present as distinct contamination in the form of a two-phase system.

Surface-active substances (detergents or surfactants) favor emulsification. The water pollution with oily and fatty substances causes odor and technical problems. Such a contamination may also involve a health hazard. The possibility of concrete being attacked by free fatty acids resulting from saponification must be particularly emphasized.

Various methods of analyzing water samples for oils and fats have become known. Three of them have been selected and are described below.

1. Gravimetric determination after extraction with n-hexane.
   In this process, only parts of the low-boiling substances are covered. On the other hand, disturbance may result from the co-extraction of surface-active substances.
2. Carbohydrate determination with infrared spectroscopy.
   After extraction of the carbohydrates with carbon tetrachloride, the spectroscopic analysis is carried out in the wavelength range of 3.2 to 3.6 µm.
3. Gas-chromatographic analysis.
   This method provides for an identification of the individual substances. The head space analysis can be applied to determine volatile organic compounds.

7.3.3.5.1 Gravimetric Determination after Extraction with n-Hexane

The extraction of a water sample with n-hexane at pH 6, the pH range common in water analysis, covers:
mineral oils and mineral fats,
vegetable and animal oils and fats (triglycerides),
free fatty acids.

This method does not cover:
fatty acids present as compounds of or combined with fatty acids (soaps). This fraction can be covered after the extraction with n-hexane by acidification to a pH range from 1 to 2, and by an additional extraction.
The pH value of the water sample must therefore always be indicated in the analysis.
Depending on the sample volume used, which may vary from 1 l to ca. 10 l, the method can be applied to determine concentrations up to ca. 0.1 mg/l.

Easily volatile oils and fats, which become volatile in drying as a function of their vapor pressure and the drying time, are not covered.

Proportions of other organic substances are covered as far as they can be extracted with a homopolar solvent, e.g. certain surface-active substances (detergents, surfactants).

When emulsions are present, they may interfere with the extraction process. The emulsion can be separated by the addition of salts (sodium chloride or sodium sulfate). When the fatty acids from soaps are to be included in the analysis, and when the water sample is acidified to a pH of 1 to 2, the emulsions are separated by the addition of the acid. If the phase separation is still poor, the emulsion must be centrifuged.

The introduction of traces of oils and fats from sampling vessels must be definitely avoided. A loss of these substances from the water sample through the cover of the bottle must also be considered.

Sampling

Only glass bottles with ground glass plugs may be used to take water samples for the determination of oily and fatty substances.

These bottles, including the ground sections and the plugs, are thoroughly cleaned as usual, then sufficiently extracted with n-hexane, and dried. The thus prepared bottles are wetted with 50 ml n-hexane (for bottles with a volume up to 5 l) or 100 ml n-hexane (for bottles with a volume of 10 l). At the sampling site, the plug is opened without touching the ground sections with the hands, and the water is filled into the bottle. The water quantity can be either measured in a fat-free measuring cylinder on the site, or an optional water quantity is filled into the bottle charged with n-hexane, and the water level is marked. The added water quantity can then be measured in the laboratory, considering the n-hexane volume. Finally, it is also possible to weigh the sampling bottles with the n-hexane volume, and to perform a differential weighing in the laboratory after the filling of an optional water quantity.

In sampling, it should be noted that oily and fatty substances frequently concentrate as a thin surface film, or that separation processes may have occurred when emulsions are present. And, finally, the sediment may also contain a concentration of oily and fatty substances as a result of adsorption processes. Water sampling must therefore be related to the respective analysis problem. It must be indicated whether the sample was taken from the surface or from a defined depth. Analogous sediment samples must be taken, but they must be analyzed separately from the water.

The ground sections must be so secured after sampling that the bottles will not open during transport.

Equipment and chemicals

Sampling bottle, volume up to 10 l
separating funnel, volume ca. 1 l
glass funnel, diameter ca. 7 cm
filter paper
water bath with thermostat, 80° C
drying cabinet, 80° C
desiccator with silica gel
platinum or glass dish, fat-free
n-hexane, p.a.
alcoholic potassium hydroxide solution, 0.1 N
sulfuric or hydrochloric acid, 2 N
buret
ERLENMEYER flask with ground section, 250 ml
reflux condenser
centrifuge
semimicro weighing apparatus

Procedure

The pH value of the water to be analyzed is electrometrically determined in a separate sample.

The volume of the water sample added to the n-hexane solution at the sampling site is marked and determined later on by measuring, or the water volume is weighed.

The extraction is carried out by thoroughly shaking the bottle (1 minute). After the separation of the phases, the aqueous phase is transferred with a siphon tube into a second glass sampling bottle, which also contains 50 or 100 ml n-hexane. There, the second extraction is carried out. After the separation of the phases in the second bottle, the water is removed with a siphon tube and discarded. Both bottles now contain residues of the water sample and the hexane phases. The contents in both sampling bottles are transferred into a fat-free separating funnel (1 l), and both sampling bottles are then washed twice with n-hexane quantities of 20 to 50 ml each. The total extraction solutions and the residual water in the separating funnel are vigorously agitated (1 minute) and after the separation of the phases, the aqueous phase is drained. The n-hexane phase, which contains all the organic substances that are accessible to extraction in this pH range, is transferred through a fat-free filter into a weighed platinum dish at constant weight. When more than 5 mg is to be weighed, glass dishes of constant weight can also be used.

The n-hexane in the platinum dish is evaporated on the water bath at 80° C at maximum, then the platinum dish with the extractable substances is dried in the drying cabinet at 80° C for 15 minutes. After a certain cooling time in the desiccator, the dish is weighed on a semi-micro weighing equipment. This operation requires that the usual precautionary measures must be taken that apply to working in the semi-micro range.

The weighed extract can be processed. The determination of the saponifiable and the nonsaponifiable fraction allows for a distinction by mineral oils and fats as well as triglycerides or fatty acids.

The saponifiable fraction can be subjected to a gas-chromatographic analysis after the production of the methyl ester in order to determine the types of fatty acids present.

The nonsaponifiable fraction can be further analyzed by infrared spectroscopy or UV fluorescence determination. These analyses may allow conclusions as to the types of carbohydrates present (paraffins or aromatic substances or mixtures).

If this analysis is also to cover the fatty acids present in a water sample in the form of compounds of or combined with fatty acids (soaps), the water samples, extracted with n-hexane, and washings (see method described above) are not to be discarded but are to be collected in a sampling bottle where they are acidified with sulfuric or hydrochloric acid to set them at a pH of 1 to 2. This process decomposes the soaps and the fatty acids. These acids must be extracted with n-hexane and then the method is carried out in analogy to the procedure described in the foregoing.

When the determined weights of the directly extracted matter and the matter extracted after acidification are below 1 mg, the analysis must be repeated with a higher water volume. Experience has shown that the upper limit is reached when 10 l water are used.

Blind values must be determined and correspondingly considered.

Evaluation:

In consideration of the water volume used, the weighed oily and fatty substances are corrected by application of the blind value and related to 1 l water (the pH value at which extraction was carried out must be indicated).

When the compounds of fatty acids are covered by the extraction in an acidic solution, this fact must be noted in the result.

Determination of the saponifiable and nonsaponifiable fraction. The weighed matter is transferred with ethanol from the platinum dish into a ground glass flask to separate the saponifiable from the nonsaponifiable fraction. The fraction insoluble in ethanol must also be flushed off. After the addition of alcoholic potassium hydroxide solution, the saponification takes place at the reflux condenser.

The mixture is cooled and the ethanolic potassium hydroxide solution is transferred into a separating funnel. The saponification flask is subsequently washed with n-hexane (ca. 10 to 50 ml). These n-hexane portions are also transferred into the separating funnel. The extraction is then carried out by vigorous shaking. The ethanolic phase is separated from the n-hexane phase, and another extraction takes place with n-hexane. The combined n-hexane extracts are transferred into a weighed platinum dish of constant weight through a fat-free filter, and the treatment is continued as described for the main procedure.

The saponified fractions in the ethanolic potassium hydroxide solution are set to pH 1 to 2 by the addition of sulfuric or hydrochloric acid, with a double extraction with n-hexane. Both separated n-hexane extracts are filtered into a weighed platinum dish, and are processed by analogy to the main procedure. The following weighed matter is obtained:

1. Nonsaponifiable mineral oils and fats, given in mg/l.
2. the fatty acids of the saponifiable fractions, given in mg/l.

7.3.3.5.2 Infrared Analysis

When infrared analysis is applied to evaluate the nonsaponifiable fraction, the following procedure applies.

As a preparation for the infrared analysis of the obtained nonsaponifiable residue, this residue is dissolved in n-hexane. The n-hexane solution is transferred into a fat-free buret. Approximately 300 mg potassium bromide are placed on a watch glass under the discharge from the buret, and the n-hexane solution with the dissolved nonsaponifiable fraction is slowly dropped onto the potassium bromide. At the same time, an infrared lamp is placed at a suitable distance from the potassium bromide mixture to heat it. The dripping rate and the distance between the buret tip and the potassium bromide are so selected that the n-hexane evaporates soon after dropping onto the potassium bromide. This is a quick and simple method of applying the total n-hexane extract to the potassium bromide with simultaneous evaporation of the n-hexane. The buret is to be flushed with some milliliters n-hexane, which are also dropped and evaporated in an analogous way. The thus prepared potassium bromide is mixed and a molded potassium bromide cake is formed for use in the infrared analysis. Some relevant evaluation bands are given below.

a) Paraffins show characteristic absorption bands.

C-H stretching frequencies between 3.33 and 3.57 μm (3000 - 2800 cm^{-1}),

C-H deformation frequencies CH_2 and CH_3 6.89 - 7.29 μm (1450 - 1350 cm^{-1})

for long-chain paraffins with more than four CH_2 groups:
framework frequency at ca. 13.89 µm (ca. 720 cm^{-1}).

b) Double bonds are mainly indicated by:
C=C stretching frequency at 6.25 µm (1600 cm^{-1}),
by C-H stretching frequency at ca. 3.29 µ (3040 cm^{-1})

c) Aromatic substances show characteristic absorption bands.
C-H stretching frequencies between 3.23 and 3.33 µ (3100 - 3000 cm^{-1}),
frequencies of the aromatic framework
  at ca. 6.18 µm (1600 cm^{-1})
       6.66 µm (1500 cm^{-1})
       6.89 µm (1450 cm^{-1})

Attention must be paid to the aromatic C-H deformation frequencies in the range from 11.1 to 15.4 µ (900 - 650 cm^{-1}). Their position in the spectrum is characteristic of the substitution and the substituents.

7.3.3.5.3 Gas-Chromatographic Analysis

When the saponifiable fraction is to be analyzed for its fatty acid spectrum, its esterification to methyl esters is recommended. The weighed saponifiable fraction is dissolved with ca. 50 to 60 ml methanol containing ca. 10 % by weight of sulfuric acid. This solution is transferred into a glass flask and boiled at the reflux condenser for ca. 1 hour. This operation results in the conversion of the fatty acids to their methyl esters. (The esterification can also be carried out with diazo methane according to the directions in textbooks.)

The methanolic solution is allowed to cool, is thoroughly extracted with n-hexane, and the n-hexane extract is reduced to a volume of ca. 5 ml on the water bath at 80° C. This residual volume is transferred into a 10 ml measuring cylinder and topped up to the calibration mark with n-hexane at 20° C. Aliquot shares are taken out, using a microliter syringe, for the gas-chromatographic analysis.

The following working program can be recommended:
column: 25 - 50 m glass capillary column with OV1 (SE 30)
Temperature program: start 60° C, 4° C/min, up to 250° C
Carrier gas: helium, pressure bias $1 \cdot 10^5$ Pa, splitting ratio 50 : 1

injector: temperature 250° C,
detector: ionizing flame detector 250° C, H_2 30 ml/min, air 300 ml/min
secondary acceleration: N_2 as make-up gas 30 ml/min

Blind values are to be subjected to analogous analyses, and the retention times are determined by calibration with mixtures of the methyl esters of the palmitic acid, the oleic acid, and the stearic acid.

7.3.3.6 Detergents (Surfactants)

The term detergent or surfactant applies to substances that are active at the interface or surface-active. They are frequently used in detergents, washing powders and emulsions, cosmetics, and as auxiliaries in industrial products. They have a very wide range of applications.

A considerable number of detergents or technical products include not only anion-active detergents but also other types of agents, e.g. cation-active detergents, ampholytic detergents, or non-ionogenic detergents. At present, anion-active detergents are predominantly used.

7.3.3.6.1 Quick Test for Detergents

Many detergents are soluble in ethanol, so that the detergent concentration can be approximated from the concentration of substances soluble in ethanol in the evaporation residue of a water sample.

Most detergents are substances that are sensitive to drying. They are prone to decompose during the determination of the evaporation residue.

<u>Equipment and chemicals</u>

ethanol, absolute
funnel
fiber filter
weighed glass dish

Procedure

The evaporation residue of the water sample is dried at 105° C and mixed with ca. 50 ml hot absolute alcohol at boiling temperature. Adhering residues in the dish are loosened with a glass rod. After ca. 30 minutes on the water bath, the solution is filtered in hot condition through the fiber filter into a previously weighed glass dish. Then the filtrate is evaporated in a drying cabinet at 85° C, dried overnight, and weighed the next morning.

Since inorganic salts may also be dissolved in this process, the method is repeated with the evaporation residue of the ethanol extract. This time, the extract is filtered into a platinum dish, with a second evaporation and weighing step.

7.3.3.6.2 Determination of Anion-Active Detergents

Special reagents having the property of a cation-active detergent, form complexes with anion-active detergents. The complexes are used for photometric determination, for the determination of the end point in titration, or for a gravimetric determination.

When cation-active substances are present in the water, in addition to anion-active detergents, these cation-active substances are disturbing. They form compounds with the anion-active substances. Moreover, disturbances have already been noted in determination that are an indication of a natural water pollution by compounds having detergent properties. It is assumed that humic acids must be considered to be the cause of such disturbances.

Sampling:

The used glass flasks must be thoroughly flushed with 10 % alcoholic hydrochloric acid and then with trichloromethane before sampling. It should be noted that it is a specific property of detergents to accumulate on the water surface and to cause foaming, depending on their foaming activity.

One part of the detergents is accessible to biological decomposition. The consequence for the chemical analyst is therefore the quickest possible transport of the samples to the laboratory since the analysis must commence upon arrival.

If necessary, samples free of hydrogen sulfide may be preserved with some ml of a mercury chloride solution (10 mg mercury (II) chloride per 1 l distilled water). Waters containing hydrogen chlorides may be treated with sulfuric acid as the preserving agent. In this case, however, the sample must be neutralized before any analysis.

The disturbance resulting from sulfide ions can be eliminated when a hydrogen peroxide solution is added. The fact must be considered that the oxidation may convert sensitive detergents.

Methods Available:

Many different methods have been developed for a quantitative determination of detergents. The majority of these methods of determining anion-active detergents mainly react to sulfonates. Two specific methods have been selected here:
the photometric determination of anion-active detergents with methylene blue,
the titration with cetyl trimethyl ammonium bromide against methylene blue.

7.3.3.6.2.1 Photometric Determination of Anion-Active Detergents with Methylene Blue

The concentration of anion-active substance should range between 20 and 150 µg in 100 ml of the sample. Waters with a higher concentration of anion-active detergents need an appropriate dilution.

Methylene blue has the property of a cation-active detergent, and when combined with an anion-active surfactant it forms a colored complex. This complex is photometrically evaluated. This method is also known by the designation "methylene blue method according to LONGWELL and MANIECE".

Equipment and chemicals

separating funnel, volume ca. 250 ml
measuring cylinder, volume 50 ml

phosphate solution: 12.52 g $Na_2HPO_4 \cdot 2\,H_2O$ (buffer substance according to SOERENSEN) are dissolved in 500 ml distilled water. Approximately 3 ml 0.5 N sodium hydroxide solution are added to set the solution to a pH value of 10, then the solution is topped to 1 l with distilled water. When the solution is stored for an extended period, the pH value must be regularly checked and reset, if necessary.

methylene blue solution neutral: 0.35 g methylene blue 6 are dissolved with distilled water to achieve 1 l. The freshly prepared solution must be allowed to stand for at least 24 hours before the calibration plot is established. The extinction of the trichloromethane phase of the blind sample, measured against trichloromethane, must not exceed the value of 0.015 per 1.0 cm of thickness of the layer. When a fresh methylene blue supply is used, the calibration plot must be established again.

methylene blue solution, acidic: 0.35 g methylene blue 6 is dissolved in 500 ml distilled water and mixed with 6.5 ml H_2SO_4 (d = 1.84 g/ml). This solution is topped with distilled water up to the 1 l mark. The freshly prepared solution must be allowed to stand for at least 24 hours before the calibration plot is established. The extinction of the trichloromethane phase of the blind sample, measured against trichloromethane, must not exceed the value of 0.015 per 1.0 cm of thickness of layer. When a fresh methylene blue supply is used, the calibration plot must be established again.

hydrogen peroxide solution, 30 % by weight H_2O_2

trichloromethane p.a., freshly distilled

cotton wool

spectrophotometer

Procedure

Turbid waters are filtered, the first 100 ml of the filtrate are to be discarded; 100 ml of the filtrate are transferred into a separating funnel.

When less than 100 ml water are used, the mixing volume is topped with distilled water. It should be emphasized here that a blind value with 100 ml dist. water must be analyzed by the same method in any case.

Ten milliliters of alkaline phosphate solution, 5 ml of neutral methylene blue solution, and 15 ml trichloromethane are added. The separating funnel is closed and shaken well for ca. 1 minute. Then the trichloromethane phase is allowed to separate from the water phase. The separation of the two phases may take a considerable length of time. The trichloromethane phase sediments in the bottom of the separating funnel and is discharged into a second separating funnel that has previously been charged with 100 ml distilled water and 5 ml acidic methylene blue solution. Then the second separating funnel is shaken again, as described above. Finally, the trichloromethane is filtered into a 50 ml measuring cylinder through a cotton wool filter wetted with trichloromethane. The whole procedure, i.e. the extraction of the water sample initially with trichloromethane and then reextraction of the trichloromethane phase with distilled water, is repeated two more times, using 10 ml trichloromethane each time. Phosphate solution and methylene blue solution are not added again. Finally, the measuring clyinder is washed by filling.

The complex of the anion-active detergent with methylene blue is now totally in the trichloromethane phase. The extinction of the trichloromethane phase is measured in a cuvette, 1 cm thickness, using a photometer at 650 nm. The blind value is also measured and correspondingly considered.

Evaluation:

The calibration plot, considering the blind value and the used water volume or the prepared dilution, is used to evaluate the extinction. The results are indicated in mg/l.

Establishing the calibration plot:

The most frequently used anion-active detergents are based on an alkyl or alkyl-aryl sulfonate type. Examples are sodium lauryl sulfate, sodium laurylether sulfate, dodecyl benzene sulfonate, tetrapropylene benzene sulfonate. Any of the listed detergents may be used to establish the calibration plot. The concentration of anion-active substance in the reference sample must be exactly known. A series of dilutions is prepared to establish the calibration plot.

Since the various detergents have different molecular weights, the reference detergent must be indicated together with the mg/l concentration.

7.3.3.6.2.2 Titration with Cetyl Trimethyl Ammonium Bromide against Methylene Blue (EPTON Titration)

The anion-active detergent of the sulfonate type, when combined with methylene blue, forms a dye complex that penetrates as a dye salt into the trichloromethane layer. It is titrated with cetyl trimethyl ammonium bromide, which has the effect of a strong cation-active detergent.

The water sample should contain ca. 480 to 880 µg/l anion-active detergents. Waters with a higher concentration must be correspondingly diluted with distilled water.

Equipment and chemicals

separating funnel, volume ca. 125 to 150 ml
micro-buret
0.005 N cetyl trimethyl ammonium bromide (CTAB): 1.80 g of the substance are dissolved in water, the solution is diluted to obtain 1 l.
methylene blue solution: 0.1 g methylene blue hydrochloride p.a. are dissolved in 50 ml of the water and diluted to 100 ml in the measuring cylinder; 30 ml of this solution are transferred with a pipette into a 1 l measuring cylinder. 50 g sodium sulfate are added, the substance is dissolved in sufficient water, 6.8 ml concentrated sulfuric acid are added, and the solution is topped to 1 l with distilled water.
ethanol, 96 % by vol.
trichloromethane p.a.

Procedure

The following is a description of the method modified by BLANK.

Turbid waters are filtered, and the first 100 ml are discarded; 25 ml of the water sample are transferred into the 125 ml mixing cylinder (with glass plug) with a pipette, 15 ml trichloromethane are added, as well as 25 ml methylene blue solution and 15 ml water. The separating funnel is closed and vigorously shaken for ca. 1 minute. Then the phases are allowed to separate. At first, 5 ml of the CTAB solution are added from a micro-buret, and the mixture is vigorously shaken. Then the titration is continued with smaller volumes of the solution, ca. 0.1 to 0.2 ml. Shaking is necessary between the individual steps of titration. With titration progressing, the intensive blue coloring of the trichloromethane phase migrates into the aqueous phase.

Toward the end of titration, 0.1 ml of the CTAB solution are added in each step. The end of titration is achieved when the bottom trichloromethane layer has a less intensive coloring than the water phase above. Normally, the trichloromethane layer is greenish in color. It is difficult for an untrained operator to recognize the end of titration. It is therefore recommended that the untrained analyst should be trained with titration of a solution having a known concentration of anion-active detergents.

A tenth of a milliliter (empirical value) is deducted from the result of titration. The surfactant concentration in the water sample is calculated according to the following formula:

$$G = \frac{b \cdot N \cdot M}{25} \; 1000 \quad [mg/l]$$

where:

G = concentration of the anion-active surfactant, in mg/l
b = number of ml CTAB solution used for titration, reduced by 0.1 ml
N = normality of the CTAB solution
M = molecular weight of the reference detergent. The reference detergent employed must be indicated with the result of the analysis.

Two grams of one of the aforementioned detergents of the sulfonate type are exactly weighed, dissolved in water, and diluted to 1 l in the measuring cylinder for standardization of the CTAB solution. Ten milliliters of this dilution are transferred with a pipette into a 125 ml separating funnel with glass plug. 30 ml water, 25 ml methylene blue solution, and 15 ml trichloromethane are added to the dilution. Then the dilution is titrated as described above. The measurement is repeated three times, the mean value of the results is applied. The normality of the CTAB solution is calculated from the following formula:

$$N_{CTAB} = \frac{a \cdot 1000}{b \cdot M}$$

where:
- a = weight of the dosed detergent in 10 ml of the stock solution. The detergent must be a salt-free product without neutral oil.
- b = ml of CTAB solution consumed, reduced by 0.1 ml
- M = molecular weight of the reference detergent used

7.3.3.6.3 Determination of Non-Ionogenic Detergents (Preliminary Test)

DRAGENDORFF's reagent, potassium iodine bismuthate $KBiJ_4$, quickly forms a well visible intensively red inclusion complex with a large number of polyalkylene oxide compounds and free polyethylene glycols, from the aqueous solutions.

Anion-active and cation-active detergents comprising polyethylene oxide functions in the molecules lead to a positive detection reaction.

Application

A positive reaction is achieved up to a concentration limit of 0.1 mg/kg (0.1 ppm) with the following substances:

oxalkylated fatty acids, alcohols, fatty amines, alkylphenols and amides of fatty acids, polyethylene and polypropylene glycols down to pentaethylene or pentapropylene glycol.

Low-molecular alkylene glycols such as tetra, tri, and di-alkylene glycol, do not result in a precipitation. Proteins and their decomposition products down to the amino acids and other agents active at the interface do not lead to a positive reaction when the molecule does not include ethylene oxide fractions.

Equipment and chemicals:

Solution A: 1.7 g base bismuth nitrate ($BiO\,NO_3\,H_2O$) are dissolved in 20 ml glacial acetic acid, and the solution is topped to 100 ml with distilled water.

Solution B: 40 g potassium iodide are dissolved in 100 ml distilled water.

Solutions A and B are combined in a 1 l measuring cylinder, mixed with 200 ml glacial acetic acid, and topped to 1 l with distilled water; 100 ml each of this solution are mixed with 50 ml 20 % barium chloride solution in distilled water. The reagent is ready for use and is stored in a brown bottle with ground glass plug. It is stable in the dark for ca. 2 weeks. Over longer periods, the originally orange-red color changes to brown, so that the reagent can no longer be used.

The water sample to be analyzed is filtered and 5 ml of the filtrate are mixed with 5 l of the reagent in a test tube and well agitated. The presence of polyalkylene oxide compounds is immediately indicated by the formation of the intensively orange-red precipitate. Very low concentrations result in turbidity of the initially clear solution. Centrifuging produces an orange-red precipitate at the bottom. This method clearly indicates concentrations as low as 0.1 mg/kg.

7.3.3.6.3.1 Determination of Non-Ionogenic Detergents (Quantitative Determination)

DRAGENDORFF's reagent, potassium iodine bismuthate $KBiJ_4$, forms an intensively orange-red inclusion complex from aqueous solutions with a large number of polyalkylene oxide compounds and free polyethylene glycols. The formation of the inclusion complex is a quantitative reaction. The inclusion complex is filtered off and dissolved, and the bismuth contained in the inclusion complex is potentiometrically titrated for a quantitative determination.

This method is suited to determine the following types of non-ionogenic detergents:

Oxalkylated fatty acids, alcohols, fatty amines, alkyl phenols and amides of fatty acids, polyethylene and polypropylene glycols down to the pentaethylene or pentapropylene glycol.

Low-molecular weight alkylene glycols such as tetra, tri and di-alkylenglycol do not result in a precipitation. Proteins and their products of decomposition down to the amino acids and other agents active at the interface, having a cationic or anionic reactivity, do not lead to a positive reaction when ethylene oxide groups are not present in the molecules.

Anion-active detergents are not disturbing in the concentration range below 5 mg/l. Anion-active and cation-active detergents having polyethylene oxide functions in the molecules, however, form the inclusion complex and are determined as well. Such detergents can be separated by ion exchange.

The concentration of non-ionogenic detergents should range between 250 and 800 µg/l. Waters with a higher concentration need a corresponding dilution.

Sampling

The used glass instruments must be thoroughly flushed with 10 % alcoholic hydrochloric acid and then with trichloromethane before sampling. It should be noted that it is a property of the detergents to accumulate at the water surface and to result in foaming, depending on their foaming activity.

One part of the detergents is accessible to biological decomposition. This means for the chemical analyst that the samples must be transported as quickly as possible into the laboratory and that the analysis must begin immediately upon arrival.

Equipment and chemicals

acetic acid ethyl ester p.a., freshly distilled

sodium hydrogen carbonate $NaHCO_3$ p.a.

hydrochloric acid 1 %, prepared from concentrated hydrochloric acid with bidistilled water

methanol p.a., freshly distilled, stored in glass bottles

bromium cresol purple solution: 0.1 g bromium cresol purple are dissolved in 100 ml methanol p.a., distilled fresh.

precipitating reagent: the precipitation reagent is a mixture of 2 parts by volume of Solution A and 1 part by volume of Solution B. The mixture must be stored in a brown bottle in the dark. It remains stable for ca. 1 week.

Solution A: 1.7 g base bismuth (III) nitrate, purest quality available, are dissolved in 20 ml glacial acetic acid p.a., the solution is topped to 100 ml with bidistilled water; 65 g potassium iodide p.a. are dissolved in ca. 200 ml bidistilled water. Then both solutions are combined in a 1000 l measuring cylinder, 200 ml glacial acetic acid p.a. are added, and the solution is topped to the calibration mark with bidistilled water.

Solution B: 290 g barium chloride ($BaCl_2 \cdot 2\ H_2O$) p.a. are dissolved in bidistilled water, and the solution is topped to 1000 ml.

glacial acetic acid, purest quality available

ammonium tartrate solution: 12.4 g tartaric acid p.a. and 18 ml ammonia solution 25 % p.a. are combined and topped to 1000 ml with bidistilled water.

ammonia solution, 1 %, prepared from ammonia solution 25 % p.a. by dilution with bidistilled water.

standard acetate buffer: 120 ml glacial acetic acid (purest quality available) are combined with 40 g sodium hydroxide p.a., both after dilution or dissolution in bidistilled water, and the mixture is topped to 1000 ml.

0.0005 N pyrrolidine dithiocarbamate solution (carbate solution): 103.0 mg pyrrolidine dithiocarbonic acid sodium salt (purest quality if possible) are dissolved in bidistilled water, 10 ml n-amyl alcohol p.a. and 0.5 g sodium hydrogen carbonate are added, then the solution is topped to 1000 ml with bidistilled water.

0.0005 N copper sulfate solution

solution I: 1.249 g copper (II) sulfate p.a. ($CuSO_4 \cdot 5\ H_2O$) and 50 ml 1 N sulfuric acid are topped to 1000 ml with bidistilled water. (Do not use disintegrated crystals!)

solution II: 50 ml of Solution I and 10 ml 1 N sulfuric acid are topped to 1000 ml with bidistilled water. This is exactly an 0.0005 N solution.

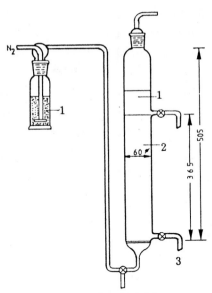

Fig. 7.3.-7: Surfactant blower apparatus
1 - ethyl acetate; 2 - water sample; 3 - glass filter frit 61

1 l apparatus for surfactant accumulation by blowing, according to WICKBOLD, as illustrated in Fig. 7.3.-7 (all dimensions in mm)

magnetic stirrer with magnetic bar (25 - 30 mm)

separating funnel 250 ml

porcelain filter crucible, size 2 (upper diameter 40 mm, height 42 mm, porosity 2)

2 aspirator bottles with adapter for filter crucible (500 and 250 ml)

round glass fiber paper filter, fiber diameter 0.5 to 1.5 µm

polyethylene squeeze bottle, volume 500 ml, for glacial acetic acid

recording potentiometer with platinum/calomel or platinum/silver chloride measuring chain, measuring range 250 mV, with automatic buret, 20 to 25 ml volume, or alternatively a manual potentiometric equipment

methanolic hydrochloric acid, 10 %

Procedure

a) Isolation of the non-ionogenic detergents

Approximately 1 l of the water sample, filtered through a soft paper filter, is necessary to perform this method. Water samples with a concentration of non-ionogenic detergents higher than 800 µg/l are correspondingly diluted. Acidic and alkaline water samples are neutralized. Then 100 g sodium chloride and 5 g sodium hydrogen carbonate are dissolved in the water sample or the dilution. The surfactant blower equipment (see Fig. 7.3.-7) is filled with the water sample up to the upper drain cock. Then 100 ml acetic acid ethyl ester are carefully topped onto the water. The frit washing bottle in the gas supply line is filled by ca. two-thirds with acetic acid ethyl ester. A gas flow, nitrogen or air, is blown through the blower equipment at a rate of ca. 50 to 60 l/h. The flow rate should be so dimensioned that turbulence will definitely not occur at the phase boundary. The phases must remain distinctly separated. This procedure avoids an intermixture which results in emulsification. The gas flow is stopped after 5 minutes.

The acetic acid ethyl ester phase is completely drained into a 250 ml separating funnel. The aqueous phase separated in the separating funnel is returned into the blower equipment.

Another volume of 100 ml acetic acid ethyl ester is topped into the surfactant blower equipment, and nitrogen or air is again passed through the equipment for 5 minutes again. The organic phase is again drained into the separating funnel. The aqueous phase in the separating funnel is discarded and the acetic acid ethyl ester phase is filtered through a fiber filter. The separating funnel and the filter are flushed with ca. 20 ml acetic acid ethyl ester. The acetic acid ethyl ester extracts and the washing solution are combined in a 250 ml beaker.

As a result of the gas blowing procedure, the detergents are accumulated at the surface of the water phase and are quantitatively absorbed into the acetic acid ethyl ester phase.

The combined extracts are carefully evaporated in the 250 ml beaker on the water bath. A slight air flow above the beaker accelerates the evaporation process, while the risk of a conversion of the detergents sensitive to heat treatment is reduced.

b) Precipitation and filtration

The evaporation residue is absorbed in 5 ml methanol, 40 ml bidistilled water and 0.5 ml of 1 N hydrochloric acid, and thoroughly stirred with a magnetic stirrer. Then 3 to 5 drops of bromium cresol purple solution are added. The indicator should immediately change over to yellow. 30 ml Dragendorff's precipitation reagent are added into the clear solution, with continued stirring. After 10 minutes, stirring is stopped and the solution is allowed to stand for at least 5 minutes.

When the water sample contains non-ionogenic detergents, an orange-red precipitate has formed. The precipitate is filtered into a 500 ml aspirator bottle through a porcelain filter crucible with aspiration effected by a water jet pump. The beaker, the magnetic stirrer, and the crucible are thoroughly washed with glacial acetic acid, using ca. 150 to 200 ml. A polyethylene squeeze bottle facilitates washing. A quantitative transfer of the precipitate from the beaker into the filter crucible is not necessary since the beaker is used again later on to receive the solution of the precipitate.

c) Dissolution of the precipitate

The filter crucible and the rubber sleeve are attached to a glass adapter and arranged on a 200 ml aspirator bottle. Spraying of the solution into the aspirator should be prevented. To preclude errors by the transfer of the precipitation reagent, the rubber sleeve used in the filtration step must not be used here again. The precipitate is dissolved by the addition of hot ammonium tartrate solution in three batches of 10 ml each. The solution may foam as a result of the detergent concentration. Additionally, 20 ml of the ammonium tartrate solution are introduced into the precipitation beaker. The glass beaker is rotated to dissolve the remaining precipitate. The filter crucible, the adapter, and the aspirator bottle are thoroughly washed with bidistilled water, and the thus obtained solution is returned into the precipitation beaker, together with the washing. The total volume of the solution in the precipitation beaker should be 150 to 200 ml.

Titration:

The solution may still be slightly acidic due to acetic acid residues resulting from washing. Therefore, some drops of bromium cresol purple solution are added to the solution. The solution is stirred with the magnetic stirrer, and a 1% ammonia solution is added to set the color change of the indicator to violet. Then 10 ml standard acetate buffer are added to achieve a pH of 4.6. The beaker is placed on the potentiograph and the electrodes are immersed into the solution. Then the potentiometric titration is carried out with 0.0005 N pyrrolidine dithiocarbamate solution beyond the potential jump; 2 ml/min are the prescribed titration rate. The paper feed should be 2 cm/min.

The bismuth in combination with the pyrrolidine dithiocarbamate forms an almost insoluble metal complex. The potentiometric titration method is highly sensitive for small bismuth quantities. With a standard solution as low as N/10000 an observable jump is still achieved. The question why a potential jump occurs has not yet been clarified. Presumably a small quantity of a reversible oxidation product is present during the potentiometric titration so that a measurable redox voltage may arise at the electrode arrangement, which changes when the dithiocarbamate excess occurs in the solution when the end of titration is exceeded.

The end of titration is defined as shown in the diagram below.

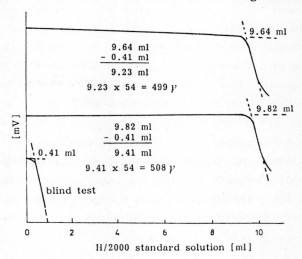

Fig. 7.3.-8: N-surfactant determination through potentiometric bi-determination with pyrrolidine dithiocarbamate standard solution, determination of 500 µg nonyl phenol with 10 ÄO

The point of intersection of the tangents of the two branches of the potential curve is defined to be the end point of titration. Occasionally, flattening of the potential jump is noted, but this phenomenon can be eliminated by carefully polishing the platinum electrode.

d) Blind test

A blind test must be carried out in analogy to the previously described procedure; 5 ml methanol and 40 ml bidistilled water are used in the blind analysis. Then the process is continued as described in item (b) (precipitation and filtration). The process is continued as described above. In the blind analysis, the volume of 0.0005 N-pyrrolidine dithiocarbamate solution should be below 1 ml. Otherwise, the reagents used are to be checked for their purity in relation to heavy metals.
The purity of the reagents can be checked by a spectral analysis of the evaporation residue.

e) Factor check of the 0.0005 N-pyrrolidine dithiocarbamate solution

10 ml of the copper sulfate calibration solution are mixed with 100 ml bidistilled water and 10 ml standard acetate buffer solution. Then the titration is carried out as described above. The factor is

$$f = 10/a$$

where a = volume of pyrrolidine dithiocarbamate solution (ml) consumed. The factor of the solution must be considered in the evaluation of the titration result.

f) Calculation of the results

Since there are many different types of non-ionogenic detergents and since the individual types may include ethylene oxides of different chain lengths, reference must be made to a standard substance with a known gram-molecular weight, unless the specific non-ionogenic surfactant is known which was present in the water sample.

Nonyl-phenol decaglycolether is defined as the standard substance. This compound comprises a phenol body, substituted by a paraffin residue with 9 carbon atoms, with the OH group of the phenol being ethered with 10 ethylene oxide units. A conversion factor of 54 has empirically been determined for this chemical compound.

The mg/l concentration of non-ionogenic detergents in the water sample is calculated according to the following formula:

$$g = (b - c) \cdot f \cdot 0.054 \qquad [mg/l]$$

where:
- g = concentration of non-ionogenic detergents in the water sample (in mg/l), related to nonyl-phenol decaglycol ether
- b = volume of pyrrolidine dithiocarbamate solution (in ml) used for the titration of the precipitate of the acetic acid ethylester extract
- c = volume of pyrrolidine dithiocarbamate solution (in ml) consumed in the blind analysis
- f = factor of the 0.0005 N-pyrrolidine dithiocarbamate solution

7.3.3.6.4 Separation of Detergents by the Foaming Method

When air is blown through the water sample, many detergents accumulate at the surface where they are collected by ethylacetate.

Not all the surfactants have a high foaming activity. Moreover, the foaming capability depends on the pH value in the water. It is the best solution to set the water to a pH value of 7.

This method covers many compounds soluble in ethyl acetate.

Procedure

Nitrogen saturated with ethyl acetate is bubbled through 400 ml water at a rate of 50 to 60 l per hour. The water sample is topped by a layer of 100 ml ethyl acetate. After 5 minutes, the gas flow is stopped and the organic phase is separated. Then another 100 ml volume of ethyl acetate is added and nitrogen is passed through for 5 minutes. Then the water sample should practically be free of surfactants. The detergent concentration can be determined by evaporation of the ethyl acetate extract on the water bath.

This method is disputed, as can also be concluded from the list of disturbances. Not only surfactants but also fatty acids may be extracted from acidic solutions, whereas the anion-active detergents are inactive in acidic solutions. The foaming activity of non-ionogenic detergents is generally rather low.

The most favorable solution is to introduce a measured quantity of a strongly foaming detergent with an exactly known concentration of active matter into a water sample. As a result of the high foaming activity, the detergents initially present in the water sample are also foamed out. The evaporation residue of the ethyl acetate extract must then be corrected. Then the analysis can be continued with the various methods described in this Section.

7.3.3.7 Extractable Organic Halogen Compounds (EOX)

The concentration of extractable organic halogen compounds in a particular water denotes the volume-related mass of organically bound halogen, which is determined under the operating conditions described here and which is indicated as EOX.

The organic halogen compounds are extracted with n-pentane or n-hexane and di-isopropyl ether. The extract is incinerated in an oxygen hydrogen flame. The inorganic mineralization products are then subjected to coulometric titration and indicated as chloride. The description of the procedure conforms to the DIN Standard draft of August 1981.

Equipment and chemicals

Measuring cylinders (1000 ml with low volume marks), nominal volume 50 ml, 100 ml and 1000 ml
magnetic stirrer, applicable for speeds up to 1100 rpm
PTFE, sheathed
transfer pipettes, nominal volumes 25 ml and 50 l
pH meter

WICKBOLD incinerator, e.g. acc. to DIN Standard 51 409

coulometer or micro-coulometer with potentiometric indication

standard solution I: 150.3 mg penta-chlorophenol are weighed into a 100 ml measuring cylinder. Then the substance is dissolved in n-pentane or n-hexane, and topped to 100 ml with the respectively used solvent. (The standard solution I contains 1000 mg/l "organically bound chlorine".)

standard solution II: 1 ml of the standard solution I is topped with n-pentane or n-hexane to 100 ml. (Standard solution II contains 10 mg/l of "organically bound chlorine".)

oxygen of sufficient purity

hydrogen of sufficient purity

acetone

distilled water, free of extractable organic halogen compounds (EOX concentration below 20 µg/l)

n-pentane, C_5H_{12}, alternatively: n-hexane, C_6H_{14}

di-isopropyl ether, $C_6H_{14}O$. This substance must be permanently stored in brown bottles on NaOH (solid) to bind any formed ether peroxides.

sulfuric acid, H_2SO_4 = 1.84 g/ml

sodium hydroxide solution, NaOH: 30 g sodium hydroxide are dissolved in 100 ml water.

sodium sulfate, Na_2SO_4, preheated at 600° C for 1 h

penta-chlorophenol, C_6Cl_5OH

pH paper, alternatively: pH meter

Procedure

Twenty grams of sodium sulfate and ca. 950 ml of the sample are introduced into a 1 l measuring cylinder. Sulfuric acid or sodium hydroxide solution is added to set the pH value of the analyzed sample to 6.0 to 8.0. Then the measuring cylinder is topped to 1 l. The solution is cooled to 4° C.

The magnetic stirring bar is introduced into the measuring cylinder; 25 ml n-pentane or 25 ml n-hexane are added to the analyzed sample. Then the cylinder is closed and the sample is stirred for 10 minutes at 1100

immediately upon sampling, mixed with 0.3 ml of the methylene blue solution, and closed with the glass plug greased with petroleum jelly. An air bubble should be carefully avoided. The sample is stored at 20° C in the incubator. When the pollution is considerable, the sample is compared every hour to a sample of distilled water of equal size, treated in the same manner. When the pollution is not too strong, the comparisons can be made at longer intervals. The time of complete decolorization should be determined as exactly as possible, provided that it occurs within 48 hours. The test is continued through a total of 120 days = 5 days.

Evaluation

The observed changes and the time of decolorization are indicated in hours after sampling. When the decolorization or any change does not occur within 5 days, the result is designated as negative.

Indication of the result

Examples: "Methylene blue test:
        a) decolorized after 48 h,
        b) lighter coloring within 48 hours,
        c) decolorized after 3 days,
        d) negative".

7.3.3.9 Waste Water Toxicity

Bacteria Test

Toxic substances dissolved in waste water inhibit the formation of organic acids from glucose by the bacterium Pseudomonas fluorescens. Thus, as a result of the bacterial activity after a certain time, the pH value in a receiving water not influenced by waste water will be lower than the pH value of a water sample containing dissolved toxic substances, which is maintained under the same conditions.

The quantitative determination of the acid formation inhibition by pH metering requires the adaptation of the pH value and the acid consumption of the waste water sample to the corresponding values of the receiving water before the analysis begins.

The indication of the ratio between the waste water volume and the receiving water/waste water total volume of that particular dilution that has the lowest deviation from the pH value of the sample taken from the receiving water, after expiration of the testing time, is used as a measure of the biological toxic effect of a toxic waste water.

The method is applicable to all commercial and industrial waste waters.

No interference or disturbance

Equipment and chemicals

incubator
autoclave,
agitator equipment with attachment for 300 ml ERLENMEYER flasks
pressure filtering equipment for diaphragms
diaphragm, pore gauge 0.2 µm (1 µm = 0.001 mm)
photometer with nephelometer attachment
filter Hg 436 nm
cuvettes (thickness of layer 10 mm) of special optical glass
electrical pH meter
Kapsenberg culture tubes (diameter 18 mm, height 180 mm) with Kapsen-
    berg caps as closures, dry sterilized
Erlenmeyer flask (volume 300 ml), dry sterilized, with plastic caps filled
    with cotton wool, as closures
transfer pipettes (volume 1 ml, graduation by 0.01 ml)
transfer pipettes (volume 10 ml, graduation by 0.1 ml)
measuring cylinders (volume 100 ml, 1000 ml)
glass beads (diameter ca. 2 mm)
inoculation eye
sodium hydroxide solution (5 N, 1 N, 0.1 N, 0.01 N)
hydrochloric acid (5 N, 1 N, 0.1 N, 0.01 N)
hydrochloric acid (25 % by mass)

Culture media:
nutrient medium for stock and preparatory cultures
  1.060 g sodium nitrate, $NaNO_3$, p.a.

0.600 g di-potassium hydrogen phosphate, K_2HPO_4, waterfree
0.300 g potassium dihydrogen phosphate, KH_2PO_4, p.a.
0.200 g magnesium sulfate, $MgSO_4 \cdot 7\ H_2O$, p.a.
10.000 g D(+) glucose (for biochemical and microbiological applications)

trace element solution (quantities indicated in g/l bidistilled water):
0.055 $Al_2(SO_4)_3 \cdot 18\ H_2O$, p.a.
0.028 KJ, p.a.
0.028 KBr, p.a.
0.055 TiO_2, p.a.
0.028 $SnCl_2 \cdot 2\ H_2O$, p.a.
0.028 LiCl, p.a.
0.389 $MnCl_2 \cdot 4\ H_2O$, p.a.
0.614 H_3BO_3, p.a.
0.055 $ZnSO_4 \cdot 7\ H_2O$, p.a.
0.055 $CuSO_4 \cdot 5\ H_2O$, p.a.
0.059 $NiSO_4 \cdot 6\ H_2O$, p.a.
0.055 $Co(NO_3)_2 \cdot 6\ H_2O$, p.a.

vitamin solutions (quantities indicated in mg/100 ml bidistilled water):
0.2 biotin
2.0 nicotinic acid
1.0 thiamin
1.0 p-aminobenzoic acid
0.5 panthothenic acid
5.0 pyridoxamin
2.0 cyanocobalamin

The prepared nutrient medium is charged in batches of 6 ml each into culture tubes. The culture tubes are then distilled in fractions in the autoclave three times for 30 minutes each. Then the nutrient medium is caused to solidify in an oblique position.

Stock solution I
20.000 g D(+) glucose (for biochemical and microbiological purposes)
 4.240 g sodium nitrate, $NaNO_3$, p.a.
 2.400 g di-potassium hydrogen phosphate, K_2HPO_4, waterfree

1.200 g potassium di-hydrogen phosphate, KH_2PO_4, p.a., and
30 ml trace element solution

The glucose or nutritive salts are separately dissolved in 500 ml bidistilled water each, sterilized in the autoclave for 30 minutes, and the fractions are combined after cooling.

Stock solution II

0.200 g iron sulfate, $FeSO_4 \cdot 7\ H_2O$, p.a.
4.000 g magnesium sulfate, $MgSO_4 \cdot 7\ H_2O$, p.a.
are dissolved in 1000 ml sterile bidistilled water.

NaCl solution: 0.500 g sodium chloride, NaCl, p.a. are dissolved in 1000 ml bidistilled water. The solution is sterilized in the autoclave for 30 minutes.

Procedure

Stock cultures of a Pseudomonas test strain (recommended: American Type Culture Collection - ATCC -, Rockville, USA, Strain No. 13525) are maintained on the culture medium for stock and preliminary cultures in oblique agar culture tubes. To continue the breeding of the test strain, new stock cultures are prepared at 1 week intervals. The inoculated stock cultures are incubated for 20 + 4 hours at 25° C and then stocked.

Preliminary cultures are taken from the stock cultures as required. They are established on the aforementioned culture medium in oblique agar test tubes and incubated at + 25° C for 20 + 4 hours. Then the cell material is flushed off with sterile NaCl solution. The turbidity value of the bacteria suspension is determined from the photoelectric measurement of the level of transmission of the monochromatic measuring radiation Hg 436 nm. On the basis of this value, the bacteria suspension is diluted with sterile NaCl solution and then the final turbidity value of the suspension is set for 10 mm thickness of layer, to the transmission level T = 37 % of the monochromatic measuring radiation Hg 436 nm (nm = nanometer).

Before the test cultures are prepared, the waste water to be tested is neutralized. The concentration of the neutralizing acid or base must be selected so that the added volume is as small as possible. Then the waste water is filtered through a diaphragm (pore size 0.2 µm) by a pressure filter equipment.

Four parallel dilution series are prepared from the thus pretreated waste water. These series are kept in 300 ml Erlenmeyer flasks closed by plastic caps that are filled with cotton wool. The dilutions contain one part by volume each of the waste water in 2, 4, 8, 16, 32, 64, 100, 200, 400, etc. parts by volume of the dilution. To prepare the dilution series, bidistilled water is added to each test flask in a quantity smaller by 20 ml than the respectively calculated dilution ratio between the waste water and the bidistilled water. Thus, each test flask initially contains 80 ml.

Then 5 ml each of the stock solution I, 5 ml of the stock solution II, and 10 ml each of the prepared bacteria suspension of the preliminary culture with a known set transmission level (T = 37 %) are added to top each test flask of the three waste water dilution series to be inoculated. Thus, the nominal volume of 100 l is reached. The initial turbidity of the test cultures after inoculation corresponds to a transmission level of T = 89 % of the monochromatic measuring radiation (Hg 436 nm) for 10 mm thickness of layer.

Each test flask containing the waste water samples in the dilution series, which is not to be inoculated, is topped with 5 ml each of the stock solution I, 5 ml of the stock solution II, and 10 ml NaCl solution.

Reference cultures are simultaneously prepared in order to check the biological reaction standard of the test organisms in the test medium not influenced by the waste water. To this end, 80 ml each of bidistilled water, 5 ml of stock solution I, 5 ml of stock solution II, and 10 ml of the prepared bacteria suspension of the preliminary culture of a known set transmission level (T = 37 %) are added in five ERLENMEYER flasks.

The waste water dilution series and the reference cultures are kept at 25° C for 16 hours. Then, after the addition of 1 ml hydrochloric acid (d = 1.125 g/ml) and after the addition of glass beads, and after 30 minutes shaking on a vibrating equipment with ca. 250 torsional oscillations per minute, the transmission level of the monochromatic measuring radiation (Hg 436 nm) is measured in the reference cultures and in the waste water dilution series in 10 mm thickness of layer.

When a coloring or a chemicophysically caused turbidity, after acidification, occurs in the waste water in the dilution series, the analogous concentration levels of the non-inoculated dilution series are used as nephelometric vacant values for the turbidity measurement of the inoculated dilution series.

Evaluation

The lowest waste water dilution level is not deemed toxic, for which the mean transmission level is not higher than the mean transmission level of the reference cultures, after the expiration of the test time. The corresponding dilution factor is indicated.

7.4 Fish Test

General Information

Fish are the final members in the aquatic food chain. Fish react very sensitively to changes in their environment. When the water quality deteriorates the fish may be damaged, which can be recognized by various symptoms, e.g. gasping for breath, reeling, loss of orientation, lethargy. These types of damage may be lethal.

The method is applicable to any kind of waste water. It is employed to determine the acute lethal effects on test fish, causing death within 48 hours. The ide or orfe (Leuciscus idus) is employed as test fish.

The fish may be damaged by the substances contained in the waste water, e.g. resulting in sealed gill lamellae or their obstruction with phlegm because the respiratory epithelium is swollen and destroyed. Apart from the immediate effect on the gills, a resorption of pollutants through the gills, the skin, or the digestive system is possible in many cases before toxication occurs.

The toxic effects are determined in dilutions of the waste water sample with diluting water in integer volume ratios. The toxic effects of a particular waste water on fish are characterized by the indication of the dilution factor G_F.

The mixture of waste water and diluting water, which is used for the fish test, is termed test water.

The dilution factor G indicates the multiple of the initial volume to which one part by volume of waste water is diluted with the diluting water to prepare the test water.

Example: when 1 part by volume of waste water is diluted with 4 parts by volume of diluting water to a total of 5 parts by volume of test water, the dilution factor G is 5.

The smallest G value of the test water in which all fish survive under the standardized conditions, is designated as dilution factor G_F.

Equipment and chemicals

all-glass aquariums
measuring pipettes
transfer pipettes
measuring flasks
measuring cylinders
ERLENMEYER flasks, nominal volume 250 ml,
beakers, nominal volume 250 ml
glass capillaries for aeration (PASTEUR pipettes)
thermometer
oxygen meter
pH meter
deionized water or water of the same purity level
calcium chloride solution, c = 0.5 mole/l: 109.55 g calcium chloride-
    6-hydrate, $CaCl_2 \cdot 6\ H_2O$, pure, crystallized, are dissolved in deionized water. The solution is topped to 1 l with deionized water; 1 ml of the solution contains 0.5 mmole calcium ions.
magnesium sulfate solution, c = 0.5 mole/l: 123.25 g magnesium sulfate-
    7-hydrate, $MgSO_4 \cdot 7\ H_2O$, pure, crystallized, are dissolved in deionized water. The solution is topped to 1 l with deionized water. 1 ml of the solution contains 0.5 mmole magnesium ions.

sodium hydrogen carbonate solution, c = 0.1 mole/l: 8.401 g sodium hydrogen carbonate, $NaHCO_3$, highest purity, are dissolved in deionized water. The solution is topped to 1 l in deionized water. 1 ml of the solution, when added to 1 l water, increases the acid capacity of the water $K_{S\ 4.3}$ by 0.1 mmole/l.

hydrochloric acid (HCl), c = 1 mole/l: 1 ml of this solution, when added to 1 l water, reduces the acid capacity $K_{S\ 4.3}$ of the water by 1 mmole/l.

diluting water prepared from chlorine-free drinking water: Drinking water free from chlorine, with a calcium ion concentration of (2.2 ± 0.4) mmole/l and a magnesium ion concentration of (0.5 ± 0.1) mmole/l, can be used to dilute the waste water sample. The mole ratio between the calcium and magnesium ions should range at 4 : 1.

synthetic diluting water: 22 ml calcium chloride solution, 5 ml magnesium sulfate solution and 5.0 ml sodium hydrogen carbonate solution are diluted with deionized water to a total volume of 5 l. The diluting water is aerated until a constant pH value is achieved.

Test fish and test fish breeding

The orfe [Leuciscus idus (L.), gold variant = gold orfe] is used as test fish, an orange-yellow colored variety of the ide (Idus idus) occurring in rivers and lakes in central and northern Europe. The orfe reaches a length of 30 to 40 cm. Its spawning time exends from April to June.

The test fish should have an overall length of 5 to 8 cm, and a corpulence factor C of 0.8 to 1.1 g/cm³. Only healthy fish may be bred. The corpulence factor is calculated according to the equation:

$$C = \frac{100 \cdot m}{l^3} \qquad [g/cm^3]$$

where:

C = corpulence factor, in g/cm³

m = live weight of the fish, in g

l = length of the fish, measured from the tip of the mouth to the end of the caudal fine, in cm

Note: C does not correspond here to the density of the fish.

The test fish are not yet mature and not discernibly differentiated by sex. The sex of the animals is therefore not considered in this method.

The fish are kept in aerated drinking water free from chlorine, at 10 to 20° C. They are best kept in through-flowing water. Five animals at maximum should be kept in 1 l of water. When a stagnant water is used, it must be sufficiently recirculated, filtered, and frequently exchanged.

The fish should be fed with a suitable dry feedstuff (e.g. dry breeding foodstuff, grain size 0) as long as they are kept in this water.

When the temperature of the breeding water is below 18° C, the fish must be adapted for a period of 48 hours or more at 20° C.

The fish should have been bred for at least one week. Then their mortality rate is normally not higher than 1 % per week.

Procedure

The test water is prepared from the waste water and the required volume of diluting water as given in the table below. It thus contains 1 part by volume of waste water and x parts by volume of diluting water.

parts by volume diluting water x	diluting factor
1	2
2	3
3	4
4	5
5	6
7	8
9	10
11	12
15	16
19	20
23	24
31	32
39	40
47	48
63	64
79	80

The aquarium is filled with 5 l of the test water. The pH value of the test water is set to pH = 7.0 ± 0.2 by the addition of hydrochloric acid or sodium hydroxide solution. Then five orfes are set into each aquarium. During the test, the fish are no longer fed.

The test water temperature is (20 ± 1)° C. During the test, a minimum oxygen concentration of 4 mg/l O_2 must be maintained in the test water. This can be frequently achieved without aeration throughout the testing period. The testing period is 48 hours. A fish is deemed dead when no movement can be detected, even after touching the fish.

A reference test with 5 l diluting water is analogously made. When one or several fish die in the reference sample, the test must not be evaluated.

Fish surviving the test must not be used for any further test.

Evaluation

The smallest dilution factor G_F of the test water, at which all fish survive, is given as the test result.

Only integer dilution factors are indicated.
Example: dilution factor $G_F = 8$.

7.5 Assessment of the Analyses

The evaluation and assessment of the many test results obtained with waste water analyses requires a high degree of practical experience.

The established analytical data, in combination with the observations and measurements, must furnish information for the harmless elimination of the respective waste water.

To this end, the waste water occurring at the various discharge points must be primarily sampled and analyzed. The analysis of waste water aims at describing the physical, physicochemical, and microbiological composition and nature of the waste water and at determining substances or their concentrations or loadings that are harmful for the environment or that may govern the issuance of a licence to discharge waste waters into sewerages, clarification facilities, sewage plants, and natural waters. This also includes statements on other types of waste water removal measures, e.g. the spreading of waste water and sewage sludges on the soil or the introduction of sludges into sanitary landfills.

When the waste water analyses find substances that may interfere with the drainage, the clarification, or the spreading of the waste water for land treatment, internal measures must be taken at the point of waste water discharge. The respective recommendations are then to be made. Particularly, those pollutants must be withheld from the waste water that lead to corrosion in the waste water system or whose toxicity or gas release may disturb the waste water system. Finally, the waste water analysis must specifically identify those substances that must be expected to affect the biological waste water purification. These substances must then be particularly considered in the assessment of the results of the analyses.

7.5.1 Concentration or Load

The concentrations of the analyzed substances, which are established in the waste water samples taken, are generally given in mg/l or similar dimension. These concentrations alone are well suited to furnish information about batch loadings. The quantitative waste water measurement during sampling may establish a relationship with the pollution load of the waste water.

The established concentrations in the waste water and the determination of the pollution load may result in conditions made for the emission source or emitter, e.g. in the form of financial contributions or limits for the emission of certain substances.

These conditions in turn require internal measures of waste water treatment upstream of the point of discharge into the sewage plant or a receiving water. This is the reason why the establishment of both concentrations and loadings is particularly important in waste water technology. On the basis of the results of the analyses, certain substances in certain concentrations must be precluded from the discharge through waste water.

7.5.2 General Load of Waste Water

The load caused by organic and inorganic nutritive substances is particularly important in terms of an optimum biological clarification. Here, conclusions can be drawn from the established global parameters COD, BOD_5, TOC, DOC, nitrogen, and phosphorus. A varying load with organic and inorganic substances in the waste water is normally balanced by the various industrial and, above all, domestic discharges upstream of the sewage plant.

7.5.3 Pollution Load of Waste Water

When pollutants in relevant concentrations have been determined by sampling and analysis, measures are required to cope with these pollutants and to manage their effects on clarification and waste water drainage. Pollutants may be summarized as certain inorganic loads, e.g. heavy metals, high salt load, or load with organic pollutants such as pesticides or cyanides, with organic solvents or halogenized compounds. The toxicity analysis in the fish test and bacteria test also furnishes information about the pollution load of the particular waste water.

Sampling and analysis, especially the volumetric waste water measurement at various points up to the sewage plant, must determine whether dilution, e.g. with domestic waste water, will reduce the pollution of industrial discharges to an extent that a particular enterprise will not disturb the operation of a biological sewage plant or a third purification stage.

If such nondisturbance is not warranted, conditions must be made for the pollution emitter. The pollutants that interfere with a biological clarification must be removed from the waste water at the point of their discharge. In view of the comparatively lower waste water quantities, internal company measures at the pollutant discharge points are easier to modify and improve than those in the sewage plant, which is bound to manage the total of all pollutants of discharged waste waters.

7.6 Sewage Sludges

Sludges from waste water treatment plants, e.g. stabilized or nonstabilized sewage sludges, digested sludges, and sludges from industrial treatment facilities, must be eliminated without causing any harm. The application of sludges from municipal sewage plants with a low commercial waste water share in land treatment is still very important. With such applications, however, the standards for heavy metal concentrations must be definitely observed. When the sludges are ploughed into the soil, the heavy metals are very tightly bound so that a long-term accumulation must be expected from an excessively frequent application of loaded sludges. Sewage sludges must therefore be more frequently analyzed. Following various recommendations for a practically applicable routine analysis, covering zinc, copper, boron, molybdenum, and cobalt as so-called micro-nutritive substances and chromium, lead, nickel, arsenic, selenium, cadmium, and mercury as so-called pollutants, as the relevant elements, the new draft of a regulation governing the application of sewage sludges (to be promulgated by the German Federal Ministry of the Interior, version of August 1, 1981), the elements lead, cadmium, chromium, copper, nickel, mercury, and zinc are listed as relevant elements for the analysis.

7.6.1 Directions for Sludge Sampling and Decompositon for Heavy Metal Determination

Sampling

In consideration of the inhomogenous composition of sludges, larger quantities, normally 5 to 10 kg, are to be sampled. When homogenization prior to sampling, e.g. by thorough mixing, is not possible or when the sample cannot be taken from the flowing sludge, several partial samples must be taken. These partial samples are taken at different points of sludge occurrence or from the corresponding storage containers, using a suitable sampling container (sampling ladle or puncturing probe), depending on the consistency of the sludges. When the phases separate in liquid or semi-liquid sludges during sampling, the sludge must either be homogenized before sampling, or the ratio between the sedimented sludge and the water phase on top must be estimated. Then the partial samples must be taken in this ratio, and combined later on into a composite sample.

The samples must be transferred to the laboratory as quickly as possible in steamtight vessels, avoiding any heating. Experience has shown that it is favorable to analyze liquid or semi-liquid sludges one to two days after sampling at the latest. To inhibit or reduce digestion processes that may be accompanied by the formation of hydrogen sulfide or methane, it is recommended to store the sludge samples taken in a cooled condition (4° C).

When the sludges are also to be rated on their hygienic conditions, the sludge samples are to be filled into sterile vessels for the microbiological analyses. Sterile 500 or 1000 g wide-neck bottles have proven successful for this purpose.

Homogenization before sampling is normally not possible with very solid sludges. Therefore, the total puncture-proof sludge occurring must be sampled. A sampling scoop of ca. 1 m length is best suited for this application. It is made of stainless steel, slotted throughout its length, and ends in an acute tip at the lower end for better puncturing. A thread should be provided at the other end so that extension rods can be threaded with a matching connector when the sludge deposit is so thick that one puncturing operation is not sufficient to cover the total cross section through the sludge. After the extension of the puncturing probe, several sampling operations are to be carried out in order to take a sample throughout the cross section of the sludge deposit. The partial samples taken with the puncturing probe are also combined into a composite sample. A hand shovel should be used to homogenize these partial samples until parts of the sample can be filled for the laboratory analyses. According to experience, 5 to 10 kg of puncture-proof sludges must also be available for the analyses.

Decomposition

The treatment with aqua regia or nitro-hydrochloric acid is often applied to decompose sludges. In this process, the dry sludge residue is milled in a grinder to a grain size below 0.1 mm. Approximately 3 g of the prepared sample are subsequently dried in the drying cabinet for ca. 30 minutes, and after cooling in the desiccator, this quantity is weighed in the reaction vessel with an accuracy of 0.01 g. The weighed quantity is wetted with a few milliliters of distilled water and mixed first with 21 ml HCl, concentrated, and then 7 ml HNO_3, concentrated. The reaction solution is left at room temperature overnight and is then heated to boiling temperature for 2 hours at the reflux condenser.

Then, the volume is transferred into the measuring cylinder and topped up to 100 ml. The solution is filtered through an acid-proof paper filter, the first part of the filtrate is discarded, and the remaining filtrate is analyzed by suitable methods.

Another decomposition method has proven successful. It has also the advantage that only small quantities of chemicals are necessary that are of highest purity and are thus expensive:

The sludges are dried at 105° C (at 50° C in the analyses for mercury). Then 1 - 2 g are weighed onto a platinum dish, mixed with ca. 5 g p-toluene sulfonic acid, and left on the sand bath until the development of smoke has stopped. The mixture is incinerated in the muffle furnace at 450° C. Then the residue is extracted with ca. 10 ml HCl conc., transferred into a beaker, and filtered off. The filter is incinerated at 450° C, the residue is treated with a few ml HF/H_2SO_4 (1 + 1) and evaporated. The residue is taken up with HCl and the total solution is topped to 100 ml.

Particularly for zinc determination, a modification of the sludge analysis method has been developed, since this element is likely to cause high blind values resulting from chemicals and the air in the laboratory:
0.5 - 1 g of dry weighed substance are filled into a 100 ml polyethylene bottle. A mixture of 10 parts of hydrochloric acid, 3 ml hydrofluoric acid, and 10 ml bidistilled water is introduced into the bottle. The mixture is heated in the water bath to 80 - 90° C for 1 - 2 hours, with the plug inserted, is cooled, and a total of 100 g is set on the weighing equipment. An undissolved residue mostly remains in the bottle. Analyses have shown that these residues do not contain any zinc.

Sludges to be deposited

Municipal sewage sludges with heavy metal concentrations higher than permissible according to the regulation on agricultural application, and industrial sludges are normally stocked in sanitary landfills for domestic garbage or special wastes.

Deposited sludges may give rise to effects on surface waters and groundwater, and the seepages originating from the sludges have often a very unfavorable composition, which is normally even beyond the composition of waste waters. A harmless deposition therefore depends on the composition of these wastes, and also on specific soluble constituents that, under certain conditions, may result in water loading. When waste materials are deposited, seepage is always produced that is loaded with soluble inorganic and organic substances from the waste body. It is therefore necessary to know the water-soluble substances occurring under practical conditions, so that an assessment can be made as to whether deposition is possible without harm to the environment, and whether the seepage can be eliminated without harm.

As a result of the inhomogenous composition of sludges, considerable sample volumes are required. Normally, weighed quantities of 1 to 10 kg are processed to analyze the water extracts. Directions for the orienting analysis of such wastes and sludges are given below, with a description of how to produce rainwater and seepage extracts.

Directions for the analysis of wastes and sludges:

The following scheme is conceivable to analyze a homogenized sample:

drying loss at 105° C

total organic substance as loss at red heat at 450° C

total carbon, analyzed by incineration

decomposable organic substance, determined by oxidation with potassium dichromate

total nitrogen

calculation of the carbon/nitrogen ratio

total phosphorus

qualitative spectral analysis or qualitative X-ray fluorescence analysis of the residue from glowing for heavy metal determination

When potentially harmful heavy metals are determined in corresponding concentrations, they must be quantitatively determined as total heavy metals concentration, if necessary /55/.

Further analyses that may become necessary for wastes or sludges:

phenol-type substances

cyanides

extractable substances (oily and fatty substances)

surfactants (detergents)

sodium

potassium

calcium

magnesium

iron and manganese

ammonium

nitrite

nitrate

chloride

sulfate

total sulfur

In addition to these orientation analyses for wastes, the assessment of a particular water load by emanating seepage from waste deposits demands elution tests.

General

This test also requires an appropriate average sample volume, possibly a fractioned average sample.

The material to be analyzed is generally not crushed or ground but inspected in its original condition as initially deposited. In exceptional cases, a coarse crushing to 10 mm grain size at maximum may be necessary.

The residues are extracted with water in a quantity 10 times the original volume. Various modifications of methods have proven successful in practice, which are suggested below. The analysis of the water extracts is then made according to the directions given in the General Analytical Methods /55/.

1. General Principles of the Method:

1 to 10 kg of the wastes are to be extracted once or several times with rainwater and/or seepage in a quantity 10 times the original volume. The extractions are carried out at 20° C. For one extract, the wastes remain in contact with the elution solution for 24 hours. During the first 8 hours, the preparation with the solid wastes and the tenfold water quantity is agitated for 10 minutes every hour. When higher quantities are prepared, e.g. 5 kg wastes and 50 kg water or 10 kg wastes and 100 kg water, the extracts are prepared in suitable plastic containers that are rolled for 10 minutes every hour.

After 24 hours contact time, the excess water is separated, filtered through a coarse filter, and analyzed.

When a second, and if necessary a third extract, of the residues must be prepared to determine the extraction gradients the residues remaining undissolved in the first elution, must be filtered off and treated, without further drying, again under analogous conditions with the same quantity of elution liquid for 24 hours.

2. Elution liquids:

a) Rainwater

Deionized or distilled water is generally used to simulate rainwater.

b) Seepage (as it may occur in landfills)

Here, rainwater or distilled or deionized water is used when almost saturated with carbon dioxide. CO_2 is passed through water of ca. 4° C in a plastic barrel to saturate it with carbonic acid. After saturation, the water is slowly heated overnight to 20° C. Then the concentration of free dissolved carbon dioxide is determined according to the method described in the Collection of methods of water analysis /55/. Carbon dioxide concentrations from ca. 1500 to 2000 mg/l water are achieved provided that the test conditions described will be observed.

c) Diluted seepage

In this test, the water saturated with carbonic acid, prepared by the method described above, is diluted with distilled or deionized water to ca. 500 mg/l concentration of free dissolved carbon dioxide.

d) Synthetic seepages

The composition of such seepages should roughly correspond to genuine seepages. Such preparations have recently been suggested. Sufficient experience has not yet been gained with the use of such preparations.

Test program for the analysis of the water extracts from solid, liquid, or sludge wastes:

The aqueous extracts prepared by the methods described above are separately analyzed (rainwater extract, seepage extract with high carbon dioxide concentration, and seepage extract with low carbon dioxide concentration), and the elution gradient is determined if necessary.

The following parameters are determined in the aqueous extracts:
pH value

electrical conductivity at 20° C
redox potential
evaporation residue at 180° C
residue from glowing of the evaporation residue at 450° C
acid capacity up to pH 4.3 (m value)
acid capacity up to pH 8.2 (p value)
total hardness
carbonate hardness
non-carbonate hardness
oxidizability with potassium permanganate
    a) as potassium permanganate volume consumed
    b) as oxygen volume consumed, in mg O_2/l
oxidizability with potassium dichromate (chemical oxygen demand, COD), given in mg/l
biochemical oxygen demand (BOD_5), given in mg/l
chloride
sulfate
phosphate
nitrate
nitrite
ammonium
Qualitative spectral analyses of the evaporation residue to check whether toxic heavy metals are present in corresponding concentrations. If necessary, the determined heavy metals must be subjected to a quantitative analysis in the seepage extracts.

cyanides
    a) total cyanide
    b) easily releasable cyanide
phenols
    a) total phenols
    b) phenols volatile with steam
surfactants (detergents)
oily and fatty substances extractable with petroleum ether
polycyclic aromatic substances
Further analyses as required in view of the origin of the wastes and certain pollutants suspected.

The first water extract from hydrated sludges is obtained by separation of the water by way of filtration or centrifuging. Then the sludges are extracted a second or third time, as described above.

The analysis results are the basis on which a judgement can be made as to whether the groundwater or surface water might be loaded by the elution of such wastes. At the same time, forecasts are possible about the risk of precipitations caused by substances with an alkaline reaction, in the course of the flow path, e.g. precipitations of iron, manganese, calcium, magnesium, or heavy metals.

Moreover, the COD/BOD_5 relationship permits statements about the presumable accessibility of dissolved organic substances to a biological decomposition.

The same demands as those applying to drinking water can, of course, not be made in relation to seepage from waste dumps. A comparison to the parameters permissible for waste water could be conceivable. In certain cases, a higher seepage load may be permitted when such a seepage is drained into the groundwater and when a biological decomposition in the aerobic or anaerobic range or precipitations and sorptions can be expected in consideration of long duration resulting from a low flow rate (less than 1 m/day). Sources of drinking water from groundwater should not be located in the catchment area of such seepage. A minimum distance of ca. 1000 m can be estimated.

When wastes are dumped in a sanitary landfill with a sealed bottom, an analysis following the above-described scheme leads to the conclusion of a prediction whether the organic seepage fractions are accessible to biological decomposition in a sewage plant over the duration there. Particular attention must also be given to the heavy metal concentration that may disturb the sewage plant operation.

Reference is made to the specialized literature for further information about this highly complicated problem.

7.6.2 Sludge Volume and Sludge Index

Apart from the concentration of dry sludge substance and the sludge volume index, the sludge volume (sludge sedimentation volume) VS_R is one of the physical characteristics of activated sludge.

The term sludge volume applies to that volume in ml that the activated sludge produces from 1 l of sludge/water mixture (generally taken from an activation basin) after undisturbed sedimentation in a transparent cylinder jar for 30 minutes.

The sludge sedimentation volume permits only a highly conditional statement about the dry substance concentration in the sludge. But this value is more relevant as an auxiliary parameter for the calculation of the sludge index.

The sludge index I_{SV} is a measure of the volume which 1 g of dry substance reaches as sludge in 1 l of a sludge/water mixture.

The index proceeds from the equation:

$$I_{SV} = \frac{S}{T} \qquad [mg/l]$$

where:
- S = sludge volume fraction of the activated sludge, in ml/l
- T = mass concentration of the dry residue from the activated sludge, in g/l
- I_{SV} = sludge index of the sludge/water mixture (activated sludge), in ml/g

Particularly with high sludge concentrations (ca. 250 ml/l) the sedimentation process may be disturbed by the wall of the vessel and by the mutual blocking of the individual flakes. The sample is then diluted as described below (Procedure).

Considerable differences in the temperature of the sample and the environment may also result in disturbance (convection, formation of gas bubbles). When the differences are higher than 2° C, it is recommended to place the sedimentation cylinder with the sample into a bucket that is also filled with the sample liquid.

Equipment

measuring cylinder, nominal volume 1 l, internal diameter 6 to 7 cm, made of glass or transparent plastic material

Procedure

The measuring cylinder is filled with the sample of the sludge/water mixture up to the 1 l mark. A ladling instrument is recommended that has a volume of 1 l up to the rim. Such an instrument avoids errors resulting from separation processes. When the sample has been left in one place, without vibrations, for 30 minutes, the sludge volume is read by the height of the sludge level (boundary between the sludge and the excess liquid). The determination must be repeated when the sludge volume exceeds 250 ml. For this second analysis, the sample must be diluted with water produced from the topping liquid of the same sludge/water mixture, or from the discharge of the final clarification basin, in a ratio of 1 : 1, 1 : 2 or 1 : 3. The sludge volume then red is multiplied with the dilution factor 2, 3, or 4

The dry residue of the sludge/water mixture is determined then.

Indication of the results

The values are rounded to 10 ml/l.

When the sample had to be diluted, the indicated value must be followed by the volume read from the diluted sample and by the dilution factor, both given in brackets.

Example:

Sludge volume (VS_R)   180 ml/l
sludge volume   510 ml/l (170 ml x 3)

8 REFERENCES

1. ABWASSERTECHNISCHE VEREINIGUNG (ATV): Lehr- und Handbuch der Abwassertechnik. Bd. I, 2. Auflage, Verlag Wilh. Ernst und Sohn, Berlin, München, Düsseldorf, 1973.
2. ABWASSERTECHNISCHE VEREINIGUNG (ATV): Lehr- und Handbuch der Abwassertechnik. Bd. II., 2. Auflage, Verlag Wilh. Ernst und Sohn, Berlin, München, Düsseldorf, 1975.
3. ABWASSERTECHNISCHE VEREINIGUNG (ATV): Lehr- und Handbuch der Abwassertechnik. Bd. III, 2. Auflage, Verlag Wilh. Ernst und Sohn, Berlin, München, Düsseldorf, 1978.
4. ABWASSERTECHNISCHE VEREINIGUNG (ATV): Arbeitsblätter:
   - A 102: Industrie- und Gewerbebetriebe, Allgemeine Planung einer Abwasserableitung und Abwasserbehandlung, 1971.
   - A 104: Lederbetrieb. Planung einer Abwasserableitung und Abwasserbehandlung (Grundlagen und Hinweise), 1957.
   - A 105: Hinweise auf die Abwasserableitung im Hinblick auf die Reinhaltung der Gewässer, Mischverfahren/Trennverfahren, 1957.
   - A 107: Hinweise für das Ableiten von Schlachthofabwasser in ein öffentliches Kanalnetz
   - A 108: Maßnahmen zum Schutz der Abwasseranlagen gegen gefährdete Stoffe, 1971.
   - A 112: Hinweise für das Ableiten von Abwasser aus fleisch- und fischverarbeitenden Betrieben in ein öffentliches Kanalnetz, 1966.
   - A 115: Hinweise für das Einleiten von Abwasser aus gewerblichen und industriellen Betrieben in eine öffentliche Abwasseranlage, 1970 (Neuer Entwurf 1980).
   - A 116: Abwasser aus landwirtschaftlichen Betrieben, 1970.
   - A 117: Richtlinien für die Bemessung, die Gestaltung und den Betrieb von Regenrückhaltebecken, 1977.
   - A 118: Richtlinien für die hydraulische Berechnung von Schmutz-, Regen- und Mischwasserkanälen, 1977.
   - A 127: Richtlinien für die statische Berechnung von Entwässerungskanälen und -leitungen, (Entwurf 1978).
   - A 128: Richtlinien für die Bemessung und Gestaltung von Regenentlastungen in Mischwasserkanälen, 1977.
   - A 129: Abwasserbeseitigung aus Erholungs- und Fremdenverkehrseinrichtungen (Entwurf 1978).
   - A 134: Planung und Bau von Abwasserpumpwerken mit kleinen Zuflüssen (Entwurf 1980).
   - A 135: Grundsätze für die Bemessung von Tropfkörpern und Scheibentauchkörpern in einstufigen biologischen Kläranlagen, (Entwurf 1980).
5. AFRICAN ENVIRONMENT, Special Report 1. Problems and Perspectives. International African Institute in association with the Environmental Training.
6. ALTER, AMOS, J.: Sewerage and Sewage Disposal in Cold Regions. USA Army Publications, 1969.
7. ANDERSON, G.K./WHO: Slaughterhouse Wastes. (Industrial Wastes Guide Nr. 12). WHO/WD. 73. 12 Geneva.

8. BABBIT, H.E., BAUMANN, E.R.: Sewerage and Sewage treatment. New-York, John Wiley] Sons, Inc. 1958.
9. BAYERISCHES LANDESAMT FÜR WASSERWIRTSCHAFT: Grundwerte für die Bemessung von Begutachtung der biologischen Stufe von Kläranlagen mit Ausbaugrößen über 500 EGW (Merkblatt II -2/77).
10. BAHR, LUTZ, A.: Verstädterung in der dritten Welt - Herausforderung an Politik und Verwaltung. Bonn, Dissertation 1974.
11. BERG, B.: An Integrated Approach to the Problem of Viruses in Water. Pre Publication Copy, Cincinnati 1970.
12. BERNHARD, H., WILHELM, A.: Einfluß von Algenstoffwechsel- und Abbauprodukten auf Flockungsprozeß. GWA 8 (1972), Schriftenreihe des Inst. für Siedlungswasserwirtschaft der RWTH Aachen, 1972.
13. BERNHARD, H.: Die Phosphoreliminierung durch chemische Abwasserbehandlung. GWA 17 Schriftenreihe des Instituts für Siedlungswasserwirtschaft der RWTH Aachen, 1975.
14. BHASKARAN, T.R./WHO: Tannery Wastes (Industrial Wastes Guide Nr. 7). WHO/WD/ 73. 14 Geneva.
15. BHASKARAN, T.R./WHO: Brewery Wastes (Industrial Wastes Guide Nr. 8). WHO/WD/ 73. 15 Geneva.
16. BHASKARAN, T.R./WHO: Cotton Textile Wastes Industrial Wastes Guide Nr. 9). WHO/WD/ 73. 16 Geneva.
17. BISCHOFSBERGER, W.: Neue Bemessungswerte der aeroben biologischen Abwasserbehandlung. GWA 15. Schriftenreihe des Inst. f. Siedlungswasserwirtschaft der RWTH Aachen, 1974.
18. BISCHOFSBERGER, W.: Weitergehende Reinigungsleistung und Kosten verschiedener Behandlungsanlagen. GWA 29, Schriftenreihe des Inst. f. Siedlungswasserwirtschaft der RWTH Aachen.
19. BILLMEIER, E.: Verbesserte Bemessungsvorschläge für horizontal durchströmte Nachklärbecken von Belebungsanlagen. Berichte aus Wassergütewirtschaft und Gesundheitsingenieurwesen, TU München, H. 21, 1978.
20. BILLMEIER, E.: Verbesserte Sedimentation. ATV-Fortbildungskurs A 4, 1978.
21. BILLMEIER, E.: Nachklärbecken. ATV-Fortbildungskurs B 1, Laasphe 1979.
22. BILLMEIER, E.: Leistungssteigerung der Nachklärung von Belebungsanlagen. Korrespondenz Abwasser, 26. Jg., H. 1, 1979.
23. BILLMEIER, E.:Feststoffretentionen in horizontal durchströmten Zwischen- und Nachklärbecken. GWA Bd. 42, Aachen 1980.
24. BILLMEIER, E.: Wirkungsweise und Bemessung der Nachklärung bei der biologischen Abwasserreinigung. Österreichische Abwasser-Rundschau, Folge 3, 1980.
25. BLITZ, E.: Epurarea apelor uzate menajere si orăşenşti (Die Behandlung der häuslichen und kommunalen Abwässer). Editura Didactică, Bukarest, 1966.
26. BLITZ, E.: Proiectarea Canalizărilor (Entwurf der Abwasserbeseitigungsanlagen). Bukarest, Editura Didactică, 1971.
27. BLITZ, E.: Cours d'Assainissement Rural. Ecole Mohammadia d'Ingénieurs, Rabat, 1978.
28. BÖHNKE, B.: Abwasserabgabe als Instrument zur Erreichung einer ausreichenden Wassergüte. GWA 19. Schriftenreihe des Institutes für Siedlungswasserwirtschaft der RWTH Aachen, 1975.

29. BÖHNKE, B.: Möglichkeiten der Abwasserreinigung durch das "Adsorptions-Belebungsverfahren". Verfahrenssystematik Versuchsergebnisse. GWA 29. Schriftenreihe des Instituts für Siedlungswasserwirtschaft der RWTH Aachen, 1978.
30. BÖHNKE, B.: "Grundzüge der Abwasserbehandlung". Skriptum zur Vorlesung, Aachen, 1977.
31. BÖHNKE, B.: Technische und gesetzliche Entwicklung der Abwassertechnik in Deutschland. Wissenschaft und Umwelt Nr. 1/1980.
32. BÖHNKE, B., DOETSCH, P.: Entwicklung von Daten für die Planung und Ausführung von Wasserversorgungsanlagen, Teil 2: Ermittlung von Kostendaten. Inst. für Siedlungswasserwirtschaft der RWTH Aachen, 1981.
33. BÖHNKE, B., DOETSCH, P., STROHMEYER,: Untersuchung über die Erweiterungsmöglichkeit von kommunalen Abwasserbehandlungsanlagen im Hinblick auf die Mindestanforderungen der Verwaltungsvorschrift nach § 7a WHG. Ministerium für Ernährung, Landwirtschaft und Forsten des Landes Nordrhein-Westfalen, 1981.
34. BORCHERT, O.: Abwässer aus der chemischen Oberflächenbehandlung. Köln-Braunsfeld, Vlg. R. Müller, 1966.
35. BORN, R.: Schlammkonditionierung. ATV-Fortbildungskurs A 2, 1977 in Eppingen.
36. BREER, C.: Die Hygienisierung von Klärschlamm mit Hilfe verschiedener Behandlungsverfahren. Wasser - Energie - Luft 72, 1960.
37. BREER, C., HESS, E., KELLER, U.: Soll Klärschlamm vor oder nach dem Ausfaulen pasteurisiert werden? Gas - Wasser - Abwasser 59, 1979.
38. BUCKSTEEG, W.: Die Abwässer von Brauereien, Brennereien und Hefefabriken. "Spezielle Abwässer aus kommunalen, gewerblichen und landwirtschaftlichen Betrieben". Heft 70, Essen 1960.
39. CALLELY, A.G., FORSTER, G.F., STAFFORD, D.A.: Treatment of Industrial Effluents. Hodder and Stoughton, London, Sydney, Auckland, Toronto, 1977.
40. CLARKE, N.A., STEVENSON, R.E., CHANG, S.L., KABLER, P.W.: Removal of Enteric Viruses from Sewage by derivated Sludge Treatment. Am. J. Publ. Hlth 51 118, 1961.
41. COUSIN, B.: Die Druckentwässerung, eine neue Technik der Abwasserbeseitigung. Wasser und Boden, 10. 1972.
42. DEGREMONT HANDBUCH: Wasseraufbereitung, Abwasserreinigung. Bauverlag GMBH, Wiesbaden, Berlin 1974.
43. DIN: - 1986, Blatt 3, 1963; 2425; 4032; 4040; 4041; 4045; 4263; 19540; 19650
44. DÖNGES, H.J.: Die Schlammbeseitigung auf der Emscherflußkläranlage. Münchner Beiträge zur Abwasser-, Fischerei- und Flußbiologie Bd. 13. Vlg. R. Oldenbourg, München 1966.
45. DÖNGES, H.J.: Bau- und Betriebskosten von Kläranlagen. GWA 4. Schriftenreihe des Instituts für Siedlungswasserwirtschaft der RWTH Aachen, 1971.
46. DRAPEAUX, A., JANKOVIC, S.: Manuel des Microbiologie de l'environnement. Organisation Mondiale de la Santé, Genève, 1977.
47. ENGELBART, F.: Diskussionsbeitrag über die Reinigung von Restabwässern der Zuckerfabriken in Stapelteichen. Münchner Beiträge zur Abwasser-, Fischerei- und Flußbiologie Bd. 11, 1964, Verlag R. Oldenbourg.
48. FAIR, M.G., GEYER, G.J.: Wasserversorgung und Abwasserbeseitigung. Vlg. R. Oldenbourg, München, 1962.

49. FEACHEM, R., MC. GARRY, M.: Water, Wastes and Health in hot climates. John Wiley] Sons, 1978.
50. FINCH, J.: The Planing and Organisation of Industrial Wastes Control Programmes (Industrial Wastes Guide Nr. 1). WHO/WD/ 70. 6 Geneva
51. FLINTOFF, F.: Management of Solid Wastes in Developing Countries. WHO Regional Publications South-East Asia Series Nr. 1., WHO, New-Delhi, 1976.
52. FORSTNER, M.J.: Probleme bei der Abwasserbeseitigung aus fleischverarbeitenden Betrieben. Münchner Beiträge zur Abwasser-, Fischerei- und Flußbiologie, Bd. 13, 1966.
53. FRANKE, G.: Abwässer, Schlämme sowie feste Rückstände der chemischen Industrie - Aufbereiten, Verbrennen, Deponieren. GWA 21. Schriftenreihe des Instituts für Siedlungswasserwirtschaft der RWTH Aachen, 1976.
54. FRAUNHOFER GESELLSCHAFT - INSTITUT FÜR SYSTEMTECHNIK UND INNOVATIONSFORSCHUNG: Entwicklung von Kriterien zur Prüfung des Umweltschutzes. Schlußbericht Teil I, Juli-Dez. 1975.
55. FRESENIUS, W., SCHNEIDER, W.: Methodensammlung zur Wasseruntersuchung. Im Auftrag der Gesellschaft für Technische Zusammenarbeit (GTZ) 1977 ("Méthodes pour l'Analyse des Eaux". 1977).
56. GEISLER, W.: Kanalisation und Abwasserbeseitigung. Berlin, Springer Vlg., 1933.
57. GEYER, J.C.: Industrial Wastes - Textile Industrie. Industr. Engineering Chemistry 39, 1947.
58. GLOYNA, E.F.: Stabilization Ponds for Waste Waters. WHO Geneva, 1972.
59. GOULD, R.H.: New Developments in Sewage Disposal. New-York City Sewage Works, Journal 13, 1942.
60. GRAU, A., KÖHLHOFF, D., RITTER, K.: Untersuchung zur Aufstellung technischer Bestimmungen für den Bau, den Betrieb und die Unterhaltung von Abwasseranlagen in Nordrhein-Westfalen. Abschlußbericht, Teil II. Institut für Siedlungswasserwirtschaft der RWTH Aachen, 1978.
61. GUEREE, H.: Pratique de l'Assainissement des Agglomerations Urbanes et Rurales. Paris, Ed. Eyrolles, 1965.
62. HALBACH, A.J. u.a.: Industrialisierung in Tropisch-Afrika. Weltform Verlag, München, 1975.
63. HAMM, A.: Die Abwasserbeseitigung bei Lederfabriken. Münchner Beiträge zur Abwasser-, Fischerei- und Flußbiologie, Bd. 10, München, Vlg. R. Oldenbourg, 1964.
64. HARTINGER, L.: Abwasserreinigung in der metallverarbeitenden Industrie. Abtrennung und Beseitigung von Schlämmen galvanischer Betriebe. IWL-Forum 64, Köln 1965.
65. HARTINGER, L.: Abwasserreinigung in der metallverarbeitenden Industrie. Entstehung, Abtrennung und Beseitigung von Schlämmen galvanotechnischer Betriebe. IWL Forum 65.
66. HEGMANN, W.: Übersicht über Entwässerungsverfahren. ATV-Fortbildungskurs c/2 in Kaiserslautern, 1981.
67. HELMER, R., SEKOULOV, I.: Weitergehende Abwasserreinigung. Mainz-Wiesbaden, Deutscher Fachschriften-Verlag, 1977.
68. HENNIG, A., POPPE, S.: Abprodukte tierischer Herkunft als Futtermittel. Stuttgart, F. Enke Verlag, 1976.

69. HÖRLER, A.: Canalisation. Zürich, EPF Zürich, 1966.
70. HUBER, L.: Neuere Entwicklungen bei der biologischen Reinigung von Erdölraffinerieabwässern. Münchner Beiträge, Bd. 11, 1964.
71. HUSMANN, W.: Der heutige Stand der Abwasserreinigung der Kohlenindustrie. Münchner Beiträge zur Abwasser-, Fischerei- und Flußbiologie, Bd. 11, Vlg. R. Oldenburg, München 1964.
72. HWWA-INSTITUT FÜR WIRTSCHAFTSFORSCHUNG HAMBURG: Promotion of Small Scale Industries in Developing Countries. Report Nr. 4/1978.
73. IMHOFF, K. u. R.R.: Taschenbuch der Stadtentwässerung. 25. Auflage R. Oldenburg Verlag, München, Wien.
74. IMHOFF, K.R.: Betriebserfahrung mit Schönungsteichen. GWA 25. Schriftenreihe des Instituts für Siedlungswasserwirtschaft der RWTH Aachen, 1978.
75. IMHOFF, K.R., KALBSKOPF, K.H.: Bemessungsgrundlagen für biologische Kläranlagen. Technische Akademie Wuppertal, e.V. 1977.
76. JACOBITZ, K.N.: Die Ableitung industrieller Abwässer in öffentliche Entwässerungen und die gemeinsame Behandlung mit häuslichem Abwasser. Dissertation, TH Darmstadt, 1965.
77. KALBSKOPF, K.H.: Über den Absetzvorgang in Sandfängen. H. 24 der Veröffentl. des Instituts für Siedlungswasserwirtschaft, Hannover, 1966.
78. KALBSKOPF, K.H., LONDONG, D.: Entwicklungstendenzen bei Kläranlagen nach dem Belebungsverfahren für kleine Gemeinden. GWF 111, 455, 1970.
79. KAUFHOLD, W.: Thermische Konditionierung, auch unter Berücksichtigung der Wirtschaftlichkeit. ATV-Fortbildungskurs c/2 in Kaiserslautern, 1981.
80. KEHR, D., RÜFFER, H., MÜHLE, K.A.: Vier Jahrzehnte biologische Reinigung des Abwassers eines Margarinewerkes. GWF 110 44, 1969.
81. KEMPA, E.S.: Verfahrensverbesserungen überlasteter Kläranlagen in der VR Polen. GWA 19, Schriftenreihe des Instituts für Siedlungswasserwirtschaft der RWTH Aachen, 1975.
82. KLING, W.: Phosphate als Bausteine moderner Waschmittel. Münchner Beiträge, Bd. 12, 1965.
83. KNOLLMANN, R.: Die biologische Abwasserreinigung im Versuch, Entwurf und Betrieb. H. 28 der Veröffentlichung des Institutes für Siedlungswasserwirtschaft, 1968.
84. KÜHLHOFF, D.: Möglichkeiten der Deponien (unter Berücksichtigung bodenmechanischer Aspekte). ATV-Fortbildungskurs c/2 in Kaiserslautern, 1981.
85. KOZIOROWSKI, B., KUCHARSKI, J.: Industrial Waste Disposal. Pergamon Press, 1972.
86. KRAFFCZYK, K.: Abwässer aus der Metallverarbeitung und Metalloberflächenbehandlung. Köln, IWL Forum 64.
87. KRAUSE: Schadstoffminderung durch innerbetriebliche Möglichkeiten. Fortbildungskurs für Führungskräfte der Siedlungswasserwirtschaft der ATV Tagung Eppingen.
88. KRAUTH, K.: Beschaffenheit von Straßenoberflächenwässern. Stuttgarter Berichte zur Siedlungswasserwirtschaft, München, R. Oldenbourg Verlag 1979.
89. LAMB, J.C.III/WHO: Pulp and Paper Manufacturing Wastes (Industrial Wastes Guide Nr. 5). WHO/WD/ 72. 11 Geneva.

Barrel cleaning water, 840
Basin rettery, 111
Bath water, 222
Bathing waters, 449, 451-452
Batter boards, 367, 368
Battery charging station, 603
Battery keeping, 199
Beaches, 6, 22
Bedding of pipes, 385-397
Beechwood, 129, 137
Beer production, 166-169
Beet flume and washing waters, 154, 156, 831, 832
Biological slime (bacterial substrate), 583, 592, 596, 597, 639
Biological waste water treatment methods, 474, 476, 575-737, 805, 811, 839
   artificial, 611
   combined, 613, 809
   natural, 610
Biomass, 582, 592, 597, 682, 748
Biotope, 669
Blast furnace gas, 60, 64, 776
Bleaching plants, 432
Blood, 5, 104
Blood processing, 175
Blow-out columns, 808
Blue algae, 576

Blue water (eau celeste), 119
Bottle cleaning, 166, 834, 840
Bottle material for sampling, 939-942
Bottling methods, samples, 921
   composite samples, 922
   equipment, 934
   random samples, 921-922
Brace spacing, 376-377
Bracing, vertical, 382
Brake lining wear, 209
Branch sewer, 253, 260
Breakpoint chlorinating, 763-764
Breweries, 148, 166-169, 428-429, 558
Brick sewer, 396-397
Bridge clearing device, 507
   mobile, 528
Bridge mounted sludge clearers, 508
Brush aerator, 566, 689
Buffer capacity (see Acid binding capacity)
Builder, 94, 95
Buna rubber, 101
Buoyancy of suspended particles, 500
Butter production, 159
Butter washing water, 5, 159
Buttermilk, 159

C

CNP relation, 418
COD, 3, 102, 192, 201, 445-447, 475, 711
Cadmium, 565, 791
Cadmium red, 88
Cadmium yellow, 88
Calcium, 466
Calcium chloride, 820
Calcium hydroxide, 574-575, 782, 820
Calcium soda process, 770
Calculation,
   of commercial and industrial waste water, 267
   of domestic (municipal) sewage volume, 265-266
   of domestic and public water demand, 25-28
   of dry weather flow, 267-268
   of earth pressure against

   trench supports, 372-375
   of flow volume in a sewer, 269, 276
   of foreign water volume, 266-267
   of irrigation systems, 621-625
   of overflows, 309-318
   of pipelines, 284-321
   of precipitation water, 270-275
   procedure for a sewerage system, 276-277
   of pumps, 362
   of settling basins, 513-514, 522, 530, 538
   of sheeting support elements, 375-378
   of slow filters, 490-491
   of solids content of recir-

culated sludge 727
of trench width, 365-367
Calculatory section, 273
Camping and tourist grounds,
22, 245, 247, 248
Candle filter, 489, 847
Cane sugar processing, 156-158
Canning factories, 53, 179-183,
411, 486, 838
Capacity of purification methods, 475
Carbon determination, 1018, 1022
Carbonic acid, gassing, 567
Carnallite, 82, 139
Carrousel method, 878-879
Cast iron, 64
Cast iron piping, 343
Catch grate, 480
Catching device, 236
Cation exchanger, 564-565
Cell filter, (see Suction filter)
Cell production, 713
Cellophane, 812
Cells, (see Bacteria)
Cellulose, 130, 859
acetate membranes, 757
fiber substances, 116
viscose method, 117
Cement production, 61-62
Centrifloc sludge blanket system, 549
Centrifuges, 819, 836, 839, 895, 898-900
Ceramic industry, 62, 486, 550, 774, 775
Chain remover, 507, 529
Chamber filter presses, 903
Chamois dressing, 105
Chamois leather, 105
Channel rettery, 111, 112
Channels, open artificial, 298
Checking of prescribed parameters, 460-463
Cheese production, 147, 160, 429
Chemical industry, 79-85, 486, 550, 565, 601
Chemical oxygen demand, (see COD)
Chemical pulp production, 130, 433, 492, 550, 821-822
Chemical water purification, (see waste water treatment)
Chemicals,
utilization of, 568-579

Chemosorption, 555
Chlorination of waste water,
92, 501, 809, 810, 830,
837, 838, 842
Chlorine, 563, 777, 808, 837, 838
compounds of, 145
Chlorocresoles 145
Chlorophenols, 145
Cholera, 8
Chromate, 784
detoxication of, 784-787, 793
Chrome, 565
Chrome passivation, 65
Chrome tanning, 105
Chrome yellow, 88
Chromic acid, 563
Cigarette paper, 133
Ciliates, 595-596, 639
Circular cross section, 339
Circular profiles, 290, 291
Circulation basin, 878
City drainage, (see sewerage system)
Clarification ponds, (see Polishing ponds)
Classification of settling basins, 525
Clay soil, 466
Cleaning, manual, 480
Cleaning of sewers, 407
Climatic conditions, 468, 469
Cloth production, 122, 820-821
Clouds, 204
Coal,
dressing, 76, 502
mining, 74-76, 229, 798
pulp method, 612
washing, 76, 436, 799
Coarse separation, 479-495
Cocoon processing, 115
Coke dust water, 78
Coking works, 77, 219, 559, 601, 799
Colcothar, 89
Coli bacteria, 17, 616
Collecting channels, 501
Collecting pits, 235, 248
Collection basin, 833, 835
Collection of waste water, 234
Collector (see sewer)
Collector basin, 253, 254
Collector funnels, 822
Collieries, 218
Combined sewerage systems, 211, 237, 241-244

Combined two-step waste water treatment, 702-713
Comminutors, 480, 484, 844
Comparison, separated vs combined method, 242-243
Composition,
    of basic materials for viscose silk or rayon manufacture, 117
    of coal pit waters, 75
    of cyanide bath liquors, 72
    of digestion waters, 873
    of domestic sewage, 19, 41-42, 113
    of filtered waters from thermally treated sludges, 897
    of flax retting wastes, 114
    of hydrochloric pickling wastes, 69
    of milk, 158
    of oil well waters, 139
    of retting waters, 112-113
    of silage waste water, 195
    of slaughterhouse waters, 175
    of soap production waters, 93
    of sewage sludges, 855
    of textile waste waters, 124-125
    of waste water
        from production of non-alcoholic beverages, 165
        from artifical silk production, 120
        from bottle cleaning, 166
        from brass works, 69
        from breweries, 169
        from brown mechanical pulp mills, 129
        from canning factories, 178-183
        from condensation of artificial resins, 101
        from different food processing industries, 147-149
        from electroplating plants, 73
        from fish canning, 178
        from fish meal production, 203
        from food and luxury item industry, 150-151
        from fruit and vegetable canning, 179-182
        from industry and commerce, 58-60
        from livestock raising in stalls, 194
        from margarine production, 172
        from oil refineries, 143
        from perlon production, 102
        from potash final liquors, 81
        from potato flour production, 185
        from potato starch production, 186
        from public buildings, 19, 20
        from rice starch production, 189
        from sugar beet processing, 157
        from synthetic detergent production, 96
        from wheat starch production, 186
        from white mechanical pulp mills, 129
        from wine production, 192
        from wool washing, 122
Composting,
    in bio-reactors, 909
    of digested sludge, 908-909, 914
    in silos, 908
    of sludge in combination with garbage, 914
Compressibility of liquids, 279
Concentration equilibration, (see Equilibration basins)
Concrete, 394-396
    application of, 396
    encasements, 387-388
Concrete pipes,
    reinforced, 338-340
    types of, 321
Condensation water, 811, 821, 835, 838
Conduits,
    closed (material), 338-344
    of in situ concrete, 341
    open, 337-338
Confectionery industry, 149
Connecting sewers, 260
Connections, conduits,
    to manholes, 394
    to other structures, 389
Constriction, 315-317
Construction costs, 242, 247
    of settling basins, 531

Construction drawings, 257, 330-335
Construction material for sewerage systems, 335, 336
Construction stages,
    brick masonry piping, 396
    reinforced concrete piping, 395
Construction work,
    begin of, 364-369
    regulations of, 365
Consumer goods industry, 51
Consuming substances, 419
Contour maps, 257
Control basins, 539
Control,
    of direct and indirect discharge, 411
    of emissions values,
        inlet to treatment plants, 460-462
        outlet of treatment plants, 462
    of immission values, 462-463
    of natural waters, 411, 419-420, 462-463
    of pollution parameters, 411, 460
        of prescribed parameters, 460-463
Convertor, 778
Cool waste water, (see Cooling water)
Coolants, 66
Cooling towers, 777
Cooling water, 53, 117, 267, 419, 427, 558, 807, 810, 811, 813, 838
Copper, 3, 71-72, 120, 573, 565, 818
    content of sludge, 910-911
Corrugated plate separator, 510
Cosmetic industry, 87

Costs, 234, 242, 251, 253, 336, 262
    construction of sewer trench, 262
Cotton bleaching, 120-121
Cracked gases, 144
Criteria,
    for choice of drainage method, 241, 254-255
    for choice of drainage system, 251-254
    for choice of receiving water, 258
    for design of waste water disposal system, 251
    for pollution of waste water, 3
    for sewerage system planning, 250
Cross network, 239
Cross section, sewers,
    egg form, 287, 292
    horseshoe form, 287, 292
    normal, 296
Crude oil production, 138-139, 828
Crude sludge, organic matter contained, 860
Crystallization, 567
Culverts, 236, 237, 238, 307-309
Cumulonimbus clouds, 204
Cupola blast furnace, 778
Cuprammonium process, 119, 818
Cutting oils, 560
Cyanic complexes with heavy metals, 788-794
Cyanide, 71, 73, 78, 604, 777, 787-794
    detoxication of, 792-793
Cyanogen chloride, 789
Cyanogen compounds, toxic, 73
Cyanosis, 620
Cylindrical aeration, 689
Cymenes, 130

D

DDT, 620
DIN 19540 (Dimensioning program of circular profiles), 288
DIN 19650, 226
DIN 2425, 330-334
DIN 4032 (Types of concrete pipes), 321, 404

DIN 4040, 4041 (Fat separators, construction principles), 503-504
DIN 4045, 52
DIN 4263 (Standard cross section forms), 288
DOC (dissolved organic carbon) 960

DORRCO method, 693
Daily peak flow rate, 265
Daily waste water flow, 30
Dairies, 159-163, 429, 833
De-oiling, 66, 801, 806
Dechlorination, 834
Decoloring, 541, 755, 806, 815, 838
Decomposition, 609, 673, 750
   in activation facilities, 586
   aerobic, 586, 595-713
   anaerobic, 713-737, 859, 861
   in biological disk filters, 586
   of biological substances (BOD5), 353, 499-500, 560, 561, 562, 586, 596,
   in biological trickle filters, 586
   capacity of,
      in activation basins, 697
      in biological filters, 652
      in treatment plants, 475
   of carbon compounds, 591, 601
   of detergents, 775
   of flavor, 557
   of nitrogen in municipal sewage, 593
   of odorous substances, 92, 557, 809
   of organic compounds, 591, 601
   of phenols, 799-800
   sludges, 859-863
   reactions, 586-597
Deep filters, 488, 491
Definitions, 235-240
Degermination, (see Disinfection)
Degreasing, 65-66, 805, 820, 840
Dekanter centrifuge, 899
Delay coefficient for rainwater runoff, 273
Denitrification, 591-593, 841
   retardation of, 920
   sludge, 592, 841
Density,
   measurement of, 948-950
   of a homogeneous body 279
   of waste water, 523
Density flows, 601
Deodorization, 541, 842
Dephenolizing, 566, 612, 799, 801-802, 813
Deposits,
   of clarification sludge, 912-913
   of screened material, 843
Depot, 843, 903
Depth of sewers, 324, 325
Desalination, 769-772
   by distillation methods, 769
   by electrodialysis, 770
   by hyperfiltration, 771-772
   by ion exchange methods, 770
   by lime-soda method, 769
Design alternatives, 250
Design criteria, 250
Design drawings, 251, 330-334
Design plans, 330-333
Desorption, 556, 756
Destabilization, 544
Destillation, 562
Desulfurization, 593-594, 680-681
Detergents, 94-97, 418
   anion active, 94
   cation active, 94
   decomposition of, 755
   determination of, 1048
   not ionized, 94
   production of, 94-97, 418
Determinants for biological processes, 584
Determination,
   of closed pipe cross section, 300-305
   of cross section for pressure lines, 307-309
   of head loss in culverts, 308-309
   of open channel cross section, 306-307
   of waste water amount, 13-37
Detoxication, 541, 568, 806, 808, 830
   of concentrates with chrome and chromic acid, 784-787
   of cyanogen containing waters, 787-794
Developing countries, 234, 271, 422, 425, 467, 468, 469, 614
Dewatering, (see also sludge dewatering),
   artificial, 897
   by belt filters, 900-902
   in centrifuges, 898-900
   by filter presses, 902-904
   natural, 890-893
   potential of sludge, 882

of tannery sludge, 815
by vacuum filtration, 904-905
Dialysis, 568
Diameter variation, 326, 327, 328, 349
Digested sludge, 678, 827, 836, 849, 873-874, 886
Digester gas, 715, 844, 856, 864, 868, 869-872
  container, 606, 864
  generation of, 869
  utilization of, 872
Digestion limit, engineering, 874
Digestion time, sludges, 534, 866
Digestion water, 832, 869, 872-873
Dilution method, BOD determination, 1025-1031
Dimensioned flow-through, 522
Dimensioning,
  of activation installations, 694-702
  of active coal demand, 755
  of biological disk filters, 664-666
  of biological filters, 650-658
  of closed channels, 284-296
    completely filled, 303, 305
    partly filled, 303-304, 305
  of digesters, 875-877
  of Emscher tanks, 534
  of excess sludge, 699
  of grating installations, 481, 483
  of long sand traps, 515-516
  of microscreens, 487
  of oil and fat separators, 503-506
  of open channels, 306-307
  of overflows, 308-318
  of oxygen input, 701
  of polishing ponds, 637
  of pressure lines, 307-309
  of sand traps, 515-516
  of screens, 487
  of secondary clarification basins, 716-735
    after activation basin, 720-735
    after biological filter, 716-719
  of settling basins, 513-514, 522, 530, 538
  of settling ponds, 634
  of sludge thickeners, 886-890
  of vertical-flow filters, 744-745
  of waste water ponds, 634-638
Dimensioning length, 273
Dimensioning programmes for closed profiles, 290-296
Dimensioning rainfall, 265, 270
Dimensioning schemes, (see flow diagrams)
Dimensions of normal cross section forms, 296
Direct discharge, 410, 419-459
Direct precipitation, 550
Direct spinning method, 102
Discharge,
  of treated waters, 237
  of waste waters, 234
  of rainwater, 234, 261-263
  of surface waters, 259
Diseases,
  contagious to humans, 8-10
Disemulgators, (see Emulsion splitting)
Disinfecting sludge, 906
Disinfection, 541, 678, 755, 815
Disposal conditions for waste water and sludge, 410-466
Distilleries, 148
Distinguishing criteria, purification methods, 470
Distribution network, 627
District network, 239
District planning, 249
Disturbance,
  of aerobic decomposition caused by chromate, 607
  in biological purification processes, 607-608
  of methanogenic fermentation 606-607
  of purification processes caused by heavy metals, 608-609
Domestic garbage, 778
Dosing station for chemicals, 569
Double culvert, 308
Drainage area, 29

of sewer network, 237, 256, 270
of sewers, 263-264, 270
Drainage method, 241-249
Drainage network, 626, 628
Drainage systems, 240, 249
   building components of, 335-363
Draining irrigated fields, 628-629
Drawing oil, 560
Drilling oil, 560, 779
Drinking water, 1, 83, 222-223, 615, 619
   supply, 6
   treatment works, 440

Drop manholes, 329, 335, 344, 351, 352
Drum filter, 489, 492-493
Dry weather
   inflow, 724, 725, 726
   outflow, 29, 30, 243, 728
Drying, thermal, 877
Drying beds, 815, 880, 890-892
Duck breeding, 632
Dumping ponds, 715, 831
Dyestuffs, 87-90, 124-126, 135, 557, 561, 807-809, 820
   inorganic, 88-90
   organic, 90
Dyeworks, 110, 432, 558
Dying stage, 583, 672

E

E.C. (European Community) directives, 448-457
   Water pollution control regulations,
      for bathing water, 451-452
      for fish waters, 453-454
      for schellfish waters, 456-457
      for surface waters which serve as source of drinking water, 450
Economical values of sludge application in agriculture, 466
Edible fat and oil production, 170, 430, 503
Efficiency of purification methods for municipal (city) waste water, 471, 475
Egg cross section (egg profile), 339
Electrical conductivity of water, 952-954
Electrodialysis method, 568
Electrosubmersion-painting, 750
Elevation station, (see Pump station)
Elimination,
   of dissolved inorganic substances, 769-772
   of dissolved organic substances, 750-757
   of nitrogen 759-764
      in activation systems 760, 841

   by ammonia desorption, 762
   in fixed bed reactors, 760-761
   in microbiological processes, 760-761
   by ion exchange, 763
   in oxidation ponds, 765
   by turning point chlorination, 763-764
   of phosphates by algae, 766
   of phosphorus, 764-769
      by aluminum phosphate precipitation, 768
      by biological treatment, 765-766
      by conventional activation methods, 765
      by iron phosphate precipitation, 767
      by physical-chemical methods, 767-769
      by pond treatment, 765
   of pickling agents, 779
Elimination processes, purification of industrial waste water, 738
Eloxal plants, 70-71, 438
Emission
   limit values, 424-425, 425-462
   standards, (emission principle), 424-425
Emissions from motor vehicles, 209
Emscher well (Emscher tank,

IMHOFF tank), 532-534,
611, 863-864
Emulsions,
   splitting of, 779
   used, 779
Enameling works, 67
Endogeneous respiration, 583, 674
Energy consumption,
   activation plants, 687-688, 691
   pump stations, 360-362
Energy metabolism, 580
Energy respiration, 674
Engineering digestion limit, 873
Environmental protection, 467-469
Enzymes, 17, 860

Equilibration basins, 539-540, 803, 804, 807, 810, 818, 819, 820
Equilizing ponds, 539
Ethylene, 101
Eukaryotes, 576, 577, 579, 706
Eutrophication, 758
Evaporation, 567, 805, 829
Excavation, 369-371
Excavation width, 366
Excess sludge, 663, 699-700, 848
   in biological filters, 656
Exchange adsorption, 802
Excreta of animal, 199, 209, 210
Expansion slit, 396
Exponential phase, 583, 672
Extraction, 558-559, 802, 830

F

FROUD number, 524
Factors,
   effecting sewage water occurrence, 25
   for settling basin calculations, 525
   of waste water occurrence and water demand in industry, 45-46
Fast filters, 491-492
Fat films, oil films, 413
Fat production, worldwide, 170
Fat recovery, 231
Fat separator, (see oil and fat separator)
Fats, 91-93, 95, 170
Fatty acids, 91, 495, 594, 810
Feces, 246, 860
Feeding,
   of digester, 874
   of sludge drying bed, 891
   of thickener, 888
Fermentability of water, testing of, 1069-1071
Fermentation processes, 594-595, 862
Fermenters, 17
Fertilizer, 82-85, 199, 620-621
   nitrogen containing, 83-84
   phosphate containing, 84-85
   production of, 82-85, 804

Fertilizing effect, 11
Fiber substances, 116-120, 414, 492
Filter, 479, 488-495
   with coarse sand or fine gravel filling, 489-492
Filter bed, 489-490, 557
   open, 557, 745-746
Filter cake, 488, 494
Filter capacity, 491
Filter plates for compressed air, 684, 685
Filter presses, 898, 902-904
Filterability index, 487
Filtermeter, 487-488
Filtrate, 494
Filtration, 541
Filtration with a neutralizing filter material, 543
Final clarification (see secondary clarification basins)
Fine paper, 133
Fir wood, 129, 137
Fish canning factories, 176-178
Fish endangerment, 588, 620
Fish meal production, 202-203, 427-428, 841-842
Fish ponds, (see Waste water fish ponds)
Fish processing industries, 147, 176-178, 503
Fish test, 937, 1076-1080

Fish waters, 449, 453-455
Fixed bed absorber, 753
Flax retteries, 110-114, 816-817
Floating layer, 480
Floating sludge, 512, 592, 841, 848
Flocculation,
  agents, 415, 512, 546, 553, 569, 739, 774, 778, 835, 903
    polymers, 573-574
  aids, 569, 774, 781
  basin, 605
  chemical, 415, 523, 541, 544-555, 568, 605, 826, 830
  preliminary, 820
  transport processes during, 544-545
Flocculation-precipitation, 545, 550
Flotation, 499, 502, 541
  fine bubble, 502
  of heavy particles, 502
  of light particles, 502
  by sludge thickening, 889
  systems of, 496, 498-511, 546
Flow characteristic value, 270-272
Flow diagram,
  of beet sugar production, 155
  of breweries, 168
  of detergent production, 97
  of fish marinade production, 178
  of fruit juice production, 165
  of maize starch production, 188
  for mechanical, chemical and biological purification plants, 471-475
  of milk processing, 160
  of olive oil extraction, 171
  of paper manufacture, 134
  of pectin production, 197
  of phenol synthesis, 100
  of potato starch production, 184
  of rayon staple mill, 118
  for round basin, 530
  of sewer network in chemical industry plant, 244
  of sewerage system, 253, 254
  of slaughterhouse, 174
  of sugar production from cane, 158
  of sulfite pulp production, 131
  of tanneries, 106
  for utilization of industrial waste waters, 216
  of waste water disposal plant, 238
  of waste water in oil refineries, 140
  of wheat flour processing, 187
  of white mechanical pulp mill, 128
  of wool washing, 123
Flow direction
  change of, 328-329
Flow inlet,
  in a city sewerage system, 833, 836, 837, 838, 842
  in a receiving water, 237
Flow inlet conditions,
  natural waters, 419-459
  public sewerage system, 413-414
  public treatment plant, 414-418
Flow velocity,
  in closed pipelines, 301-305
  in sedimentation basins, 523
Flow volume calculations,
  of commercial waste water, 267
  of domestic sewage, 265-266
  by dry weather, 267-268
  of foreign water, 266-267
  of industrial waste waters, 267
  of rainwater, 270-275
  for outlying areas, 275
  of surface waters, 275
Flow-through time,
  of biological trickle filters (contact time), 651
  of final clarification basins of biological filter systems, 117
  of settling basins, 522
Fluctuations,
  in waste concentration, 41-42
  in waste water amount, 29-30
Flue gas, 567
Fluidized bed absorber, 755

Fluidized bed furnace, 915
Fluoride concentration, 781-782
Flushing devices, 352
Flushing of pipelines, 241, 407-408
Flushing screens, 485
Flushing station, 246
Fly larvae, 541
Foam formation, 95, 108, 778, 811, 812
Foamer, 498
Food industry, 52, 145-203, 487, 560, 596, 749
Foreign water, 2, 42-44, 235, 237, 241, 265
Formwork, special, 400
Foundries, 218
Freezing method in sludge dewatering, 567
French Emission Values, 458-459

Fresh milk dairies, 159
Fresh sludge, 848, 855, 867-869, 870, 873, 876, 886
Frictional losses, 284, 481
　full flow, 286-296
　grating, 481
　in open artifical channels, 298-300
　partial flow, 297-298
　in pipes, 284-300
Fruit canning factories, 53, 179-183, 838
Fruit juice beverages, 164-166
Fuel storage facilities, 510
Full flow, 286, 303
Fulling, 122
Fungus, 576
Fur finishing plants, 435
Furfurol, 130
Furrow (bed) irrigation, 623, 624

G

G value, drinking water supply, 450
GEIGER aerator, 690, 692
GOULD system, 677
Galvanizing plants, 71, 219-220, 437-438, 565
Garages, 510
Gas,
　purification of, 776
　scrubber water, 84, 835
　utilization, 872
Gas volumes for commercial waste waters, 871
Gas-air mixture, explosive, 510
Gasoline separator, 502, 510-511
Gasoline stations, 144, 510, 831
Gassing, 567, 681, 830
Gate shaft, 352
Gelatin factories, 499
Geodetic suction head, (see Pump stations)
German standard method for water, waste water and sludge investigation, 444
Glass processing industry, 439
Glucose, 4, 189
Glycerinated waste waters, 810
Gold, 565

Gradient
　change of, 326, 327
　of pipelines, 245, 246, 302, 324-326, 364, 389
　of sewer, adjustment to terrain, 324
　in sewerage system, 326, 327
Graphic representation of sewerage system, 330-334
Grated material, 483, 484, 843-844, 846
　comminutor of, 479, 482, 844
　presses of, 844
Grating, 414, 479, 480-484
　manual cleaning, 482
　mechanical cleaning, 482
Gravel washing, 436
Gravity pipes, 237
Gravity separation, 495-540
Ground digester basin, 612, 864, 833, 837
Ground pressure calculation, 372-378
Groundwater, 43, 74, 138, 468
　lowering of, 234
　recharging of, 224-225
Growth curve (cell numbers), (see Growth stages)
Growth metabolism, 580
Growth rate, microorganisms, 578

Growth stages, 583, 672
   dying stage, 583, 672
   exponential stage, 583, 672
   initial stage, 583, 672
   stationary stage, 583, 672

Guidelines for water protection, (see E.C. Standards)
Guidelines, French, 457-459
Gutters, 337
Gyrox aerator, 690, 692

H

Halogen compounds (EOX), determination of, 1065-1068
Hamburg basin, 528, 529
Hard salt production, 82
Hardening works, 439
Harmful effects of waste water, 413
   direct damage, 6
   indirect damage, 7
   on natural waters, 6-11
Harmful substances, 1, 63, 419
Head loss in grating facilities, 481
Heating of sludge digesters, 866
Heavy metals
   balance of, 911-912
   concentration of, limit values
     in raw sewage, 465
     in sludges, 464
     in the soil, 464
   content of in sludge, 910, 911
   ions of, 565
   problems with, 909-912
Heavy-duty detergent, 94
Hemp retteries, 110-114, 816-817
Herbicides, 619-620
Heterotrophy, 580, 581
High pressure flushing devices for sewer networks, 352
Homogenization basins, 539-540
Hops, 834

Horizontal trench sheeting, 371-372
Horseshoe cross section (Horseshoe profile) 339
Hotels, 21-22
House connections, 236, 242, 260, 344, 345
Housing structure, 469
Hydraulic characteristics for closed lines, 305
Hydraulic conditions for conduits, 327-329, 336
   calculation of sewer system, 278-320
   dimensioning culverts, 307-309
   dimensioning pipelines, 300-309
   waterhammer, 310-315
Hydraulic data for normal cross sections, 296
Hydraulic radius, 283
Hydraulic waterhammer, 318-320
Hydro-cyclone, 520-521
Hydroanthracite, 488
Hydrochloric acid, 3, 80, 803
Hydrocyanic acid, 787
Hydrogen ion concentration, (see pH value)
Hydrogen sulfide, 554, 594, 826
Hydrolytic stage, 862
Hyperfiltration, 559, 784, 756-757

I

I values, drinking water supply, 450
IMHOFF tank (see Emscher well)
IMMEDIUM filter, 492

IMPERIAL filter, 493
INKA aeration, 686
Ignition loss in digestion processes, 854
Immission,

limit values of, 423-424
limitation method of, 423
regulations of, 424
In-situ concrete sewers, 394-396
Incineration,
of animal excreta, 199
of screenings, 844
of sludge, 612, 886, 903, 914
of sulfite liquor, 823
Indirect discharge, 410-418
Industrial water, 1, 421, 467, 488
treatment works, 440
Infiltration, 614, 616, 618
Infiltration basin, 237, 238
Infiltration water, 2, 235, 237, 241, 264
Inhibiting substances, 603-609
Inlet zones,
in rectangulair basins, 526
in waste water ponds, 631
Inorganic dyestuffs, 88-89
Inorganic fertilizer, 82-85
Inorganic industrial waste water, 60-85
Insect larvae, 639
Insecticide, 641
Inspection,
of pump stations, 405
of sewers, 404-405

Intake structures, 335, 344, 347-348
Intercepting system, 239
Intestinal diseases, 8, 9
Intestine usage (intestine processing plants), 198
Ion exchange method, 564-566
Ion exchangers, 564-566
anion exchanger, 564
cation exchanger, 564, 565
with macroporous resinstructure, 564
Iron (II) sulfate heptahydrate, 571-572
Iron (III) chloride, 571
Iron (III) sulfate 571
Iron pickling plant, 230, 437
Iron salts, 554, 569-573, 801, 811, 894
Iron works, 64, 219, 776
waste water of, 64
Irrigated land, 625-626
Irrigation, 467, 623-624, 802, 833, 838, 839
surface, 614-627
underground, 864
Irrigation equipment, agricultural, 623-629
Isopropyl ether, 559

J

JAR test, 549
Juice production, 147, 164, 834

Junctions, 335, 344, 349, 351

K

KESSENER basin, 689-690
KORTE process, 114, 817
KROPP fiber collector system, 822

KUNZE connectors, 394
Ketone, 594
Kieselgur, 86

L

LAWA values
  for cooling and condensation water, 427
  for potable and industrial water treatment works, 440
  for waste water,
    from breweries and malthouses, 428-429
    from canning factories, 441
    from cleaning cloth laundries, 440-441
    from coal washing, 436
    from dairies, cheese factories and canned milk production, 429-430
    domestic, dry weather, 426
    domestic, rainwater mixed, 426-427
    from eloxal plants, 438-439
    from fish meal and animal carcass disposal plants, 427-428
    from galvanizing plants, 437-438
    from glass processing, 439
    from hardening works, 439
    from iron pickling, 437
    from margarine, edible oil and fat production, 430-431
    from mineral water, table water and limonade production, 439
    from non-metallic minerals industry, gravel washing, 436-437
    from paper and cardboard production, 434-435
    from petroleum industry, 436
    from potato processing, 441-442
    from press yeast production, 429
    from pulp industry, wood basis, 433-434
    from slaughterhouses and meat processing plants, 427
    from soap production, 431
    from starch production, 442-443
    from sugar processing, 431-432
    from tanneries, fur and leather handling, 435-436
    from textile fabrication, bleaching and dyeing facilities, 432-433
    from workshops, 443
Laboratory analysis, 959-961
Lamella separator, 739, 740
Lamella settling basin, 536-537
Lampblack, 89
Large-scale catering establishments, 846
Layer clouds, 204
Layer filters, 489, 494
Layered structures, open filter beds, 557
Layout for hydraulic calculation of sewerage system, 331
Layout schemes, 239
Leaching processes, 114
Leather making, 103, 109, 435, 815
Leptospirosis, 8
Light sludge, 595, 608, 838
Lignin, 129, 824
Lime briquetting, 823
Lime wash (lime water), 574, 575, 631, 804, 813, 820
Limit concentrations,
  for discharge, (see E.C. normes)
  for heavy metals, 464, 465, 603
  for metals, 604
  in the soil, 464
  for untreated waste water, 465
Limit of drainage area, 239
Lining open channels, 337
Linseed oil, 505
Liquid manure, 5
Lithophone, 89
Livestock feed production, 199, 560
Livestock raising, 193, 194
Local bylaws, 413
Local conditions, sewage treatment plant, 237
Long sand trap, 515-519
Longitudinal section of sewage treatment plant 239
Longitudinal settling basins, 513, 525-529
Low pressure aeration, 501
Lubricants, 65, 66
Lye baths, 117

M

Machinery production, 66, 779
Machines, in sewer system, 236
Magnesium, 466
Main collector (see main sewer)
Main sewer, 238, 240, 253, 259
Maintenance of sewer networks, 400
Maintenance of sewers, 406-408
Maize starch production, 188
Malthouses, 148, 166-168, 428-429, 834
Mammoth pumps, 359
Manhole connections, 344, 345
Manholes, 235, 237, 344, 348-349, 394
Maps, 259
Margarine production, 148, 170-172, 430, 503
Marinade production, 177
Material separation,
    coarse, 479-495
    methods of, 477-479
Materials, in sewer system, 336-337
Maw worm eggs, maw worms, 622
Meat processing plants, 173-176
Meat residues, 860
Mechanical purification, 477-540
Mechanical workshops, 144-145, 443, 779
Membrane separation, 559-562, 568, 749
Membranes, 559, 568, 748
Mercury chloride, 920
Metabolism,
    of aerobic microorganisms, 599
    of bacteria, 578
    of eukaryotes, 576-580
    in aerobic stabilization, 861
    in anaerobic stabilization, 861
    processes of, 582-584
    of prokaryotes, 578-580, 595
Metal ions, 565
Metal oxides, 859
Metal pickling plants, 766-70
Metal precipitation, 793
Methane, 576, 613
Methanogenic bacteria, 613, 859
Methanogenic fermentation, 783, 856

Methanogenic stage, 862
Methanol, 130
Methods of waste water treatment, 470, 471
    combined, 702-713
    convential, 471
Microorganisms,
    autotrophic 580, 581
    decomposition processes, 586-596
    heterotrophic, 580, 581
    metabolism of, 582-586
    nature of, 576-582
Microscreens, 486-487, 741-742
Milk
    canning facility, 429
    composition of, 161
    processing plants, 159-163, 833
Minamata disease, 7
Mineral oil, 495, 506
Mineral oil separator, 502, 506-510
    corrugated plates, 510
    parallel plates, 509
    round, 508
    separation chambers, 507-508
Mineral water processing, 147, 164, 439, 834
Mineralization, 581
Mines and ore dressing plants, 74-78
Minimum depth, sewers, 325
Minimum diameter, sewers, 339
Minimum requirements (emission values), 444-447
    for different industries, 446-447
Minimum width of trenchs, 366
Mining, 49, 565
Mixing equipment, 547
Moisture content, (see Water content)
Molasses, 154, 155, 156, 158
Mortar production, 61, 773
Movement of water, theoretical principles, 281-283
Multiple use of industrial waste water, 217-221
Multiple-layer filters, 492, 743
Multistage waste water treatment, 702-713
Must slime, 840
Mycelium sludge, 86

N

N : P_2O_5 : BOD_5 ratio, 113
NIERS method, 809
Natural purification methods, 610, 613
Nematodes, 596, 639
Neutralization, 541, 542-544, 780, 782, 783-784, 793, 794-798, 806, 809, 811, 813, 818, 820, 830, 836, 839, 841
  in basins, 605, 781
Newsprint, 133
Nickel, 565
Nimbostratus or layer clouds, 204
Nitrate determination, 968-970
Nitration, 90
Nitric acid, 3, 80, 803
Nitrification, 587-591, 704
  in artificial, biological plants, 588
  in receiving waters, 588-589
  in treatment plants, 589
  inhibition of during sampling, 920

Nitrite determination, 965-968
Nitrobacteria, 581, 587, 590
Nitrogen, 19, 582, 601, 615, 620, 748, 758
Nitrogen content, 591, 602
Nitrogen fertilizer, 83-84
Nitrogen gas bubbles, 592
Nitrogen salts, 82
Nitrosomonas, 581, 587, 590
Non-alcoholic beverage production, 164-166
Non-ionic products, 574
Normal cross section shapes, 289, 296
Normal requirements for waste water discharge, 249
Nuisance substances, 419
Nutrient balance, 418, 585
Nutrient content,
  in city sewage, 582
  in sludge, 912
Nutrient elimination, 758
Nutrient salts, 811, 813, 818
Nutrients, 419, 827, 835

O

OC load, 702
Obligatory points, 325
Odor elimination, 541, 810, 844, 863
Odor nuisance, 483, 554, 640, 837, 838, 880
Odor-protective covering, 504
Odorous substances, 557
Oil and fat films in sewers, 413
Oil and fat separators, 479, 496, 500, 502, 503-506, 811, 829, 835, 837, 841, 843
  matter collected in, 843, 846-847
Oil containing waste waters, 141, 828
Oil dissociation, 779
Oil emulsions, 141, 779
Oil industry, 139-142, 436, 829-830
Oil recovery, 231
Oil regeneration, used, 846-847

Oil residues, 778
Oil separators, 479, 502, 506-510, 829
Oil storage tanks, 144
Oilfields, 138, 828
Oils and fats,
  determination, 1040-1048
    by gas chromatographic analysis, 1047-1048
    by infrared analysis, 1046-1047
  sampling, 1042
Oily fish, 176-177
Olefins, 143
Olive oil production, 171
Open channels, 337-338
Open storm water basins, 353-354
Operation and maintenance,
  of Emscher tank, 864
  of sewerage system, 242-243, 406-409
  of sludge drying bed, 891

Operational conditions in secondary clarification basins, 724
Operational costs, 6, 251
Operations personnel, 410
Organic industrial waste water, 85-203
Organisms,
    autotrophic, 633
    heterotrophic, 633
Organized beaches, 22
Osmosis, reverse (see also Hyperfiltration) 559-560
Outfall structure, 237
Outlet channel, 238
Outlet zone, sedimentation basin, 526
Outlying areas, 275
Over-beaten substances, 135
Overflow water of digester, (see digestion water)
Overflows, 238, 309-318
Oxidation, 90, 541, 562-563, 781
    hydrogen sulfide, 804
    of $Fe(OH)_2$ to $Fe(OH)_3$, 793
    with ozone, 755-756
Oxidation basin, 781, 833, 835, 836
Oxidation ditches, 679, 833
Oxidation ponds, 816, 833
Oxidation processes, 562
Oxygen, 584, 630, 672, 955
    consumption of for bio-content of, 674
    in activation plants, 680-681
    in water, 421
    degradation, 583, 609, 673
    demand of, (see BOD)
    determination of, 955-958
    gassing with, 680-684
    load, 702
    supply with, 590, 674, 700
    dimensioning parameters for, 701
    recommended values for, 686-688
Ozone, 562, 755-756

P

pH
    measurement, 950-952
    value, 96, 109, 192, 201, 418, 565, 585, 590, 594, 597-599, 606, 608, 834, 854, 857, 869, 923
PE (Population equivalent values), 32-39, 108, 112, 115, 125, 127, 135, 150-151, 852, 866, 877
PORTEOUS method, 896
PRANDTL-COLEBROOK dimensioning formula, 286, 287
Packing paper, 133
Paddle wheel aeration, 566
Paper and cardboard industry, 126, 132-135, 220, 434-435, 486, 492, 550, 601
Paper, used 133
Paraffins, 143, 812
Parallel plate separators, 509
Paris green, 89
Partial flow, 297, 303
Partial system, 239
Partial-flow curves for pipes, 298
Passage loss, grating, 481
Passivation of metal surfaces, 71
Pasteurization, 874, 882, 906-907
Pathogenic agents, 10, 17
Paved surface drainage, 236
Pectin production, 196-197
Perforated sheet metal, 484
Perlon, 102
Permanganate consumption ($KMnO_4$), 93, 108, 109, 112, 157, 166
Permeability coefficient, filter layer, 491
Permeate, 560
Permission, grant of, (sanitation laws) 413
Personnel waste water, industry, 266
Pesticides, 615, 619-620
Petrochemical industry, 142-144, 566, 830
Pharmaceutical industry, 86-87, 566, 601, 806-807

Phases,
   aerobic stabilization, 861
   anaerobic decomposition, 861, 862
Phenols, 5, 78, 799
   decomposition of, 801
   determination of, 920, 1036-1040
   extraction of, 7
   synthesis of, 100
Phenol-decomposing bacteria, 799
Phenosolvan method, 230, 799
Phosphates, 114
   determination of, 921, 970-975
   precipitation of, 488, 547
Phosphoric acid, 602
Phosphorus, 11, 466, 582, 601, 748, 758
   balance, 602
   content, 554, 602
   determination of, total, 971
Photographic laboratories and establishments, 603
Photosynthesis, 581
Pickling plants, 66-70
Pine wood, 129
Pipe anchorage, 389
Pipe bedding, 385-397
   cohesive soil, 386
   concrete, 387
   for prefabricated pipes, 385-386
   non-cohesive soil, 386
   other bedding, 386
   sand, gravel, 386
Pipe concrete encasement, 387-388
Pipe connections, 339, 340
   in manholes, 389, 394
Pipe cross sections, 287-289
Pipe frictional losses, 284-299
Pipe jacking methods, 384-385
Pipe laying, 363-400
Pipe sealing, 392, 393
Pipe support, 389
Pipelines, 241, 337-344
   in pump station, 358
   in sewer network, 237
Pipes,
   asbestos cement, 342
   brick, 341
   cast iron, 343
   ceramic, see stoneware,
   concrete, reinforced concrete, 338-340, 395
   heavy, light, 390
   in-situ concrete, 341, 394
   mortise concrete, 393
   plastic and plastic coated, 342-343
   socket, 391-392
   special constructions, 388
   steel, 343
   stoneware, 340-341
Piping in pump station, 358
Pit coal, 74-78, 218-219, 799-802
Pit waters, 74-75, 798
Pithead bath waste waters, 75-76, 798
Planks, boards, 371, 372, 375, 376, 378-379, 381
Planning of drainage system, 249-334
Planning,
   of drainage systems, 249
   of municipal sewerage system, 255-257
   of waste water disposal system, 251-259
   of water management in developed areas, 249
Plans, 257
Plant growth, 758
Plastic pipes, 342-343
Plastics industry, 98-103
   cellulose, 98-99, 812
   condensed products, 99-101, 812-813
   polymerized products, 101-103, 813
Polishing ponds, 633, 637, 710-713, 748, 761
Pollution,
   artificial, 10
   atmospheric, 206-207
   in combined sewerage systems, 211-212
   determination of, 923
   in natural waters, 10, 468
   not traffic related, 207
   in rainwater runoff, 211-213
   in separated sewerage systems, 212-213
   strongest, 10
   traffic related, 207
   values,
      in precipitations, 207
      in street deposits, 210
Polyacrylamide fiber (Perlon), 102
Polyacrylnitril fiber (Dralon), 102, 116

Polyamide fiber, 102, 116
Polyelectrolyte, 574
Polyester fiber, 116
Polyethylene, 101
Polyglycol, 101
Polymerization products, 101
Polymers, 116
Polypropylene, 101
Polystyrene, 101
Polyvinylchloride, 101, 116
Polyvinylester, 101
Pond stages, 631
Population equivalent values, (see PE)
Porcelain production, 62, 774
Pore volume, slow filter, 489
Postclarification (see secondary clarification basins)
Potash,
 final liquors, 80, 81
 industry, 80-81
 waste waters, 80
Potassium, 466, 582, 114
 salts, 82
Potato chip production, 190
Potato distillery, 148
Potato processing plants, 189, 190, 441-442
Potato starch production, 183-185
Potatoes, dried products, 189
Potential evaporative transpiration, 621
Power generation with digester gas, 872
Pre-coating method, 902-903
Precipitation, 415, 541, 568, 739, 783, 809, 853
 preliminary, 550-551, 768-769, 820
 sludge of, 853
 using lime, 782
Precipitation, types of, 204-206
Precipitation/Flocculation, 415, 568
Preclarification, 415, 521, 522, 523, 676, 825
Press yeast production, 429
Pressure drainage, (see pressurized sewer system)
Pressure filtration, 495
 with active coal, 558
Pressure oxidation, 562

Pressure relaxation flotation, 498-499
Pressure water,
 from fish meal production, 153, 203
 from sugar beet pulp presses, 52, 154-156, 831
Pressurized pipelines, 237, 244-246, 288
Pressurized sewer system, 245-246
Pretreatment of industrial waste water, (see Indirect discharge)
Primary product and capital goods industries, 50
Private sanitary installations, 236
Process water, 249
Producer gas (see digester gas)
Profile of a main sewer, 302
Profiles (Egg profile, Circular profile, Horseshoe profile), 339
Profiles of the sewers, 329
Prokaryotes, 576, 577, 579, 580
 growth rate of, 578
Propeller pump, 359
Properties,
 of construction material, 336
 of waste water, 3, 59, 124, 237, 335, 586
Protection basin, 539
Proteins, 810
Protista, 576-577
Protozoa, 576, 595-596, 612
Prussian blue, 88, 791
Prussian white, 791
Pseudomonas, 591
Psychoda fly, 640-641
Pump diagram, 362-363
Pump head and capacity, 356-362
Pump house, 357
Pumping stations, 238, 255, 356-363, 478, 626, 627-628, 629
Pumps,
 number of, 361
 power demand of, 360-361
 types of, 358, 359
Purification capacity of natural waters, 7
Purification, third stage, 737

R

REYNOLDS number, 283, 524
Radioactivity, 541
   measurement of, 921
Rain, 204-206, 242, 243
   frequency of, 206, 271
Rainfall
   duration of, 206, 271, 272-273
   intensity of, 205
   per second per area (r), 205, 270-272
Rainwater, 2, 3, 142, 204-213, 234, 235, 236, 237, 241, 264, 270, 807
   outflow, 728-729
Random samples, 920, 921-922
Rayon staple mills, 116-120, 432
Reactions, with and without material conversion, 541
Receiving waters, 237-240, 243, 253, 255, 258, 420, 810, 840
   inflow, 397
   pollution of, 258
Recirculated sludge, 668
   amount of, 696
   flow of, 722
   solids content in, 698
Recirculated water,
   in iron works, 491, 777
   in paper and cardboard production, 826
   in sugar production, 831
Recirculation of industrial waste waters, 217
Recovery,
   of acids and metal salts, 230, 780-781
   of caustic soda, 819, 824
   of dyes, 820
   of fats and oil, 231
   of fertilizer, 820
   of fibers, 231, 820
   of magnesia cement, 803
   of magnesite and magnesium oxide, 803
   of magnesium chloride, 803
   of spinning oil, 820
   of valuable metals, 565
   of valuable substances, industrial waste water, 228-232, 484, 820
   of wool grease, 819

Recreation areas, 6
Rectangular basins, 506-507, 525-529, 530
Red water (synthetic resin condensation), 101
Redox potential, 786, 954-955
Reduction, 541, 563
Reduction,
   of fresh water consumption, in food industry, 152
   in fruit and vegetable processing, 182-183
   of substances in food industry discharges, 153
Reed, 893
Refineries, 502, 506, 508, 829
Refinery gases, 142
Regeneration basins, 678
Regeneration, active coal, 556
Regional planning, 249, 256
Regulations,
   for direct discharge, 419-459
   for indirect discharge, 410-418
   for public sewage treatment plants, 414-419
   for sewerage system, 413-414
   for sludge utilization in agriculture, 463-466
Regulators, 353, 355-356
Relationship, trench depth and costs, 262
Removal of deposits, waste water network, 257
Representation, graphic, 330
Requirements and regulations,
   for public water works, 414-418
   for surface waters, 421
Residues,
   treatment of, 842-915
Restaurants, 21, 503, 846
Retaining walls,
   lowering of, 380-381
Retention basins, 539
Retention of substances, 144, 168, 470, 836
Retention profiles, 516
Retting waters, 112-114
Reuse of waste water, 214-232, 237, 488
   for agricultural purposes, 226-228

in coal dressing, 799
direct, 215
for drinking and bathing purposes, 222-223
in dual purpose systems, water supply, 224
for groundwater recharging, 224-225
indirect, 216-217, 223
in other commercial branches, 214, 215-217
pit water, 798
for recovery of valuable residues, 228-232
in recreation areas, 225-226
in same factory, 214, 215
Reverse osmosis (see Hyperfiltration)
Rice starch production, 189
Rinsing waters, 67, 70
Rising motion, 496
Rising velocity, 496
Road surface wear-off, 208
Roll sinter, 778
Rolling, 65, 778-779
Rolling mills, 60, 64-65, 219, 489, 557-558, 778-779
Root dehydration, 893
Rotary aerator, 566, 690-691, 693
Rotary drills, 138
Rotary process sludge, 138
Rotation-equilibration basin, 540
Rotatoria, 596, 639
Roughness parameter (frictional losses), 284-288
Round basins, 529-532
Round sand traps, 519-520
Routing of sewers, 322-324
Running water retteries, 111
Runoff delay coefficient, 273

S

SHELL parallel separator, 509, 829
STOKES law, 496
Safety basin, 539
Safety measures,
  in metal working industry, 783-784
  at sampling, 918, 919
  in sewer network, 408-409
Safety ponds, 638
Safety zone, in settling basins, 526
Sale of digester gas, 872
Salt water, 138, 803-804, 828
Sample preservation, 921, 935-938
Sampling,
  report, 916, 918
  of sludge, 916, 942-943, 1084-1085
  of waste water, 916
Sand, 488, 859
  classifier, 517, 845
  drying areas, 845
  filter, 489, 743, 745-747
  filtration, 555, 834
  removal of, 520
  separation chamber, 518
Sand trap, 237, 414, 479, 496, 515-521, 843
Sand trappings, 845-846
Sandy limestone production, 61, 773
Sandy soil, 466
Saponification, 92
Screen belt presses, 900
Screen belts, 485
Screen cylinder, 486
Screen drum, 486
Screen mesh, 486
Screen panels, 485
Screen plate, 486
Screening facilities, 484-487, 818
Screenings, 485, 486
Screens, 479, 839
Screw presses, 836
Scrubber water, 802
Scum layer, 866, 867, 868, 869
Sealing methods for pipes, 391-394
Seasonal plants, 154, 179, 192, 838, 839, 840
Seawater, 213
Secondary clarification basins, 521, 523, 592, 667, 716-737
  after activation plants, 720-722
  after biological filters, 716-719
  horizontal flow systems,

722-730
  with pipe filters, 740-741
  settling process in, 720-721
  vertical flow systems,
    730-735
Secondary clarification ponds,
  (see Polishing ponds)
Secondary precipitation 551-553
Sections for pressure pipes,
  307
Sedimentary substances,
  determination of, 958-959
Sedimentation and digestion
  tank (combined), 532
Sedimentation,
  accelerator, 569
  basins, 244, 415, 479, 496,
    503, 506, 514, 515, 521-539,
    611, 626-627, 705, 706, 707,
    714, 776, 804, 805, 808, 812,
    819, 822
    horizontal flow, 525-529
    lamella, 536-537
    mechanical cleaning, 528
      intermediate floor, 532,
        535
      trenched floor, 527, 528
    other types, 538-539
    two-level systems, 532-535
    vertical flow, 537-538
  container, 845
  facilities, 415, 512-539
  ponds, 633, 822
  process of, 513, 415, 541,
    605, 781, 830
  sludge of, 521
  velocity of, 415
  zone of, 526, 631
Seepage, silos, 5
Selection of flow velocity in
  pipelines, 301
Sensory tests in situ, 945-947
Separated sewerage system, 44,
  242, 243, 244, 254, 300
Separation zone (in settling
  basins), 526
Septic tanks, 863-864
Series or parallel systems, 239
Settling basins,
  length of, 524-525
  types of, 525
Settling tube, 740-741
Sewage sludge regulations, 857
Sewer bridges, 238
Sewer cleaning, 407-408
Sewer construction,
  advanceable formwork, 395
  in situ pipes, 394-396
Sewer construction, 363-400
  finished pipes, 389
Sewer cross sections,
  closed, 300-305
  open, 306-307
Sewer inspection, 406-407
Sewer joint sealing, 392-394
Sewer network, 236, 335
  crossings, 344, 363
  operation, maintenance,
    400-406
Sewer outfalls, 397, 398
Sewer position,
  depth of, 324, 325
  in street profile, 323
Sewer profile,
  brick, 397
  concrete, 338
Sewer repair, 408
Sewer shutter, 352
Sewer storage section, 355
Sewer trench, 369
Sewer walls, construction
  method, 381, 382
Sewerage system, 1, 204, 235,
  240-244
  city (municipal), 237, 241
  industrial, 237
  official acceptance of,
    403-406
  parts of, 236-238
  structures of, 236, 237,
    335, 344-363
Sewers, 242, 251, 240, 260, 336
  alignment of (routing), 259,
    261, 322-324, 364-369
Sheep's wool, 121
Sheet piling, 379, 380
Sheeting, 371-385
  Berlin method, 382
  horizontal, 371-378
  in tunnel method, 382-383
  large trenches, 398
  other types, 381
  with sewer walls, 381
  sheet piling, 379-381
  shields, 383-385
  special formwork, 400
  tiered, 379
  vertical, 378-385
  with wooden planks, 378, 381
Shellfish waters, 449, 456-457
Shield advancing,
  sewer construction, 385

Shields, 383
Shigellosis, 9
Ships, at sea, 1, 248
Side weirs, 311-316
Sight rails, 367-369
Silages, 195
Silk boiling, 109, 114-115, 817
Silo composting, 908
Simcar aerator, 690, 692
Simultaneous precipitation, 553-555, 769
Sinking motion, 497
Sinking velocity, 496, 512, 514
Sinter, 778, 779
Site plan (diagram),
  agricultural irrigation, 629
  sewer system, chemical works, 244
Slag stores, 777
Slaughterhouses, 147, 173-176, 427, 499, 503, 846
Slope irrigation, 624
Slot-screens, 484
Slow filters, 489-491
Sludge, 238, 470, 521, 612, 657, 782, 873
  activated, 555, 580, 592, 667-671, 721
  from activation plants, 870
  age of, 663, 670, 880
  agricultural use of, 463-466, 892, 905-906
  analysis of, 1083-1093
  from biological filters, 662-663
  biological purification, 848
  calorific value of, 855, 857
  collection sump, 520
  completely digested, 869, 873-874, 853
  composition of, 853-859, 884
  composting of, 914
  conditioning of, 756, 893-897
    chemical, 894-895
    thermal, 894, 895-897
  decomposability of, 859
  deposition of,
    landfills, 912-913
    open sea, 914
  dewatering of, 612, 782, 824, 826, 857, 882, 883-905
    with belt filters, 900-902
    with centrifuges, 898-900
    on drying beds, 890-892
    with filter presses, 902-904
    potential of, 613, 857
    process stages of, 885
    with vacuum filtration, 904-905
  digesters, 863-877
    heated, 865-877
    two-stage, 876
    uncovered, 612, 864-865
    unheated, 863-865
  digestion of, 480, 595, 612, 853, 868, 876
  discharge of, 534
  drying, 886, 914-915
  fine flaked, 497
  fresh, 848, 855, 859, 867, 870, 873, 876, 886
  gas, (see digester gas)
  incineration, 612, 886, 914
  index, 670-671, 683, 722
  initial dewatering of, 486
  load, 590, 671, 699, 877
  mycelium, 86
  occurrence,
    anodizing plants, 782
    biological filters, 657
  parasite eggs in, 859
  pasteurization of, 906-907
  pipe line, 244
  ponds (lagoons), 863, 892-893
  from preliminary clarification basins, 870
  presses, 902
  regulations, 463-466, 857
  from regulator basins, 870
  removal of,
    hydraulic, 527
    mechanical, 528
  separation of, 794
  silos, 880
  stabilization, 612, 676, 859
    aerobic, 861, 862, 877-882
    anaerobic, 861, 863-877
    of fresh sludge, 859, 879
  stabilized, 853
  thickening of, 521, 782, 886-890
  toxic substances in, 858
  trap, 538
  treated, 237, 238
  treatment 470, 521, 611, 626-627, 847
    aerobic thermophile, 907-908
  types of, 848-849
  volume of, 849-853
  water, 520, 594
  water content of, 670, 849-851
  zone, 526, 528, 631

Sludge and sand pump, 520
Snow melt water, 204
Snow-melt runoff, 44
Soap production, 91-93, 431, 499
Soda lye, 132
Soda production, 81-82
Sodium aluminate, 570
Softening, 798
Soil,
    dust, 209
    filtration, 616, 747, 809, 833, 836
    types of, 373
Soldier piles, 382
Solids,
    concentration of, reachable in sludges, 888
    dewatering of, 492
    separation of, 541
Solubility of metal ions, 795
Solvents, 558
Special head losses, 299
Specific gravity, homogeneous body, 279
Spectrophotometric determination,
    of cyanide with barbituric acid pyridine, 991-994, 998
    of nitrite with sulfanilic acid, 965-968
Spinning mills, 110
Spray irrigation, 623, 624-625, 627, 833, 838
Stabilization basins, 879
Stabilizing time, 881
Stand rettery, 111
Standard values, german (see LAWA values) 425-443
Standards,
    French, 458-459
    Swiss, 505
Starch production, 183-189, 442, 486, 838
Starch syrup production, 189
Start-up,
    of biological filters, 639, 646
    of digesters, 867
Start-up phase, 583
Static calculations, 251
    of ground pressure, 373-375
    of sewer lines, 320-321
    of sheet piling, 381
    of trench sheeting, 375-378
Static thickening, 890
Stationary stage, 583, 672

Steam condensate, (see Condensate)
Steel pipes, 344
Steel works, 64-65, 778-779
Step aeration (distributed water inlet), 677
Stirring, 868-869, 873
Storage basin, 539
Storage room at pump stations, 357
Storm clouds (cumulonimbus), 204
Storm water, (see also rainwater), 204-213
    atmospheric impurities in, 206-207
    ground surface impurities, 207-210
    intensity of, 205
Storm-water drainage, 204, 234, 261-263, 322
Storm-water overflows, 239, 240, 243, 335, 344, 353
Storm-water retention tank, 237, 335, 344
Storm-water surplusing works, 253
Storm-water tanks, 354
Story furnaces, 915
Straw fiber board production, 135-136, 827
Street deposits, 210
Street inlets, 246, 345-346
Street profile, 323
Street spreading, salt spreading, 208
Stripping, 566, 756, 757, 829
Strongyloidiasis, 9
Structures, special, 397-400
Styropor slices, 660
Submersible pumps, 359
Substances contained in industrial and commercial waste waters, 58, 59
Substances, sewer clogging, 414
Substrate respiration, Substrate phosphorylation, 583
Suction filter, 489, 494
Suction head, 248
Suction pipe in pump station, 358
Sugar, 823
    extraction plants, 156
    production of, 154-158, 221, 431-432, 608, 715
Sulfate pulp process, 130-132, 824

Sulfate pulp production, 132, 824
Sulfate, determination, 982-984
Sulfide, 814
Sulfite determination, 980-982
Sulfite pulp process, 130-132
Sulfite pulp production, 130-132
Sulfite waste liquor, 7, 823
Sulfites, 11
Sulfonation, 90
Sulfur, 40, 84
  bacteria, 554, 608
  dyes, 124
Sulfuric acid, 3, 79, 501, 803
Surface aeration, 688-693
Surface charge,
  of biological filters, 651, 654-655
  definition of, 642
  of preliminary clarification basins, 523
  of secondary clarification basins 523
    of activation systems 728-729
  of sedimentation basins, 514, 520
Surface filters, 488

Surface tension of water, 498
Surface waters, 3, 449-450
  quality classification, 420-423
Surfactants (see also Detergents), 95, 921
Surge discharges, 418
Suspended matter, removal, 739-750
Suspended particles, 498
Suspended substance,
  in sludge, 860
  in waste water, 120, 512, 513
Suspension adsorption, 555
Suspension load, 523
  of industrial waste water, 852
Suspensions, 414, 512, 867
  separator of, 496
Swimmer foam trap, 501
Swimmer for aerator, 690
Synthetic detergent production, 93-97, 811-812
Synthetic packing elements, 655
Synthetic resins, 770, 802
Syrup production, 189

T

TOC (total organic carbon), 3, 5, 960
TOD (Total oxygen demand) 3, 6
Table water production, 164, 439
Tanneries, 103-108, 435-436, 814-815
Taste limits for drinking water, 83
Temperature, 599-601, 865, 869, 923
  of air, 944
  compensation factors for conductivity measurement, 953
  in digester, 869-870
  by sludge digestion, 865
  of waste water, 418, 523, 524, 562, 584-585, 590, 599-601, 640, 821
  of water, 945

Testing on the site, waste water, 917-919
Tetanus, 9
Textile industry, 109-126, 432-433, 492
Textiles, 806, 809
Thermal insulation of digesters, 867
Thermal pollution, 419
Thermobacter mix, 824
Thermophilic bacteria, 865
Thickening, 826, 874, 880, 886-890
  of sludges, 886-890
Thread twisting facilities, 561
Tilting troughs, 648, 649
Tire abrasion, 208-209
Titanium white, 89, 808
Toilets, 236
  mobile, 248
Total discharge, determination, 300

Supplements

Supplements

Federal Ministry for Economic Cooperation
Federal Republic of Germany

WATER SUPPLY AND SANITATION PROJECTS
IN DEVELOPING COUNTRIES

"Sector Paper"

Guidelines for the Planning and Implementation
of Bilateral Cooperation Projects
of the Federal Republic of Germany
in the Drinking Water and Sanitation Sector

Bonn, May 22, 1984

TABLE OF CONTENTS

I.   INTRODUCTION 4

II. IMPORTANCE OF THE "WATER SUPPLY AND SANITATION"
     SECTOR IN THE DEVELOPMENT PROCESS. 5

III. TARGETS AND OPPORTUNITIES OF THE "WATER DECADE". . . 8

IV. NECESSARY ADAPTION OF THE TARGETS. 10

V.   CRITERIA FOR PROJECT SELECTION AND DESIGN. 12

   V.1.   Basis for Project Planning and
          Implementation 13
   V.1.1 General. 13
   V.1.2 Preservation of resources. 14
   V.1.3 Drinking Water Supply. 16
   V.1.4 Disposal of Waste Water and of Faeces. . . . 19
   V.1.5 Solid Waste Disposal 20

   V.2.   Preparatory and Accompanying Measures. . . . 21
   V.2.1 Hygiene Education Campaigns. 21
   V.2.2 Project Executing Agencies 22
   V.2.3 Tariff Policy. 23

VI. CONCLUSION 25

I. INTRODUCTION

The present sector paper replaces the one entitled "Communal Water Supply in Third World Countries", which was issued in May 1974. It incorporates the experiences acquired in project work during the last ten years, and now also includes rural water supply and sanitation.

A number of project evaluations carried out in the last few years have helped to identify bottlenecks and weaknesses in the planning, implementation and operation of projects. Taking into account these evaluation results and the experience of other donors, indicative figures and guidelines were elaborated for future development cooperation in the "drinking water supply and sanitation" sector.[1]

The new sector paper addresses the relative importance within the development policy of water supply and sanitation and defines the principal criteria for project selection and design.

This sector paper is supplemented[2] by a "Summarized checklist of Project Assessment Criteria" (target group: Regional Sections of the BMZ) and a more detailed "Project Assessment Manual for Drinking Water and Sanitation Projects in Developing Countries". The latter is based on the principles contained in the present paper and will contain a specific cat-

---

[1] "Sanitation" means all measures connected with the disposal of liquid and solid waste, excreta disposal, and hygiene education.

[2] The "Summarized Checklist of Project Assessment Criteria" and the detailed "Project Assessment Manual" will be finalized by BMZ, GTZ and KfW on the basis of the present Sector Paper in the next few months. In addition, KfW will adapt its internal project appraisal guidelines "Drinking Water Supply" to this Sector Paper and its assessment criteria, and will supplement it by incorporating also Sanitation.

alogue for collecting and evaluating project data. It is intended as an aid, in particular for consulting engineers and project-executing organisations, to systematise and simplify the planning and assessment of projects.

II. IMPORTANCE OF THE "WATER SUPPLY AND SANITATION" SECTOR IN THE DEVELOPMENT PROCESS

The provision of qualitative safe water[3] is one of the most important basic human needs. For health reasons, water supply projects must always be planned with a view to water resources assessment and protection and to the disposal of the resulting waste water, of faeces and solid waste; wherever necessary, measures related to these aspects should be included into the scope of the project.

The Federal German Government has set the supra-sectoral target of meeting basic needs (food, drinking water, health, housing, clothing, education) and therefore promotes measures in this sector.

Being particularly orientated to basic needs, water supply and sanitation have acquired special importance as factors of economic and social development and as a result of the increasing population pressure in the developing countries. The traditional way of life and type of economy of the population used to be heavily dependent on locally and regionally available water resources as a result of the clearly apparent interdependence between the natural conditions and the peoples' behaviour patterns. As a result of far-reaching changes in the socio-economic structure of the developing

---

[3] "Safe Water Supply" (World Health Statistical Report, 1976, Volume 29, X, p. 546). Guidelines for safe water are provided in Volume I of "WHO Guidelines for Drinking Water Quality", WHO, Geneva, 1983 (EFP/82.39).

countries, which were either deliberately promoted (colonisation; monocultivation; industrialisation etc.) or which resulted as a consequence (population growth; urbanisation; environmental impacts; dwindling resources; changes in political structures etc.) temporary shortages, with originally only limited effects in terms of duration and affected areas, have become lasting phenomena all over the world. Therefore, the supply of drinking water acquires increasingly a key role in the ability of urban conglomerations and rural regions to survive.

Although they cater to basic needs and have a direct impact on health and performance, water supply, and, in particular, complementary sanitary measures, have not been accorded the appropriate priority in the development process of many developing countries. According to estimates of the World Health Organisation (WHO) in 1981 only

- 73% of urban population and
- 32% of rural population in developing countries

had access to safe drinking water and only

- 53% of urban population and
- 15% of rural population

had adequate sanitation facilities. Some experts consider these data to be even overestimations.

The fundamental hazards associated with unsafe drinking water and poor sanitation often lead to intolerable living conditions in urban areas.

This extremely unsatisfactory state of affairs is normally demonstrated by the following circumstances:

- Since only a part of the population in developing countries had access to hygienically safe drinking water, the other population is forced to use unsuitable, often heavily polluted and contaminated water for drinking. This leads to serious health hazards and damages, which are often not recognised, underestimated, or their origin not understood.

- Town centres and priviledged residential areas usually are equipped with water supply systems - even if the water quality and quantity are often inadequate - or have been equipped in the last two decades with such systems, frequently with external support. The urban problem zones and rural areas often cannot obtain sufficient support to improve their drinking water supplies and to promote sanitation.

- The health hazards affecting a large number of people because of the absence of complementary sanitation measures (liquid and solid waste disposal; hygiene education), frequently become apparent only some time after the water supply improvements. Even today the often disastrous effects of these health hazards are largely underestimated both by the people concerned and by the decision-makers. Moreover, the activities between the executing agencies of water and sanitation projects and the health authorities are frequently not sufficiently coordinated, and curative health measures receive usually higher priority than preventive measures.

Yet, according to findings of WHO, 80% of all diseases in the developing countries are "water-related", of which a significant proportion is attributable to inadequate drinking water supplies and sanitation installations, and to the absence of hygiene education.

III. TARGETS AND OPPORTUNITIES OF THE "WATER DECADE"

Targets:

On account of the health hazards resulting from the inadequate supply and disposal of water, the United Nations conference of 1977 in Mar del Plata declared the period from 1981 to 1990 as the "International Decade for Drinking Water Supply and Sanitation", setting the following targets to be achieved for over 2 billion people in the Third World by the year 1990:

- to provide all people with a minimum quantity of safe drinking water in urban and rural areas and to

- improve the sanitary conditions by implementing measures for the disposal of waste water, faeces, and solid waste.

That means that a quarter of the urban population and two thirds of the rural population in these countries are to be supplied with drinking water, and about half of the urban population and about three quarters of the rural population are to be provided with basic adequate sanitation facilities. In addition, the existing water supply and sanitation facilities must be maintained and rehabilitated.

Attainability of these Targets:

Assuming the continued application of the non-adapted technologies, frequently found in countries of the Third World, between US$ 500 and 600 billion would be required until 1990 to meet these Decade targets, as estimated by the World Bank and WHO. However, even using low-cost adapted technologies, about US$ 300 billion would be necessary, that is, about US$ 30 billion in each year of the Decade; or that, according to WHO, about US$ 6 billion would have to be raised through external support (actuals in 1981 only US$ 2,175 billion).

Unless appropriate alternatives are chosen in future in the planning and implementation of water supply and sanitation projects, these figures mean that at best 30% of the Decade targets can be achieved. This does not take account of the fact that a substantial part of the available funds is actually used for re-investment, in other words for the maintenance of existing facilities. In reality, therefore, we are in fact further away from reaching the Decade targets today than we were in 1977.

Deterioration in the Situation:

In many developing countries the situation has deteriorated further as a result of

- the general population increase in conjunction with increasing population concentrations into urban centres;

- the unexpectedly high increase in the overall demand for water, caused in part by a high level of water losses and wastage, and the consequent premature capacity utilisation of existing water supply facilities;

- the sometimes dramatic reduction or even depletion of available resources suitable for drinking water supplies (e.g. as a result of desertification and deforestation; overexploitation and contamination; the absence of water resource management);

- concentration of projects on high-cost, non-adapted supply and disposal technologies; neglect of measures designed to maintain or make better use of such installations; failure to set up efficient executing agencies (including cost-related progressive tariffs, training and payment of qualified personnel);

- the frequently isolated planning of individual projects without the necessary supplementary measures and without involving the beneficiaries concerned (e.g. drinking water supply without liquid and solid waste disposal; water supply and sanitation works without hygiene education; housing and industrial settlement programs without water supply and sanitation).

Conclusions:

Taking into account that the currently available funds cannot realistically be expected to increase significantly in the years ahead, these must be used more effectively in order nevertheless to bring about a significant improvement in water supplies and sanitation in the developing countries, to reduce health hazards and to meet basic needs. Among other measures, it is therefore necessary to ensure a greater participation of the population in the planning, construction, operation and maintenance of the projects.

For economic as well as technical reasons it is necessary to apply technologies that are as simple as possible and appropriate to the situation in the respective developing country and project region and to reduce the technical standards and water consumption rates (frequently still oriented according to the industrial nations). It is in this context that the process of rethinking that has already begun both in the developing countries and at the donor organisations must be intensified.

IV. NECESSARY ADAPTION OF THE TARGETS

Assuming the continued provision of full conventional water supply systems, as usually implemented in the past, the Decade targets would imply a level of investment for which nowhere near enough financing could be found. This also ap-

planning, constructing and operating water supply projects.

Management and preservation of the water potential are elements of integrated environment planning and are, in the long run, more effective and, as a rule, less expensive than implementing only new water supply and sanitation systems or having to implement later corrective measures of damage repair.

All water resources development schemes must in general be in accordance with national water laws and rights. The drinking water supply must be given priority over all other forms of water utilisation. The steps necessary for the management and preservation of water resources comprise:

- exploration, survey and evaluation of water resources,

- water use planning and actual water utilisation.

Exploration serves to obtain an overall inventory of the available and exploitable potential of surface water (streams and lakes), groundwater (including springs) and other water sources (e.g. precipitation). In order to survey water resources and to prepare water balance plans, stock-taking and sometimes supplementary hydrogeological and hydrological investigations are necessary. The assessment of water resources should furnish data on their location, the economic possiblities and limitations of their utilisation in quantitative and qualitative terms.

Water use planning sets out the possibilities and limitations of resource exploitation in the project region for all types of water uses (drinking water supply, irrigation, livestock, commerce and industry, hydropower, etc.). At the same time, the risks of contamination through domestic sewage/solid wastes, industry, mining, agriculture, etc. must also be taken into account as a limiting factor. Water use planning should define the technical priorities for the management and preservation of the water resources.

Water use planning must, in principle, include water protection measures (e.g. protective areas for drinking water). In addition to the technical measures, legislative and administrative action must also be provided (for e.g. at water capture structure). The utilisation of surface water for drinking water supplies is to be considered as equivalent to that of groundwater, provided its quality meets adequate standards of hygiene and/or allows application of simple technologies for its treatment and conveyance.

The decision to utilise either the surface water or ground water resource must be made primarily dependent upon economic aspects, taking into account, however, socio-cultural water consumption habits. In order to prevent overexploitation, actual water use must in general be compatible with the long-term renewal of the resources.

When planning and designing water development and supply schemes long-term negative impacts to water resources (e.g. through dangerous sewage) must be prevented.

V.1.3 Drinking Water Supply

Domestic Water Demand:

The principal objective of water supply projects is to ensure the supply of a sufficient amount of hygienically safe drinking water for basic human needs (drinking, cooking, washing) for all strata of the population.

What is to be regarded as a sufficient and adequate basic demand depends on the water use habits as well as on climatic and cultural conditions. A sufficient basic supply can be achieved with 20 to 40 litres per inhabitant and day (20-40 l/cd) (WHO: 30 l/cd).

In general, only those projects should be promoted where the target group of the poorer population strata represents the major part of the beneficiaries. Therefore, projects with a high proportion of public standpipes and individual yard connections enjoy special priority.

When defining supply standards for a specific project area, a supply solely based on public standpipes should normally first be considered. If higher standards (yard connections and house connections) are envisaged for certain consumers or for parts or all of the project area, this must be justified by specific local conditions. The following consumption rates are adequate and eligible from a development policy point of view:

  up to 40 l/cd for public standpipes
  up to 60 l/cd for yard connections
  up to 120 l/cd for house connections.

The distance between public standpipes or other public or semi-public water taps depends on the population density and on reasonable waiting times. In urban areas walking distances from the dwelling of the user should not exceed 300 m. In rural areas a greater distance to the water tap or well may have to be accepted. Even for settlements in arid regions drinking water should be available within one hour's walking distance. In areas of extreme water shortage the water supply must be adapted to the available resources and limited to the basic demand and may, if necessary, have to be rationed.

Drinking water projects must include all necessary project components, i.e. the water catchment works, treatment facilities (if necessary) and the conveyance to the consumers, as well as preparation of systematic maintenance and the necessary hygiene education. Partial measures are eligible for financial assistance only if the other necessary system components have already been implemented or if the financing of their, preferably simultaneous, implementation is ensured through other sources.

Public Demand:

Water demand of public institutions (community needs), especially in urban areas, must always be covered and the corresponding measures must therefore also be a component of the project. The danger of uncontrollable and excessive consumption of those consumers, however, makes it necessary to employ all appropriate technical measures in the design of the works and installations that will ensure low water consumption volumes and contribute to a reduction of cost.

Commercial and Industrial Demand:

Water supply for commercial and industrial enterprises is indispensable for regional economic development and should therefore be taken into account in the planning of water supply projects.

Because of the higher production cost of water with safe quality, only those commercial water needs which really require potable water quality (food production, water for employees, etc.) should normally be covered by the public drinking water system, provided there are alternative supply possibilities. The production water demand without such quality requirements should, if possible, be covered by a separate water system of the company concerned; nonetheless, requirements of resource preservation should also be considered when granting permission for such separate private water supply systems.

Special Considerations for Rural Areas:

The principle of aiming at a maximum simplicity in the design of water supply projects is of particular importance in rural areas, not only with a view to cost-saving but because of the special difficulties to ensure adequate operation and maintenance of the facilities. This implies depending upon the local conditions construction of simple dug wells or drilled wells - equipped with easy-to-maintain hand-pumps -, of rain water cisterns, or of spring water catchments with gravity lines into the supply area. Motor-driven pumps should be used only in exceptional and justified cases not only because of their higher operating and maintenance costs (usually representing foreign exchange requirements) but above all for reasons of environmental protection and safeguarding of resources (excessive grazing in the well catchment area, excessive use of ground water).

In cases of small individual supply systems, such as simple wells or rain water cisterns, distribution networks should not be installed. Only under special circumstances (spring water chambers connected to gravity lines, single wells outside settlements) simple supply lines can be installed to convey the drinking water to the necessary public standpipes. In very small villages, it may even be sufficient to install a single water tap.

V.1.4 Disposal of Waste Water and of Faeces

Under every water supply project, the planning should include provisions for suitable disposal facilities for waste water (domestic waste water, possibly also rain water) and for faeces, as well as for the necessary finance. Without proper sewage disposal and sanitation facilities no lasting improvement in the hygiene conditions can normally be attained.

The selection of the appropriate sanitation technology (e.g. latrine, cesspit or conventional sewerage) is determined essentially by the volume of drinking water consumption and the resulting sewage volume. Decentralised solutions should be given preference for reasons of lower cost as well as smaller environmental impact. Installation of conventional sewerage systems is justified only in urban areas with high population densities and large sewage volumes because of the extremely high construction and operating costs. The construction of a conventional sewerage system requires the simultaneous construction of sewage treatment facilities, unless the safe disposal of the sewage can be achieved otherwise. For the treatment of sewage simple and natural purification processes (e.g. bio-oxydation ponds) should be applied as far as possible.

Latrinisation programmes, as developed by the World Bank/UNDP, are particularly suitable for promotion as they are typical low-cost solutions adapted to local conditions and provide opportunities for users' participation in the construction, operation and maintenance of the sanitation facilities. Water water disposal and sanitation programmes, too, should be accompanied by supplementary hygiene education campaigns, specifically tailored to the project.

V.1.5 Solid Waste Disposal

In the interest of environmental proection as well as housing and drinking water hygiene, solid waste disposal requirements should be taken into account as an integral part of the planning for any sanitary engineering project and should cover all strata of the population and settlements in the project. Similarly, storm water disposal projects must take into account adequate disposal of solid waste, even if no drinking water and sanitation measures are planned or carried out simultaneously.

The very different composition of solid waste in developing countries and the traditional processes of collection and treatment (labour-intensive recycling of raw materials; composting) must form part of the planning. A reduction of solid waste disposal costs should be sought by recycling and subsequent reuse of raw materials (e.g. composting).

V.2. Preparatory and Accompanying Measures

Hygiene education campaigns, institution-building support activities and training of technical staff must be included in line with local requirements as integral part of all water supply and sanitation projects.

V.2.1 Hygiene Education Campaigns

Supplying drinking water and implementing sanitary measures have a direct impact on public health in the project areas.

In this connection, one of the most important aspects is the timely and comprehensive information of the target groups, especially the women, about

- the importance of safe water for their health;

- the necessity to actually and exclusively use the new supplies of safe water;

- the proper hygienic use of the drinking water (protection of the new water catchment works against human and animal pollution; utilisation of clean vessels for drawing and storing water; observing the danger of qualitative deterioration of stored water);

- the required supplementary installation and use of sanitation facilities.

Without hygiene education campaigns, drinking water projects may even have negative effects on the health situation of the target group, because the consumption of safe water reduces the traditional immunity against certain diseases. Thus, when again drinking contamined water afterwards, the risk of acquiring a serious disease will be much higher than before. Hygiene education campaigns should be conducted by qualified persons, preferably coming from existing institutions within the primary health care system.

The participation of the people in the planning, construction, operation and maintenance of water supply and sanitation projects is designed to ensure in the long run their acceptance of the new facilities and to contribute towards decisively improving the efficiency and prolonging the life of these facilities.

V.2.2 Project-Executing Agencies

Qualified and efficient executing agencies are an essential prerequisite for the success of water supply and sanitation projects. Because of the interdependence of water supply and sanitation both should, if possible, be under the responsibility of a single executing agency; this does not necessarily also apply to solid waste disposal.

The executing agencies should be invested with clearly defined responsibilities and should have a technical capacity - with a potential for further improvement - to organise the planning, implementation, operation and maintenance of the facilities. Should, however, the responsibility for planning and construction be vested with other organisations, the operating institutions should be involved in the design and construction of their future installations. The executing agencies should be as independent as possible and able to employ local operating staff on appropriate wages. It must be ensured that water/sewage charges are in fact collected and, to the maximum extent, used by the executing agency under its own responsibility.

Where no suitable executing agencies are available or not yet sufficiently effective, this must be remedied, prior to the commencement of construction, if necessary by means of advisory and training measures. Particular attention should be given to practice-oriented institution-building advisory services to executing agencies on organisation, technology and financial administration. This applies especially to advisory measures on operation and maintenance.

V.2.3 Tariff Policy

As a matter of principle, cost-covering charges should be aimed at for water supply as well as for sewerage-sanitation. At least, actual effective income must cover all running expenses (operating and maintenance costs, small replacement expenditures) of the project and/or of the executing agency.

Projects in which present revenues do not comply with this requirement because tariffs are too low or collection systems inefficient, are eligible for financial assistance only if the necessary tariff increases and improvements of collection systems and financial administration management can realistically be expected to be enforced.

If no charges have so far been levied for the provision of drinking water, a flat rate should at least be introduced in a first stage (e.g. fixed charge per inhabitant or household supplied with water) as a contribution towards the project cost. Here, too, however, consumption-related charges (per volume of consumption unit) should be sought as this is an economic prerequisite for a sound financial management by the executing agency which would ensure cost recovery. Furthermore, also at standpipes water should be delivered only against payment of a charge. However, especially in rural areas, the consumption charges can partly be substituted by an appropriate contribution (in money or labour) by the consumer to the construction cost.

Above-average tariffs should be levied for water consumption exceeding basic requirements. Through application of progressive tariffs the marginal costs of supply and disposal of these water quantities can thus be borne fully by the users; in addition, such a tariff system can provide the basic for cross-subsidising insufficient cost recovery from problem areas. Tariffs for large industrial/commercial consumers must ensure full cost coverage of their consumption.

The rationale for aiming at steeply progressive tariffs follows from several reasons:

- they ensure a more economical utilisation of water resources which are scarce in many developing countries;

- they reduce undesirably large waste water volumes and contribute thus to avoiding rather costly conventional sewage collection and treatment systems and to protecting the environment; they also prevent a further deterioration in the hygienic conditions resulting from excessive amounts of waste water;

- they permit a "cross-subsidisation" for socially legitimate cost recovery shortfalls resulting from provision of basic water quantities.

In some developing countries (in particular LLDCs) neither a full cost recovery (operation and maintenance, depreciation and adequate interest return) nor even a covering of running expenses through collection of charges will be economically feasible in the short and medium run. In these cases it must be ensured that external financial support covers, during an initial phase (up to five years), also the cost of supply of spares and operational requirements, as well as part of maintenance costs, provided the executing agency is willing and able to gradually assume an increasing financial share itself.

Cost-covering sewerage tariffs must be introduced if the level of water consumption requires waste water disposal via a conventional sewerage system. Here, also tariff structures should be developed along the same lines as for water supply. It is desirable that the water supply agency is also responsible for collecting the sewerage charges.

Even if, in the case of individual private sanitation facilities (by means of a latrine, cesspits, etc.), the executing agency incurs no expenditures for construction it should perform - in the interest of public health - advisory and supervisory functions during planning, execution and maintenance of the facilities.

Tariffs for solid waste disposal should be levied per inhabitant and dependent upon the type of disposal. In countries where, because of the low per capita income, charging of each household is unrealistic, other types of revenue, geared to the interest of the general public (e.g. in the form of land tax) must be ensured for the executing agency.

VI. CONCLUSION

In view of the limited financial means and especially the marked shortage in foreign exchange of most developing countries, and also of the limited funds available to this sector under external development assistance, it is apparent that only a part of the "Decade" targets can be achieved.

This implies, that in future only those projects can be assisted, which are in line with the targets and criteria laid down in this Sector Paper and thus attain the largest possible developmental impact.

In the future, the guiding figures presented in this Sector Paper and in the list of Project Assessment Criteria should, as far as possible, also be enforced - for economic reasons - vis-à-vis the decision-making bodies in the developing country and should be pursued vis-à-vis other donors in the case of joint negotiations.

Finally, reference is again made to the "Project Assessment Manual for Drinking Water and Sanitation Projects in Developing Countries", which should be taken into account additionally in project preparation and implementation.

INTERNATIONAL DRINKING WATER SUPPLY AND SANITATION DECADE

1981-1990

Global Sector Concepts for Water Supply and Sanitation

March 1987

This publication was prepared under the Interregional Cooperation Programme between the World Health Organization and the German Federal Ministry for Economic Cooperation.

CONTENTS

	Page
Background	2
Why Global Concepts are Needed	3
Concept No. 1 **INSTITUTIONAL CHANGES**	4-5
Concept No. 2 **COST RECOVERY**	6-7
Concept No. 3 **BALANCED DEVELOPMENT**	8-9
Concept No. 4 **OPERATION, MAINTENANCE & REHABILITATION**	10-11
Concept No. 5 **COMMUNITY PARTICIPATION AND HYGIENE EDUCATION**	12-13
Concept No. 6 **COORDINATION AND COOPERATION**	14-15
The Way Forward	16

Background

The **Global Concepts** described in this document have emerged from a series of meetings of multilateral and bilateral donor organizations aimed at improving coordination and resource mobilization activities in support of the International Drinking Water Supply and Sanitation Decade (IDWSSD).

From 1978 through 1984, WHO carried out an Interregional Cooperation Programme aiming primarily at promoting Decade goals and formulating water supply and sanitation sector development plans in a number of developing countries, with support from the German Federal Ministry for Economic Cooperation (BMZ) through the German Agency for Technical Cooperation (GTZ). The next phase of the WHO/GTZ programme involved organization of a series of meetings for regional and country-level donor coordination and resource mobilization, which sought assistance from external support agencies for implementation of Decade plans and projects.

This programme of meetings was launched by a *European Donor Consultation* held in Koenigswinter, near Bonn, Federal Republic of Germany, in October 1984. There followed a meeting, in May 1985, of the *Development Assistance Committee (DAC)* of the Organization for Economic Cooperation and Development (OECD), which concentrated on improving aid effectiveness in the drinking water supply and sanitation sector.

To follow up on the results of these global meetings, the Interregional Cooperation Programme organized and carried out, in collaboration with the three regional development banks, three *Regional External Support Consultations* to formulate new sector concepts. The consultations included representatives of external support agencies and water supply an sanitation sector experts from the respective regions. The Regional Consultation for Asia, co-sponsored by the Asian Development Bank, took place in Manila, Philippines, in October 1985; the one for Africa, co-sponsored by the African Development Bank, was organized in Abidjan, Cote d'Ivoire, in November 1985; and the Consultation for the Americas, co-sponsored by the Inter-American Development Bank, was held in Washington DC, USA, in April 1986. In parallel, and as a continuous activity, a number of country-level *Decade Consultative Meetings* were organized to promote resource mobilization and coordinate water supply and sanitation sector activities in recipient countries with the participation of the donor community.

All these activities have been directed at means of coordinating activities of external support agencies within the framework of the established **Decade Approaches**.

DECADE APPROACHES

Complementarity in developing water supply and sanitation.

Strategies giving precedence to underserved rural and urban populations.

Programmes promoting self-reliant, self-sustained action.

Community involvement in all stages of project implementation.

Socially relevant systems that people can afford, using technologies appropriate to specific projects.

Association of water supply and sanitation with relevant programmes in other sectors, particularly with primary health care, concentrating on hygiene education, human resources development, and the strengthening of institutional performance.

Why Global Concepts are Needed

The *International Drinking Water Supply and Sanitation Decade (1981–1990)* has focussed attention on the plight of about two billion people in developing countries who lack adequate sanitation facilities, including access to a safe supply of potable water. Considerable progress was made during the first half of the Decade, but major challenges remain.

External support agencies have been looking closely at the achievements and the disappointments of recent activities in the water supply and sanitation sector. In two multilateral and three regional consultations (see opposite) six major constraints have been identified and analysed.

THE CONSTRAINTS

1. **Institutions** responsible for water supply and sanitation sector activities in developing countries are frequently inefficient and financially weak.

2. **Cost recovery** is generally ineffective.

3. **Imbalances** exist between the provision of water supply and of sanitation facilities;d between sector inputs in central urban areas and those in urban-fringe and rural areas.

4. **Operation, maintenance and rehabilitation** receive insufficient attention, and the problem is aggravated by application of inappropriate and often too sophisticated technologies, which are neither affordable nor manageable.

5. **Community participation and hygiene education** efforts are inadequate.

6. **Coordination and cooperation** is inadequate among external support agencies, between these agencies and the national water supply and sanitation sector agencies, among the sector agencies themselves, and between the water and sanitation sector and related sector programmes.

Discussions have shown that the relative importance of each constraint varies from region to region. In **Africa**, the central problems are the shortage of qualified, adequately trained people, and poor institutional performance. In **Asia**, the emphasis is on a need to improve coordination among external support agencies and on ways of establishing satisfactory cost-recovery schemes. In the **Americas**, major attention is being given to operation, maintenance and rehabilitation, as a means of optimizing the use of existing resources in a difficult economic environment.

Analysis of the six constraints has led to development of six **Global Concepts** for improving the performance of the water supply and sanitation sector.

CONCEPT No. 1:

Constraint

Institutions *responsible for water supply and sanitation sector activities in developing countries are frequently inefficient and financially weak.*

Actions Needed

1. Institutional Structures

The International Drinking Water Supply and Sanitation Decade strategy involves emphasis on underserved populations, particularly those in rural and urban fringe areas. Institutional structures and modes of operation need to be adapted to suit this new emphasis. A major change, for many countries, will be **decentralization** and/or privatization of responsibilities for water supply and sanitation activities, and substantial participation of communities in all stages of projects.

External support agencies should, therefore, substantially increase resources devoted to "software" inputs, such as institutional reforms, management and staff training, awareness campaigns and hygiene education. Relevant models for **integrated sector programmes** at district level should be pursued with ministries responsible for coordinating rural development (e.g. Ministry of Community or Rural Development or similar), with technical input from existing or strengthened water supply and sanitation sector ministries.

2. Technical Cooperation

Strengthening of institutions' sector management planning and project preparation capacities is necessarily a long term process. It requires uniform development strategies from external support agencies, as well as reliable and continuous performance by managerial and technical staff in the institutions. It is therefore essential that the external support agencies provide national institutions with support for human resources development, including management training, through financial and technical cooperation. In addition to providing their own support programmes, the extenal support agencies should encourage activities of Technical Cooperation among Developing Countries (TCDC).

3. Appropriate Technology Research and Application

Participation of benefitting communities has been shown to be vital, if water supply and sanitation projects are to be sustainable and replicable. Appropriate and socially acceptable technologies can save investment and bring down operation and maintenance costs, so making it more viable for communities to take a direct part in all project activities. External support agencies are urged to expand their **research and development** (and application) programmes in that direction, and to help, to the extent possible, to promote **local manufacture** of water supply and sanitation equipment. It is recommended that the leading UN agencies active in the sector (UNDP, The World Bank, WHO) should intensify the elaboration of **engineering design criteria** guidelines for the sector, emphasizing regional differences where these are applicable.

INSTITUTIONAL CHANGES

The Role of External Support Agencies

- Increase resources for public awareness campaigns and hygiene education.

- Encourage decentralization and/or privatization of water supply and sanitation institutions, or certain functions of these institutions, and promote collaboration with rural development agencies on integrated programmes.

- Involve benefitting communities in project identification, planning, design, implementation, operation and maintenance.

- Establish harmonized strategies to be adopted by all agencies active in particular countries or regions.

- Provide support for institutions' management and staff training (as well as for education in community participation and hygiene awareness) through technical cooperation. TCDC should be encouraged.

- Expand R&D programmes and encourage local manufacture. Press for standardized engineering design criteria relating to appropriate water supply and sanitation technologies.

CONCEPT No. 2:

Constraint

Cost recovery is generally ineffective

Actions needed

1. Cost Recovery Objectives

The cost of water supply and sanitation services must be borne, or at least shared by the beneficiaries, to ensure adequate operation, maintenance and expansion of installed facilities. Cost recovery is a crucial step towards the **financial viability** and, eventually, autonomy of sector agencies. Full cost recovery involves recuperation of investment costs as well as those for operation and maintenance. To achieve any degree of cost recovery, developing country governments must have the **political will** to require consumers to pay for water supply and sanitation services. The population's **willingness to pay** must be motivated where necessary, by public awareness campaigns which make clear the benefits deriving from the services provided. In **dialogues** with recipient countries, extenal support agencies need to emphasize the need for maximum cost recovery. Project **designs and technical cooperation** should be based on the principle of cost recovery.

2. Urban Policy

In urban areas, developing countries, with the aid of external support agencies where required, should establish a **cost-recovery strategy** based on the criteria of: making drinking water and sanitation accessible to all segments of the population; ensuring the gradual financial autonomy of the water supply and sanitation agency; and discouraging the waste of water. Full cost recovery (operation and maintenance, depreciation of equipment, and debt servicing) is a long-term objective, to be reached preferably by cross-subsidizing tariffs. No single group of the population should be privileged by external subsidies (e.g. for household or yard connections) while other groups in the project area have no access to any reliable water supply. In the short run, operation and maintenance costs, including replacement of equipment, should be recovered as a minimum target. In all cases, water supply and sanitation costs should be **affordable** by all consumer income groups. Revenues of water and sanitation agencies should remain in the sector.

3. Rural Policy

In rural areas, income levels are generally low. Wherever possible, beneficiaries should contribute towards construction, operation and maintenance costs of new services, through a mixture of cash payments, labour, and the supply of local materials, as part of the process of **community participation**. Before projects are prepared, governments and donor agencies should discuss with communities the implications of operation and maintenance costs and provision of labour, and the choice of technology should be appropriate for available resources. In some special cases, particularly in Africa, a **transition** period may be necessary, during which operation and maintenance costs are co-funded by external support agencies. However, the objective should be that beneficiaries should gradually assume responsibility for the full costs of operation and maintenance. Financial contributions for replacement of equipment is a longer term objective.

COST RECOVERY

The Role of External Support Agencies

- Emphasize in all dialogues with recipient country governments, the crucial importance of cost recovery in sustainable and replicable programs.

- Encourage the establishment of strongly progressive, cross-subsidizing tariffs.

- Support public awareness campaigns which stress the benefits of water supply and sanitation services and so promote willingness to pay.

- Promote and support urban project designs based on full cost recovery from affordable technologies. Back sector agencies in strategies to achieve self-sufficiency and financial autonomy.

- Use early community participation in rural areas to establish commitments to contribute cash, labour and materials for construction, operation and maintenance of appropriately designed facilities.

- Extend support where necessary into the operation and maintenance phase of projects, but always with the long-term aim of establishing community responsibility for recurrent costs.

CONCEPT No. 3:

Constraint

Imbalances *exist between the provision of water supply and sanitation; and between sector inputs in central urban areas and those in urban-fringe and rural areas.*

Actions Needed

1. Promotion and Education

The severe neglect of sanitation services in comparison with water supply reflects insufficient appreciation of the **value** of sanitation. This in turn results from a lack of hygiene education. Sanitation also lags behind because sanitation projects have a **lower prestige** value than those for water supply, and because **traditional design** standards for sanitation result in prohibitively high investment and running costs. In many ways, similar factors have caused rural sector developments to trail behind those in the urban sector. Correcting these imbalances calls for the application of appropriate technology and for emphasis in public awareness campaigns of the **complementarity** of water supply and sanitation in the achievement of health benefits — one of the fundamental elements of the International Drinking Water Supply and Sanitation Decade.

2. Project Planning

Maximum benefits are obtained when water supply, sanitation and hygiene education form part of **integrated** programmes, preferably under the responsibility of a single executing agency. Water supply and sanitation agencies need to strengthen their resources, with the help of external support agencies, to equip their managerial and technical staff to promote, design and implement sanitation components of projects. Development of appropriate and socially acceptable sanitation **technologies** has progressed a long way in recent years, and efforts are now needed on all sides to see that suitable sanitation components are incorporated in future urban and rural water supply programmes.

BALANCED DEVELOPMENT

The Role of External Support Agencies

- **Ensure that hygiene education campaigns emphasizing the complementarity of water supply and sanitation are included in sector programmes receiving donor support.**

- **Bring to the attention of programme planners and designers the sources of information on low-cost and socially acceptable sanitation technologies.**

- **Raise the proportion of technical cooperation and funding support given to integrated projects, and to the expansion of national water supply agencies' capacities, to enable them to cope with liquid and solid waste disposal activities.**

- **Re-emphasize the key Decade concept of precedence for the *underserved* urban and *rural* populations, and encourage recipient countries to balance investments accordingly.**

CONCEPT No. 4:

Constraint

Operation, maintenance and rehabilitation *receive insufficient attention, and the problem is aggravated by application of inappropriate and often too sophisticated technologies (which are neither affordable nor manageable).*

Actions Needed

1. Optimising Use of Resources
Premature failure or poor performance of existing water supply and sanitation systems sets back progress towards Decade goals and represents wasted investment. In a worldwide economic climate unfavourable to social sector investments, it is of utmost importance that developing countries and external support agencies can point to successful programmes which bring **long-term** benefits. More attention to the needs of operation and maintenance is vital, and begins with selection of technologies and management systems which are appropriate for **available resources**. Rehabilitation of existing systems should be considered as a necessary precedent of major investments. It may often serve as a substitute for new installations, or a way of postponing them.

2. Policies and Budget Provisions
With the encouragement and support of external support agencies, water supply and sanitation sector agencies need to review policies and **staffing resources**, to ensure that they cater for the operation and maintenance needs of existing and future systems. Assignment of O&M responsibilities to adequately equipped and trained communities will usually be a desirable policy change, but must be accompanied by the right internal structure, including **decentralization**.

Operation and maintenance needs and costs must be evaluated in the project planning and design stages, with due budgetary allowances made in project costings. External support agencies may be willing in some circumstances to continue support into the operation and maintenance phase, particularly in the field of **training** and **institutional development**, but programmes must be designed with the long-term aim of self-sufficiency.

Programme planning and project appraisal by external support agencies should include comparison of proposed new projects with alternative (or supplementary) investments in rehabilitation.

OPERATION, MAINTENANCE & REHABILITATION

The Role of External Support Agencies

- Ensure that project or programme proposals take account of operation and maintenance needs, and that financial and human resources are available.

- Compare proposed investments in new projects with alternatives for rehabilitation of existing systems which are disused or underperforming.

- Assist sector agencies in developing countries to establish policies and institutional structures which provide for adequate operation and maintenance of existing and proposed new facilities.

- Extend programme support, where necessary, beyond completion of construction, to help equip agencies and communities for their O&M tasks.

CONCEPT No. 5:

Constraint

Community participation and hygiene education efforts are inadequate.

Actions Needed

1. Community Participation
The International Drinking Water Supply and Sanitation Decade has produced compelling evidence that participation of benefitting communities in all stages of water supply and sanitation projects is a prerequisite of success. Too many projects prove unsustainable when central agencies assume all decision-making and managerial responsibilities and then prove unable to meet the long-term commitments. A sense of **ownership**, engendered by full involvement of the community in planning, design, construction, operation and maintenance, is the best way to provide for satisfactory upkeep of installed facilities.

Involving **women** in each project stage is particularly important. As the prime users and beneficiaries of improved water and sanitation services, women have continually proved also to be the most diligent in ensuring that those services are properly maintained.

2. Hygiene Education
Motivation of communitites to participate in water supply and sanitation activities is most readily accomplished through hygiene education programmes and public awareness campaigns which stress the benefits to be achieved from such improvements. Hygiene education is clearly also important in its own right, as a method for **maximizing health benefits** from the provision of improved water and sanitation facilities.

3. Software and Training
Community participation in water supply and sanitation activities can only be fully effective if it is supported by measures to equip community members to undertake tasks and duties expected of them. Software programmes or components need to include both training for community workers to give them the capacity to take on responsibility for the upkeep and management of water supply and sanitation systems, and the provision of necessary **support** structures (spare parts supplies, power/fuel availability, technical advice). External support agencies are committed (Concept No. 1) to increasing resources for hygiene education and public awareness campaigns, and to providing technical cooperation for training at all levels.

COMMUNITY INVOLVEMENT

The Role of External Support Agencies

- In providing programme support, ensure that the balance of "software" and "hardware" is correct, and that training of community workers is part of the package.

- Use hygiene education programmes to motivate community members to participate in all project phases, with special emphasis on the role of women. Bring the benefits of water supply and sanitation investments into health education messages promoted through other sector agencies.

- Provide technical cooperation to establish — where possible — the support system necessary for community management of completed installations to function effectively.

- Ensure that project proposals have considered and properly reflected the views of the community on technology choice, service level, affordability, and operation and maintenance commitments.

CONCEPT No. 6:

Constraint

Coordination and cooperation *is inadequate among external support agencies, between these agencies and the national water supply and sanitation sector agencies, among the sector agencies themselves, and between the water and sanitation sector and related sector programmes.*

Actions Needed

1. Country-level Coordination

It is the prime responsibility of the developing country itself to coordinate sector activities. The International Drinking Water Supply and Sanitation Decade has helped through the concept of **National Action Committees**, which are performing this role successfully in a number of countries. As the Decade focal point at the country level, the **UNDP** Resident Representative should also assist the government, through regular meetings with the locally-represented donor community to discuss sector issues. The aim should be to have a single water supply and sanitation strategy for each of the urban and rural subsectors, which is known to each agency operating in the sector, and to the external support community. The subsectoral strategies should be formulated so as to complement one another.

2. Intersectoral Coordination

Just as water supply and sanitation improvements produce benefits in other sectors — most notably the **health** sector, but also **agricultural** and **industrial** production — so, it is helpful to coordinate activities with other sector programmes such as **housing** and urban and rural development, where water and sanitation components may be introduced into investments with other prime purposes. Integration should mean better use of scarce resources, and, with proper planning, can bring enhanced benefits in all sectors. The process of coordination between sectors can be facilitated by external support agencies, who commonly have dealings in more than one sector.

3. Standardization

One symptom of uncoordinated activities between governments and donors is a proliferation of different types of equipment and services, often the result of **tied aid**. External support agencies have an important role to play by placing more emphasis in discussions among themselves and with governments of developing countries on arrangements for standardizing on equipment and services supplied as well as for the introduction of appropriate technologies. The Development Assistance Committee (DAC) of the Organization for Economic Cooperation and Development (OECD) has available draft guidelines entitled *Minimum Conditions for Effective International Competitive Bidding (DAC 86-23)*, which contain useful advice.

4. Information Exchange

The World Health Organization is in the process of establishing a *Country External Support System (CESI)*, which will collect from and disseminate to donors and recipient governments information on ongoing and planned projects in the water supply and sanitation sector. The system will depend on accurate and timely inputs, and external support agencies are urged to collaborate fully in the build-up of the system, which aims at **streamlining** sector inputs and so maximizing their benefits.